ADAPTIVE PROCESSING OF BRAIN SIGNALS

ADAPTIVE PROCESSING OF BRAIN SIGNALS

Saeid Sanei

Reader in Neurocomputing
University of Surrey, UK

Library of Congress Cataloging-in-Publication Data

Sanei, Saeid.
   Adaptive processing of brain signals / Dr. Saeid Sanei.
      pages   cm
   Includes bibliographical references and index.
   ISBN 978-0-470-68613-3 (hardback)
   1. Neural networks (Neurobiology).   2. Brain–Physiology.   3. Signal processing–Digital techniques.
I. Title.
   QP363.3.S26 2013
   573.8′5–dc23

                                                                                    2013005190

A catalogue record for this book is available from the British Library

Typeset in 10/12pt Times by Aptara Inc., New Delhi, India
Printed and bound in Singapore by Markono Print Media Pte Ltd

Contents

Preface

Since the book, *EEG Signal Processing* (Sanei, Chambers 2007) was published, there have been more demands for research and development of signal processing tools and algorithms for the much wider exploration of EEG and other neuroimaging systems. More recent advances in digital signal processing are expected to underpin key aspects of the future progress in biomedical research and technology, particularly on measurements and assessment of brain activity and its response to various stimulations.

Although most of the concepts in multi-channel EEG digital signal processing, particularly the use of adaptive techniques and iterative learning algorithms, have their origin in distinct application areas, such as acoustics, communications engineering, speech and biometrics, together with the processing of other physiological signals; it is shown in this book that new approaches stem from many recent neurological, psychological and clinical neuroscience findings.

As well as some fundamental concepts in both the modelling and processing of brain activities, a number of new signal processing topics are explored, including new definitions and algorithms. Multichannel, multidimensional and multiway signal processing, which cover a more inclusive and dynamic description of brain information, are presented in this book. This extends to multimodal data analysis in the context of simultaneous EEG–fMRI recordings and information fusion.

Motivated by research in the field over more than two decades, techniques specifically related to EEG processing, such as brain source localization, detection and classification of event-related potentials, sleep signal analysis, emotion effects, mental fatigue, brain connectivity, seizure detection and prediction, together with brain–computer interfacing are all explained in detail. A comprehensive illustration of new signal processing results in the form of signals, graphs, images, and tables has been provided for a better understanding of the concepts.

Following the history and basic definitions and physiological brain-related concepts, the first few chapters cover a comprehensive overview of the tools and algorithms currently being used for processing EEG and magnetoencephalography (MEG) signals. There is more emphasis put on algorithms utilised for the analysis of scalp recordings.

Detection and tracking of event-related potentials (ERPs), particularly for single-trial estimation, using recent techniques are provided next. ERPs are the brain responses to audio, visual, and tactile stimuli. Therefore, many neurological and psychiatric brain disorders as well as movement-related abnormalities are diagnosed and monitored using these techniques.

Brain connectivity and its dynamics change significantly for various physiological and neurological states of the brain. A comprehensive overview of the techniques plus new directions in the assessment and exploitation of brain connectivity are provided next.

Seizure and epileptic brain discharges are still investigated by many researchers by looking at both scalp EEG signals and intracranial spike recordings using subdural electrodes. Some very recent methods in seizure prediction are demonstrated. This area of research is later extended to a new multimodal methodology employing simultaneous EEG–fMRI data recording and analysis.

The state of the art in the localization of brain signal sources has been followed by two new inclusive forward methods. In places where the desired source signals can be defined using known waveforms forward models result in accurate localization of the sources. A deflation-based localization approach presented here can localize multiple brain signal sources.

Mental fatigue assessment and analysis from EEG is another new research area covered in this book. Combining signal processing and machine learning algorithms leads to establishing a robust and clinically proven method for detection of the stages of mental fatigue. This includes the changes in both brain responses to stimuli and the changes in brain normal rhythms.

Despite a variety of approaches for evaluating emotions from facial expressions and body biometrics, researchers are focussing on the changes in brain rhythms, mainly in terms of source localization and connectivity, using EEG, MEG, and fMRI neuroimaging techniques. A comprehensive overview of the methods and algorithms developed for this purpose is given.

The book also covers very recent machine learning tools in brain–computer interfacing (BCI). Although, the principles of BCI have not changed significantly over the last five years, a number of extensions to well established algorithms, such as common spatial patterns, have been proposed very recently. These approaches are covered in this book.

Unfortunately, the signals are affected by noise and interferences. Much effort, therefore, has to be made to restore the signals. In some applications, such as joint EEG–fMRI recording and analysis, this problem is significant. In this book artefact removal has been extensively addressed and the results of new algorithms discussed and illustrated.

In the preparation of the content of this book, it is assumed that the reader has a background in the fundamentals of digital signal processing and wishes to focus on the processing of brain signals, particularly EEG and MEG. It is hoped that the concepts covered in each chapter provide a foundation for future research and development in the field.

In conclusion, I wish to stress that there is no attempt to challenge any clinical or diagnostic knowledge. Instead, the tools and algorithms described here can, I believe, enhance the clinically-related information within EEG signals significantly and thereby aid physicians in better diagnosis, treatment and monitoring of brain abnormalities.

Before ending this preface I wish to thank most sincerely my friends and colleagues who reviewed this book for their valuable and very useful comments: Ahmad Reza Hosseini-Yazdi, Jonathan Clark, Tracey Kah Mein Lee, and Clive Cheong-Took. Next, my appreciation to my recent PhD students in the University of Surrey, who contributed to provision of the materials in this book, especially, Delaram Jarchi, Hamid Mohseni, Javier Escodero, Foad Ghaderi, Bahador Makkiabadi, Saideh Ferdowsi, and Kostas Eftaxias.

Last but not least, I appreciate the support and patience of my wife Maryam and my sweethearts Erfan, Ideen and Shaghayegh to whom I dedicate this book.

Saeid Sanei
June 2013

1

Brain Signals, Their Generation, Acquisition and Properties

1.1 Introduction

The brain is the most astonishing and complicated part of the human body and is naturally responsible for controlling all other organs. The neural activity of the human brain starts between the 17th and 23rd weeks of prenatal development. It is believed that from this early stage and throughout life electrical signals generated by the brain represent not only the brain function but also the status of the whole body. This assumption provides the motivation to apply advanced digital signal processing methods to the brain functional data, including electroencephalogram (EEG), magnetoencephalogram (MEG), and functional magnetic resonance image (fMRI) sequences. Although the emphasis in this book is on EEG and MEG, there will be some analysis of simultaneously recorded EEG-fMRI sequences too. Other functional brain information, such as that obtained by near-infrared spectroscopy (NIRS) recently developed for recording movement=related cortical potentials and some under-developing imaging systems, such as ultrawideband or microwave brain imaging are rarely referred to.

Nowhere in this book does the author attempt to comment on the physiological aspects of brain activities. However, there are several issues related to the nature of the original sources, their generation, their actual patterns, and the characteristics of the propagating environment.

Understanding of neuronal functions and neurophysiological properties of the brain, together with the mechanisms underlying the generation of signals and their recordings is, however, vital for those who deal with these signals for detection, diagnosis, and treatment of brain disorders and the related diseases. We begin by providing a brief history of EEG recording.

1.2 Historical Review of the Brain

EEG history goes back to the time when, for the first time, some activity of the brain was recorded or displayed. Carlo Matteucci (1811–1868) and Emil Du Bois-Reymond (1818–1896) were the first people to register the electrical signals emitted from muscle nerves using

Adaptive Processing of Brain Signals, First Edition. Saeid Sanei.
© 2013 John Wiley & Sons, Ltd. Published 2013 by John Wiley & Sons, Ltd.

a galvanometer and establish the concept of neurophysiology [1, 2]. However, the concept of *action current,* introduced by Hermann Von Helmholz [3], clarified and confirmed the negative variations which occur during muscle contraction.

Richard Caton (1842–1926) a scientist from Liverpool, England, used a galvanometer and placed two electrodes over the scalp of a human subject and thereby first recorded brain activity in the form of electrical signals in 1875. Since then, the concepts of electro-(referring to registration of brain electrical activities) encephal-(referring to emitting the signals from the head) and gram (or graphy), meaning drawing or writing, were combined so that the term EEG was henceforth used to denote electrical neural activity of the brain.

Fritsch (1838–1927) and Hitzig (1838–1907) discovered that the human cerebrum can be electrically stimulated. Vasili Yakovlevich Danilevsky (1852–1939) followed Caton's work and finished his PhD thesis in the investigation of brain physiology in 1877 [4]. In this work he investigated the activity of the brain following electrical stimulation as well as spontaneous electrical activity in the brain of animals.

The cerebral electrical activity observed over the visual cortex of different species of animals was reported by Ernst von Fleischl-Marxow (1845–1891). Napoleon Cybulski (1854–1919) provided EEG evidence of an epileptic seizure in a dog caused by electrical stimulation.

The idea of association of epileptic attacks with abnormal electrical discharges was expressed by Kaufman [5]. Pravidch-Neminsky (1879–1952) a Russian physiologist, recorded the EEG from the brain, termed dura, and the intact skull of a dog in 1912. He observed a 12–14 cycle s^{-1} rhythm under normal conditions which slowed under asphyxia. He later called it the *electrocerebrogram.*

The discoverer of the existence of human EEG signals was Hans Berger (1873–1941); see Figure 1.1. He began his study of human EEGs in 1920 [6]. Berger is well known by almost all electroencephalographers. He started working with a string galvanometer in 1910, then migrated to a smaller Edelmann model and, after 1924, to a larger Edelmann model. In 1926, Berger started to use a more powerful Siemens double coil galvanometer (attaining a

Figure 1.1   Hans Berger (image from http://www.s9.com/Biography/Berger-Hans)

sensitivity of 130 μV cm^{-1}) [7]. His first report of human EEG recordings of 1–3 min duration on photographic paper was in 1929. In this recording he only used a one-channel bipolar method with fronto-occipital leads. Recording of the EEG became popular in 1924. The first report of 1929 by Berger included the alpha rhythm, as the major component of the EEG signals, as described later in this chapter, and the alpha blocking response.

During the 1930s the first EEG recording of sleep spindles was undertaken by Berger. He then reported the effect of hypoxia on the human brain, the nature of several diffuse and localized brain disorders, and gave an inkling of epileptic discharges [8]. During this time another group, established in Berlin-Buch and led by Kornmüller, provided more precise recording of the EEG [9]. Berger was also interested in cerebral localization and particularly in the localization of brain tumours. He also found some correlation between mental activities and the changes in the EEG signals.

Toennies (1902–1970) from the group in Berlin built the first biological amplifier for recording brain potentials. A differential amplifier for recording EEGs was later produced by the Rockefeller foundation in 1932.

The importance of multichannel recordings and using a large number of electrodes to cover a wider brain region was recognised by Kornmüller [10]. The first EEG work focusing on epileptic manifestation and the first demonstration of epileptic spikes were presented by Fischer and Löwenbach [11–13].

In England, W. Gray Walter became the pioneer of clinical EEG. He discovered the foci of slow brain activity (delta waves), which initiated enormous clinical interest in the diagnosis of brain abnormalities. In Brussels, Fredric Bremer (1892–1982) discovered the influence of afferent signals on the state of vigilance [14].

Research activities related to EEGs started in North America in around 1934. In this year, Hallowell Davis illustrated a good alpha rhythm for himself. A cathode ray oscilloscope was used around this date by the group in St. Louis University in Washington, in the study of peripheral nerve potentials. The work on human EEGs started at Harward in Boston and the University of Iowa in the 1930s. The study of epileptic seizure, developed by Fredric Gibbs, was the major work on EEGs during these years, as the realm of epileptic seizure disorders was the domain of their greatest effectiveness. Epileptology may be divided historically into two periods [15]: before and after the advent of EEG. Gibbs and Lennox applied Fischer's idea and the effect of picrotoxin on the cortical EEG in animal and human epileptology. Berger [16] showed a few examples of paroxysmal EEG discharges in a case of presumed petit mal attacks and during a focal motor seizure in a patient with general paresis. For a couple of decades the EEG work focused on the study of epilepsy.

As the other great pioneers of EEG in North America, Hallowel and Pauline Davis were the earliest investigators of the nature of EEG during human sleep. A. L. Loomis, E. N. Harvey, and G. A. Hobart were the first who mathematically studied the human sleep EEG patterns and the stages of sleep. At McGill University, Jasper studied the related behavioural disorder before he found his niche in basic and clinical epileptology [17].

The American EEG society was founded in 1947 and the first international EEG Congress was held in London, U K, around this time. While the EEG studies in Germany were still limited to Berlin, Japan gained attention by the work of Motokawa, a researcher of EEG rhythms [18]. During these years the neurophysiologists demonstrated the thalamocortical relationship through anatomical methods. This led to the development of the concept of centrencephalic epilepsy [19].

Throughout the 1950s the work on EEGs expanded in many different places. During this time surgical operation to remove the epileptic foci became popular and the book entitled *Epilepsy and the Functional Anatomy of the Human Brain* (Penfield and Jasper) was published. During this time microelectrodes were invented. They were made of metals such as tungsten, or glass, filled with electrolytes, such as potassium chloride, with diameters of less than 3 μm.

Recordings of deep brain EEG sources of a human were first obtained with implanted intracerebral electrodes by Mayer and Hayne (1948). Invention of intracellular microelectrode technology revolutionarized this method and was used in the spinal cord by Brock *et al.* in 1952 [20], and in the cortex by Phillips in 1961 [21].

Analysis of EEG signals started during the early days of EEG measurement. Berger assisted by Dietch (1932) applied Fourier analysis to EEG sequences which was rapidly developed during the 1950s. Analysis of sleep disorders with EEGs started its development in the 1950s through the work of Kleitman at the University of Chicago.

In the 1960s analysis of the EEGs of full-term and premature newborns began its development [22]. Investigation of evoked potentials (EPs), especially visual EPs, as commonly used for monitoring mental illnesses, progressed during the 1970s.

MEG signals, on the other hand were first measured by David Cohen, a University of Illinois physicist, in 1968 [23] before the availability of the superconducting quantum interference device (SQUID), using a copper induction coil as the detector. To reduce the magnetic background noise, the measurements were made in a magnetically shielded room. The coil detector was barely sensitive enough, resulting in poor, noisy MEG measurements that were difficult to use. Later, Cohen built a better shielded room at Massachusetts Institute of Technology (MIT), and used one of the first SQUID detectors, just developed by James E. Zimmerman, a researcher at Ford Motor Company [24], to again measure MEG signals [1003]. This time the signals were almost as clear as those of EEG. Subsequently, various types of spontaneous and evoked MEGs began to be measured.

Using a single SQUID detector to successively measure the magnetic field at a number of points around the human head was cumbersome. Therefore, in the 1980s, MEG manufacturers began to fabricate multiple sensors into arrays to cover a larger area of the head. Present-day MEG arrays are set in a helmet-shaped dewar that typically contains 300 sensors, covering most of the head. In this way, MEG signals can be recorded much faster.

One advantage of MEG over EEG signals is their much lower sensitivity to the nonlinearity and non-uniformity in brain tissues. EEG on the other hand, is less noisy and much cheaper than MEG.

The history of EEG and MEG, however, has been a continuous process which started from the early 1800s and has brought daily development of clinical, experimental, and computational studies for discovery, recognition, diagnosis, and treatment of a vast number of neurological and physiological abnormalities of the brain and the rest of the central neural system (CNS) of human beings. Nowadays, EEGs are recorded invasively and non-invasively using fully computerised systems. The EEG machines are equipped with many signal processing tools, delicate and accurate measurement electrodes, and enough memory for very long-term recordings of several hours. Although the MEG machines are expensive due to the technology involved and the requirement for low noise SQUIDs and amplifiers, they are being used as an effective tool for brain source localization. EEG or MEG machines may be integrated with other neuroimaging systems, such as fMRI. Very delicate needle-type electrodes can also be used for recording the EEGs from over the cortex (electrocorticogram), thereby avoiding the attenuation and nonlinearity effects induced by the skull. We next proceed to describe the nature of neural activities within the human brain.

1.3 Neural Activities

The central nervous system (CNS) generally consists of nerve cells and glia cells, which are located between neurons. Each nerve cell consists of axons, dendrites and cell bodies. Nerve cells respond to stimuli and transmit information over long distances. A nerve cell body has a single nucleus, and contains most of the nerve cell metabolism, especially that related to protein synthesis. The proteins created in the cell body are delivered to other parts of the nerve. An axon is a long cylinder, which transmits an electrical impulse and can be several meters long in vertebrates (giraffe axons go from the head to the tip of spine). In humans the length can be a percentage of a millimetre to more than a metre. An axonal transport system for delivering proteins to the ends of the cell exists and the transport system has "molecular motors" which ride upon tubulin rails.

Dendrites are connected to either the axons or dendrites of other cells and receive impulses from other nerves or relay the signals to other nerves. In the human brain each nerve is connected to approximately 10 000 other nerves, mostly through dendritic connections.

The activities in the CNS are mainly related to the synaptic currents transferred between the junctions (called synapses) of axons and dendrites, or dendrites and dendrites of cells. A potential of 60–70 mV with negative polarity may be recorded under the membrane of the cell body. This potential changes with variations in synaptic activities. If an action potential travels along the fibre, which ends in an *excitatory* synapse, an excitatory following neuron. If two action potentials travel along the same fibre over a short distance, there will be a summation of EPSPs producing an action potential on the postsynaptic neuron providing a certain threshold of membrane potential is reached. If the fibre ends in an *inhibitory* synapse, then hyperpolarization will occur, indicating an inhibitory postsynaptic potential (IPSP) [25, 26]. Figure 1.2 shows the above activities schematically.

Following the generation of an IPSP, there is an overflow of cations from the nerve cell or an inflow of anions into the nerve cell. This flow ultimately causes a change in potential along the nerve cell membrane. Primary transmembranous currents generate secondary inonal currents along the cell membranes in the intra- and extra-cellular space. The portion of these currents that flows through the extracellular space is directly responsible for the generation of field potentials. These field potentials, usually with less than 100 Hz frequency, are called EEGs when there are no changes in the signal average, and called DC potential if there are slow drifts in the average signals, which may mask the actual EEG signals. A combination of EEG and DC potentials is often observed for some abnormalities in the brain, such as seizure (induced by pentylenetetrazol), hypercapnia, and asphyxia [27]. We next focus on the nature of action potentials.

1.4 Action Potentials

The information transmitted by a nerve is called an action potential (AP). APs are caused by an exchange of ions across the neuron membrane and an AP is a temporary change in the membrane potential that is transmitted along the axon. It is usually initiated in the cell

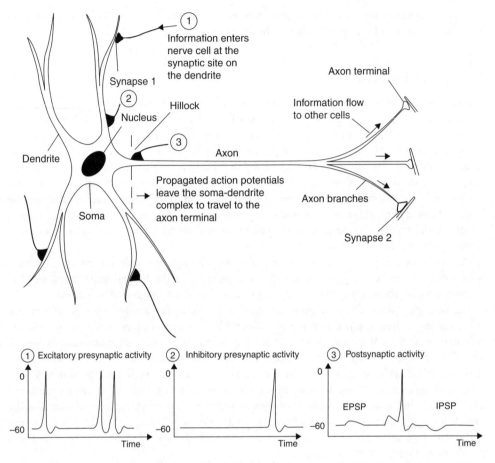

Figure 1.2 The neuron membrane potential changes and current flow during synaptic activation recorded by means of intracellular microelectrodes. Action potentials in the excitatory and inhibitory presynaptic fibre, respectively, lead to EPSP and IPSP in the postsynaptic neuron

body and normally travels in one direction. The membrane potential depolarizes (becomes more positive) producing a spike. After the spike reaches its peak amplitude the membrane repolarizes (becomes more negative). The potential becomes more negative than the resting potential and then returns to normal. The action potentials of most nerves last between 5 and 10 milliseconds.

The conduction velocity of action potentials lies between 1 and 100 m s^{-1}. APs are initiated by many different types of stimuli; sensory nerves respond to many types of stimuli, such as chemical, light, electricity, pressure, touch and stretching. On the other hand, the nerves within the CNS (brain and spinal cord) are mostly stimulated by chemical activity at synapses.

A stimulus must be above a threshold level to set off an AP. Very weak stimuli cause a small local electrical disturbance, but do not produce a transmitted AP. As soon as the stimulus strength goes above the threshold, an action potential appears and travels down the nerve.

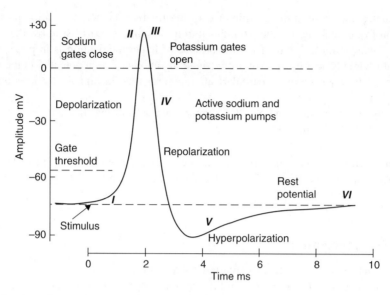

Figure 1.3 An action potential (membrane potential) for a giant squid by closing the Na channels and opening K channels. Taken from [28], © John Benjamins Publishing Co.

The spike of the AP is mainly caused by opening of Na (sodium) channels. The Na pump produces gradients of both Na and K (potassium) ions – both are used to produce the action potential; Na is high outside the cell and low inside. Excitable cells have special Na and K channels with gates that open and close in response to the membrane voltage (voltage-gated channels). Opening the gates of Na channels allows Na to rush into the cell, carrying +Ve charge. This makes the membrane potential positive (depolarization), producing the spike. Figure 1.3 shows the stages of the process during evolution of an action potential for a giant squid. For a human being the amplitude of the AP ranges between approximately −60 and 10 mV. During this process [28]:

I. When the dendrites of a nerve cell receive the stimulus the Na^+ channels will open. If the opening is sufficient to drive the interior potential from −70 to −55 mV, the process continues.

II. As soon as the action threshold is reached, additional Na^+ channels (sometimes called voltage-gated channels) open. The Na^+ influx drives the interior of the cell membrane up to about +30 mV. The process to this point is called depolarization.

III. Then Na^+ channels close and the K^+ channels open. Since the K^+ channels are much slower to open, the depolarization has time to be completed. Having both Na^+ and K^+ channels open at the same time would drive the system towards neutrality and prevent the creation of the action potential.

IV. Having the K^+ channels open, the membrane begins to repolarize back towards its rest potential.

V. The repolarization typically overshoots the rest potential to a level of approximately −90 mV. This is called hyperpolarization, and would seem to be counterproductive, but it is actually important in the transmission of information. Hyperpolarization prevents the neuron from receiving another stimulus during this time, or at least raises the threshold for any new stimulus. Part of the importance of hyperpolarization is in preventing any stimulus already sent up an axon from triggering another action potential in the opposite direction. In other words, hyperpolarization assures that the signal is proceeding in one direction.

VI. After hyperpolarization, the Na^+/K^+ pumps eventually bring the membrane back to its resting state of -70 mV.

The nerve requires approximately 2 ms before another stimulus is presented. During this time no AP can be generated. This is called the refractory period. The generation of EEG signals is next described.

1.5   EEG Generation

An EEG signal is a measurement of currents that flow during synaptic excitations of the dendrites of many pyramidal neurons in the cerebral cortex. When brain cells (neurons) are activated, the synaptic currents are produced within the dendrites. These currents generate a magnetic field measurable by EMG machines and a secondary electrical field over the scalp measurable by EEG systems.

Differences in electrical potentials are caused by summed postsynaptic graded potentials from pyramidal cells that create electric dipoles between the soma (body of a neuron) and apical dendrites which branch from neurons (Figure 1.4). The current in the brain is generated

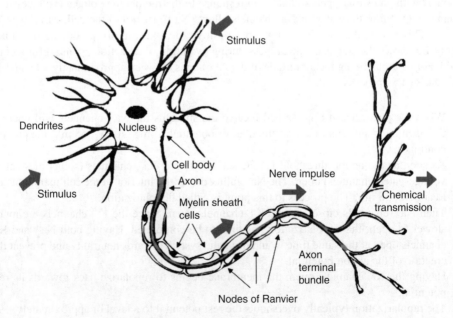

Figure 1.4   Structure of a neuron, adapted from [29]

mostly due to pumping the positive ions of sodium, Na^+, potassium, K^+, calcium, or Ca^{++}, and the negative ion of Cl^-, through the neuron membranes in the direction governed by the membrane potential [29].

The human head consists of different layers, including scalp, skull, brain and many other thin layers in between. The head layers have different thickness and current resistivities; the scalp has a thickness of approximately 0.2–0.5 cm and resistivity of 300–400 Ω, the skull has a thickness of 0.3–0.7 cm and resistivity of 10–25 kΩ when measured *in vivo*. In addition, the scalp consists of different layers, such as skin, connective tissue, which is a thin layer of fat and fibrous tissue lying beneath the skin, the loose areolar connective tissue, and the pericranium which is the periosteum of the skull bones and provides nutrition to bone and capacity for repair. Thebrain is covered by a thin layer called the cortex, which encompasses the entire brain lobes. The cortex has a thickness of 0.1–0.3 cm and an *in vivo* resistivity of 50–150 Ω. The cortex includes arachnoid, meninges, dura, epidural, and subarachnoid space. The skull attenuates the signals approximately one hundred times more than the soft tissue.

Since the layers have different electrical properties, EEG signals are generally a nonlinear sum of the brain sources. However, since the majority of the sources are cortical, that is, very close to the cortex, this nonlinearity does not significantly affect the common source separation processes. For this reason MEG is more popular for brain source localization.

On the other hand, most of the noise is generated either within the brain (internal noise) or over the scalp (system noise or external noise). Therefore, only large populations of active neurons can generate enough potential to be recordable using the scalp electrodes. These signals are later amplified greatly for display purposes. Approximately 10^{11} neurons are developed at birth when the CNS becomes complete and functional [30]. This makes an average of 10^4 neurons per mm^3. Neurons are interconnected into neural nets through synapses. Adults have approximately 5.10^{14} synapses. The number of synapses per neuron increases with age, whereas, the number of neurons decreases with age. From an anatomical point of view the brain may be divided into three parts; the cerebrum, cerebellum, and brain stem (Figure 1.5). The cerebrum consists of both left and right lobes of the brain with highly convoluted surface layers called the cerebral cortex.

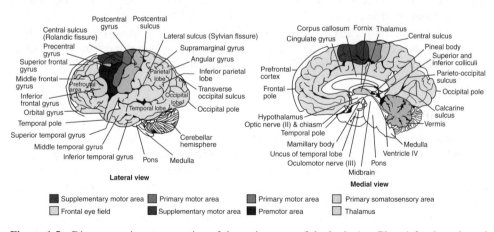

Figure 1.5 Diagrammatic representation of the major parts of the brain (see Plate 1 for the coloured version)

The cerebrum includes the regions for movement initiation, conscious awareness of sensation, complex analysis, and expression of emotions and behaviour. The cerebellum coordinates voluntary movements of muscles and maintenance of balance.

The brain stem controls involuntary functions, such as respiration, heart regulation, biorythms, neurohormone and hormone secretion [31].

Based on the above section it is clear that the study of EEGs paves the path for diagnosis of many neurological disorders and other abnormalities in the human body. The acquired EEG signals from a human (and also from animals) may for example be used for investigation of the following clinical problems [31, 32]:

1. Monitoring alertness, coma, and brain death
2. Locating areas of damage following head injury, stroke, and tumour
3. Testing afferent pathways (by evoked potentials)
4. Monitoring cognitive engagement (alpha rhythm)
5. Producing biofeedback situations
6. Controlling anaesthesia depth (servo anaesthesia)
7. Investigating epilepsy and locating seizure origin
8. Testing epilepsy drug effects
9. Assisting in experimental cortical excision of epileptic focus
10. Monitoring the brain development
11. Testing drugs for convulsive effects
12. Investigating sleep disorders and physiology
13. Investigating mental disorders
14. Providing a hybrid data recording system together with other imaging modalities.

This list confirms the rich potential for EEG analysis and motivates the need for advanced signal processing techniques to aid the clinician in their interpretation. We next proceed to describe the brain rhythms which are expected to be measured within EEG signals.

Some of the mechanisms that generate the EEG signals are known at the cellular level and rest on a balance of excitatory and inhibitory interactions within and between populations of neurons. Although it is well established that the EEG (or MEG) signals result mainly from extracellular current flow, associated with summed postsynaptic potentials in synchronously activated and vertically oriented neurons, the exact neurophysiological mechanisms resulting in such synchronisation to a given frequency band, remain obscure.

1.6 Brain Rhythms

Many brain disorders are diagnosed by visual inspection of EEG signals. Clinical experts in the field are familiar with the manifestation of brain rhythms in the EEG signals. In healthy adults, the amplitudes and frequencies of such signals change from one state of a human to another, such as wakefulness and sleep. The characteristics of the waves also change with age. There are five major brain waves distinguished by their different frequency ranges. These frequency bands from low to high frequencies, respectively, are called alpha (α), theta (θ), beta (β), delta (δ) and gamma (γ). The alpha and beta waves were introduced by Berger in 1929. Jasper and Andrews (1938) used the term "gamma" to refer to the waves of above 30 Hz. The delta rhythm was introduced by Walter (1936) to designate all frequencies below the alpha range. He also introduced theta waves as those having frequencies within the range 4–7.5 Hz. The notion of a theta wave was introduced by Wolter and Dovey in 1944 [33].

Delta waves lie within the range 0.5 to 4 Hz. These waves are primarily associated with deep sleep and may be present in the waking state. It is very easy to confuse artefact signals caused by the large muscles of the neck and jaw with the genuine delta response. This is because the muscles are near the surface of the skin and produce large signals, whereas the signal of interest originates from deep within the brain and is severely attenuated in passing through the skull. Nevertheless, by applying simple signal analysis methods to the EEG, it is very easy to see when the response is caused by excessive movement.

Theta waves lie within the range 4 to 7.5 Hz. The term theta might be chosen to allude to its presumed thalamic origin. Theta waves appear as consciousness slips towards drowsiness. Theta waves have been associated with access to unconscious material, creative inspiration and deep meditation. A theta wave is often accompanied by other frequencies and seems to be related to level of arousal. We know that healers and experienced mediators have an alpha wave which gradually lowers in frequency over long periods of time. The theta wave plays an important role in infancy and childhood. Larger contingents of theta wave activity in the waking adult are abnormal and are caused by various pathological problems. The changes in the rhythm of theta waves are examined for maturational and emotional studies [34].

The alpha waves appear in the posterior half of the head, are usually found over the occipital region of the brain, and can be detected in all parts of brain posterior lobes. For alpha waves the frequency lies within the range 8–13 Hz, and commonly appears as a round or sinusoidal-shaped signal. However, in rare cases it may manifest itself as sharp waves. In such cases, the negative component appears to be sharp and the positive component appears to be rounded, similar to the wave morphology of the rolandic mu (μ) rhythm. Alpha waves have been thought to indicate both a relaxed awareness without any attention or concentration. The alpha wave is the most prominent rhythm in the whole realm of brain activity and possibly covers a greater range than has been previously accepted. You can regularly see a peak in the beta wave range in frequencies even up to 20 Hz, which has the characteristics of an alpha wave state rather those of a beta wave. Again, we very often see a response at 75 Hz which appears in an alpha setting. Most subjects produce some alpha waves with their eyes closed and this is why it has been claimed that it is nothing but a waiting or scanning pattern produced by the visual regions of the brain. It is reduced or eliminated by opening the eyes, by hearing unfamiliar sounds, by anxiety or mental concentration or attention. Albert Einstein could solve complex mathematical problems while remaining in the alpha state; though generally, beta and theta waves are also present. An alpha wave has a higher amplitude over the occipital areas and has an amplitude of normally less than 50 μV. The origin and physiological significance of an alpha wave is still unknown and yet more research has to be undertaken to understand how this phenomenon originates from cortical cells [35].

A beta wave is the electrical activity of the brain varying within the range 14–26 Hz (though in some literature no upper bound is given). A beta wave is the usual waking rhythm of the brain associated with active thinking, active attention, focus on the outside world or solving concrete problems, and is found in normal adults. A high level beta wave may be acquired when a human is in a panic state. Rhythmical beta activity is encountered chiefly over the frontal and central regions. Importantly, a central beta rhythm is related to the rolandic mu rhythm and can be blocked by motor activity or tactile stimulation. The amplitude of the beta rhythm is normally under 30 μV. Similar to the mu rhythm the beta wave may also be enhanced because of a bone defect [33] and also around tumoural regions.

The frequencies above 30 Hz (mainly up to 45 Hz) correspond to the gamma range (sometimes called fast beta wave). Although the amplitudes of these rhythms are very low and their occurrence is rare, detection of these rhythms can be used for confirmation of certain brain diseases. The regions of high EEG frequencies and the highest levels of cerebral blood flow (as well as oxygen and glucose uptake) are located in the frontocentral area. The gamma wave band has also been proved to be a good indication of event-related synchronization (ERS) of the brain and can be used to demonstrate the locus for right and left index finger movement, right toes and the rather broad and bilateral area for tongue movement [36].

Waves of frequencies much higher than the normal activity range of EEG, mostly in the range 200–300 Hz have been found in cerebellar structures of animals but they have not played any role in clinical neurophysiology [37, 38].

Figure 1.6 shows the typical normal brain rhythms with their usual amplitude levels. In general the EEG signals are the projection of neural activities which are attenuated by

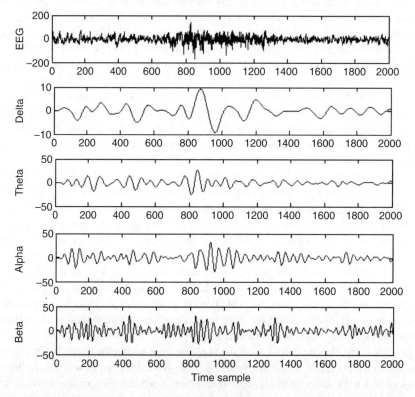

Figure 1.6 The EEG signal on the top and four typical dominant brain normal rhythms, from low to high frequencies; the delta wave is observed in infants and sleeping adults, the theta wave in children and sleeping adults, the alpha wave is detected in the occipital brain region when there is no attention, and the beta wave appears frontally and parietally with low amplitude

leptomeninges, cerebrospinal fluid, dura matter, bone, galea, and the scalp. Cortiographic discharges show amplitudes of 0.5–1.5 mV in range and up to several millivolts for spikes. However, on the scalp the amplitudes commonly lie within 10–100 μV.

The above rhythms may last if the state of the subject does not change and therefore they are approximately cyclic in nature. On the other hand, there are other brain waveforms, which may:

1. have a wide frequency range or appear as spiky type signals such as k-complexes, vertex waves (which happen during sleep), or a breach rhythm, which is an alpha-type rhythm due to a cranial bone defect [39], which does not respond to movement, and is found mainly over the midtemporal region (under electrodes T3 or T4), and some seizure signals,
2. be a transient such as an event related potential (ERP) and contain positive occipital sharp transient (POST) signals (also called rho (ρ)) waves,
3. originate from the defected regions of the brain, such as tumoural brain lesions,
4. be spatially localised and considered as cyclic in nature, but can be easily blocked by physical movement, such as mu rhythm. Mu denotes motor and is strongly related to the motor cortex. Rolandic (central) mu is related to posterior alpha in terms of amplitude and frequency. However, the topography and physiological significance are quite different. From the mu rhythm one can investigate the cortical functioning and the changes in brain (mostly bilateral) activities subject to physical and imaginary movements. The mu rhythm has also been used in feedback training for several purposes, such as treatment of epileptic seizure disorder [33].

Also, there are other rhythms introduced by researchers such as:

5. Phi (φ) rhythm (less than 4 Hz) occurring within two seconds of eye closure. The phi rhythm was introduced by Daly [35].
6. The kappa (κ) rhythm, which is an anterior temporal alpha-like rhythm and is believed to be the result of discrete lateral oscillations of the eyeballs and considered to be an artefact signal.
7. The sleep spindles (also called sigma (σ) activity) within the 11–15 Hz frequency range.
8. Tau (τ) rhythm which represents the alpha activity in the temporal region.
9. Eyelid flutter with closed eyes which gives rise to frontal artefacts in the alpha band.
10. Chi rhythm is a mu-like activity believed to be a specific rolandic pattern of 11–17 Hz. This wave has been observed during the course of Hatha Yoga exercises [40].
11. Lambda (λ) waves are most prominent in waking patients, but are not very common. They are sharp transients occurring over the occipital region of the head of walking subjects during visual exploration. They are positive and time-locked to saccadic eye movement with varying amplitude, generally below 90 μV [41].

Often it is difficult to understand and detect the brain rhythms from scalp EEGs, even with trained eyes. Application of advanced signal processing tools, however, should enable separation and analysis of the desired waveforms from within the EEGs. Therefore, definition of foreground and background EEG is very subjective and entirely depends on the abnormalities and applications. We next consider the development in the recording and measurement of EEG signals.

An early model to generate the brain rhythms is the model of Jansen and Rit [42]. This model uses a set of parameters to produce alpha activity through an interaction between inhibitory and excitatory signal generation mechanisms in a single area. The basic idea behind these models is to make excitatory and inhibitory populations interact such that oscillations emerge. This model was later modified and extended to generate and emulate the other main brain rhythms, that is, delta, theta, beta, and gamma, too [43]. The assumptions and mathematics involved in building the Jansen model and its extension are explained in Chapter 3. Application of such models in the generation of postsynaptic potentials and using them as the template to detect, separate, or extract ERPs is of great importance. In Chapter 9 of this book, we can see the use of such a template in the extraction of the ERPs.

1.7   EEG Recording and Measurement

Acquiring signals and images from the human body has become vital for early diagnosis of a variety of diseases. Such data can be in the form of electrobiological signals such as electrocardiogram (ECG) from the heart, electromyography (EMG) from muscles, EEG and MEG from the brain, electrogastrography (EGG) from the stomach, and electro-occlugraphy (electrooptigraphy, EOG) from eye nerves. Measurements can also be taken using ultrasound or radiographs, such as a sonograph (or ultrasound image), computerised tomography (CT), magnetic resonance imaging (MRI) or functional MRI (fMRI), positron emission tomography (PET), single photon emission tomography (SPET), or near-infrared spectroscopy (NIRS).

Functional and physiological changes within the brain may be registered by either EEG, MEG or fMRI. Application of fMRI is, however, very limited in comparison with EEG or MEG due to a number of important reasons:

1. The time resolution of fMRI image sequences is very low (for example approximately two frames/s), whereas complete EEG bandwidth can be viewed using EEG or MEG signals.
2. Many types of mental activities, brain disorders and malfunctions of the brain cannot be registered using fMRI since their effect on the level of blood oxygenation is low.
3. The accessibility to fMRI (and currently to MEG) systems is limited and costly.
4. The spatial resolution of EEG, however, is limited to the number of recording electrodes (or number of coils for MEG).

The first electrical neural activities were registered using simple galvanometers. In order to magnify very fine variations of the pointer a mirror was used to reflect the light projected to the galvanometer on the wall. The d'Arsonval galvanometer later featured a mirror mounted on a movable coil and the light focused on the mirror was reflected when a current passed the coil. The capillary electrometer was introduced by Lippmann and Marey [44]. The string galvanometer, as a very sensitive and more accurate measuring instrument, was introduced by Einthoven in 1903. This became a standard instrument for a few decades and enabled photographic recording.

More recent EEG systems consist of a number of delicate electrodes, a set of differential amplifiers (one for each channel) followed by filters [31], and needle (pen) type registers. The multichannel EEGs could be plotted on plane paper or paper with a grid. Soon after this system came to the market, researchers started looking for a computerised system, which could digitise and store the signals. Therefore, to analyse EEG signals it was soon understood that the signals must be in digital form. This required sampling, quantization, and encoding of the signals. As the number of electrodes grows the data volume, in terms of the number of bits, increases.

The computerised systems allow variable settings, stimulations, and sampling frequency, and some are equipped with simple or advanced signal processing tools for processing the signals.

The conversion from analogue to digital EEG is performed by means of multichannel analogue-to-digital converters (ADCs). Fortunately, the effective bandwidth for EEG signals is limited to approximately 100 Hz. For many applications this bandwidth may be considered even half of this value. Therefore, a minimum frequency of 200 Hz (to satisfy the Nyquist criterion) is often enough for sampling the EEG signals. In some applications where a higher resolution is required for representation of brain activities in the frequency domain, sampling frequencies of up to 2000 sample/s may be used.

In order to maintain the diagnostic information the quantization of EEG signals is normally very fine. Representation of each signal sample with up to 16 bits is very popular for the EEG recording systems. This makes the necessary memory volume for archiving the signals massive, especially for sleep EEG and epileptic seizure monitoring records. However, the memory size for archiving the images is often much larger than that used for archiving the EEG signals and with the help of new technology this is a minor problem.

A simple calculation shows that for a one hour recording from 128-electrode EEG signals sampled at 500 samples/s a memory size of $128 \times 60 \times 60 \times 500 \times 16 \approx 3.68$ Gbits $\approx$ 0.45 Gbyte is required. Therefore, for longer recordings of a large number of patients there should be enough storage facilities such as in today's technology Zip disks, CDs, DVDs, large removable hard drives, optical disks, and share servers or the cloud.

Although the format of reading the EEG data may be different for different EEG machines, these formats are easily convertible to spreadsheets readable by most signal processing software packages, such as MATLAB and Python.

The EEG recording electrodes and their proper function are crucial for acquiring high quality data. There are different types of electrodes often used in the EEG recording systems such as:

- Disposable (gel-less, and pre-gelled types)
- Reusable disc electrodes (gold, silver, stainless steel or tin)
- Headbands and electrode caps
- Saline-based electrodes
- Needle electrodes

For multichannel recordings with a large number of electrodes, electrode caps are often used. Commonly used scalp electrodes consist of Ag-AgCl disks, less than 3 mm in diameter, with long flexible leads that can be plugged into an amplifier. Needle (cortical) electrodes are those which have to be implanted under the skull with minimally invasive operations. High impedance between the cortex and the electrodes as well as electrodes with high impedances can lead to distortion, which can even mask the actual EEG signals. Commercial EEG recording systems are often equipped with impedance monitors. To enable a satisfactory recording the electrode impedances should read less than 5 kΩ and be balanced to within 1 kΩ of each other. For more accurate measurement the impedances are checked after each trial.

Due to the layered and spiral structure of the brain, however, distribution of the potentials over the scalp (or cortex) is not uniform [45]. This may affect some of the results of source localization using the EEG signals.

1.7.1   Conventional EEG Electrode Positioning

The International Federation of Societies for Electroencephalography and Clinical Neurophysiology has recommended the conventional electrode setting (also called 10–20) for 21 electrodes (excluding the earlobe electrodes) as depicted in Figure 1.7 [17]. Often, the earlobe electrodes called A1 and A2, connected respectively to the left and right earlobes, are used as the reference electrodes. The 10–20 system avoids both eyeball placement and considers some constant distances by using specific anatomical landmarks from which the measurement would be made and then uses 10 or 20% of that specified distance as the electrode interval. The odd electrodes are on the left and the even ones on the right.

For setting a larger number of electrodes using the above conventional system, the rest of the electrodes are placed equidistantly in between the above electrodes. For example C_1 is placed between C_3 and C_z. Figure 1.8 represents a larger setting for 75 electrodes, including the reference electrodes based on the guidelines by the American EEG Society. Extra electrodes are sometimes used for the measurement of EOC, ECG, and EMG of the eyelid and eye surrounding muscles. In some applications, such as ERP analysis and brain computer interfacing, a single channel may be used. In such applications, however, the position of the corresponding electrode has to be well determined. For example C_3 and C_4 can be used to record, respectively, the right and left finger movement related signals for BCI applications. Also F_3, F_4, P_3, and P_4 can be used for recordings of the ERP P300 signals.

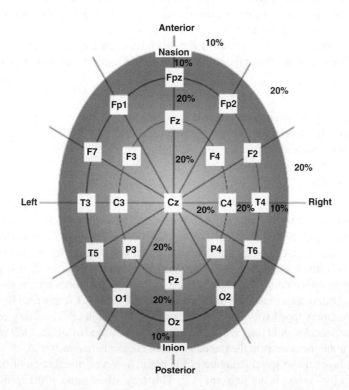

Figure 1.7   Conventional 10–20 EEG electrode positions for the placement of 21 electrodes

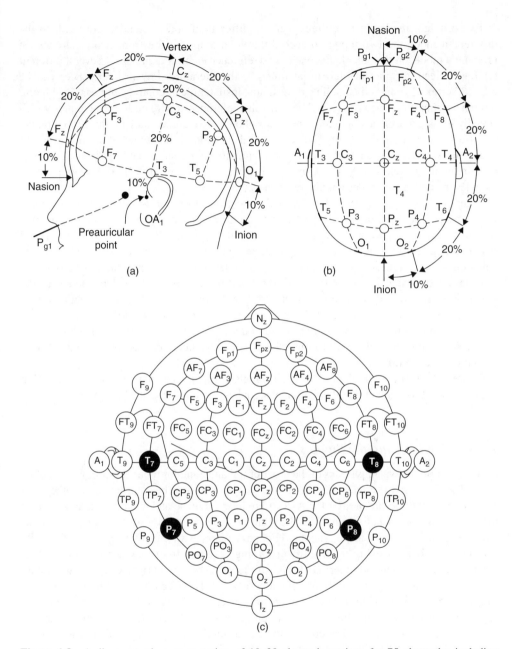

Figure 1.8 A diagrammatic representation of 10–20 electrode settings for 75 electrodes including the reference electrodes; (a) and (b) represent the three-dimensional measures and (c) indicates a two-dimensional view of the electrode set-up configuration

Two different modes of recordings, namely differential and referential, are used. In the differential mode the two inputs to each differential amplifier are from two electrodes. In referential mode, on the other hand, one or two reference electrodes are used. Several different reference electrode placements can be found in the literature. Physical references can be used as vertex (C_z), linked-ears, linked-mastoids, ipsilateral ear, contralateral ear, C_7, bipolar references, and tip of the nose [32]. There are also reference-free recording techniques which actually use a common average reference. The choice of reference may produce topographic distortion if the reference is not relatively neutral. In modern instrumentation, however, the choice of a reference does not play an important role in the measurement [46]. In such systems other references such as FP_z, hand, or leg electrodes may be used [47]. The overall setting includes the active electrodes and the references.

In another similar setting called the Maudsley electrode positioning system the conventional 10–20 system has been modified to better capture the signals from epileptic foci in epileptic seizure recordings. The only difference between this system and the 10–20 conventional system is that the outer electrodes are slightly lowered to enable better capturing of the medial temporal seizure signals. The advantage of this system over the conventional one is that it provides a more extensive coverage of the lower part of the cerebral convexity, increasing the sensitivity for the recording from basal sub-temporal structures [48]. Other deviations from the international 10–20 system as used by researchers are found in [49, 50].

In many applications such as brain-computer interfacing (BCI) and study of mental activity, often a small number of electrodes around the movement-related regions are selected and used from the 10–20 setting system.

Figure 1.9 illustrates a typical set of EEG signals during approximately 7 s of normal adult brain activity. It is evident that the dominant signal cycles are those of alpha rhythm.

1.7.2   Conditioning the Signals

Raw EEG signals have amplitudes of the order of μV and contain frequency components of up to 300 Hz. To retain the effective information the signals have to be amplified before the ADC and filtered, either before or after the ADC, to reduce the noise and make the signals suitable for processing and visualization. The filters are designed in such a way as not to introduce any change or distortion to the signals. Highpass filters with cut-off frequency of usually less than 0.5 Hz are used to remove the disturbing very low frequency components such as those of breathing. On the other hand, high frequency noise is mitigated by using lowpass filters with cut-off frequency of approximately 50 to 70 Hz. Use of notch filters with a null frequency of 50 Hz is often necessary to ensure perfect rejection of the strong 50 Hz power supply. In this case the sampling frequency can be as low as twice the bandwidth commonly used by most EEG systems. The commonly used sampling frequencies for EEG recordings are 100, 250, 500, 1000, and 2000 samples/s. The main artefacts can be divided into patient-related (physiological) and system noise. The patient-related or internal artefacts are body movement-related, EMG, ECG (and pulsation), EOG, ballistocardiogram, and sweating. The system artefacts are 50/60 Hz power supply interference, impedance fluctuation, cable defects, electrical noise from the electronic components, and unbalanced impedances of the electrodes. Often in the preprocessing stage these artefacts are highly mitigated and the informative information is restored. Some methods for removing the EEG artefacts will be discussed in

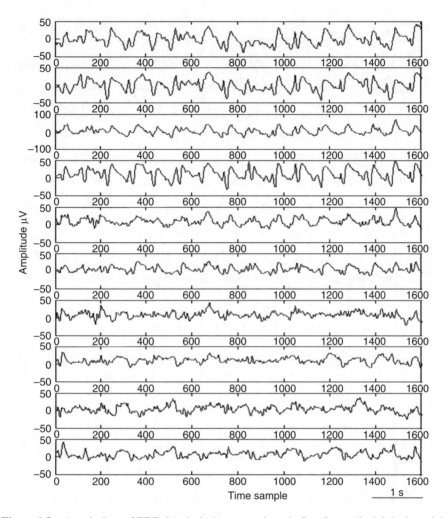

Figure 1.9 A typical set of EEG signals during approximately 7 s of normal adult brain activity

the related chapters of this book. Figure 1.10 shows a set of normal EEG signals affected by an eye-blinking artefact. Similarly, Figure 1.11 represents a multichannel EEG set with clear appearance of ECG signals over the electrodes in the occipital region.

We continue in the next section to highlight the changes in EEG measurements which correlate with physiological and mental abnormalities in the brain.

1.8 Abnormal EEG Patterns

Variations in the EEG patterns for certain states of the subject indicate abnormality. This may be due to distortion and disappearance of abnormal patterns, appearance and increase of abnormal patterns, or disappearance of all patterns. Sharbrough [51] divided the

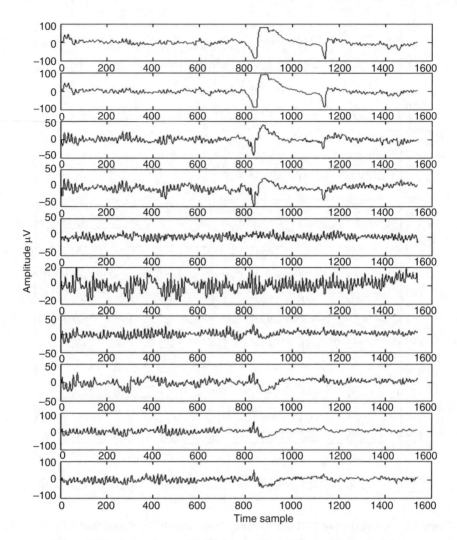

Figure 1.10   A set of normal EEG signals affected by eye-blinking artefact

non-specific abnormalities in the EEGs into three categories; (i) widespread intermittent slow wave abnormalities, often in the delta wave range and associated with brain dysfunction, (ii) bilateral persistent EEG usually associated with impaired conscious cerebral reactions, and (iii) focal persistent EEG usually associated with focal cerebral disturbance.

The first category is a burst type signal, which is attenuated by alerting the individual and eye opening, and accentuated with eye closure, hyperventilation, or drowsiness. The peak amplitude in adults is usually localized in the frontal region and influenced by age. In children, however, it appears over the occipital or posterior head region. Early findings showed that this abnormal pattern frequently appears with an increased intracranial pressure with tumour or aqueductal stenosis. Also, it correlates with grey matter disease, both in cortical and subcortical

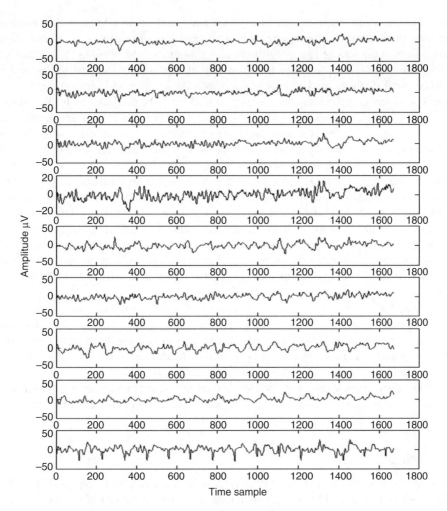

Figure 1.11 A multichannel EEG set with clear appearance of ECG signals over the electrodes in the occipital region

locations. However, it can be seen in association with a wide variety of pathological processes, varying from systemic toxic or metabolic disturbances to focal intracranial lesions.

Regarding the second category, that is, bilateral persistent EEG, the phenomenon in different stages of impaired, conscious, purposeful responsiveness are aetiologically non-specific and the mechanisms responsible for their generation are only partially understood. However, the findings in connection with other information concerning aetiology and chronicity may be helpful in arriving more quickly at an accurate prognosis concerning the patient's chance of recovering his previous conscious life.

As for the third category that is, focal persistent EEG, these abnormalities may be in the form of distortion and disappearance of normal patterns, appearance and increase of abnormal patterns, or disappearance of all patterns, but such changes are seldom seen at the cerebral cortex. The focal distortion of normal rhythms may produce an asymmetry of amplitude, frequency, or reactivity of the rhythm. The unilateral loss of reactivity of a physiological rhythm, such as the loss of reactivity of the alpha rhythm to eye opening [52] or to mental alerting [53], may reliably identify the focal side of abnormality. A focal lesion may also distort or eliminate the normal activity of sleep-inducing spindles and vertex waves.

Focal persistent non-rhythmic delta activity (PNRD) may be produced by focal abnormalities. This is one of the most reliable findings of a focal cerebral disturbance. The more persistent, the less reactive, and the more nonrhythmic and polymorphic is such focal slowing, the more reliable an indicator it becomes for the appearance of a focal cerebral disturbance [54–56]. There are other cases, such as focal inflammation, trauma, vascular disease, brain tumour, or almost any other cause of focal cortical disturbance, including an asymmetrical onset of CNS degenerative diseases that may result in similar abnormalities in the brain signal patterns.

The scalp EEG amplitude from cerebral cortical generators underlying a skull defect is also likely to increase unless acute or chronic injury has resulted in significant depression of underlying generator activity. The distortions in cerebral activities are because focal abnormalities may alter the interconnections, number, frequency, synchronicity, voltage output, and access orientation of individual neuron generators, as well as the location and amplitude of the source signal itself.

With regards to the three categories of abnormal EEGs, their identification and classification requires a dynamic tool for various neurological conditions and any other available information. A precise characterization of the abnormal patterns leads to a clearer insight into some specific pathophysiologic reactions, such as epilepsy, or specific disease processes, such as subacute sclerosing panencephalitis (SSPE) or Creutzfeldt-Jakob (CJD) [51].

Over and above the reasons mentioned above there are many other causes for abnormal EEG patterns. The most common abnormalities are briefly described in the following sections.

1.9   Aging

The aging process affects the normal cerebral activity in waking and sleep, and changes the brain response to stimuli. The changes stem from reducing the number of neurons and due to a general change in the brain pathology. This pathology indicates that the frontal and temporal lobes of the brain are more affected than the parietal lobes, resulting in shrinkage of large neurons and increasing the number of small neurons and glia [57]. A diminished cortical volume indicates that there is age-related neuronal loss. A general cause for aging of the brain may be the decrease in cerebral blood flow [57].

A reduction in the alpha frequency is probably the most frequent abnormality in EEG. This often introduces a greater anterior spread to frontal regions in the elderly and reduces the alpha wave blocking response and reactivity. The diminished mental function is somehow related to the degree of bilateral slowing in the theta and delta waves [57].

Although the changes in high frequency brain rhythms have not been well established, some researchers have reported an increase in beta wave activity. This change in beta wave activity may be considered as an early indication of intellectual loss [57].

As for the sleep EEG pattern, older adults enter into drowsiness with a more gradual decrease in EEG amplitude. Over the age of 60, the frontocentral waves become slower, the frequency of the temporal rhythms also decreases, frequency lowering with slow eye movements become more prominent, and spindles appear in the wave pattern after the dropout of the alpha rhythm. The amplitudes of both phasic and tonic NREM sleep EEG [57] reduce with age. There is also significant change in REM sleep organization with age; the REM duration decreases during the night and there is significant increase in the sleep disruption [57].

Dementia is the most frequent mental disorder which occurs predominantly in the elderly. Therefore, the prevalence of dementia increases dramatically with aging of the society. Generally, EEG is a valuable diagnostic tool in differentiation between organic brain syndromes (OBSs) and functional psychiatric disorders [57], and together with evoked potentials (EPs) plays an important role in the assessment of normal and pathological aging. Aging is expected to change most neurophysiological parameters. However, the variability of these parameters must exceed the normal degree of spontaneous variability to become a diagnostic factor in acute and chronic disease conditions. Automatic analysis of the EEG during sleep and wakefulness may provide a better contrast in the data and enable a robust diagnostic tool. We next describe particular and very common mental disorders whose early onset may be diagnosed and their treatment or progress monitored with EEG measurements.

1.10   Mental Disorders

1.10.1   Dementia

Dementia is a syndrome that consists of a decline in intellectual and cognitive abilities. This consequently affects the normal social activities, mode, and the relationship and interaction with other people [58]. EEG is often used to study the effect and progress of dementia. In most cases such as in primary degenerative dementia, for example, Alzheimer's, and psychiatric disorder, for example, depression with cognitive impairment, EEG can be used to detect the abnormality [59].

Dementia is classified into cortical and subcortical forms [59]. The most important cortical dementia is Alzheimer's disease (AD), which accounts for approximately 50% of the cases. Other known cortical abnormalities are Pick's and CJD diseases. They are characterised clinically by findings such as aphasia, apraxia, and agnosia. CJD can often be diagnosed using the EEG signals. Figure 1.12 shows a set of EEG signals from a CJD patient. On the other hand, the most common subcortical diseases are Parkinson's disease, Huntington's disease, lacunar state, normal pressure hydrocephalus, and progressive supranuclear palsy. These diseases are characterised by forgetfulness, slowing of thought processes, apathy, and depression. Generally, subcortical dementias introduce less abnormality to the EEG patterns than the cortical ones [59].

In AD the posterior EEG (alpha) rhythm slows down and the delta and theta wave activities increase. On the other hand, beta wave activity may decrease. In the severe cases epileptiform discharges and triphasic waves can appear. In such cases, cognitive impairment often results. The spectral power also changes; the power increases in delta and theta bands and decreases in beta and alpha bands and also in mean frequency. New research shows the impairment of connectivity for AD patients too.

The EEG wave morphology is almost the same for AD and Pick's disease. Pick's disease involves the frontal and temporal lobes. An accurate analysis followed by an efficient classification of the cases may discriminate these two diseases. CJD is a mixed cortical and subcortical dementia. This causes slowing of the delta and theta wave activities and, after

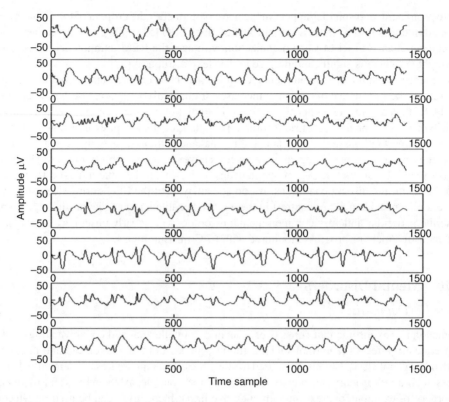

Figure 1.12   A set of multichannel EEG signals from a patient suffering from CJD

approximately three months from the onset of the disease, periodic sharp wave complexes are generated which occur almost every second, together with decrease in the background activity [59]. Parkinson's disease is a subcortical dementia, which causes slowing down of the background activity and an increase in the theta and delta wave activities. Some work has been undertaken using spectral analysis to confirm the above changes [60]. Some other disorders, such as depression, have a lesser effect on the EEGs and more accurate analysis of the EEGs has to be performed to detect the signal abnormalities for these brain disorders.

Generally, EEG is usually used in the diagnosis and evaluation of many cortical and sub-cortical dementias. Often it can help to differentiate between a degenerative disorder such as AD, and pseudo-dementia due to psychiatric illness [59]. The EEG may also show whether the process is focal or diffuse (i.e. involves the background delta and theta wave activities). In addition, EEG often can reveal the early CJD-related abnormalities. However, more advanced signal processing and quantitative techniques may be implemented to achieve robust diagnostic and monitoring performance.

1.10.2   Epileptic Seizure and Nonepileptic Attacks

Often the onset of a clinical seizure is characterised by a sudden change of frequency in the EEG measurement. It is normally within the alpha wave frequency band with slow decrease in frequency (but increase in amplitude) during the seizure period. It may or may not be spiky in shape. Sudden desynchronization of electrical activity is found in electrodecremental seizures. The transition from preictal to ictal state, for a focal epileptic seizure, consists of gradual change from chaotic to ordered waveforms. The amplitude of the spikes does not necessarily represent the severity of the seizure. Rolandic spikes in a child of 4–10 years for example, are very prominent; however, the seizure disorder is usually quite benign or there may not be clinical seizure [61].

In terms of spatial distribution, in childhood the occipital spikes are very common. Rolandic central-midtemporal-parietal spikes are normally benign, whereas frontal spikes or multifocal spikes are more epileptogenic. The morphology of the spikes varies significantly with age. However, the spikes may occur in any level of awareness including wakefulness and deep sleep.

Distinction of seizure from common artefacts is not difficult. Seizure artefacts within an EEG measurement have a prominent spiky but repetitive (rhythmical) nature, whereas the majority of other artefacts are transients or noise-like in shape. For the case of the ECG, the frequency of occurrence of the QRS waveforms is approximately 1 Hz. These waveforms have a certain shape which is very different from that of seizure signals.

The morphology of an epileptic seizure signal changes slightly from one type to another. The seizure may appear in different frequency ranges. For example, a petit mal discharge often has a slow spike around 3 Hz, lasting for approximately 70 ms, and normally has its maximum amplitude around the frontal midline. On the other hand, higher frequency spike wave complexes occur for patients over 15 years old. Complexes at 4 and 6 Hz may appear in the frontal region of the brain of epileptic patients. As for the 6 Hz complex (also called benign EEG variants and patterns), patients with anterior 6 Hz spike waves are more likely to have epileptic seizures and those with posterior discharges tend to have neuro-autonomic disturbances [62]. The experiments do not always result in the same conclusion [61]. It was also found that the occipital 6 Hz spikes can be seen and are often drug related (due to hypoanalgetics or barbiturates) and associated with drug withdrawal [63].

Amongst nonepileptics, the discharges may occur in patients with cerebrovascular disorder, syncopal attacks, and psychiatric problems [61]. Fast and needle-like spike discharges may be seen over the occipital region in most congenitally blind children. These spikes are unrelated to epilepsy and normally disappear in older age patients.

Bursts of 13–16 Hz or 5–7 Hz, as shown in Figure 1.13, (also called 14 and 6 Hz waves) with amplitudes less than 75 μV and arch shapes may be seen over the posterior temporal and the nearby regions of the head during sleep. These waves are positive with respect to the background waves. The 6 and 14 Hz waves may appear independently and be found, respectively, in younger and older children. These waves may be confined to the regions lying beneath a skull defect. Despite the 6 Hz wave, there are rhythmical theta bursts of wave activities relating to drowsiness around the mid-temporal region with a morphology very similar to ictal patterns. In old age patients other similar patterns, such as *subclinical rhythmic EEG discharges of adults* (SREDA) over the 4–7 Hz frequency band around the centroparietal region, and a wide frequency range (2–120 Hz) *temporal minor sharp transient* and *wicket spikes* over anterior temporal and midtemporal lobes of the brain may occur. These waves are also nonepileptic but with seizure-type waveform [61].

The epileptic seizure patterns, called ictal wave patterns appear during the onset of epilepsy. Although Chapter 15 of this book focuses on analysis of these waveforms from a signal processing point of view, here a brief explanation of the morphology of these waveforms is given. Researchers in signal processing may exploit these concepts in the development of

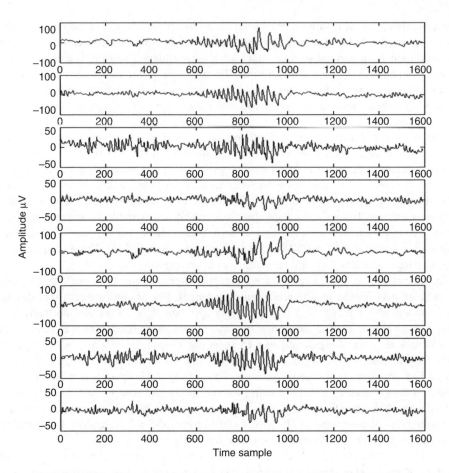

Figure 1.13   Bursts of 3–7 Hz seizure activity in a set of adult EEG signals

their algorithms. Although these waveform patterns are often highly obscured by the muscle movements, they normally maintain certain key characteristics.

Tonic-clonic seizure (also called grand mal) is the most common type of epileptic seizure. It appears in all electrodes but more toward the frontal electrodes (Figure 1.14). It has a rhythmic but spiky pattern in the EEG and occurs within the frequency range 6–12 Hz. Petit mal is another interictal paroxysmal seizure pattern which occurs at approximately 3 Hz with a generalized synchronous spike wave complex of prolonged bursts. A temporal lobe seizure (also called a psychomotor seizure or complex partial seizure) is presented by bursts of serrated slow waves with relatively high amplitude of above 60 μV and frequencies of 4–6 Hz. Cortical (focal) seizures have contralateral distribution with rising amplitude and diminishing frequency during the ictal period. The attack is usually initiated by local desynchronization, that is, very fast and very low voltage spiky activity, which gradually rises in amplitude with diminishing frequency. Myoclonic seizures have concomitant polyspikes seen clearly in the EEG signals. They can have generalized or bilateral spatial distribution more dominant in the frontal region

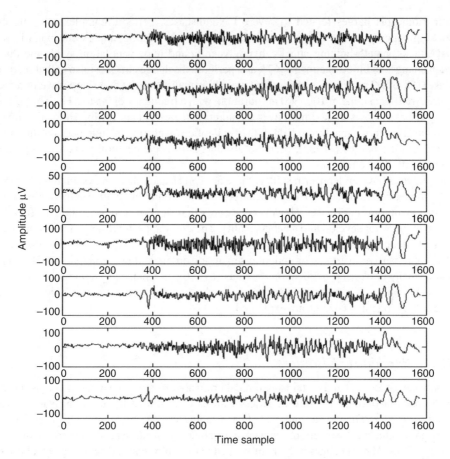

Figure 1.14   Generalized tonic-clinic (grand mal) seizure. The seizure appears in almost all the electrodes

[64]. Tonic seizures occur in patients with Lennox–Gastaut syndrome [65] and have spikes which repeat with frequency approximately 10 Hz. Atonic seizures may appear in the form of a few seconds drop attack or be inhibitory, lasting for a few minutes. They show a few polyspike waves or spike waves with generalized spatial distribution of approximately 10 Hz, followed by large slow waves of 1.5–2 Hz [66]. Akinetic seizures are rare and characterised by arrest of all motion, which, however, is not caused by sudden loss of tone as in atonic seizure and the patient is in an absent-like state. They are rhythmic with frequency of 1–2 Hz. Jackknife seizures also called salaam attacks, are common in children with hypsarrhythmia (infantile spasms, West syndrome) and are either in the form of sudden generalised flattening desynchronization or have rapid spike discharges [65].

There are generally several varieties of recurring or quasi-recurring discharges, which may or may not be related to epileptic seizure. These abnormalities may be due to psychogenic changes, variation in body metabolism, circulatory insufficiency (which appears often as acute cerebral ischaemia). Of these, the most important ones are: periodic or quasiperiodic discharges related to severe CNS diseases, periodic complexes in subacute sclerosing panencepalities (SSPE), periodic complexes in herpes simplex encephalitis, syncopal attacks, breath-holding attacks, hypoglycemia and hyperventilation syndrome due to sudden changes in blood chemistry [67], and periodic discharges in CJD (also called mad cow disease) [68,69]. The waveforms for this latter abnormality consist of a sharp wave or a sharp triphasic transient signal of 100–300 ms duration, with a frequency of 0.5 to 2 Hz. The periodic activity usually shows a maximum over the anterior region except for the Heidenhain form, which has a posterior maximum [61]. Other epileptic waveforms include periodic literalised epileptiform discharges (PLED), periodic discharges in acute cerebral anoxia, and periodic discharges of other etiologies.

Despite the above epileptiform signals there are spikes and other paroxysmal discharges in healthy nonepileptic persons. These discharges may be found in healthy individuals without any other symptoms of diseases. However, they are often signs of certain cerebral dysfunctions that may or may not develop into an abnormality. They may appear during periods of particular mental challenge on individuals, such as soldiers in the war front line, pilots, and prisoners.

A comprehensive overview of epileptic seizure disorders and non-epileptic attacks can be found in many books and publications such as [67, 70]. In Chapter 15 some recent attempts in the application of advanced signal processing techniques to the automatic detection and prediction of epileptic seizures are explained.

1.10.3   Psychiatric Disorders

Not only can functional and certain anatomical brain abnormalities be investigated using EEG signals, pathophysiological brain disorders can also be studied by analysing such signals. According to the "Diagnostic and Statistical Manual (DSM) of Mental Disorders" of the American Psychiatric Association, changes in psychiatric education have evolved considerably since the 1970s. These changes have mainly resulted from physical and neurological laboratory studies based upon EEG signals [71].

There have been evidences from EEG coherence measures suggesting differential patterns of maturation between normal and learning disabled children [72]. This finding can lead to the establishment of some methodology in monitoring learning disorders.

Several psychiatric disorders are diagnosed by analysis of evoked potentials (EPs) achieved by primarily averaging a number of consecutive trials having the same stimuli. New approaches for single trial analysis and trial-by-trial tracking have considerably enhanced the clinical diagnosis and treatment.

A number of pervasive mental disorders, such as: dyslexia, which is a developmental reading disorder; autistic disorder which is related to abnormal social interaction, communication, and restricted interests and activities, and starts appearing from the age of three; Rett's disorder, characterized by the development of multiple deficits following a period of normal post-natal functioning; and Asperger's disorder which leads to severe and sustained impairments in social interaction and restricted repetitive patterns of behaviour, interests, and activities; cause significant losses in multiple functioning areas [71].

Attention deficit hyperactivity disorder (ADHD) and attention deficit disorder (ADD), conduct disorder, oppositional defiant disorder, and disruptive behaviour disorder have also been under investigation and considered within the DSM. Most of these abnormalities appear during childhood and often prevent children from learning and socialising well. The associated EEG features have been rarely analytically investigated, but the EEG observations are often reported in the literature [73–77]. However, most of such abnormalities tend to disappear with advancing age.

EEG has also been analysed recently for the study of delirium [78, 79], dementia [80, 81], and many other cognitive disorders [82]. In EEGs, characteristics of delirium include slowing or dropout of the posterior dominant rhythm, generalized theta or delta slow-wave activity, poor organization of the background rhythm, and loss of reactivity of the EEG to eye opening and closing. In parallel with that, the quantitative EEG (QEEG) shows increased absolute and relative slow-wave (theta and delta) power, reduced ratio of fast-to-slow band power, reduced mean frequency, and reduced occipital peak frequency [79].

Dementia includes a group of neurodegenerative diseases that cause acquired cognitive and behavioural impairment of sufficient severity to interfere significantly with social and occupational functioning. AD is the most common of the diseases that cause dementia. At present, the disorder afflicts approximately 5 million people in the United States and more than 30 million people worldwide. A larger number of individuals have lesser levels of cognitive impairment, which frequently evolve into full-blown dementia. Prevalence of dementia is expected to nearly triple by 2050, since the disorder preferentially affects the elderly, who constitute the fastest-growing age bracket in many countries, especially in industrialized nations [81].

Amongst other psychiatric and mental disorders, amnestic disorder (or amnesia), mental disorder due to general medical condition, substance-related disorder, schizophrenia, mood disorder, anxiety disorder, somatoform disorder, dissociative disorder, sexual and gender identity disorder, eating disorders, sleep disorders, impulse-controlled disorder, and personality disorders have often been addressed in the literature [71]. However, in many of these cases the corresponding EEGs have seldom been analysed by means of advanced signal processing tools.

1.10.4 External Effects

EEG signal patterns may significantly change when using drugs for the treatment and suppression of various mental and CNS abnormalities. Variation in the EEG patterns may also arise by just looking at the TV screen or listening to music without any attention. However, amongst the external effects the most significant ones are the pharmachological and drug effects. Therefore, it is important to know the effects of these drugs on the changes of EEG waveforms due to chronic overdosage, and the patterns of overt intoxication [83]. The changes in these waveforms may be monitored if taken during drug infusion.

The effect of administration of drugs for anaesthesia on EEGs is of interest to clinicians. The related studies attempt to find the correlation between the EEG changes and the stages of anaesthesia. It has been shown that in the initial stage of anaesthesia a fast frontal activity appears. In deep anaesthesia this activity becomes slower with higher amplitude. In the last stage, a burst-suppression pattern indicates the involvement of brainstem functions, including respiration and finally the EEG activity ceases [83]. In cases of acute intoxication, the EEG patterns are similar to those of anaesthesia [83].

As an example, barbiturates are commonly used as anticonvulsant and antiepileptic drug. With small dosage of barbiturate the activities within the 25–35 Hz frequency band around the frontal cortex increase. This changes to 15–25 Hz and spreads to the parietal and occipital regions. Dependence and addiction to barbiturates are common. Therefore, after a long-term ingestion of barbiturates, its abrupt withdrawal leads to paroxysmal abnormalities. The major complications are myoclonic jerks, generalized tonic-clonic seizures, and delirium [83].

Many other drugs are used in addition to barbiturates as sleeping pills, such as melatonin and bromides. Very pronounced EEG slowing is found in chronic bromide encephalopathies [83]. Antipsychotic drugs also influence the EEG patterns. For example, neuroleptics increase the alpha wave activity but reduce the duration of beta wave bursts and their average frequency. As another example, clozapine increases the delta, theta, and above 21 Hz beta wave activities. As another antipsychotic drug, tricyclic antidepressants, such as imipramine, amitriptyline, doxepin, desipramine, notryptiline, and protriptyline, increase the amount of slow and fast activity along with instability of frequency and voltage, and also slow down the alpha wave rhythm. After administration of tricyclic antidepressants the seizure frequency in chronic epileptic patients may increase. With high dosage, this may further lead to single or multiple seizures occurring in nonepileptic patients [83].

During acute intoxication, a widespread, poorly reactive, irregular 8–10 Hz activity and paroxysmal abnormalities including spikes, as well as unspecific coma patterns, are observed in the EEGs [83]. Lithium is often used in the prophylactic treatment of bipolar mood disorder. The related changes in the EEG pattern consist of slowing of the beta rhythm and of paroxysmal generalised slowing, occasionally accompanied by spikes. Focal slowing also occurs, which is not necessarily a sign of a focal brain lesion. Therefore, the changes in the EEG are markedly abnormal with lithium administration [83]. The beta wave activity is highly activated by using benzodiazepines, as an anxiolytic drug. These activities persist in the EEG as long as two weeks after ingestion. Benzodiazepine leads to a decrease in an alpha wave activity and its amplitude, and slightly increases the 4–7 Hz frequency band activity. In acute intoxication the EEG shows prominent fast activity with no response to stimuli [83]. The sychotogenic drugs such as lysergic acid diethylamide and mescaline decrease the amplitude and possibly depress the slow waves [83].

The CNS stimulants increase the alpha and beta wave activities and reduce the amplitude and the amount of slow waves and background EEGs [83].

The effect of many other drugs, especially antiepileptic drugs, is investigated and new achievements are published frequently. One of the significant changes in the EEG of epileptic patients with valproic acid consists of reduction or even disappearance of generalized spikes along with seizure reduction. Lamotrigine is another antiepileptic agent that blocks voltage-gated sodium channels, thereby preventing excitatory transmitter glutamate release. With the intake of lamotrigine a widespread EEG attenuation occurs [83]. Penicillin if administered in high dosage may produce jerks, generalized seizures, or even status epilepticus [83].

1.11 Memory and Content Retrieval

Recognition of human memory is a very complicated subject to follow. Memory refers to the processes by which information is encoded, stored, and retrieved. Encoding allows information from the outside world to stimulate the brain in the forms of chemical and physical stimuli. This involves three stages. In the first stage we must change the information so that we may put the memory into the encoding process. Storage, as the second memory process or stage, entails that we maintain information over periods of time. Finally, the third process/stage concerns the retrieval of information stored previously.

Memory is initially recognized as either recognition or recall memory. Recognition memory tasks require individuals to indicate whether they have encountered a stimulus (such as an audio or visual one) before. Recall memory tasks require participants to retrieve previously learned information. For example, individuals may perform a sequence of actions they have seen before or repeat a set of words they have heard before.

Memory may be separated into declarative, requiring conscious retrieva, and non-declarative, related to skills and habits [84]. Recognition memory is a subcategory of declarative memory. Declarative memory is sub-divided into semantic (facts) and episodic (events). Semantic memory refers to facts which are available to everyone while episodic memory concerns events particular to the person's experiences. Memory can further be classified into sensory, short- and long-term memories [85]. Sensory memory corresponds approximately to the first 200–500 ms after an item is perceived. The ability to look at an item and remember what it looked like with just a glance or memorization is an example of sensory memory. Sensory information which comes from the environment is archived into short-term memory of the human brain. Hence, the temporary storage system called working memory, as part of short-term memory, transfers information to the long-term memory. The level of attention/interest to a specific stimulus and the amount of neuronal inter-connections information can be kept for a longer time [85].

A number of measurement procedures have been proposed for memory evaluation. For example Old–New recognition is a protocol used to assess recognition memory based on the pattern of yes–no responses [86]. In this simple test a participant is given an item and asked to indicate 'yes' if it is old and 'no' if it is a new item. This method of recognition testing makes the retrieval process easy to record and analyse [87]. Another testing protocol is forced choice recognition. Following this protocol, the participants are asked to identify out of several items which one is old [86]. In the presented items, one is the target, which is the actual previously presented item. The other items can be very similar and act as distractors. This allows the experimenter some degree of manipulation and control in item similarity or item resemblance. This helps provide a better understanding of retrieval and what kinds of existing knowledge are used to make decisions based on memory [86].

From a signal processing perspective, detection and tracking the changes in the ERP characteristics is the main ground for research in the assessment of memory. The ERP components N100, P200, N400, P600 and so-called later ERPs change with the change in the memory task [85]. Therefore, evaluation of these components has been the main agenda in memory processing.

In a number of memory studies, for example, the Old–New effect has been hypothesized as related to some specific ERP components, like N400 in the anterior parts of the brain, P600 in the posterior parts of the brain and the components within 800–1800 ms called late positive complex. For instance, in [88], the early visual ERP component at the frontal lobe (frontal-N400) and later components were related to familiarity and recollection in a dual process framework, respectively. Also in [89], the early ERP component (100–175 ms) was found to be sensitive to novelty of individual fractal (abstract images) but later components hypothesized to be indicative for the novelty of the association between fractals. In another EEG/ERP source retrieval study in recognition memory, an early parietal ERP Old–New effect in the time interval (300–700 ms) and a late posterior negativity (LPN) in time interval

(300–700 msec) was larger for Old (already studied) than New (not studied before) responses, irrespective of source accuracy [90]. Similarly, using the Remember–Know routine [91] a more positive late positive complex associated with "Remember" compared to "Know" responses has been reported.

1.12   MEG Signals and Their Generation

The MEG signals, though noisier, look similar to EEG signals. Synchronized neuronal currents induce weak magnetic fields of 10 femtotesla (fT) for cortical activity and 10^3 fT for the human alpha rhythm. This is considerably smaller than the ambient magnetic noise in an urban environment, which is of the order of 10^8 fT or 0.1 µT. The essential problem of biomagnetism is thus the weakness of the signal relative to the sensitivity of the detectors, and to the competing environmental noise.

Like EEG signals, The MEG is generated from the net effect of ionic currents flowing in the dendrites of neurons during synaptic transmission. Any electrical current, consequently, produces an orthogonally oriented magnetic field which is measured as an MEG field. The net currents can be thought of as electric dipoles, that is, currents with a position, orientation, and magnitude, but no spatial extent. According to the right-hand rule, a current dipole gives rise to a magnetic field that flows around the axis of its vector component.

To generate a signal that is detectable, approximately 50 000 active neurons are needed [92]. Since the current dipoles must have similar orientations to generate magnetic fields that reinforce each other, it is often the layer of pyramidal cells, which are situated perpendicular to the cortical surface, that give rise to measurable magnetic fields. Bundles of these neurons that are orientated tangentially to the scalp surface project measurable portions of their magnetic fields outside the head, and these bundles are typically located in the sulci [92].

MEG is not attenuated by brain tissue. Therefore, the head is a more homogenous medium for MEG compared with EEG. This allows estimation of a better model of the head and consequently more accurate localization of the brain sources as dipoles. In addition, deep brain sources are more likely to be detected from these signals. This, however, has not yet become practical and no clinically useful method is currently available to use MEG for this purpose.

It is worth noting that action potentials do not usually produce an observable field, mainly because the currents associated with action potentials flow in opposite directions and the magnetic fields cancel out. However, action fields have been measured from peripheral nerves.

In some MEG research laboratories a multichannel EEG set can be integrated within the MEG system. This allows detection of tangential currents whose fields cannot be detected by MEG SQUIDs. Such a set-up can also be used for estimation of the real head model for EEG analysis.

1.13   Conclusions

In this chapter the fundamental concepts in the generation of action potentials, and consequently the EEG signals, have been briefly explained. The conventional measurement set-ups for EEG recording and the brain rhythms present in normal or abnormal EEGs have also been described. In addition the effects of popular brain abnormalities, such as mental diseases, aging, and epileptic and non-epileptic attacks have been pointed out. Despite the known neurological, physiological, pathological, and mental abnormalities mentioned in this chapter, there are many other brain disorders and dysfunctions which may or may not manifest some kinds of abnormalities in the related EEG signals. Degenerative disorders of the CNS [93], such as a variety of lysosomal disorders, several peroxisomal disorders, a number of mitochondrial disorders, inborn disturbances of the urea cycle, many aminoacidurias, and other metabolic and degenerative diseases, as well as chromosomal aberrations have to be evaluated and their symptoms correlated with the changes in the EEG patterns. The similarities and differences within the EEGs of these diseases have to be well understood. On the other hand, the developed mathematical algorithms need to take the clinical observations and findings into account to further enhance the outcome of such processing. Although a number of technical methods have been well established for processing the EEGs with relation to the above abnormalities, there is still a long way to go and many questions to be answered.

The following chapters of this book introduce new digital signal processing techniques employed mainly for analysis of EEG signals, followed by a number of examples in the applications of such methods.

References

[1] Caton, R. (1875) The electric currents of the brain. *Br. Med. J.*, **2**, 278.

[2] Walter, W.G. (1964) Slow potential waves in the human brain associated with expectancy, attention and decision. *Arch. Psychiat. Nervenkr.*, **206**, 309–322.

[3] Cobb, M. (2002) Exorcizing the animal spirits: Jan Swammerdam on nerve function. *Neuroscience*, **3**, 395–400.

[4] Danilevsky, V.D. (1877) Investigation into the Physiology of the Brain. Doctoral Thesis, University Charkov. Quoted after Brazier.

[5] Brazier, M.A.B. (1961) *A History of the Electrical Activity of the Brain; The First Half-Century*, Macmillan, New York.

[6] Massimo, A. (2004) In Memoriam Pierre Gloor (1923–2003): An appreciation. *Epilepsia*, **45**(7), 882.

[7] Grass, A.M. and Gibbs, F.A. (1938) A Fourier transform of the electroencephalogram. *J. Neurophysiol.*, **1**, 521–526.

[8] Haas, L.F. (2003) Hans Berger (1873–1941), Richard Caton (1842–1926), and electroencephalography. *J. Neurol. Neurosurg. Psychiatry*, **74**, 9.

[9] Spear, J.H. (2004) Cumulative change in scientific production: Research technologies and the structuring of new knowledge. *Persp. Sci.*, **12**(1), 55–85.

[10] Shipton, H.W. (1975) EEG analysis: A history and prospectus. Annual Reviews, University of Iowa, USA, pp. 1–15.

[11] Fischer, M.H. (1933) Elektrobiologische Auswirkungen von Krampfgiften am Zentralnervensystem. *Med. Klin.*, **29**, 15–19.

[12] Fischer, M.H. and Lowenbach, H. (1934) Aktionsstrome des Zentralnervensystems unter der Einwirkung von Krampfgiften, "1. Mitteilung Strychnin und Pikrotoxin". *Arch. F. Exp. Pathol. Und Pharmakol.*, **174**, 357–382.

[13] Kornmuller, A.E. (1935) Der Mechanismus des epileptischen anfalles auf grund bioelektrischer untersuchungen am zentralnervensystem. *Fortschr. Neurol. Psychiatry*, **7**, 391–400; 414–432.

[14] Bremer, F. (1935) Cerveau isole' et physiologie du sommeil. *C.R. Soc. Biol. (Paris)*, **118**, 1235–1241.

[15] Niedermeyer, E. (1999) Historical aspects, Chapter 1, in *Electroencephalography, Basic Principles, Clinical Applications, and Related Fields*, 4th edn (eds E. Niedermeyer and F. Lopes da Silva), Lippincott Williams & Wilkins, pp. 1–14.

[16] Berger, H. (1933) Über das Elektrenkephalogramm des menschen, 7th report. *Arch. Psychiat. Nervenkr.*, **100**, 301–320.

[17] Jasper, H. (1958) Report of committee on methods of clinical exam in EEG. *Electroencephalogr. Clin. Neurophysiol.*, **10**, 370–375.

[18] Motokawa, K. (1949) Electroencephalogram of man in the generalization and differentiation of condition reflexes. *Tohoku J. Expl. Medicine*, **50**, 225.

[19] Niedermeyer, E. (1973) Common generalized epilepsy. The so-called idiopathic or centrencephalic epilepsy. *Eur. Neurol.*, **9**(3), 133–156.

[20] Brock, L.G., Coombs, J.S. and Eccles, J.C. (1952) The recordings of potentials from motor neurons with an intracellular electrode. *J. Physiol.*, **117**, 431–460.

[21] Phillips, C.G. (1961) Some properties of pyramidal neurones of the motor cortex, in *The Nature of Sleep* (eds G.E.W. Wolstenholme and M. O'Conner), Little, Brown, Boston, pp. 4–24.

[22] Aserinsky, E. and Kleitman, N. (1953) Regularly occurring periods of eye motility, and concomitant phenomena, during sleep. *Science*, **118**, 273–274.

[23] Cohen, D. (1968) Magnetoencephalography: evidence of magnetic fields produced by alpha rhythm currents. *Science*, **161**, 784–786.

[24] Zimmerman, J.E., Theine, P. and Harding, J.T. (1970) Design and operation of stable rf-biased superconducting point-contact quantum devices, etc. *J. Appl. Phys.*, **41**, 1572–1580.

[25] Speckmann, E.-J. and Elger, C.E. (1999) Introduction to the neurophysiological basis of the EEG and DC potentials, in *Electroencephalography*, 4th edn (eds E. Niedermeyer and F. Lopes da Silva), Lippincott Williams and Wilkins, pp. 15–27.

[26] Shepherd, G.M. (1974) *The Synaptic Organization of the Brain*, Oxford University Press, London.

[27] Caspers, H., Speckmann, E.-J. and Lehmenkühler, A. (1986) DC potentials of the cerebral cortex, Seizure activity and changes in gas pressures. *Rev. Physiol. Biochem. Pharmacol.*, **106**, 127–176.

[28] Perry, E.K., Ashton, H. and Young, A.H (eds) (2002) *Neurochemistry of Consciousness*, John Benjamins Publishing Co.

[29] Attwood, H.L. and MacKay, W.A. (1989) *Essentials of Neurophysiology*, B.C. Decker, Hamilton, Canada.

[30] Nunez, P.L. (1995) *Neocortical Dynamics and Human EEG Rhythms*, Oxford University Press, New York.

[31] Teplan, M. (2002) Fundamentals of EEG measurements. *Meas. Sci. Rev.*, **2**, Sec. 2.

[32] Bickford, R.D. (1987) Electroencephalography, in *Encyclopedia of Neuroscience* (ed. G. Adelman), Birkhauser, Cambridge (USA), pp. 371–373.

[33] Sterman, M.B., MacDonald, L.R. and Stone, R.K. (1974) Biofeedback training of sensorimotor EEG in man and its effect on epilepsy. *Epilepsia*, **15**, 395–416.

[34] Ashwal, S. and Rust, R. (2003) Child neurology in the 20th century. *Pediatr. Res.*, **53**, 345–361.

[35] Niedermeyer, E. (1999) The normal EEG of the waking adult, Chapter 10, in *Electroencephalography, Basic Principles, Clinical Applications, and Related Fields*, 4th edn, (eds E. Niedermeyer and F. Lopes da Silva), Lippincott Williams & Wilkins, pp. 174–188.

[36] Pfurtscheller, G., Flotzinger, D. and Neuper, C. (1994) Differentiation between finger, toe and tongue movement in man based on 40 Hz EEG. *Electroencepalogr. Clin. Neurophysiol.*, **90**, 456–460.

[37] Adrian, E.D. and Mattews, B.H.C. (1934) The Berger rhythm, potential changes from the occipital lob in man. *Brain*, **57**, 345–359.

[38] Trabka, J. (1963) High frequency components in brain waves. *Electroencephalogr. Clin. Neurophysiol.*, **14**, 453–464.

[39] Cobb, W.A., Guiloff, R.J. and Cast, J. (1979) Breach rhythm: The EEG related to skull defects. *Electroencephalogr. Clin. Neurophysiol.*, **47**, 251–271.

[40] Roldan, E., Lepicovska, V., Dostalek, C. and Hrudova, L. (1981) Mu-like EEG rhythm generation in the course of Hatha-yogi exercises. *Electroencephalor. Clin. Neurophysiol.*, **52**, 13.

[41] IFSECN (1974) A glossary of terms commonly used by clinical electroencephalographers. *Electroencephalogr. Clin. Neurophysiol.*, **37**, 538–548.

[42] Jansen, B.H. and Rit, V.G. (1995) Electroencephalogram and visual evoked potential generation in a mathematical model of coupled cortical columns. *Biol. Cybern.*, **73**, 357–366.

[43] David, O. and Friston, K.J. (2003) A neural mass model for MEG/EEG coupling and neuronal dynamics. *NeuroImage*, **20**, 1743–1755.

[44] O'Leary, J.L. and Goldring, S. (1976) *Science and Epilepsy*, Raven Press, NY, pp. 19–152.

[45] Gotman, J., Ives, J.R. and Gloor, R. (1979) Automatic recognition of interictal epileptic activity in prolonged EEG recordings. *Electroencephalogr. Clin. Neurophysiol.*, **46**, 510–520.

[46] Effects of electrode placement, http://www.focused-technology.com/electrod.htm, California, accessed January 2013.

[47] Collura, T. (1998) A Guide to Electrode selection, Location, and Application for EEG Biofeedback, Ohio.

[48] Nayak, D., Valentin, A., Alarcon, G. *et al.* (2004) Characteristics of scalp electrical fields associated with deep medial temporal epileptiform discharges. *Clin. Neurophysiol.*, **115**, 1423–1435.

[49] Barrett, G., Blumhardt, L., Halliday, L. *et al.* (1976) A paradox in the lateralization of the visual evoked responses. *Nature*, **261**, 253–255.

[50] Halliday, A.M. (1978) Evoked potentials in neurological disorders, in *Event-related Brain Potentials in Man* (eds E. Calloway, P. Tueting and S.H. Coslow), Academic Press, pp. 197–210.

[51] Sharbrough, F.W. (1999) Nonspecific abnormal EEG patterns, Chapter 12, in *Electroencephalography, Basic Principles, Clinical Applications, and Related Fields*, 4th edn (eds E. Niedermeyer and F. Lopes Da Silva) Lippincott Williams & Wilkins, pp. 215–234.

[52] Bancaud, J., Hecaen, H. and Lairy, G.C. (1955) Modification de la reactivite E.E.G., troubles des functions symboliques et troubles con fusionels dans les lesions hemispherigues localisees. *Electroencephalogr. Clin. Neurophysiol.*, **7**, 179.

[53] Westmoreland, B. and Klass, D. (1971) Asymetrical attention of alpha activity with arithmetical attention. *Electroencephalogr. Clin. Neurophysiol.*, **31**, 634–635.

[54] Cobb, W. (1976) *EEG Interpretation in Clinical Medicine, Part B*, Handbook of Electroencephalography and Clinical Neurophysiology, vol. **11** (ed. A. Remond), Elsevier.

[55] Hess, R. (1975) *Brain Tumors and Other Space Occupying Processing, Part C*, Handbook of Electroencephalography and Clinical Neurophysiology, vol. **14** (ed. A. Remond), Elsevier.

[56] Klass, D. and Daly, D. (eds) (1979) *Current Practice of Clinical Electroencephalography*, 1st edn, Raven Press.

[57] Van Sweden, B., Wauquier, A. and Niedermeyer, E. (1999) Normal aging and transient cognitive disorders in the elderly, Chapter 18, in *Electroencephalography, Basic Principles, Clinical Applications, and Related Fields*, 4th edn (eds E. Niedermeyer and F. Lopes Da Silva), Lippincott Williams & Wilkins, pp. 340–348.

[58] America Psychiatric Association (1994) *Committee on nomenclature and statistics, Diagnostic and Statistical Manual of Mental Disorder: DSM-IV*, 4th edn, American Psychiatric Association, Washington DC.

[59] Brenner, R.P. (1999) EEG and dementia, Chapter 19, in *Electroencephalography, Basic Principles, Clinical Applications, and Related Fields*, 4th edn (eds E. Niedermeyer and F. lopes Da Silva), Lippincott Williams & Wilkins, pp. 349–359.

[60] Neufeld, M.Y., Bluman, S., Aitkin, I. *et al.* (1994) EEG frequency analysis in demented and nondemented parkinsonian patients. *Dementia*, **5**, 23–28.

[61] Niedermeyer, E. (1999) Abnormal EEG patterns: Epileptic and paroxysmal, Chapter 13, in *Electroencephalography, Basic Principles, Clinical Applications, and Related Fields*, 4th edn (eds E. Niedermeyer and F. Lopes Da Silva), Lippincott Williams & Wilkins, pp. 235–260.

[62] Hughes, J.R. and Gruener, G.T. (1984) Small sharp spikes revisited: further data on this controversial pattern. *Electrencephalogr. Clin. Neurophysiol.*, **15**, 208–213.

[63] Hecker, A., Kocher, R., Ladewig, D. and Scollo-Lavizzari, G. (1979) Das Minature-Spike-Wave. *Das EEG Labor.*, **1**, 51–56.

[64] Geiger, L.R. and Harner, R.N. (1978) EEG patterns at the time of focal seizure onset. *Arch. Neurol.*, **35**, 276–286.

[65] Gastaut, H. and Broughton, R. (1972) *Epileptic Seizure*, Charles C. Thomas, Springfield, IL.

[66] Oller-Daurella, L. and Oller-Ferrer-Vidal, L. (1977) *Atlas de Crisis Epilepticas*, Geigy Division Farmaceut.

[67] Niedermeyer, E. (1999) Nonepileptic attacks, Chapter 28, in *Electroencephalography, Basic Principles, Clinical Applications, and Related Fields*, 4th edn (eds E. Niedermeyer and F. Lopes Da Silva), Lippincott Williams & Wilkins, pp. 586–594.

[68] Creutzfeldt, H.G. (1920) Uber eine eigenartige herdformige erkrankung des zentralnervensystems. *Z. Ges. Neurol. Psychiatr.*, **57**, 1. Quoted after W.R. Kirschbaum 1968.

[69] Jakob, A. (1921) Uber eigenartige erkrankung des zentralnervensystems mit bemerkenswerten anatomischen befunden (spastistische pseudosklerose, encephalomyelopathie mit disseminerten degenerationsbeschwerden). *Deutsch, Z. Nervenheilk.*, **70**, 132. Quoted after W.R. Kirschbaum 1968.

[70] Niedermeyer, E. (1999) Epileptic seizure disorders, Chapter 27, in *Electroencephalography, Basic Principles, Clinical Applications, and Related Fields*, 4th edn (eds E. Niedermeyer and F. Lopes da Silva), Lippincott Williams & Wilkins, pp. 476–585.

[71] Small, J.G. (1999) Psychiatric disorders and EEG, Chapter 30, in *Electroencephalography, Basic Principles, Clinical Applications, and Related Fields*, 4th edn (eds E. Niedermeyer and F. Lopes da Silva), Lippincott Williams & Wilkins, pp. 235–260.

[72] Marosi, E., Harmony, T., Sanchez, L. *et al.* (1992) Maturation of the coherence of EEG activity in normal and learning disabled children. *Electroencephalogr. Clin. Neurophysiol.*, **83**, 350–357.

[73] Linden, M., Habib, T. and Radojevic, V. (1996) A controlled study of the effects of EEG biofeedback on cognition and behavior of children with attention deficit disorder and learning disabilities. *Biofeedback Self Regul.*, **21**(1), 35–49.

[74] Hermens, D.F., Soei, E.X., Clarke, S.D. *et al.* (2005) Resting EEG theta activity predicts cognitive performance in attention-deficit hyperactivity disorder. *Pediatr. Neurol.*, **32**(4), 248–256.

[75] Swartwood, J.N., Swartwood, M.O., Lubar, J.F. and Timmermann, D.L. (2003) EEG differences in ADHD-combined type during baseline and cognitive tasks. *Pediatr. Neurol.*, **28**(3), 199–204.

[76] Clarke, A.R., Barry, R.J., McCarthy, R. and Selikowitz, M. (2002) EEG analysis of children with attention-deficit/hyperactivity disorder and comorbid reading disabilities. *J. Learn. Disabil.*, **35**(3), 276–285.

[77] Yordanova, J., Heinrich, H., Kolev, V. and Rothenberger, A. (2006) Increased event-related theta activity as a psychophysiological marker of comorbidity in children with tics and attention-deficit/hyperactivity disorders. *Neuroimage.*, **32**(2), 940–955.

[78] Jacobson, S. and Jerrier, H. (2000) EEG in delirium. *Semin. Clin. Neuropsychiatry*, **5**(2), 86–92.

[79] Onoe, S. and Nishigaki, T. (2004) EEG spectral analysis in children with febrile delirium. *Brain Dev.*, **26**(8), 513–518.

[80] Brunovsky, M., Matousek, M., Edman, A. *et al.* (2003) Objective assessment of the degree of dementia by means of EEG. *Neuropsychobiology*, **48**(1), 19–26.

[81] Koenig, T., Prichep, L., Dierks, T. *et al.* (2005) Decreased EEG synchronization in Alzheimer's disease and mild cognitive impairment. *Neurobiol Aging.*, **26**(2), 165–171.

[82] Babiloni, C., Binetti, G., Cassetta, E. *et al.* (2006) Sources of cortical rhythms change as a function of cognitive impairment in pathological aging: a multicenter study. *Clin. Neurophysiol.*, **117**(2), 252–268.

[83] Bauer, G. and Bauer, R. (1999) EEG, drug effects, and central nervous system poisoning, Chapter 35, in *Electroencephalography, Basic Principles, Clinical Applications, and Related Fields*, 4th edn (eds E. Niedermeyer and F. Lopes da Silva), Lippincott Williams & Wilkins, pp. 671–691.

[84] Kane, K., Picton, T.W., Moscovitch, M. and Winocur, G. (2000) Event-related potentials during conscious and automatic memory retrieval. *Brain Res. Cogn. Brain Res.*, **10**(1–2), 19–35.

[85] Cowan, N. (2008) What are the differences between long-term, short-term, and working memory? *Prog. Brain Res.*, **169**, 323–338.

[86] Radvansky, G. (2006) *Human Memory*, Pearson Education Group, Boston, MA.

[87] Finnigan, S., Humphreys, M.S., Dennis, S. and Geffen, G. (2002) ERP. *Neuropsychologia*, **40**(13), 2288–2304.

[88] Wolk, D.A., Schacter, D.L., Lygizos, M. *et al.* (2006) ERP correlates of recognition memory: effects of retention interval and false alarms. *Brain Res.*, **1096**(1), 148–162.

[89] Speer, N.K. and Curran, T. (2007) ERP correlates of familiarity and recollection processes in visual associative recognition. *Brain Res.*, **1174**, 97–109.

[90] Mecklinger, A., Johansson, M., Parra, M. and Hanslmayr, S. (2007) Source-retrieval requirements influence late ERP and EEG memory effects. *Brain Res.*, 1172, 110–23.

[91] Duarte, A., Ranganath, C., Winward, L. *et al.* (2004) Dissociable neural correlates for familiarity and recollection during the encoding and retrieval of pictures. *Els. J. Brain Res.*, **18**, 255–272.

[92] Okada, Y. (1983) Neurogenesis of evoked magnetic fields, in *Biomagnetism: An Interdisciplinary Approach* (eds S.H. Williamson, G.L. Romani, L. Kaufman and I. Modena), Plenum Press, New York, pp. 399–408.

[93] Naidu, S. and Niedermeyer, E. Degenerative disorders of the central nervous system, in *Electroencephalography, Basic Principles, Clinical Applications and Related Fields*, eds E. Niedermeyer and F. Lopes da Silva, 4th edn, Lippincott Williams and Wilkins, Chapter 20.

2

Fundamentals of EEG
Signal Processing

2.1 Introduction

EEG signals are the signatures of neural activities. They are captured by multiple-electrode EEG machines either from inside the brain, over the cortex under the skull, or certain locations over the scalp, and can be recorded in different formats. However, often the signals from the scalp electrodes are called EEG, the recordings by cortical electrodes are called electrocorticogram (ECoG) and those recorded from inside the brain, often by using subdural electrodes, are called intracranial (iEEG) signals.

Unlike scalp EEG which is noninvasive, ECoG may be performed either in the operating room during surgery (intraoperative ECoG) or outside of surgery (extraoperative ECoG). Because a craniotomy (a surgical incision into the skull) is required to implant the electrode grid, ECoG is an invasive procedure. The signals are normally presented in the time domain, however, many new EEG machines are capable of applying simple signal processing tools such as the Fourier transform to perform frequency analysis, and equipped with some imaging tools to visualise EEG topographies (spatial maps of the brain activities).

EEG signals have an effective frequency bandwidth of less than 50 Hz. This allows online processing of these signals using conventional signal processing algorithms for analysis of the data in time, frequency, space, multidimensional or multiway domains. On the other hand, different patterns of the brain rhythms can be viewed and well distinguished during different states of the brain, such as for asleep or wakeful humans. The pattern can also change from healthy individual to epileptic patient during seizure onset. Also, as discussed in later chapters there are event and movement related changes in the brain signals which can indicate the state of the brain during external (audio, visual, somatosensory) stimulations and brain motor activity during intentional body movement.

The EEG signals are called instantaneous in time since there is no time lag when the waveforms travel from the sources to the electrodes. However, by assuming distributed sources (as opposed to a dipole assumption) the sources may change in space (in analogues to synaptic currents). Therefore, multichannel analysis of EEGs may involve complexity due to the source motions.

Adaptive Processing of Brain Signals, First Edition. Saeid Sanei.
© 2013 John Wiley & Sons, Ltd. Published 2013 by John Wiley & Sons, Ltd.

Although many advanced signal processing algorithms have been used for processing the EEG signals many others have been developed to analyse and evaluate brain signals in particular. In addition, some algorithms have been developed to visualise brain activity from images reconstructed from only the EEGs.

Separation of the desired sources from the multisensor EEGs has been another research area. This can later lead to the detection of brain abnormalities, such as epilepsy, and the sources related to various physical and mental activities. Later in this book we will also see how the recent works in brain-computer interfacing (BCI) [1] have been focused upon the development of advanced signal processing tools and algorithms for this purpose.

Modelling of neural activities is probably more difficult than modelling the function of any other organ. However, some simple models for generating EEG signals have been proposed. Some of these models have also been extended to include generation of abnormal EEG signals.

Localization of brain signal sources is another very important field of research [2]. In order to provide a reliable algorithm for localization of the sources within the brain sufficient knowledge about both the propagation of electromagnetic waves, and how the information from the measured signals can be exploited in separation and localization of the sources within the brain is required. The sources might be considered as magnetic dipoles for which the well known inverse problem has to be solved, or they can be considered as distributed current sources.

Patient monitoring and sleep monitoring require real-time processing of (up to a few days) long EEG sequences. The EEG provides important and unique information about the sleeping brain. Major brain activities during sleep can be captured using the developed algorithms [3] such as the method of matching pursuits (MPs) [4].

Epilepsy monitoring, detection, and prediction have also attracted many researchers. Dynamical analysis of time series together with application of blind separation of the signal sources has enabled prediction of focal epilepsies from the scalp EEGs [5]. On the other hand, application of time–frequency domain analysis for detection of the seizure in neonates has paved the way for further research in this area.

Another problem to be solved in the EEG analysis context is classification of the signals. This is very popular in BCI where the motor activity for moving a particular body part has to be recognised. A number of methods, such as support vector machines (SVM), linear discriminant analysis [6], and neural networks have often been used for this purpose. The traditional and more BCI-related topics are explained in detail in the related chapters. Classification of brain signals requires an accurate and robust feature detection process. In a recent approach common spatial patterns (CSP) [7, 8] and their variants have been proposed. CSPs are also described in the related chapters of this book.

In the following sections some fundamental concepts used in developing EEG signal processing tools and algorithms for the above objectives are explained and the mathematical foundations discussed. The reader should also be aware of the required concepts and definitions developed in linear algebra. Further details of which can be found in [9].

2.2 Nonlinearity of the Medium

The head as a mixing medium combines EEG signals which are locally generated within the brain at the sensor positions. As a system, the head may be more or less susceptible to such sources in different situations. Generally, an EEG signal can be considered as the output of a nonlinear system, which may be characterised deterministically.

The changes in brain metabolism as a result of biological and physiological phenomena in the human body can change the mixing process. Some of these changes are influenced by the activity of the brain itself. These effects make the system nonlinear. Analysis or modelling of such a system is very complicated and as yet nobody has fully modelled the system to aid in the analysis of brain signals.

On the other hand, some measures borrowed from chaos theory and analysis of the dynamics of time serie,s such as dissimilarity, attractor dimension, and largest Lyapunov exponents (LLE) can characterise the nonlinear behaviour of EEG signals. These concepts are discussed in Chapter 5.

2.3   Nonstationarity

The statistics of the signals, such as mean, variance, and higher order statistics, may change with time. As a result the signals are generally nonstationary. Nonstationarity of the signals can be quantified by measuring some statistics of the signals at different time lags with time. The signals can be deemed stationary if there is no considerable variation in these statistics.

Although, generally, the multichannel EEG distribution is considered as a multivariate Gaussian, the mean and covariance properties generally change from segment to segment. As such, EEGs are considered stationary only within short intervals, that is, quasi-stationary.

This Gaussian assumption holds during normal brain condition, however, during mental and physical activities this assumption is not valid. Some examples of nonstationarity of the EEG signals can be observed during the change in alertness and wakefulness (where there are stronger alpha oscillations), during eye blinking, during the transitions between various ictal states, and in the event-related potential (ERP) and evoked potential (EP) signals.

The change in the distribution of the signal segments can be measured in terms of both the parameters of a Gaussian process and the deviation of the distribution from Gaussian. The non-Gaussianity of the signals can be checked by measuring or estimating some higher-order moments, such as skewness, kurtosis, negentropy, and Kulback Leibler (KL) distance.

Skewness is a measure of symmetry, or more precisely, lack of symmetry of the distribution. A distribution, or data set, is symmetric if it looks the same to the left and right of the centre point. The skewness is defined for a real signal as

$$Skewness = \frac{E\left[(x(n) - \mu)^3\right]}{\sigma^3} \tag{2.1}$$

where μ and σ are respectively, the mean and standard deviation and E denotes statistical expectation. If the distribution is more to the right of the mean point the skewness is negative, and vice versa. For a symmetric distribution such as Gaussian, the skewness is zero.

Kurtosis is a measure of how peaky (sharp), relative to a normal distribution, the signal is. That is, data sets with high kurtosis tend to have a distinct peak near the mean, decline rather rapidly, and have heavy tails. Data sets with low kurtosis tend to have a flat top near the mean rather than a sharp peak. A uniform distribution would be the extreme case. The kurtosis for a signal $x(n)$ is defined as:

$$kurt = \frac{m_4(x(n))}{m_2^2(x(n))} \tag{2.2}$$

where $m_i(x(n))$ is the ith central moment of the signal $x(n)$, that is, $m_i(x(n)) = E[(x(n) - \mu)^i]$. The kurtosis for signals with normal distributions is 3. Therefore, an excess or normalised kurtosis is often used and defined as

$$Ex\ kurt = \frac{m_4(x(n))}{m_2^2(x(n))} - 3 \tag{2.3}$$

which is zero for Gaussian distributed signals. Often the signals are considered ergodic, hence the statistical averages can be assumed identical to time averages so that they can be estimated with time averages.

The negentropy of a signal $x(n)$ [10] is defined as:

$$J_{neg}(x(n)) = H(x_{Gauss}(n)) - H(x(n)) \tag{2.4}$$

where, $x_{Gauss}(n)$ is a Gaussian random signal with the same covariance as $x(n)$ and H(.) is the differential entropy [11] defined as:

$$H(x(n)) = \int_{-\infty}^{\infty} p(x(n)) \log \frac{1}{p(x(n))} dx(n) \tag{2.5}$$

and $p(x(n))$ is the signal distribution. Negentropy is always non-negative.

The KL distance between two distributions p_1 and p_2 is defined as

$$KL = \int_{-\infty}^{\infty} p_1(z) \log \frac{p_1(z)}{p_2(z)} dz \tag{2.6}$$

It is clear that the KL distance is generally asymmetric, therefore by changing the position of p_1 and p_2 in this equation the KL distance changes. The minimum of the KL distance occurs when $p_1(z) = p_2(z)$.

2.4 Signal Segmentation

Often it is necessary to label the EEG signals into segments of similar characteristics particularly meaningful to clinicians and for assessment by neurophysiologists. Within each segment, the signals are considered statistically stationary, usually with similar time and frequency statistics. As an example, an EEG recorded from an epileptic patient may be divided into three segments of preictal, ictal, and postictal segments. Each may have a different duration. Figure 2.1 represents an EEG sequence including all the above segments.

In segmentation of EEGs time or frequency properties of the signals may be exploited. This eventually leads to a dissimilarity measurement denoted as $d(m)$ between the adjacent EEG frames, where m is an integer value indexing the frame and the difference is calculated between the m and $(m - 1)$th (consecutive) signal frames. The boundary of the two different segments is then defined as the boundary between the m and $(m - 1)$th frames provided $d(m) > \eta_T$, where η_T is an empirical threshold level. An efficient segmentation is possible by highlighting and effectively exploiting the diagnostic information within the signals with the help of expert

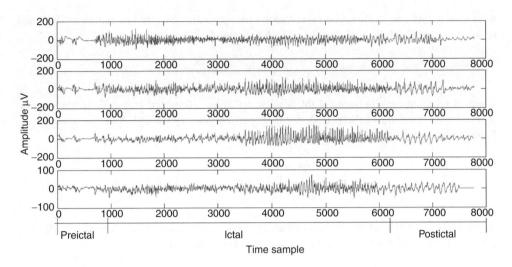

Figure 2.1 An EEG set of tonic-clonic seizure signals including three segments of preictal, ictal, and postictal behaviour

clinicians. However, access to such experts is not always possible and, therefore, algorithmic methods are required.

A number of different dissimilarity measures may be defined based on the fundamentals of signal processing. One criterion is based on the autocorrelations for segment m defined as:

$$r_x(k, m) = E[x(n, m)x(n + k, m)] \tag{2.7}$$

The autocorrelation function of the mth length N frame for an assumed time interval $n, n + 1, \dots, n + (N - 1)$, can be approximated as

$$\hat{r}_x(k, m) = \begin{cases} \dfrac{1}{N} \displaystyle\sum_{l=0}^{N-1-k} x(l + m + k)x(l + m), & k = 0, \dots, N - 1 \\ 0 & k = N, N + 1, \dots \end{cases} \tag{2.8}$$

Then the criterion is set to

$$d_1(m) = \frac{\displaystyle\sum_{k=-\infty}^{\infty} (\hat{r}_x(k, m) - \hat{r}_x(k, m - 1))^2}{\hat{r}_x(0, m)\hat{r}_x(0, m - 1)} \tag{2.9}$$

A second criterion can be based on higher order statistics. The signals with more uniform distributions, such as normal brain rhythms, have a low kurtosis, whereas, seizure signals or ERPs often have high kurtosis values. Kurtosis is defined as the fourth order cumulant at zero time lags and related to the second and fourth order moments, as given in equations (2.2) and (2.3). A second level discriminant $d_2(m)$ is then defined as:

$$d_2(m) = kurt_x(m) - kurt_x(m-1) \tag{2.10}$$

where m refers to the mth frame of the EEG signal $x(n)$. A third criterion is defined using the spectral error measure of the periodogram. A periodogram of the mth frame is obtained by Fourier transforming the correlation function of the EEG signal;

$$S_x(\omega, m) = \sum_{k=-\infty}^{\infty} \hat{r}_x(k, m)e^{-j\omega k} \quad \omega \in [-\pi, \pi] \tag{2.11}$$

where $\hat{r}_x(k, m)$ is the autocorrelation function for the mth frame as defined above. A third criterion is then defined based on the normalised periodogram as:

$$d_3(m) = \frac{\int_{-\pi}^{\pi} (S_x(\omega, m) - S_x(\omega, m-1))^2 \, d\omega}{\int_{-\pi}^{\pi} S_x(\omega, m)d\omega \int_{-\pi}^{\pi} S_x(\omega, m-1)d\omega} \tag{2.12}$$

The test window sample autocorrelation for the measurement of both $d_1(m)$ and $d_3(m)$ can be updated through the following recursive equation over the test windows of size N:

$$\hat{r}_x(k, m) = \hat{r}_x(k, m-1) + \frac{1}{N}(x(m-1+N)x(m-1+N-k) - x(m-1+k)x(m-1)) \tag{2.13}$$

and thereby computational complexity can be reduced in practice. A fourth criterion corresponds to the error energy in autoregressive (AR)-based modelling of the signals. The prediction error in the AR model of the mth frame is simply defined as:

$$e(n, m) = x(n, m) - \sum_{k=1}^{p} a_k(m)x(n-k, m) \tag{2.14}$$

where p is the prediction order and $a_k(m)$, $k = 1, 2, \ldots, p$, are the prediction coefficients. For certain p the coefficients can be found directly (for example Durbin's method) in such a way as to minimise the error (residual) signal energy. In this approach it is assumed that the frames of length N are overlapped by one sample. The prediction coefficients estimated for the $(m-1)th$ frame are then used to predict the first sample in the mth frame, which we denote as $\hat{e}(1, m)$. If this error is small, it is likely that the statistics of the mth frame are similar to those of the $(m-1)th$ frame. On the other hand, a large value is likely to indicate a change. An indicator for the fourth criterion can then be the differencing (differentiating) of this prediction signal, which gives a peak at the segment boundary that is,

$$d_4(m) = \max(\nabla_m \hat{e}(1, m)) \tag{2.15}$$

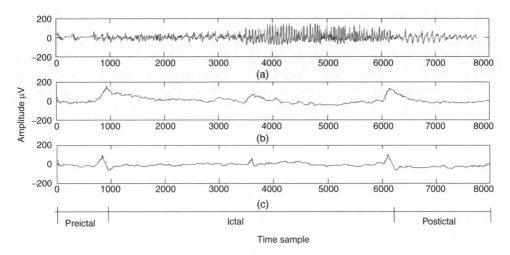

Figure 2.2 (a) An EEG seizure signal including preictal ictal, and postictal segments, (b) the error signal and (c) the approximate gradient of the signal, which exhibits a peak at the boundary between the segments. The number of prediction coefficients $p = 12$

where $\nabla_m(.)$ denotes the gradient with respect to m, approximated by a first order difference operation. Figure 2.2 shows the residual and the gradient defined in equation (2.15).

Finally, a fifth criterion $d_5(m)$ may be defined by using the AR-based spectrum of the signals in the same way as STFT for $d_3(m)$. The above AR model is a univariate model, that is, it models a single channel EEG. A similar criterion may be defined when multichannel EEGs are considered [12]. In such cases a multivariate AR (MVAR) model is analysed. The MVAR can also be used for characterisation and quantification of the signal propagation within the brain and is discussed in the next section.

Although the above criteria can be effectively used for segmentation of EEG signals, better systems may be defined for the detection of certain abnormalities. In order to do that, the features, which best describe the behaviour of the signals have to be identified and used. Therefore, the segmentation problem becomes a classification problem for which different classifiers can be used.

2.5 Other Properties of Brain Signals

A number of very important properties often used for analysis of brain dynamics have to be understood before moving to various applications. Correlation between the brain sources, for example, is the major barrier in using blind source separation. Synchronisation of different brain zones, their phase coherency, and the changes in the brain dynamics are the good indicators of brain connectivity. The connectivity can significantly change when the subject is under mental fatigue or having deterioration of the brain because of Alzheimer's disease or mild cognitive impairment. The methods for evaluation and monitoring of these properties are also provided in Chapter 8.

2.6   Conclusions

Some fundamental concepts and definitions for understanding EEG properties, characterising the signals and those useful for the processing of EEG signals have been briefly explained in this chapter. Other important concepts such as independency, signal modelling, signal transforms, multivariate modelling and direct transfer functions, chaos and dynamic analysis, independent component analysis and blind source separation, classification and clustering, and matching pursuits are presented in detail in the later chapters.

References

[1] Lebedev, M.A. and Nicolelis, M.A. (2006) Brain-machine interfaces: past, present and future. *Trends. Neurosci.*, **29**, 536–546.

[2] Lopes da Silva, F. (2004) Functional localization of brain sources using EEG and/or MEG data: volume conductor and source models. *J. Magn. Reson. Imaging.*, **22**(10), 1533–1538.

[3] Malinovska, U., Durka, P.J., Blinowska, K.J., Szelenberger, W. and Wakarow, A. (2006) Micro- and macrostructure of sleep EEG, A universal; adaptive time-frequency parametrization. *IEEE Eng. Med. Biol. Mag.*, **25**(4), 26–31.

[4] Durka, P.J., Dobieslaw, I. and Blinowska, K.J. (2001) Stochastic time-frequency dictionaries for matching pursuit. *IEEE Trans. Signal Proc.*, **49**(3) 507–510.

[5] Corsini, J., Shoker, L., Sanei, S. and Alarcon, G. (2006) Epileptic seizure predictability from scalp EEG incorporating constrained blind source separation. *IEEE Trans. Biomed. Eng.*, **53**(5), 790–799.

[6] Vapnik, V. (1998) *Statistical Learning Theory*, Wiley-Interscience.

[7] Blankertz, B., Tomioka, R., Lemm, S., Kawanabe, M. and Muller, K.-R. (2008) Optimizing spatial filters for robust EEG single-trial analysis. *IEEE Signal Proc. Mag.*, **25**(1), 41–56.

[8] Lotte, F. and Guan, C. (2011) Regularizing common spatial patterns to improve BCI designs: unified theory and new algorithms. *IEEE Trans. Biomed. Eng.*, **58**(2), 355–362.

[9] Strang, G. (1998) *Linear Algebra and Its Applications*, 3rd edn. Thomson Learning.

[10] Hyvarinen, A., Kahunen, J. and Oja, E. (2001) *Independent Component Analysis*, Wiley-Interscience.

[11] Cover, T.M. and Thomas, J.A. (2001) *Elements of Information Theory*, Wiley-Interscience.

[12] Morf, M., Vieria, A., Lee, D. and Kailath, T. (1978) Recursive multichannel maximum entropy spectral estimation. *IEEE Trans. Geosci. Elect.*, **16**, 85–94.

3

EEG Signal Modelling

Generation of electrical potentials or magnetic fields measurable from the brain is due to a nonlinear sum/distribution of electrochemical active potentials within all the neurons involved in cognitive or movement related processes. An accurate model that can link the chemical processes within corresponding neurons generating the active potentials is hard to achieve. A number of models, however, have been introduced since the 1950s.

3.1 Physiological Modelling of EEG Generation

A number of models for simulating the coupling between two or more neurons have been introduced. In [1] three models for generation of brain potentials have been introduced and compared.

3.1.1 Integrate-and-Fire Models

For coupling two neurons two integrateand-fire neurons with mutual excitatory or inhibitory coupling have been described in [1]. The neurons with activation variables x_i for $i = 1, 2$, satisfy

$$\frac{\mathrm{d}x_i}{\mathrm{d}t} = \xi - x_i + E_i(t) \tag{3.1}$$

Where $\xi > 1$ is a constant, $0 < x_i < 1$, and $E_i(t)$ is the synaptic input to neuron i. Neuron i fires when $x_i = 1$ and then resets x_i to 0. If cell $j \neq i$ fires at time t_j the function E_i is augmented to $E_i(t) + E_s(t - t_j)$, where E_s is the contribution coming from one spike [1]. In an example in [1] this function is selected as

$$E_s(t) = g\alpha^2 t e^{-\alpha t} \tag{3.2}$$

where g and α are parameters determining the strength and speed of the synapse, respectively, and the factor of α^2 in equation (3.2) normalizes the integral of E_s over time to the value g. In the considered cases the two neurons continue firing periodically when they are coupled

---

Adaptive Processing of Brain Signals, First Edition. Saeid Sanei.
© 2013 John Wiley & Sons, Ltd. Published 2013 by John Wiley & Sons, Ltd.

together. Assuming that neuron 1 fires at times $t = nT$, where T is the period and n is an integer, while neuron 2 fires at $t = (n - \varphi)T$. Therefore, both neurons are firing at the same frequency but are separated by a phase φ. We wish to determine possible values of the phase difference φ and conditions under which they arise.

3.1.2   Phase-Coupled Models

Neuronal synchronization processes, measured with brain imaging data, can be described using weakly coupled oscillator (WCO) models. Dynamic causal modelling (DCM) is used to fit the WCOs to brain imaging data and so make inferences about the structure of neuronal interactions [2]. The complex behaviours are mediated by the interaction of particular brain regions. Recent studies agreed that such interactions may be instantiated by the transient synchronization of oscillatory neuronal ensembles [3]. For example, contour detection is accompanied by gamma band synchronization in distant parts of the visual cortex, multimodal object processing by parieto-temporal synchronization in the beta band [4], and spatial memory processes by hippocampal–prefrontal synchronization in the theta band [2, 5]. DCM allows different model structures to be compared using Bayesian model selection [6]. In [2] DCM has been extended to the study of phase coupling. One direction is based on the WCO models, in which the rate of change of phase of one oscillator is related to the phase differences between itself and other oscillators [7].

The WCO theory applies to system dynamics close to limit cycles. By assuming that weak coupling leads to only small perturbations away from these cycles, one can reduce a high-dimensional system of differential equations to one based solely on the phases of the oscillators, and pairwise interactions between them [2].

Dynamics on the limit cycle are given by [2]

$$\dot{X}_0 = F(X_0)$$

$$X_0(t + T) = X_0(t) \tag{3.3}$$

$$\dot{\phi}(X_0) = f(X_0)$$

Then, for any perturbation of phase $p(\varphi)$ this changes to [2]:

$$\dot{X} = F(X) + P(X) \tag{3.4}$$

$$\dot{\phi}(X_0) = f(X_0) + z(\phi)p(\phi) \tag{3.5}$$

where

$$z(\phi) = \frac{d\phi(X_0)}{dX_0} \tag{3.6}$$

For a pair of oscillators of phases φ_1 and φ_2, using the same analysis (see Figure 3.1, assuming $p_{12} = p_{21}$):

$$\dot{\phi}_1 = f + z_1(\phi_1)p_{12}(\phi_1, \phi_2) \tag{3.7}$$

$$\dot{\phi}_2 = f + z_2(\phi_2)p_{21}(\phi_2, \phi_1) \tag{3.8}$$

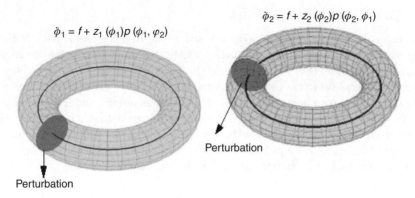

$$\dot{\phi}_2 = f + z_2 \, (\phi_2)p \, (\phi_2, \phi_1)$$

$$\dot{\phi}_1 = f + z_1 \, (\phi_1)p \, (\phi_1, \phi_2)$$

Perturbation

Perturbation

Figure 3.1 A pair of oscillators weakly coupled via the perturbation function $p(\varphi_1, \varphi_2)$. Taken from [2]

If it is further assumed that the phase difference $\varphi_2 - \varphi_1 = \varphi$ changes slowly, then [2]:

$$\dot{\phi}_1 = f + \Gamma_{12}(\phi_1 - \phi_2) \tag{3.9}$$

$$\dot{\phi}_2 = f + \Gamma_{21}(\phi_2 - \phi_1) \tag{3.10}$$

$$\Gamma_{ij}(\phi) = \frac{1}{2\pi} \int_0^{2\pi} z_i(\psi) p_{ij}(\psi, \psi + \phi) d\psi \tag{3.11}$$

$\Gamma_{ij}(\phi)$ is called phase interaction function (PIF). Similarly, for N_R regions the rate of change of phase of the ith oscillator is given by

$$\dot{\phi}_i = f_i + \sum_{j=1}^{N_R} \Gamma_{ij}((\phi_i - \phi_j) - d_{ij} \tag{3.12}$$

where f_i is the intrinsic frequency of the ith oscillator. In these formulations there are two key assumptions; the first assumes that the perturbations are sufficiently small that the differentiations can equivalently be evaluated at X_0 rather than X. The second assumption is that the relative changes in the oscillator phase are sufficiently slow with respect to the oscillation frequency, that the phase offset term can be replaced by a time average.

In [2] an extension of the above weakly coupled oscillator model is used to describe the dynamic phase changes in a network of oscillators. The use of Bayesian model comparison allows one to infer the mechanisms underlying synchronization processes in the brain. This has been applied to synthetic bimanual finger movement data from physiological models and to MEG data from a study of visual working memory. The WCO approach accommodates signal nonstationarity by using differential equations which describe how the changes in phase are driven by pairwise differences in instantaneous phase.

3.1.3   Hodgkin and Huxley Model

Most probably the earliest physical model is based on the Hodgkin and Huxley Nobel Prize winning model for the squid axon published in 1952 [8–10]. A nerve axon may be stimulated and the activated sodium (Na^+) and potassium (K^+) channels produced in the vicinity of the cell membrane may lead to the electrical excitation of the nerve axon. The excitation arises from the effect of the membrane potential on the movement of ions, and from interactions of the membrane potential with the opening and closing of voltage-activated membrane channels. The membrane potential increases when the membrane is polarised with a net negative charge lining the inner surface and an equal but opposite net positive charge on the outer surface. This potential may be simply related to the amount of electrical charge Q, using

$$E = Q/C_{\mathrm{m}} \qquad (3.13)$$

where Q has units coulomb (C) cm^{-2}, C_{m} is the measure of the capacity of the membrane and has units farad (F) cm^- and E has units of V. In practice, in order to model the action potentials (APs) the amount of charge Q^+ on the inner surface (and Q^- on the outer surface) of the cell membrane has to be mathematically related to the stimulating current I_{stim} flowing into the cell through the stimulating electrodes. Figure 3.2 illustrates how the neuron excitation results in generation of APs. Based on this the neuron acts as a signal converter [11].

The electrical potential (often called electrical force) E is then calculated using equation (3.13). The Hodgkin and Huxley model is illustrated in Figure 3.3. In this figure I_{memb} is the result of positive charges flowing out of the cell. This current consistsmainly of three currents namely, Na, K, and leak currents. The leak current is due to the fact that the inner and outer Na and K ions are not exactly equal.

Hodgkin and Huxley estimated the activation and inactivation functions for the Na and K currents and derived a mathematical model to describe an AP similar to that of a giant squid. The model is a neuron model that uses voltage-gated channels. The space-clamped version of the Hodgkin–Huxley model may be well described using four ordinary differential equations [12]. This model describes the change in the membrane potential (E) with respect to time and

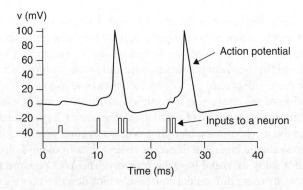

Figure 3.2   Generation of active potentials by a neuron

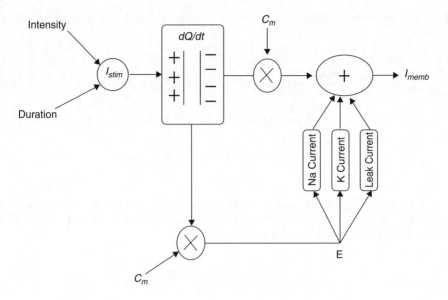

Figure 3.3   The Hodgkin–Huxley excitation model. Taken from [13]

is described in [13]. The overall membrane current is the sum of capacity current and ionic current as:

$$I_{\text{memb}} = C_{\text{m}} \frac{\mathrm{d}E}{\mathrm{d}t} + I_{\text{i}} \tag{3.14}$$

where I_{i} is the ionic current and, as indicated in Figure 3.1, can be considered as the sum of three individual components: Na, K, and leak currents.

$$I_{\text{i}} = I_{\text{Na}} + I_{\text{K}} + I_{\text{leak}} \tag{3.15}$$

I_{Na} can be related to the maximal conductance $\bar{g}_{\text{Na}}$, activation variable a_{Na}, inactivation variable h_{Na}, and a driving force $(E - E_{\text{Na}})$ through:

$$I_{\text{Na}} = \bar{g}_{\text{Na}} a_{\text{Na}}^{3} h_{\text{Na}} (E - E_{\text{Na}}) \tag{3.16}$$

Similarly I_{K} can be related to the maximal conductance $\bar{g}_{\text{K}}$, activation variable a_{K}, and a driving force $(E - E_{\text{K}})$ as:

$$I_{\text{K}} = \bar{g}_{\text{K}} a_{\text{K}} (E - E_{\text{K}}) \tag{3.17}$$

and I_{leak} is related to the maximal conductance $\bar{g}_{\text{l}}$ and a driving force $(E - E_{\text{l}})$ as:

$$I_{\text{leak}} = \bar{g}_{\text{l}} (E - E_{\text{l}}) \tag{3.18}$$

The changes in the variables a_{Na}, a_K, and h_{Na} vary from 0 to 1 according to the following equations:

$$\frac{da_{Na}}{dt} = \lambda_t \left[\alpha_{Na}(E)(1 - a_{Na}) - \beta_{Na}(E)a_{Na} \right] \tag{3.19}$$

$$\frac{dh_{Na}}{dt} = \lambda_t \left[\alpha_h(E)(1 - h_{Na}) - \beta_h(E)h_{Na} \right] \tag{3.20}$$

$$\frac{da_K}{dt} = \lambda_t \left[\alpha_K(E)(1 - a_K) - \beta_K(E)a_K \right] \tag{3.21}$$

where $\alpha(E)$ and $\beta(E)$ are, respectively, forward and backward rate functions and λ_t is a temperature-dependent factor. The forward and backward parameters depend on voltage and were empirically estimated by Hodgkin and Huxley as:

$$\alpha_{Na}(E) = \frac{3.5 + 0.1E}{1 - e^{-(3.5 + 0.1E)}} \tag{3.22}$$

$$\beta_{Na}(E) = 4e^{-(E+60)/18} \tag{3.23}$$

$$\alpha_h(E) = 0.07e^{-(E+60)/20} \tag{3.24}$$

$$\beta_h(E) = \frac{1}{1 + e^{-(3+0.1E)}} \tag{3.25}$$

$$\alpha_K(E) = \frac{0.5 + 0.01E}{1 - e^{-(5+0.1E)}} \tag{3.26}$$

$$\beta_K(E) = 0.125e^{-(E+60)/80} \tag{3.27}$$

As stated in the simulator for neural networks and action potentials (SNNAP) literature [12], the $\alpha(E)$ and $\beta(E)$ parameters have been converted from the original Hodgkin–Huxley version to agree with the present physiological practice where depolarisation of the membrane is taken to be positive. In addition, the resting potential has been shifted to –60 mV (from the original 0 mV). These equations are used in the model described in the SNNAP. In Figure 3.4 an AP has been simulated. For this model the parameters are set to $C_m = 1.1$ μF cm^{-2}, $\bar{g}_{Na} = 100$ ms cm^{-2}, $\bar{g}_K = 35$ ms cm^{-2}, $\bar{g}_{leak} = 0.35$ ms cm^{-2}, and $E_{Na} = 60$ mV.

The simulation can run to generate a series of action potentials as practically happens in the case of ERP signals. If the maximal ionic conductance of the potassium current $\bar{g}_K$, is reduced the model will show a higher resting potential. Also, for $\bar{g}_K = 16$ ms cm^{-2}, the model will begin to exhibit oscillatory behaviour. Figure 3.5 shows the result of a Hodgkin–Huxley oscillatory model with reduced maximal potassium conductance.

The SNNAP can also model bursting neurons and central pattern generators. This stems from the fact that many neurons show cyclic spiky activities followed by a period of inactivity. Several invertebrate as well as mammalian neurons are bursting cells and exhibit alternating periods of high-frequency spiking behaviour followed by a period of no spiking activity.

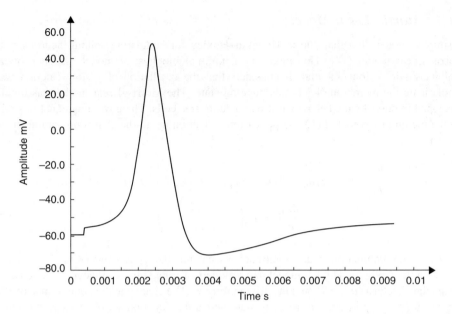

Figure 3.4 A single AP in response to a transient stimulation based on the Hodgkin–Huxley model. The initiated time is at $t = 0.4$ ms and the injected current is 80 μA cm^{-2} for a duration of 0.1 ms. The selected parameters are $C_m = 1.2$ μF cm^{-2}, $\bar{g}_{Na} = 100$ mS cm^{-2}, $\bar{g}_K = 35$ ms cm^{-2}, $\bar{g}_{leak} = 0.35$ ms cm^{-2}, and $E_{Na} = 60$ mV

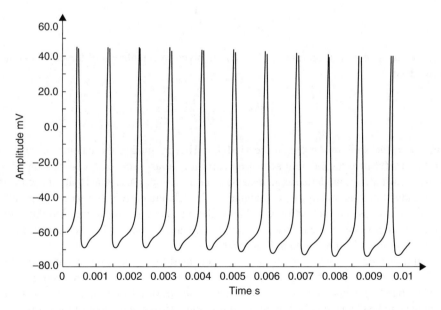

Figure 3.5 The AP from a Hodgkin–Huxley oscillatory model with reduced maximal potassium conductance. Taken from [13]

3.1.4   Morris–Lecar Model

A simpler model than that due to Hodgkin–Huxley for simulating spiking neurons is the Morris–Lecar model [14]. This model is a minimal biophysical model, which generally exhibits single action potential. It considers that the oscillation of a slow calcium wave depolarising the membrane leads to a bursting state. The Morris–Lecar model was initially developed to describe the behaviour of barnacle muscle cells. The governing equations relating the membrane potential (E) and potassium activation w_K to the activation parameters are given as:

$$C\frac{dE}{dt} = I_i - \bar{g}_{Ca}a_{Ca}(E)(E - E_{Ca}) - \bar{g}_K w_K (E - E_K) - \bar{g}_l(E - E_l) \qquad (3.28)$$

$$\frac{dw_K}{dt} = \lambda_t \left(\frac{w_\infty(E) - w_K}{\tau_K(E)} \right) \qquad (3.29)$$

where I_i is the combination of three ionic currents, calcium (Ca), potassium (K) and leak (l) that, similar to the Hodgkin–Huxley model, are products of a maximal conductance $\bar{g}$, activation components (such as a_{Ca}, w_k), and the driving force E. The change in the potassium activation variable w_K is proportional to a steady-state activation function $w_\infty(E)$ (a sigmoid curve) and a time-constant function $\tau_K(E)$ (a bell-shaped curve). These functions are respectively defined as:

$$w_\infty(E) = \frac{1}{1 + e^{-(E-h_w)/S_w}} \qquad (3.30)$$

$$\tau_K(E) = \frac{1}{e^{(E-h_w)/2S_w} + e^{-(E-h_w)/2S_w}} \qquad (3.31)$$

The steady-state activation function $a_{Ca}(E)$, involved in calculation of the calcium current, is defined as:

$$a_{Ca}(E) = \frac{1}{1 + e^{-(E-h_{Ca})/s_m}} \qquad (3.32)$$

Similar to the sodium current in the Hodgkin–Huxley model, the calcium current is an inward current. Since the calcium activation current is a fast process in comparison with the potassium current, it is modelled as an instantaneous function. This means that for each voltage E, the steady-state function $a_{Ca}(E)$ is calculated. The calcium current does not incorporate any inactivation process. The activation variable w_K here is similar to a_K in the Hodgkin–Huxley model, and finally the leak currents for both models are the same [12]. A simulation of the Morris–Lecar model is presented in Figure 3.6.

Calcium-dependent potassium channels are activated by intracellular calcium, the higher the calcium concentration the higher the channel activation [12]. For the Morris–Lecar model to exhibit bursting behaviour, the two parameters of maximal time constant and the input current have to be changed [12]. Figure 3.7 shows the bursting behaviour of the Morris–Lecar model. The basic characteristics of a bursting neuron are the duration of the spiky activity, the frequency of the action potentials during a burst, and the duration of the quiescence period. The period of an entire bursting event is the sum of both active and quiescence duration [12].

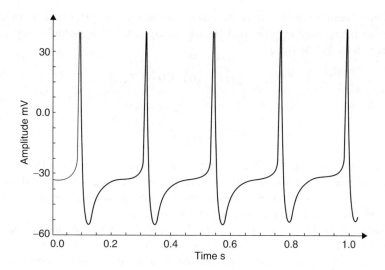

Figure 3.6   Simulation of an AP within the Morris–Lecar model. The model parameters are: $C_m = 22$ μF cm^{-2}, $\bar{g}_{Ca} = 3.8$ ms cm^-, $\bar{g}_K = 8.0$ ms cm^{-2}, $\bar{g}_{leak} = 1.6$ ms cm^{-2}, $E_{Ca} = 125$ mV, $E_K = -80$ mV, $E_{leak} = -60$ mV, $\lambda_t = 0.06$, $h_{Ca} = -1.2$, $S_m = 8.8$

Neurons communicate with each other across synapses through axon–dendrite or dendrite–dendrite connections, which can be excitatory, inhibitory, or electric [12]. By combining a number of the above models a neuronal network can be constructed. The network exhibits oscillatory behaviour due to the synaptic connection between the neurons. It is commonly assumed that excitatory synaptic coupling tends to synchronize neural firing while inhibitory coupling pushes neurons toward anti-synchrony. Such behaviour has been seen in models of neuronal circuits [1].

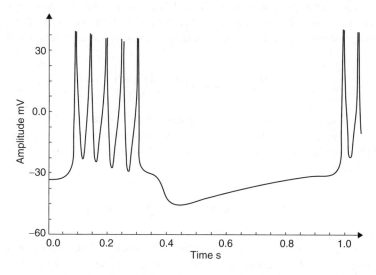

Figure 3.7   An illustration of the bursting behaviour that can be generated by the Morris–Lecar model

A synaptic current is produced as soon as a neuron fires an AP. This current stimulates the connected neuron and may be modelled by an alpha function multiplied by a maximal conductance and a driving force as:

$$I_{\text{syn}} = \bar{g}_{\text{syn}} \cdot g_{\text{syn}}(t) \left(E(t) - E_{\text{syn}} \right) \tag{3.33}$$

where

$$g_{\text{syn}}(t) = t \cdot e^{(-t/u)} \tag{3.34}$$

and t is the latency or time since the trigger of the synaptic current, u is the time to reach the peak amplitude, E_{syn} is the synaptic reversal potential, and $\bar{g}_{\text{syn}}$ is the maximal synaptic conductance. The parameter u alters the duration of the current while $\bar{g}_{\text{syn}}$ changes the strength of the current. This concludes the treatment of the modelling of APs.

As the nature of the EEG sources cannot be determined from the electrode signals directly, many researchers have tried to model these processes on the basis of information extracted using signal processing techniques. The method of linear prediction described in the later sections of this chapter is frequently used to extract a parametric description.

A tutorial on realistic neural modelling using the Hodgkin–Huxley excitation model by David Beeman has been documented at the first annual meeting of the World Association of Modelers (WAM) Biologically Accurate Modelling Meeting (BAMM) in 2005 in Texas, USA. This can be viewed at http://www.brains-minds-media.org/archive/218.

3.2   Mathematical Models

3.2.1   Linear Models

3.2.1.1   Prediction Method

The main objective of using prediction methods is to find a set of model parameters which best describe the signal generation system. Such models generally require a noise type input. In autoregressive (AR) modelling of signals each sample of a single channel EEG measurement is defined to be linearly related with respect to a number of its previous samples, that is,

$$y(n) = -\sum_{k=1}^{p} a_k y(n-k) + x(n) \tag{3.35}$$

where a_k, $k = 1, 2, \ldots, p$, are the linear parameters, n denotes the discrete sample time normalized to unity, and $x(n)$ is the noise input. In an autoregressive moving average (ARMA) linear predictive model each sample is obtained based on a number of its previous input and output sample values, that is,

$$y(n) = -\sum_{k=1}^{p} a_k y(n-k) + \sum_{k=0}^{q} b_k x(n-k) \tag{3.36}$$

where b_k, $k = 1, 2, \ldots, q$, are the additional linear parameters. The parameters p and q are the model orders. The Akaike criterion can be used to determine the order of the appropriate model of a measurement signal by minimizing the following equation [15] with respect to the model order.

$$AIC(i, j) = N \, \ln\left(\sigma_{ij}^2\right) + 2(i + j) \tag{3.37}$$

where i and j represent, respectively, the assumed AR and MA model prediction orders, N is the number of signal samples, and σ_{ij}^2 is the noise power of the ARMA model at the ith and jth stage. Later in this chapter we will see how the model parameters are estimated either directly or by employing some iterative optimisation techniques.

In a MVAR approach a multichannel scheme is considered. Therefore, each signal sample is defined versus both its previous samples and the previous samples of the signals from other channels, that is, for channel i we have:

$$y_i(n) = -\sum_{j=1}^{m}\sum_{k=1}^{p} a_{jk} y_j(n - k) + x_i(n) \tag{3.38}$$

where m represents the number of channels and $x_i(n)$ represents the noise input to channel i. Similarly, the model parameters can be calculated iteratively in order to minimize the error between the actual and predicted values [16].

There are numerous applications for linear models. These applications are discussed in other chapters of this book. Different algorithms have been developed to find efficiently the model coefficients. In the maximum likelihood estimation (MLE) method [17–19] the likelihood function is maximized over the system parameters formulated from the assumed real, Gaussian distributed, and sufficiently long input signals of approximately 10–20 s (consider a sampling frequency of $f_s = 250$ Hz as often used for EEG recordings). Using Akaike's method the gradient of the squared error is minimised using the Newton–Raphson approach applied to the resultant nonlinear equations [19, 20]. This is considered as an approximation to the MLE approach. In the Durbin method [21] the Yule–Walker equations, which relate the model coefficients to the autocorrelation of the signals, are iteratively solved. The approach and the results are equivalent to those using a least-squared-based scheme [22]. The MVAR coefficients are often calculated using the Levinson–Wiggines–Robinson (LWR) algorithm [23]. The MVAR model and its application in representation of what is called a direct transfer function (DTF), and its use in quantification of signal propagation within the brain, will be explained in detail in Chapter 8. After the parameters are estimated the synthesis filter can be excited with wide sense stationary noise to generate the EEG signal samples. Figure 3.8 illustrates the simplified system.

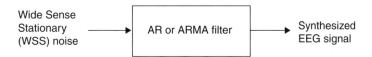

Figure 3.8   A linear model for the generation of EEG signals

3.2.1.2 Prony's Method

Prony's method has been previously used to model evoked potentials (EPs) [24, 25]. Based on this model an EP which is obtained by applying a short audio or visual brain stimulus to the brain can be considered as the impulse response (IR) of a linear infinite impulse response (IIR) system. The original attempt in this area was to fit an exponentially damped sinusoidal model to the data [26]. This method was later modified to model sinusoidal signals [27]. Prony's method is used to calculate the linear prediction (LP) parameters. The angles of the poles in the z-plane of the constructed LP filter are then referred to the frequencies of the damped sinusoids of the exponential terms used for modelling the data. Consequently, both the amplitude of the exponentials and the initial phase can be obtained following the methods used for an AR model, as follows.

Based on the original method we can consider the output of an AR system with zero excitation to be related to its IR as

$$y(n) = \sum_{k=1}^{p} a_k y(n-k) = \sum_{j=1}^{p} w_j \sum_{k=1}^{p} a_k r_j^{n-k-1} \tag{3.39}$$

where $y(n)$ represents the exponential data samples, p is the prediction order, $w_j = A_j e^{j\theta_j}$, $r_k = \exp((\alpha_k + j2\pi f_k)T_s)$, T_s is the sampling period normalized to 1, A_k is the amplitude of the exponential, α_k is the damping factor, f_k is the discrete-time sinusoidal frequency in samples/s, and θ_j is the initial phase in radians.

Therefore, the model coefficients are first calculated using one of the methods previously mentioned in this section, that is, $\mathbf{a} = -\mathbf{Y}^{-1}\breve{\mathbf{y}}$, where

$$\mathbf{a} = \begin{bmatrix} a_0 \\ a_1 \\ . \\ . \\ . \\ a_p \end{bmatrix}, \mathbf{Y} = \begin{bmatrix} y(p) & \cdots & y(1) \\ y(p-1) & \cdots & y(2) \\ & . & \\ & . & \\ y(2p-1) & \cdots & y(p) \end{bmatrix}, \text{and } \breve{\mathbf{y}} = \begin{bmatrix} y(p+1) \\ y(p+2) \\ . \\ . \\ y(2p) \end{bmatrix} \tag{3.40}$$

and $a_0 = 1$. On the basis of (3.39), $y(n)$ is calculated as the weighted sum of its p past values. $y(n)$, is then constructed and the parameters f_k and r_k are estimated. Hence, the damping factors are obtained as

$$\alpha_k = \ln |r_k| \tag{3.41}$$

and the resonance frequencies as

$$f_k = \frac{1}{2\pi} \tan^{-1} \left(\frac{\mathrm{Im}(r_k)}{\mathrm{Re}(r_k)} \right) \tag{3.42}$$

where Re(.) and Im(.) denote respectively the real and imaginary parts of a complex quantity. The w_k parameters are calculated using the fact that $y(n) = \sum_{k=1}^{p} w_k r_k^{n-1}$ or

$$
\begin{bmatrix} r_1^0 & r_2^0, & \cdots & r_p^0 \\ r_1^1 & r_2^1, & \cdots & r_p^1 \\ \cdot & \cdot & & \cdot \\ \cdot & \cdot & & \cdot \\ \cdot & \cdot & & \cdot \\ r_1^{p-1} & r_2^{p-1} & \cdots & r_p^{p-1} \end{bmatrix} \begin{bmatrix} w_1 \\ w_2 \\ \cdot \\ \cdot \\ \cdot \\ w_p \end{bmatrix} = \begin{bmatrix} y(1) \\ y(2) \\ \cdot \\ \cdot \\ \cdot \\ y(p) \end{bmatrix} \tag{3.43}
$$

In vector form this can be illustrated as $\mathbf{Rw} = \mathbf{y}$, where $[\mathbf{R}]_{k,l} = r_l^k$, $k = 0, 1, \ldots, p - 1$, $l = 1, \ldots, p$ denoting the elements of the matrix in the above equation. Therefore, $\mathbf{w} = \mathbf{R}^{-1}\mathbf{y}$, assuming R is a full-rank matrix, that is, there are no repeated poles. Often, this is simply carried out by implementing the Cholesky decomposition algorithm [28]. Finally, using w_k, the amplitude and initial phases of the exponential terms are calculated as follows:

$$
A_k = |w_k| \tag{3.44}
$$

and

$$
\theta_k = \tan^{-1}\left(\frac{\mathrm{Im}(w_k)}{\mathrm{Re}(w_k)}\right) \tag{3.45}
$$

In the above solution we considered that the number of data samples N is equal to $N = 2p$, where p is the prediction order. For the cases where $N > 2p$ a least-squares (LS) solution for $\mathbf{w}$ can be obtained as:

$$
\mathbf{w} = (\mathbf{R}^H\mathbf{R})^{-1}\mathbf{R}^H\mathbf{y} \tag{3.46}
$$

where $(.)^H$ denotes conjugate transpose. This equation can also be solved using the Cholesky decomposition method. For real data, such as EEG signals, this equation changes to $\mathbf{w} = (\mathbf{R}^T\mathbf{R})^{-1}\mathbf{R}^T\mathbf{y}$, where $(.)^T$ represents the transpose operation. A similar result can be achieved using principal component analysis (PCA) [18].

In the cases where the data are contaminated with white noise the performance of Prony's method is reasonable. However, for non-white noise the noise information is not easily separable from the data and, therefore, the method may not be sufficiently successful.

As we will see in a later chapter of this book, Prony's algorithm has been used in modelling and analysis of audio and visual evoked potentials (AEP and VEP) [29, 30].

3.2.2   Nonlinear Modelling

An approach similar to AR or MVAR modelling, in which the output samples are nonlinearly related to the previous samples, may be followed based on the methods developed for forecasting financial growth in economics studies.

In the generalised autoregressive conditional heteroskedasticity (GARCH) method [31] each sample relates to its previous samples through a nonlinear (or sum of nonlinear) function(s).

This model was originally introduced for time-varying volatility (honoured with the Nobel Prize in Economic sciences in 2003).

Nonlinearities in the time series are declared with the aid of the McLeod–Li [32] and BDS (Brock, Dechert, and Scheinkman) tests [33]. However, both tests lack the ability to reveal the actual kind of nonlinear dependence.

Generally, it is not possible to discern whether the nonlinearity is deterministic or stochastic in nature, nor can we distinguish between multiplicative and additive dependences. The type of stochastic nonlinearity may be determined on the basis of the Hseih test [34]. The additive and multiplicative dependences can be discriminated by using this test. However, the test itself is not used to obtain the model parameters.

Considering the input to a nonlinear system to be $u(n)$ and the generated signal as the output of such a system to be $x(n)$, a restricted class of nonlinear models suitable for the analysis of such a process is given by

$$x(n) = g(u(n-1), u(n-2), \cdots) + u_n.h(u(n-1), u(n-2), \cdots) \qquad (3.47)$$

Multiplicative dependence means nonlinearity in the variance, which requires the function $h(.)$ to be nonlinear; additive dependence, on the other hand, means nonlinearity in the mean, which holds if the function $g(.)$ is nonlinear. The conditional statistical mean and variance are, respectively, defined as:

$$E(x(n)|\chi_{n-1}) = g(u(n-1), u(n-2), \cdots) \qquad (3.48)$$

and

$$\mathrm{Var}(x(n)|\chi_{n-1}) = h^2(u(n-1), u(n-2), \cdots) \qquad (3.49)$$

where χ_{n-1} contains all the past information up to time $n-1$. The original GARCH(p,q) model, where p and q are the prediction orders, considers a zero mean case that is, $g(.) = 0$. If $e(n)$ represents the residual (error) signal using the above nonlinear prediction system, we have

$$\mathrm{Var}(e(n)|\chi_{n-1}) = \sigma^2(n) = \alpha_0 + \sum_{j=1}^{q} \alpha_j e^2(n-j) + \sum_{j=1}^{p} \beta_j \sigma^2(n-1) \qquad (3.50)$$

where α_j and β_j are the nonlinear model coefficients. The second term (first sum) on the right side corresponds to a qth order moving average (MA) dynamical noise term and the third term (second sum) corresponds to an autoregressive (AR) model of order p. It is seen that the current conditional variance of the residual at time sample n depends on both its previous sample values and previous variances.

Although in many practical applications, such as forecasting of stock prices, the orders p and q are set to small fixed values such as $(p,q) = (1,1)$; for a more accurate modelling of natural signals, such as EEGs, the orders have to be determined mathematically. The prediction coefficients for various GARCH models or even the nonlinear functions g and h are estimated iteratively as for the linear ARMA models [31,32].

Such simple GARCH models are only suitable for multiplicative nonlinear dependence. In addition, additive dependences can be captured by extending the modelling approach to the class of GARCH-M models [35].

Another limitation of the above simple GARCH model is failing to accommodate sign asymmetries. This is because the squared residual is used in the update equations. Moreover, the model cannot cope with rapid transitions, such as spikes. Considering these shortcomings, numerous extensions to the GARCH model have been proposed. For example, the model has been extended and refined to include the asymmetric effects of positive and negative jumps, such as the exponential GARCH model EGARCH [36], the GJR-GARCH model [37], the threshold GARCH model (TGARCH) [38], the asymmetric power GARCH model APGARCH [39], and quadratic GARCH model QGARCH [40].

In these models different functions for $g(.)$ and $h(.)$ in equations (3.48) and (3.49) are defined. For example, in the EGARCH model proposed by Glosten *et al.* [36] $h(n)$ is iteratively computed as:

$$h_n = b + \alpha_1 u^2(n-1)(1 - \eta_{t-1}) + \alpha_2 u^2(n-1)\eta_{t-1} + \kappa h_{n-1} \qquad (3.51)$$

where b, α_1, α_2, and κ are constants and η_n is an indicator function that is zero when u_n is negative and one otherwise.

Despite modelling the signals, the GARCH approach has many other applications. In some recent works [41] the concept of GARCH modelling of covariance is combined with Kalman filtering to provide a more flexible model with respect to space and time for solving the inverse problem. There are several alternatives for solution to the inverse problem. Many approaches fall into the category of constrained least-squares methods employing Tikhonov regularization [42]. Amongst numerous possible choices for the GARCH dynamics the EGARCH [36] has been used to estimate the variance parameter of the Kalman filter iteratively.

Nonlinear models have not been used for EEG processing. To enable use of these models the parameters and even the order should be adapted to the EEG properties. Also, such a model should incorporate the changes in the brain signals due to abnormalities and onset of diseases. In the next section we consider the interaction amongst various brain components to establish a more realistic model for generation of the EEG signals.

3.2.3   Gaussian Mixture Model

In this very popular modelling approach the signals are characterised using the parameters of their distributions. The distributions in terms of probability density functions are the sum of a number of Gaussian functions with different variances which are weighted and delayed differently [43]. The overall distribution subject to a set of K Gaussian components is defined as:

$$p(x|\theta_k) = \sum_{k=1}^{K} w_k \, p(x|\mu_k, \sigma_k)$$

The vector of unknown parameters $\theta_k = [w_k, \mu_k, \sigma_k]$ for $k = 1, 2, \ldots, K$. w_k is equivalent to the probability (weighting) that the data sample is generated by the kth mixture component density subject to

$$\sum_{k=1}^{K} w_k = 1$$

μ_k, and σ_k are the mean and variances of the kth Gaussian distribution and $p(x|\mu_k, \sigma_k)$ is a Gaussian function of x with parameters μ_k, and σ_k. Expectation maximization (EM) [44] is often used to estimate the above parameters by maximizing the log-likelihood of the mixture of Gaussian (MOG) for an N-sample data defined as:

$$L(\theta_k) = \sum_{i=1}^{N} \log \left(\sum_{k=1}^{K} w_k p(x(i)|\mu_k, \sigma_k) \right)$$

The EM algorithm alternates between updating the posterior probabilities used for generating each data sample by the kth mixture component (in a so-called E-step) as:

$$h_i^k = \frac{w_k p(x_i|\mu_k, \sigma_k)}{\sum_{j=1}^{K} w_j p(x_i|\mu_j, \sigma_j)}$$

and weighted maximum likelihood updates of the parameters of each mixture component (in a so-called M-step) as:

$$w_k \leftarrow \frac{1}{N} \sum_{i=1}^{N} h_i^k$$

$$\mu_k \leftarrow \frac{\sum_{i=1}^{N} h_i^k x_i}{\sum_{j=1}^{N} h_j^k}$$

$$\sigma_k \leftarrow \sqrt{\frac{\sum_{i=1}^{N} h_i^k (x_i - \mu_k^*)^2}{\sum_{j=1}^{N} h_j^k}}$$

The EM algorithm (especially in cases of high dimensional multivariate Gaussian mixtures) may converge to spurious solutions when there are singularities in the log-likelihood function due to small sample sizes, outliers, repeated data points or rank deficiencies leading to "variance collapse". Some solutions to these shortcomings have been provided by many researchers, such as those in [45–47]. Figure 3.9 demonstrates how an unknown multimodal distribution can be estimated using a weighted sum of Gaussians with different mean and variances.

Similar to prediction-based models, the model order can be estimated using the Akaike information criterion (AIC) or by iteratively minimising the error for best model order. Also, a mixture of exponential distributions can be used instead of a mixture of Gaussians where there are sharp transients within the data. In [48] it has been shown that these models can be used

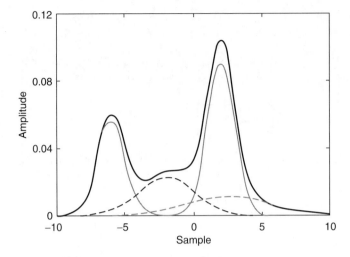

Figure 3.9 Mixture of Gaussians (dotted curves) model of a multimodal unknown distribution (bold curve)

for modelling a variety of physiological data, such as EEG, EOG, EMG, and ECG. In another work [49], the MOG model has been used for segmentation of magnetic resonance brain images. As a variant of GMM, Bayesian Gaussian mixture models have been used for partial amplitude synchronization detection in brain EEG signals [50]. This work introduces a method to detect subsets of synchronized channels that do not consider any baseline information. It is based on a Bayesian Gaussian mixture model applied at each location of a time-frequency map of the EEGs.

3.3   Generating EEG Signals Based on Modelling the Neuronal Activities

The objective in this section is to introduce some established models for generating normal and some abnormal EEGs. These models are generally nonlinear and some have been proposed [51] for modelling a normal EEG signal and some others for abnormal EEGs.

A simple distributed model consisting of a set of simulated neurons, thalamocortical relay cells, and interneurons was proposed [52, 53], that incorporates the limited physiological and histological data available at that time. The basic assumptions were sufficient to explain the generation of the alpha rhythm, that is, the EEGs within the frequency range 8–13 Hz.

A general nonlinear lumped model may take the form shown in Figure 3.10. Although the model is analogue in nature all the blocks are implemented in discrete form. This model can take into account the major characteristics of a distributed model and it is easy to investigate the result of changing the range of excitatory and inhibitory influences of thalamocortical relay cells and interneurons.

In this model [52] there is a feedback loop including the inhibitory postsynaptic potentials, the nonlinear function, and the interaction parameters C_3 and C_4. The other feedback includes mainly the excitatory potentials, nonlinear function, and the interaction parameters C_1 and

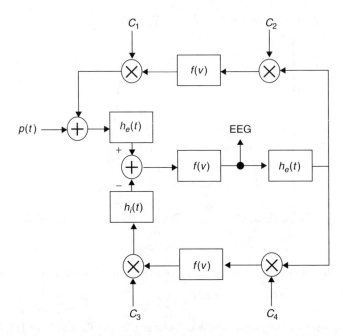

Figure 3.10 A nonlinear lumped model for generating the rhythmic activity of the EEG signals; $h_e(t)$ and $h_i(t)$ are the excitatory and inhibitory postsynaptic potentials, $f(v)$ is normally a simplified nonlinear function, and the C_is are, respectively, the interaction parameters representing the interneurons and thalamocortical neurons [52]

C_2. The role of the excitatory neurons is to excite one or two inhibitory neurons. The latter, in turn, serve to inhibit a collection of excitatory neurons. Thus, the neural circuit forms a feedback system. The input $p(t)$ is considered as a white noise signal. This is a general model; more assumptions are often needed to enable generation of the EEGs for the abnormal cases. Therefore, the function $f(v)$ may change to generate the EEG signals for different brain abnormalities. Accordingly, the C_i coefficients can be varied. In addition, the output is subject to environment and measurement noise. In some models, such as the local EEG model (LEM) [52] the noise has been considered as an additive component in the output.

Figure 3.11 shows the LEM model. This model uses the formulation by Wilson and Cowan [54] who provided a set of equations to describe the overall activity (not specifically the EGG) in a cartel of excitatory and inhibitory neurons having a large number of interconnections [55]. Similarly, in the LEM for the EEG it is assumed that the rhythms are generated by distinct neuronal populations, which possess frequency selective properties. These populations are formed by the interconnection of the individual neurons and are assumed to be driven by a random input. The model characteristics, such as the neural interconnectivity, synapse pulse response, and threshold of excitation are presented by the LEM parameters. The changes in these parameters produce the relevant EEG rhythms.

The input $p(t)$ is assumed to result from the summation of a randomly distributed series of random potentials which drive the excitatory cells of the circuit, producing the ongoing background EEG signal. Such signals originate from other deeper brain sources within the

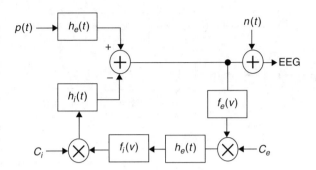

Figure 3.11 The local EEG model (LEM); the thalamocortical relay neurons are represented by two linear systems having impulse responses $h_e(t)$, on the upper branch, and the inhibitory postsynaptic potential by $h_i(t)$. The nonlinearity of this system is denoted by $f_e(v)$ representing the spike generating process. The interneuron activity is represented by another linear filter $h_e(t)$ in the lower branch, which generally can be different from the first linear system, and a nonlinearity function $f_i(v)$. C_e and C_i represent, respectively, the number of interneuron cells and the thalamocortical neurons. Taken from [54]

thalamus and brain stem and constitute part of the ongoing or spontaneous firing of the central nerve system (CNS). In the model, the average number of inputs to an inhibitory neuron from the excitatory neurons is designated by C_e and the corresponding average number from inhibitory neurons to each individual excitatory neuron is C_i. The difference of two decaying exponentials is used for modelling each post-synaptic potential h_e or h_i:

$$h_e(t) = A\left[\exp(-a_1 t) - \exp(-a_2 t)\right] \tag{3.52}$$
$$h_i(t) = B\left[\exp(-b_1 t) - \exp(-b_2 t)\right] \tag{3.53}$$

where A, B, a_k, and b_k are constant parameters, which control the shape of the pulse waveforms. The membrane potentials are related to the axonal pulse densities via the static threshold functions f_e and f_i. These functions are generally nonlinear; however, to ease the manipulations they are considered linear for each short time interval. Using this model, the normal brain rhythms, such as alpha wave, is considered as filtered noise.

A more simplified model is that presented by Jansen and Rit [56]. This model is demonstrated in Figure 3.12. The model is a neurophysiologically inspired model simulating electrical brain activity (including EEG, evoked potentials or EPs). A previously developed lumped-parameter model [52] of a single cortical column has been implemented in their work. The model could produce a large variety of EEG-like waveforms and rhythms. Coupling two models, with delays in the interconnections to simulate the synaptic connections within and between cortical areas, made it possible to replicate the spatial distribution of alpha and beta activity.

EPs were simulated by presenting pulses to the input of the coupled models. In general, the responses were more realistic than those produced using a single model. The proposed model is based on a nonlinear model of a cortical column described by Jansen and Rit [56] and also on Lopes da Silva's lumped parameter model [52, 54]. The cortical column is modelled by a population of 'feed forward' pyramidal cells, receiving inhibitory and excitatory feedback from local interneurons (i.e. other pyramidal, stellate or basket cells residing in the same

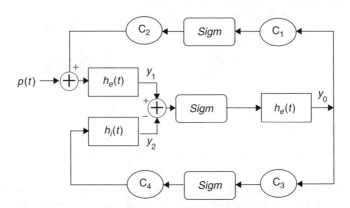

Figure 3.12   Simplified model for brain alpha generation. Taken from [56]

column) and excitatory input from neighbouring or more distant columns. The input can be a pulse, arbitrary function or noise.

Each of the neuron populations is modelled by two blocks. The first block transforms the average pulse density of action potentials coming to the population of neurons into an average postsynaptic membrane potential which can either be excitatory or inhibitory. This block is referred to as the PSP block and represents a linear transformation with an impulse response given by [56].

The main problem with such a model is due to the fact that only a single channel EEG is generated and unlike the phase coupling model explained in Section 3.1.2, there is no modelling of inter-channel relationships and the inherent connectivity of the brain zones. Therefore, a more accurate model has to be defined to enable simulation of a multichannel EEG generation system. This is still an open question and remains an area of research.

3.4   Electronic Models

These models describe the cell as an independent unit. The well established models such as the Hodgkin–Huxley one have been implemented using electronic circuits. Although for accurate models a large number of components are required, in practice it has been shown that a good approximation of such models can be achieved using simple circuits [57].

3.4.1   Models Describing the Function of the Membrane

Most of the models describing the excitation mechanism of the membrane are electronic realizations of the theoretical membrane model of Hodgkin and Huxley. In the following sections, two of these realizations are discussed.

3.4.1.1   Lewis Membrane Model

Lewis electronic membrane models are based on the Hodgkin–Huxley equations. All the components are parallel circuits connected between nodes representing the inside and outside of

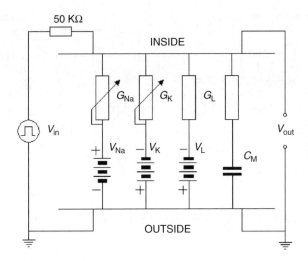

Figure 3.13   The Lewis membrane model. Reprinted from [58], © Stanford University Press

the membrane. He realized the sodium and potassium conductances using electronic hardware in the form of active filters, as shown in the block diagram of Figure 3.12. Since the output of the model is the transmembrane voltage V_m, the potassium current can be evaluated by multiplying the voltage corresponding to G_K by $(V_m - V_K)$. Figure 3.13 is consequently an accurate physical analogue to the Hodgkin–Huxley expressions, and the behaviour of the output voltage V_m corresponds to that predicted by the Hodgkin–Huxley equations. The electronic circuits in the Lewis neuromime had provision for inserting (and varying) not only such constants as $G_{K\,max}$, $G_{Na\,max}$, V_K, V_{Na}, V_{Cl}, which enter the Hodgkin–Huxley formulation, but also τ_h, τ_m, τ_n, which allow modifications from the Hodgkin–Huxley equations. In this realization the voltages of the biological membrane are multiplied by 100 to fit the electronic circuit. In other quantities, the original values of the biological membrane have been used.

3.4.1.2   Roy Membrane Model

As in Figure 3.14, Roy also introduced a model based on the Hodgkin–Huxley model [59]. He used field-effect transistors (FETs) to simulate the sodium and potassium conductances.

In the Roy model the conductance is controlled by a circuit including an operational amplifier, capacitors, and resistors. This circuit is designed to make the conductance behave according to the Hodgkin–Huxley model. Roy's main goal was to achieve a very simple model rather than to simulate accurately the Hodgkin–Huxley model.

3.4.2   Models Describing the Function of Neurons

3.4.2.1   Lewis Neuron Model

The Lewis model is based on the Hodgkin–Huxley membrane model and the theories of Eccles on synaptic transmission [60]. The model circuit is illustrated in Figure 3.15. This neuron model is divided into two sections: the synaptic section and the section generating the

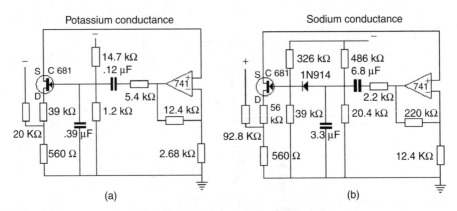

Figure 3.14 The circuits simulating (a) potassium and (b) sodium conductance in the Roy membrane model. Reprinted from [59], © IEEE

action pulse. Both sections consist of parallel circuits connected to the nodes representing the intracellular and extracellular sites of the membrane.

The section representing the synaptic junction is divided into two components; the inhibitory junction and the excitatory junction. The sensitivity of the section generating the action pulse to a stimulus introduced at the excitatory synaptic junction is reduced by the voltage introduced at the inhibitory junction. The section generating the action pulse is based on the Hodgkin–Huxley model which consists of the normal circuits simulating the sodium and potassium conductances, the leakage conductance, and the membrane capacitance. The circuit also includes an amplifier for the output signal. This model may be used in research on neural networks. However, it is actually a simplified version of Lewis's 46-transistor network having the same form. The purpose of this simplified Lewis model is to simulate the form of the action pulse with moderate accuracy following simple models.

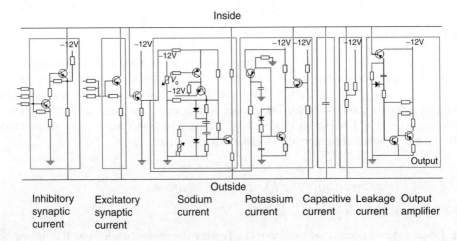

Figure 3.15 The Lewis neuron model from 1968. Reprinted from [58], © Stanford University Press

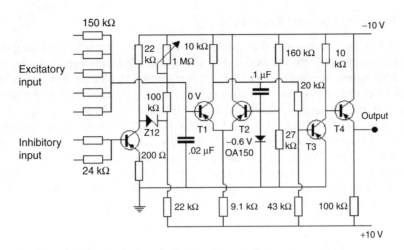

Figure 3.16 Harmon neuron model. Reprinted from [61]

3.4.2.2 The Harmon Neuron Model

The electronic realizations of the Hodgkin–Huxley model are very accurate in simulating the function of a single neuron. However, these circuits are often very complicated. Harmon managed to develop a neuron model having a very simple circuit [61]. A circuit of the Harmon neuron model is given in Figure 3.16. The model is equipped with five excitatory inputs which can be adjusted. These include diode circuits representing various synaptic functions. The signal introduced at excitatory inputs charges the 0.02 μF capacitor which, after reaching a voltage of about 1.5 V, allows the monostable multivibrator, formed by transistors T1 and T2, to generate the action pulse. This impulse is amplified by transistors T3 and T4. The output of one neuron model may drive the inputs of a large number of the neighbouring neuron models.

3.4.3 A Model Describing the Propagation of an Action Pulse in an Axon

Lewis simulated the propagation of an action pulse in a uniform axon and obtained interesting results [58]. The model structure, illustrated in Figure 3.17, can be seen to include a network of membrane elements as well as axial resistors representing the intracellular resistance. Six membrane elements are depicted in the figure. The model is an electronic realization of the linear core-conductor model with active membrane elements. An approximation of an action potential is generated in the output of each membrane element.

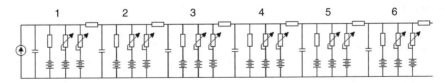

Figure 3.17 Lewis model for simulation of the propagation of the action pulse. Reprinted from [58], © Stanford University Press

3.4.4   Integrated Circuit Realizations

Mahowald [62] used electronic neuron models or neuron-like circuits as processing elements for electronic computers, called neurocomputers. Two examples of these models are as follows. The electronic neuron model developed by [63] is realized with integrated circuit technology. The circuit includes one neuron with eight synapses. The chip area of the integrated circuit is 4.5×5 mm^2. The array contains about 200 NPN and 100 PNP transistors, and about 200 of them are used. In 1991, Mahowald and Douglas [64] published an integrated circuit realization of an electronic neuron model. The model was realized with complementary metal oxide semiconductor (CMOS) circuits using very large-scale integrated (VLSI) technology. Their model accurately simulates the spikes of a neocortical neuron. The power dissipation of the circuit is 60 µW, and it occupies less than 0.1 mm^2. It is estimated that 100–200 such neurons could be fabricated on a 1 cm $\times$ 1 cm die.

3.5   Dynamic Modelling of the Neuron Action Potential Threshold

The neuron threshold is the transmembrane voltage level at which any further depolarization will activate a sufficient number of sodium channels to enter a self-generative positive feedback phase [65]. This threshold is often considered constant. Any alteration to the threshold influences the neuron spike-train temporal transformation. There are, however, evidences that the threshold is affected by the AP firing history nonlinearly [66–68]. In [65] a method for dynamically varying the threshold for intercellular activity has been proposed. The method is suitable for the systems with spikes in both their inputs and outputs.

3.6   Conclusions

The model based on phase-coupling explained here introduces the weakly coupled oscillators which can be used to model the interactions between the neurons. The Hodgkin–Huxley model on the other hand provides a detailed and accurate model for generation of active potentials. Linear and nonlinear prediction filers can be used in modelling the neuro generators. Gaussian mixtures are capable of modelling the EEG, particularly event-related, evoked and movement-related potentials. Finally, circuit models have been introduced to combine the excitatory and inhibitory postsynaptic potentials for generation of an EEG signal. The synaptic currents have also been modelled using electronic circuits. These circuits can be expanded to very accurately model either a single neuron or a membrane. They can also be used to model a large number of neurons for generation of particular brain waveforms.

References

[1] van Vreeswijk, C., Abbott, L.F. and Ermentrout, G.B. (1994) When inhibition not excitation synchronizes neural firing. *J. Comput. Neurosci.*, **1**, 313–321.
[2] Penny, W.D., Litvak, V., Fuentemilla, L., *et al.* (2009) Dynamic causal models for phase coupling. *J. Neurosci. Meth.*, **183**(1), 19–30.
[3] Ward, L. (2003) Synchronous neural oscillations and cognitive processes. *Trends Cogn. Sci.*, **7**(12), 553–559.
[4] von Stein, A., Rappelsberger, P., Sarnthein, J. and Petsche, H. (1999) Synchronization between temporal and parietal cortex during multimodal object processing in man. *Cereb. Cortex*, **9**(2), 137–150.

[5] Jones, M. and Wilson, M. (2005) Theta rhythms coordinate hippocampal–prefrontal interactions in a spatial memory task. *PLoS Biol.*, **3**(12), e402.

[6] Penny, W.D., Stephan, K.E., Mechelli, A. and Friston, K.J. (2004) Comparing dynamic causal models. *NeuroImage*, **22**(3), 1157–1172.

[7] Hoppensteadt, F., Izhikevich, E. (1997) *Weakly Connected Neural Networks*, Springer-Verlag, New York, USA.

[8] Benedek, G. and Villars, F. (2000) *Physics, with Illustrative Examples from Medicine and Biology*, Springer-Verlag, New York.

[9] Hille, B. (1992) *Ionic channels of Excitable Membranes*, Sinauer, Sunderland, MA.

[10] Hodgkin, A. and Huxley, A. (1952) A quantitative description of membrane current and its application to conduction and excitation in nerve. *J. Physiol. (Lond.)*, **117**, 500–544.

[11] Doi, S., Inoue, J., Pan, Z. and Tsumoto, K. (2010) *Computational Electrophysiology*, Springer.

[12] Simulator for Neural Networks and Action Potentials (SNNAP) (2003) Tutorial, The University of Texas-Houston Medical School, http://snnap.uth.tmc.edu.

[13] Ziv, I., Baxter, D.A. and Byrne, J.H. (1994) Simulator for neural networks and action potentials: description and application. *J. Neurophysiol.*, **71**, 294–308.

[14] Gerstner, W. and Kistler, W.M. (2002) *Spiking Neuron Models*, 1st edn. Cambridge University Press.

[15] Akaike, H. (1974) A new look at statistical model order identification. *IEEE Trans. Auto. Cont.*, **19**, 716–723.

[16] Kay, S.M. (1988) *Modern Spectral Estimation: Theory and Application*, Prentice Hall.

[17] Guegen, C. and Scharf, L. (1980) Exact maximum likelihood identification of ARMA models: a signal processing perspective, in *Signal Processing Theory Applications* (eds M. Kunt and F. de Coulon), North Holland, Amsterdam, pp. 759–769.

[18] Akay, M. (2001) *Biomedical Signal Processing*, Academic Press.

[19] Kay, S.M. (1988) *Modern Spectral Estimation, Theory and Application*, Prentice Hall, Englewood Cliffs, NJ.

[20] Akaike, H. (1974) A New Look at statistical model identification. *IEEE Trans. Auto. Cont.*, **19**, 716–723.

[21] Durbin, J. (1959) Efficient estimation of parameters in moving average models. *Biometrika*, **46**, 306–316.

[22] Trench, W.F. (1964) An algorithm for the inversion of finite Toelpitz matrices. *J. Soc. Ind. Appl. Math.*, **12**, 515–522.

[23] Morf, M., Vieira, A., Lee, D. and Kailath, T. (1978) Recursive multichannel maximum entropy spectral estimation. *IEEE Trans. Geosci. Elect.*, **16**, 85–94.

[24] Spreckelesen, M. and Bromm, B. (1988) Estimation of single-evoked cerebral potentials by means of parametric modelling and Kalman filtering. *IEEE Trans. Biomed. Eng.*, **33**, 691–700.

[25] Demiralp, T. and Ademoglu, A. (1992) Modeling of evoked potentials as decaying sinusoidal oscillations by Prony's method. Proceedings of the IEEE EMBS, Paris, 1992.

[26] De Prony, B.G.R. (1795) Essai experimental et analytique: sur les lois de la dilatabilite de fluids elastiques et sur celles de la force expansive de la vapeur de l'eau et de la vapeur de l'alkool, a differentes temperatures. *J. E. Polytech.*, **1**(2), 24–76.

[27] Marple, S.L. (1987) *Digital Spectral Analysis with Applications*, Prentice-Hall.

[28] Lawson, C.L. and Hanson, R.J. (1974) *Solving Least Squares Problems*, Prentice Hall, Englewood Cliffs, NJ.

[29] Demiralp, T. and Ademoglu, A. (1992) Modelling of evoked potentials as decaying sinusoidal oscillations by Prony method. Proceedings of the IEEE, EMBS, Paris, 1992.

[30] Bouattoura, D., Gaillard, P., Villon, P. and Langevin, F. (1996) Multilead evoked potentials modelling based on the Prony's method. Proceedings of IEEE TECON-Digital Signal Processing Applications, 1996, pp. 565–568.

[31] Dacorogna, M.M., Gencay, R., Müller, U., Olsen, R.B. and Pictet, O.V. (2001) *An Introduction to High-frequency Finance*, Academic Press, San Diego, USA.

[32] McLeod, A.J. and Li, W.K. (1983) Diagnostics checking ARMA time series models using squared residual autocorrelations. *J. Time Series Analysis*, **4**, 269–273.

[33] Brock, W.A., Hsieh, D.A. and LeBaron, B. (1992) *Nonlinear Dynamics, Chaos, and Instability: Statistical Theory and Economic Evidence*, The MIT Press, Cambridge, Massachussetts.

[34] Hsieh, D.A. (1989) Testing for nonlinear dependence in daily foreign exchange rates. *J. Business*, **62**, 339–368.

[35] Engle, R.F., Lilien, D.M. and Robin, R.P. Estimating time-varying risk premia in the term structure: the ARCH-M model. *Econometrica*, **55**, 391–407.

[36] Nelson, D.B. (1990) Stationarity and persistence in the GARCH(1,1) model. *J. Econometrics*, **45**, 7–35.

[37] Glosten, L.R., Jagannathan, R. and Runkle, D. (1995) On the relation between the expected value and the volatility of the nominal excess return on stocks. *J. Finance*, **2**, 225–251.

[38] Zakoian, J.M. (1994) Threshold heteroskedastic models. *J. Econ. Dyn. Control.*, **18**, 931–955.

[39] Ding, Z., Engle, R.F. and Granger, C.W.J. (1993) A long memory property of stock market returns and a new model. *J. Empirical Finance*, **1**, 83–106.

[40] Nelson, D.B. (1996) Modelling stock market volatility changes in *Modelling Stock Market Volatility: Bridging the Gap to Continuous Time*, ed. P.H. Rossi, Academic Press.

[41] Galka, A., Yamashita, O. and Ozaki, T. (2004) GARCH modelling of covariance in dynamical estimation of inverse solutions. *Phys. Lett. A*, **333**, 261–268.

[42] Tikhonov, A. (1992) *Ill-Posed Problems in Natural Sciences*, Coronet.

[43] Roweis, S. and Ghahramani, Z. (1999) A unifying review of linear Gaussian models. *Neural Comput.*, **11**, 305–345.

[44] Dempster, A.P., Laird, N.M. and Rubin, D.B. (1977) Maximum likelihood from incomplete data via the EM algorithm. *J. Roy. Stat. Soc. B*, **39**, 1–38.

[45] Redner, R.A. and Walker, H.F. (1984) Mixture densities, maximum likelihood and the EM algorithm. *SIAM Rev.*, **26**, 195–239.

[46] Ormoneit, D. and Tresp, V. (1998) Averaging, maximum penalized likelihood and Bayesian estimation for improving Gaussian mixture probability density estimates. *IEEE Trans. Neural Netw.*, **9**(4), 639–650.

[47] Archambeau, C., Lee, J.A. and Verleysen, M. (2003) On convergence problems of the EM algorithm for finite Gaussian mixtures. Proceedings of the European Symposium on Artificial Neural Networks (ESANN 2003), Bruges, Belgium, 23–25 April 2003, pp. 99–106.

[48] Hesse, C.W., Holtackers, D. and Heskes, T. (2006) On the use of mixtures of gaussians and mixtures of generalized exponentials for modelling and classification of biomedical signals. Belgian Day on Biomedical Engineering IEEE Benelux EMBS Symposium, December 7–8, 2006.

[49] Greenspan, H., Ruf, A. and Goldberger, J. (2006) Constrained gaussian mixture model framework for automatic segmentation of MR brain images. *IEEE Trans. Med. Imaging*, **25**(9), 1233–1245.

[50] Rio, M., Hutt, A. and Loria, B.G. (2010) Partial amplitude synchronization detection in brain signals using Bayesian Gaussian mixture models. Cinquième Conférence Plénière Française de Neurosciences Computationnelles Neurocomp'10, Lyon, France.

[51] Lagerlund, T.D., Sharbrough, F.W. and Busacker, N.E. (1997) Spatial filtering of multichannel electroencephalographic recordings through principal component analysis by singular value decomposition. *J. Clin. Neurophysiol.*, **14**(1), 73–82.

[52] Da Silva, F.H., Hoeks, A., Smits, H. and Zetterberg, L.H. (1974) Model of brain rhythmic activity: the alpharhythm of the thalamus. *Kybernetic*, **15**, 27–37.

[53] Lopes da Silva, F.H., van Rotterdam, A., Barts, P. *et al.* (1976) Models of neuronal populations: The basic mechanisms of rhythmicity, *Prog. Brain Res.*, **45**, 281–308 (Special Issue Perspective of Brain Research (eds M.A. Corner and D.F. Swaab)).

[54] Wilson, H.R. and Cowan, J.D. (1972) Excitatory and inhibitory interaction in localized populations of model neurons. *J. Biophys.*, **12**, 1–23.

[55] Zetterberg, L.H. (1973) Stochastic activity in a population of neurons—A system analysis approach. Report No. 2.3.153/1, TNO, Utrecht.

[56] Jansen, B.H. and Rit, V.G. (1995) Electroencephalogram and visual evoked potential generation in a mathematical model of coupled cortical columns. *Biol. Cybern.*, **73**, 357–366.

[57] Malmivuo, J. and Plonsey, R. (1995) *Bioelectromagnetism; Principles and Applications of Bioelectric and Biomagnetic Fields*, Oxford University Press.

[58] Lewis, E.R. (1964) An electronic model of the neuron based on the dynamics of potassium and sodium ion fluxes, in *Neural Theory and Modelling. Proceedings of the 1962 Ojai Symposium* (eds R.F. Reiss, H.J. Hamilton, L.D. Harmon *et al.*), Stanford University Press, Stanford, p. 427.

[59] Roy, G. (1972) A simple electronic analog of the squid axon membrane. The neurofet. *IEEE Trans. Biomed. Eng.*, **19**(1), 60–63.

[60] Eccles, J.C. (1964) *The Physiology of Synapses*, Springer-Verlag, Berlin.

[61] Harmon, L.D. (1961) Studies with artificial neurons, I: Properties and functions of an artificial neuron. *Kybernetik Heft*, **3**(Dez.), 89–101.

[62] Mahowald, M.A., Douglas, R.J., LeMoncheck, J.E. and Mead, C.A. (1992) An introduction to silicon neural analogs. *Semin. Neurosci.*, **4**, 83–92.

[63] Prange, S. (1990) Emulation of biology-oriented neural networks, in Proceedings of the International Conference on Parallel Processing in Neural Systems and Computers (ICNC), Düsseldorf.

[64] Mahowald, M.A. and Douglas, R.J. (1991) A silicon neuron. *Nature*, **354**, 515–518.

[65] Lu, U., Roach, S.M., Song, D. and Berger, T.W. (2012) Nonlinear dynamic modelling of neuron action potential threshold during synaptically driven broadband intercellular activity. *IEEE Tran. Biomed. Eng.*, **59**(3), 706–716.

[66] Azouz, R. and Gray, C.M. (1999) Cellular mechanisms contributing to response variability of cortical neurons in Vivo. *J. Neurosci.*, **19**(6), 2209–2223.

[67] Henze, D.A. and Buzsáki, G. (2001) Action potential threshold of hippocampal pyramidal cells in vivo is increased by recent spiking activity. *Neuroscience*, **105**(1), 121–130.

[68] Chacron, M.J., Lindner, B. and Longtin, A. (2007) Threshold fatigue and information transfer. *J. Comput. Neurosci.*, **23**(3), 301–311. Epub 2007.

4

Signal Transforms and Joint Time–Frequency Analysis

4.1 Introduction

One of the events which made a revolution in the signal processing field was the invention of the Fourier transform in the early nineteenth century. Since then signals have often been analysed in the frequency domain more effectively and efficiently.

EEG signals behave differently for different states of the brain. Normal brain rhythms during wakefulness may be considered as the sum of a limited number of sinusoids. Such signals often have a limited frequency band. During sleep, however, the waveforms change to the sum of very low frequency cyclic components and spikes. The spikes theoretically manifest as a broadband frequency component. Similarly, for some abnormalities, such as seizure, there is a sum of narrow band cortical activities and weak spikes generated within the hippocampus. Moreover, the artefacts such as eye blinks occur in random positions in time. Some other artefacts, such as those caused by magnetic fields in joint EEG-fMRI recordings, have periodic or quasi-periodic nature and are often strong. Therefore, it is worth analysing the brain signals (EEG or MEG) in the frequency or time–frequency domain. Following the presentation of some popular transforms, very recent EEG analyses in the transform domain are detailed in the following sections.

If the signals are statistically stationary it is straightforward to characterise them in either the time or the frequency domain. The frequency-domain representation of a finite-length signal can be found by using linear transforms, such as the discrete Fourier transform (DFT), cosine transform (DCT) or other semi-optimal transforms, which have kernels independent of the signal. However, the results of these transforms can be degraded by spectral smearing due to the short-term time-domain windowing of the signals and fixed transform kernels. An optimal transform, such as the Karhunen–,Loéve transform (KLT) requires complete statistical information, which may not be available in practice.

Adaptive Processing of Brain Signals, First Edition. Saeid Sanei.
© 2013 John Wiley & Sons, Ltd. Published 2013 by John Wiley & Sons, Ltd.

4.2 Parametric Spectrum Estimation and Z-Transform

Parametric spectrum estimation methods such as those based on AR or ARMA modelling, can outperform the DFT in accurately representing the frequency domain characteristics of a signal, but they may suffer from poor estimation of the model parameters, mainly due to the limited length of the measured signals. For example, in order to model the EEGs using an AR model, accurate values for the prediction order and coefficients are necessary. A high prediction order may result in splitting the true peaks in the frequency spectrum and a low prediction order results in combining peaks in close proximity in the frequency domain.

For an AR model of the signal $x(n)$ of the form

$$x(n) = \sum_{k=1}^{p} a_k x(n-k) + e(n) \tag{4.1}$$

the error or driving signal is considered to be zero mean white noise. Therefore, by applying the Z-transform to both sides of the above equation and dropping the block index m, and replacing Z by $e^{j\omega}$ we have:

$$\frac{X_p(\omega)}{E(\omega)} = \frac{1}{1 - \sum_{k=1}^{p} a_k e^{-jk\omega}} \tag{4.2}$$

where, $E(\omega) = K_\omega$ (Constant), is the power spectrum of the white noise and $X_p(\omega)$ is used to denote the signal power spectrum. Hence,

$$X_p(\omega) = \frac{K_\omega}{1 - \sum_{k=1}^{p} a_k e^{-jk\omega}} \tag{4.3}$$

and the parameters K_ω, a_k, $k = 1, \ldots, p$, are the exact values. In practical AR modelling these would be estimated from the finite length measurement, thereby degrading the estimate of the spectrum. Figure 4.1 provides a comparison of the spectrum of an EEG segment of approximately 1550 samples of a single channel EEG using both DFT analysis and AR modelling.

The fluctuations in the DFT result as shown in Figure 4.1(b) are the consequence of the statistical inconsistency of periodogram-like power spectral estimation techniques. The result from the AR technique (Figure 4.1(c)) overcomes this problem provided the model fits the actual data. EEG signals are often statistically nonstationary, particularly where there is an abnormal event captured within the signals. In these cases the frequency domain components are integrated over the observation interval and do not show the characteristics of the signals accurately. A time–frequency (TF) approach is the solution to the problem.

In the case of multichannel EEGs, where the geometrical positions of the electrodes reflect the spatial dimension, space–time–frequency (STF) analysis through multiway processing methods has also become popular [1]. The main concepts in this area together with the parallel factor analysis (PARAFAC) algorithm will be reviewed in Chapter 7 where its major applications will be discussed.

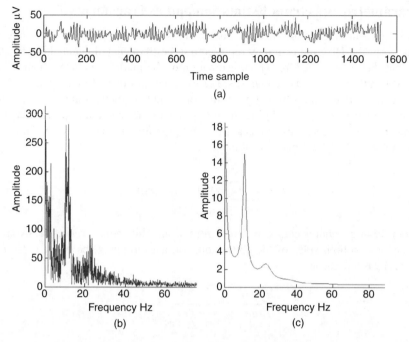

Figure 4.1 Single channel EEG spectrum: (a) a segment of EEG signal with a dominant alpha rhythm, (b) spectrum of the signal in (a) using DFT, (c) spectrum of the signal in (a) using a 12-orderAR model

4.3 Time–Frequency Domain Transforms

4.3.1 Short-Time Fourier Transform

The short-time Fourier transform (STFT) is defined as the discrete-time Fourier transform evaluated over a sliding window. The STFT can be performed as

$$X(n, \omega) = \sum_{\tau=-\infty}^{\infty} x(\tau)w(n - \tau)e^{-j\omega\tau} \qquad (4.4)$$

where the discrete time index n refers to the position of the window $w(n)$. Analogous with the periodogram a spectrogram is defined as

$$S_x(n, \omega) = |X(n, \omega)|^2 \qquad (4.5)$$

Based on the uncertainty principle that is, $\sigma_t^2\sigma_\omega^2 \geq \frac{1}{4}$, where σ_t^2 and σ_ω^2 are respectively time and frequency domain variances, perfect resolution cannot be achieved in both time- and frequency-domains. Windows are typically chosen to eliminate discontinuities at block edges and to retain positivity in the power spectrum estimate. The choice also impacts upon the spectral resolution of the resulting technique, which, put simply, corresponds to the minimum frequency separation required to resolve two equal amplitude frequency components [2].

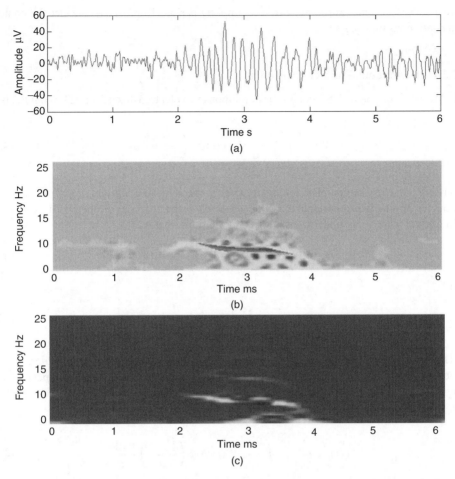

Figure 4.2 TF representation of an epileptic waveform in (a) for different time resolutions using a Hanning window of (b) 1 ms, and (c) 2 ms duration (see Plate 2 for the coloured version)

Figure 4.2 shows the TF representation of an EEG segment during the evolution from preictal to ictal and to postictal stages. In this figure the effect of time resolution has been illustrated using Hanning windows of different durations of 1 and 2 s. Importantly, in this figure the drift in frequency during the ictal period is observed clearly.

4.3.2 Wavelet Transform

The wavelet transform (WT) is another alternative for time-frequency analysis. There is already a well established literature detailing the WT, such as [3,4]. Unlike the STFT, the time-frequency kernel for the WT-based method can better localize the signal components in time–frequency space. This efficiently exploits the dependence between time and frequency components. Therefore, the main objective of introducing the WT by Morlet [3] was likely to have a coherence time proportional to the sampling period. To proceed, consider the context of a continuous time signal.

4.3.2.1   Continuous Wavelet Transform

The Morlet-Grossmann definition of the continuous wavelet transform for a 1D signal $f(t)$ is:

$$W(a, b) = \frac{1}{\sqrt{a}} \int\limits_{-\infty}^{\infty} f(t) \psi^* \left(\frac{t - b}{a} \right) dt \tag{4.6}$$

where $(.)^*$ denotes the complex conjugate, $\psi(t)$ is the analysing wavelet, a (>0) is the scale parameter (inversely proportional to frequency) and b is the position parameter. The transform is linear and is invariant under translations and dilations, that is,

$$\text{If } f(t) \rightarrow W(a, b) \text{ then } f(t - \tau) \rightarrow W(a, b - \tau) \tag{4.7}$$

and

$$f(\sigma t) \rightarrow \frac{1}{\sqrt{\sigma}} W(\sigma a, \sigma b) \tag{4.8}$$

The last property makes the wavelet transform very suitable for analysing hierarchical structures. It is similar to a mathematical microscope with properties that do not depend on the magnification. Consider a function $W(a,b)$ which is the wavelet transform of a given function $f(t)$. It has been shown [5, 6] that $f(t)$ can be recovered according to:

$$f(t) = \frac{1}{C_\phi} \int\limits_{0}^{\infty} \int\limits_{-\infty}^{\infty} \frac{1}{\sqrt{a}} W(a, b) \phi \left(\frac{t - b}{a} \right) \frac{da\,db}{a^2} \tag{4.9}$$

where

$$C_\phi = \int\limits_{0}^{\infty} \frac{\hat{\psi}^*(v) \hat{\phi}(v)}{v} dv = \int\limits_{-\infty}^{0} \frac{\hat{\psi}^*(v) \hat{\phi}(v)}{v} dv \tag{4.10}$$

Although often it is considered that $\psi(t) = \phi(t)$, other alternatives for $\phi(t)$ may enhance certain features for some specific applications [7]. The reconstruction of $f(t)$ is subject to having C_ϕ defined (admissibility condition). The case $\psi(t) = \phi(t)$ implies $\hat{\psi}(0) = 0$, that is, the mean of the wavelet function is zero.

4.3.2.2   Examples of Continuous Wavelets

Different waveforms/wavelets/kernels have been defined for the continuous wavelet transforms. The most popular ones are given below.

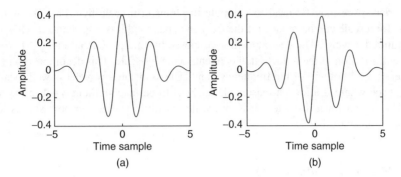

Figure 4.3 Morlet wavelet: real and imaginary parts shown, respectively, in (a) and (b)

Morlet wavelet is a complex waveform defined as:

$$\psi(t) = \frac{1}{\sqrt{2\pi}} e^{-\frac{t^2}{2} + j2\pi b_0 t} \tag{4.11}$$

This wavelet may be decomposed into its constituent real and imaginary parts as:

$$\psi_r(t) = \frac{1}{\sqrt{2\pi}} e^{-\frac{t^2}{2}} \cos(2\pi b_0 t) \tag{4.12}$$

$$\psi_i(t) = \frac{1}{\sqrt{2\pi}} e^{-\frac{t^2}{2}} \sin(2\pi b_0 t) \tag{4.13}$$

where b_0 is a constant, and it is considered that $b_0 > 0$ to satisfy the admissibility condition. Figure 4.3 shows, respectively, the real and imaginary parts.

As another wavelet, the Mexican hat defined by Murenzi [4] is given as:

$$\psi(t) = (1 - t^2) e^{-0.5t^2} \tag{4.14}$$

which is the second derivative of a Gaussian waveform.

4.3.2.3 Discrete Time Wavelet Transform

To enable computation of the wavelet coefficients for processing digital signals a discrete approximation of the wavelet coefficients is required. The discrete wavelet transform (DWT) can be derived in accordance with the sampling theorem if we process a frequency band-limited signal.

The continuous form of WT may be discretised with some simple considerations on the modification of the wavelet pattern by dilation. Since generally the wavelet function $\psi(t)$ is not band-limited, it is necessary to suppress the values outside the frequency components above half the sampling frequency to avoid aliasing (overlapping in frequency) effects.

A Fourier space may be used to compute the transform scale-by-scale. The number of elements for a scale can be reduced if the frequency bandwidth is also reduced. This requires a band-limited wavelet. The decomposition proposed by Littlewood and Paley [8] provides a very nice illustration of the reduction of elements scale-by-scale. This decomposition is based on an iterative dichotomy of the frequency band. The associated wavelet is well localized in Fourier space where it allows a reasonable analysis to be made although not in the original space. The search for a discrete transform, which is well localized in both spaces leads to multiresolution analysis.

4.3.3   Multiresolution Analysis

Multiresolution analysis results from the embedded subsets generated by interpolations (or down-sampling and filtering) of the signal at different scales. A function $f(t)$ is projected at each step j onto the subset V_j. This projection is defined by the scalar product $c_j(k)$ of $f(t)$ with the scaling function $\varphi(t)$, which is dilated and translated:

$$C_j(k) = \left\langle f(t), \, 2^{-j}\varphi\left(2^{-j}t - k\right)\right\rangle \tag{4.15}$$

where $\langle \cdot, \cdot \rangle$ denotes an inner product and $\varphi(t)$ has the property

$$\frac{1}{2}\varphi\left(\frac{t}{2}\right) = \sum_{n=-\infty}^{\infty} h(n)\varphi(t - n) \tag{4.16}$$

where the right side is a convolution of h and ϕ. By taking the Fourier transform of both sides

$$\Phi(2\omega) = H(\omega)\Phi(\omega) \tag{4.17}$$

where $H(\omega)$ and $\Phi(\omega)$ are the Fourier transforms of $h(t)$ and $\phi(t)$ respectively. For a discrete frequency space (i.e. using the DFT) the above equation permits the computation of the wavelet coefficient $C_{j+1}(k)$ from $C_j(k)$ directly. If we start from $C_0(k)$ we compute all $C_j(k)$, with $j > 0$, without directly computing any other scalar product:

$$C_{j+1}(k) = \sum_n C_j(n)h(n - 2k) \tag{4.18}$$

where k is the discrete frequency index.

At each step, the number of scalar products is divided by two and consequently the signal is smoothed. Using this procedure the first part of a filter bank is built up. In order to restore the original data, Mallat uses the properties of orthogonal wavelets, but the theory has been generalized to a large class of filters by introducing two other filters $\tilde{h}$ and $\tilde{g}$, also called conjugate filters. The restoration is performed with:

$$C_j(k) = 2 \sum_l \left[C_{j+1}(l)\tilde{h}(k + 2l) + w_{j+1}(l)\tilde{g}(k + 2l)\right] \tag{4.19}$$

where $w_{j+1}(.)$ are the wavelet coefficients at the scale $j + 1$ defined later in this section. For an exact restoration, two conditions have to be satisfied for the conjugate filters:

Anti-aliasing condition:

$$H\left(\omega + \frac{1}{2}\right)\tilde{H}(\omega) + G\left(\omega + \frac{1}{2}\right)\tilde{G}(\omega) = 0 \qquad (4.20)$$

Exact restoration:

$$H(\omega)\tilde{H}(\omega) + G(\omega)\tilde{G}(\omega) = 1 \qquad (4.21)$$

In the decomposition stage the input is successively convolved with the two filters H (low frequencies) and G (high frequencies). Each resulting function is decimated by suppression of one sample out of two. In the reconstruction, we restore the sampling by inserting a zero between each two samples. Then, we convolve the signal with the conjugate filters $\tilde{H}$ and $\tilde{G}$, the resulting outputs are added and the result multiplied by 2. The iteration can continue until the smallest scale is achieved. Orthogonal wavelets correspond to the restricted case where:

$$G(\omega) = e^{-2\pi\omega}H^*\left(\omega + \frac{1}{2}\right) \qquad (4.22)$$

$$\tilde{H}(\omega) = H^*(\omega) \qquad (4.23)$$

$$\tilde{G}(\omega) = G^*(\omega) \qquad (4.24)$$

and

$$|H(\omega)|^2 + \left|H\left(\omega + \frac{1}{2}\right)\right|^2 = 1 \quad \forall\omega \qquad (4.25)$$

We can easily see that this set satisfies the two basic relations (4.15) and (4.16). Amongst various wavelets, Daubechies wavelets are the only compact solutions to satisfy the above conditions. For biorthogonal wavelets we have the relations:

$$G(\omega) = e^{-2\pi\omega}\tilde{H}^*\left(\omega + \frac{1}{2}\right) \qquad (4.26)$$

$$\tilde{G}(\omega) = e^{2\pi\omega}H^*\left(\omega + \frac{1}{2}\right) \qquad (4.27)$$

and

$$H(\omega)\tilde{H}(\omega) + H^*\left(\omega + \frac{1}{2}\right)\tilde{H}^*\left(\omega + \frac{1}{2}\right) = 1 \qquad (4.28)$$

In addition, the relations (4.15) and (4.16) need also to be satisfied. A large class of compact wavelet functions can be used. Many sets of filters were proposed, especially for coding [9]. It was shown that the choice of these filters must be guided by the regularity of the scaling and the wavelet functions. The complexity is proportional to N. The algorithm provides a pyramid of N elements.

4.3.3.1 Wavelet Transform Using the Fourier Transform

Consider the scalar products $c_0(k) = \langle f(t).\varphi(t-k)\rangle$ for continuous wavelets. If $\varphi(t)$ is band limited to half of the sampling frequency, the data are correctly sampled. The data at the resolution $j = 1$ are:

$$c_1(k) = \left\langle f(t).\frac{1}{2}\varphi\left(\frac{t}{2}-k\right)\right\rangle \tag{4.29}$$

and we can compute the set $c_1(k)$ from $c_0(k)$ with a discrete-time filter with frequency response $H(\omega)$:

$$H(\omega) = \begin{cases} \dfrac{\Phi(2\omega)}{\Phi(\omega)} & \text{if } |\omega| < \omega_c \\[2mm] 0 & \text{if } \omega_c \leq |\omega| < \dfrac{1}{2} \end{cases} \tag{4.30}$$

and for $\forall \omega$ and $\forall$ integer m

$$H(\omega + m) = H(\omega) \tag{4.31}$$

Therefore, an estimate of the coefficients is:

$$C_{j+1}(\omega) = C_j(\omega)H(2^j\omega) \tag{4.32}$$

The cut-off frequency is reduced by a factor 2 at each step, allowing a reduction in the number of samples by this factor. The wavelet coefficients at the scale $j + 1$ are

$$w_{j+1} = \langle f(t),\ 2^{-(j+1)}\psi(2^{-(j+1)}t - k)\rangle \tag{4.33}$$

and they can be computed directly from C_j by:

$$W_{j+1}(\omega) = C_j(\omega)G(2^j\omega) \tag{4.34}$$

where G is the following discrete-time filter:

$$G(\omega) = \begin{cases} \dfrac{\Psi(2\omega)}{\Phi(\omega)} & \text{if } |\omega| < \omega_c \\[2mm] 0 & \text{if } \omega_c \leq |\omega| < \dfrac{1}{2} \end{cases} \tag{4.35}$$

and for $\forall \omega$ and $\forall$ integer m

$$G(\omega + m) = G(\omega) \tag{4.36}$$

The frequency band is also reduced by a factor of two at each step. These relationships are also valid for DWT following Section 4.2.2.3.

4.3.3.2 Reconstruction

The reconstruction of the data from the wavelet coefficients can be performed step-by-step, starting from the lowest resolution. At each scale, we compute:

$$C_{j+1} = H(2^j \omega) C_j(\omega) \tag{4.37}$$

$$W_{j+1} = G(2^j \omega) C_j(\omega) \tag{4.38}$$

we look for C_j knowing C_{j+1}, W_{j+1}, h and g. Then $C_j(\omega)$ is restored by minimising

$$P_h(2^j \omega) \left| C_{j+1}(\omega) - H(2^j \omega) C_j(\omega) \right|^2 + P_g(2^j \omega) \left| W_{j+1}(\omega) - G(2^j \omega) C_j(\omega) \right|^2 \tag{4.39}$$

using a least minimum squares estimator. $P_h(\omega)$ and $P_g(\omega)$ are weight functions which permit a general solution to the restoration of $C_j(\omega)$. The relationship of $C_j(\omega)$ is in the form of

$$C_j(\omega) = C_{j+1}(\omega) \tilde{H}(2^j \omega) + W_{j+1}(\omega) \tilde{G}(2^j \omega) \tag{4.40}$$

where the conjugate filters have the expressions:

$$\tilde{H}(\omega) = \frac{P_h(\omega) H^*(\omega)}{P_h(\omega) |H(\omega)|^2 + P_g(\omega) |G(\omega)|^2} \tag{4.41}$$

$$\tilde{H}(\omega) = \frac{P_g(\omega) G^*(\omega)}{P_h(\omega) |H(\omega)|^2 + P_g(\omega) |G(\omega)|^2} \tag{4.42}$$

It is straightforward to see that these filters satisfy the exact reconstruction condition given in equation (4.21). In fact, equations (4.41) and (4.42) give the general solutions to this equation. In this analysis, the Shannon sampling condition is always respected. No aliasing exists, so that the anti-aliasing condition (4.20) is not necessary.

The denominator is reduced if we choose:

$$G(\omega) = \sqrt{1 - |H(\omega)|^2} \tag{4.43}$$

This corresponds to the case where the wavelet is the difference between the squares of two resolutions:

$$|\Psi(2\omega)|^2 = |\Phi(\omega)|^2 - |\Phi(2\omega)|^2 \tag{4.44}$$

The reconstruction algorithm then performs the following steps:

1. Compute the FFT of the signal at the low resolution.
2. Set j to n_p; perform the following iteration steps:
3. Compute the FFT of the wavelet coefficients at the scale j.
4. Multiply the wavelet coefficients W_j by $\tilde{G}$.
5. Multiply the signal coefficients at the lower resolution C_j by $\tilde{H}$.
6. The inverse Fourier Transform of $W_j\tilde{G} + C_j\tilde{H}$ gives the coefficients C_{j-1}.
7. $j = j - 1$ and return to step 3.

The use of a band-limited scaling function allows a reduction of sampling at each scale and limits the computation complexity.

The wavelet transform has been widely used in EEG signal analysis. Its application to seizure detection, especially for neonates, modelling of the neuron potentials, and the detection of evoked potentials (EP) and event-related potentials (ERP) will be discussed in the corresponding chapters of this book.

This method has previously been shown to be useful in the evaluation of spontaneous fluctuations in the brain activity and has been used as a measure for decoding brain states [10, 11].

4.4 Ambiguity Function and the Wigner–Ville Distribution

The ambiguity function for a continuous time signal is defined as:

$$A_x(\tau, v) = \int_{-\infty}^{\infty} x^* \left(t - \frac{\tau}{2}\right) x \left(t + \frac{\tau}{2}\right) e^{jvt} dt \tag{4.45}$$

This function has its maximum value at the origin as

$$A_x(0, 0) = \int_{-\infty}^{\infty} |x(t)|^2 dt \tag{4.46}$$

As an example, if we consider a continuous time signal consisting of two modulated signals with different carrier frequencies such as

$$x(t) = x_1(t) + x_2(t)$$
$$= s_1(t)e^{j\omega_1 t} + s_2(t)e^{j\omega_2 t} \tag{4.47}$$

The ambiguity function $A_x(\tau, v)$ will be in the form of

$$A_x(\tau, v) = A_{x_1}(\tau, v) + A_{x_2}(\tau, v) + \text{cross terms} \qquad (4.48)$$

This concept is very important in the separation of signals using the TF domain. This will be addressed in the context of blind source separation (BSS) later in Chapter 7. Figure 4.4 demonstrates this concept.

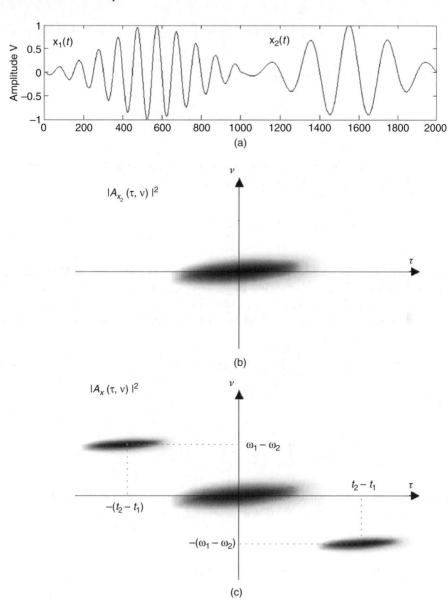

Figure 4.4 (a) A segment of a signal consisting of two modulated components, (b) ambiguity function for $x_1(t)$ only, and (c) the ambiguity function for $x(t) = x_1(t) + x_2(t)$

The Wigner–Ville frequency distribution of a signal $x(t)$ is then defined as the two-dimensional Fourier transform of the ambiguity function;

$$X_{WV}(t, \omega) = \frac{1}{2\pi} \int\limits_{-\infty}^{\infty} \int\limits_{-\infty}^{\infty} A_x(\tau, v) e^{-jvt} e^{-j\omega t} dv d\tau$$

$$= \frac{1}{2\pi} \int\limits_{-\infty}^{\infty} \int\limits_{-\infty}^{\infty} \int\limits_{-\infty}^{\infty} x^* \left(\beta - \frac{\tau}{2}\right) x \left(\beta + \frac{\tau}{2}\right) e^{-jv(t-\beta)} e^{-j\omega\tau} d\beta dv d\tau$$

(4.49)

which changes to the dual form of the ambiguity function as:

$$X_{WV}(t, \omega) = \int\limits_{-\infty}^{\infty} x^* \left(t - \frac{\tau}{2}\right) x \left(t + \frac{\tau}{2}\right) e^{-j\omega\tau} d\tau \qquad (4.50)$$

A quadratic form for the TF representation with the Wigner–Ville distribution can also be obtained using the signal in the frequency domain as:

$$X_{WV}(t, \omega) = \int\limits_{-\infty}^{\infty} X^* \left(\omega - \frac{v}{2}\right) X \left(\omega + \frac{v}{2}\right) e^{-jvt} dv \qquad (4.51)$$

The Wigner–Ville distribution is real and has very good resolution in both time- and frequency-domains. Also it has time and frequency support properties, that is, if $x(t) = 0$ for $|t| > t_0$, then $X_{WV}(t,\omega) = 0$ for $|t| > t_0$, and if $X(\omega) = 0$ for $|\omega| > \omega_0$, then $X_{WV}(t,\omega) = 0$ for $|\omega| > \omega_0$. It has also both time-marginal and frequency-marginal conditions of the form:

$$\frac{1}{2\pi} \int\limits_{-\infty}^{\infty} X_{WV}(t, \omega) dt = |X(t)|^2 \qquad (4.52)$$

and

$$\int\limits_{-\infty}^{\infty} X_{WV}(t, \omega) dt = |X(\omega)|^2 \qquad (4.53)$$

If $x(t)$ is the sum of two signals $x_1(t)$ and $x_2(t)$ that is, $x(t) = x_1(t) + x_2(t)$, the Wigner–Ville distribution of $x(t)$ with respect to the distributions of $x_1(t)$ and $x_2(t)$ will be:

$$X_{WV}(t, \omega) = X_{1WV}(t, \omega) + X_{2WV}(t, \omega) + 2\text{Re}\{X_{12WV}(t, \omega)\} \qquad (4.54)$$

where $\text{Re}\{.\}$ denotes the real part of a complex value and

$$X_{12WV}(\tau, \omega) = \int\limits_{-\infty}^{\infty} x_1^* \left(t - \frac{\tau}{2}\right) x_2 \left(t + \frac{\tau}{2}\right) e^{-j\omega\tau} d\tau \qquad (4.55)$$

It is seen that the distribution is related to the spectra of both auto- and cross-correlations. A pseudo Wigner–Ville distribution (PWVD) is defined by applying a window function, $w(\tau)$, centred at $\tau = 0$ to the time-based correlations, that is,

$$\breve{X}_{WV}(t, \omega) = \int_{-\infty}^{\infty} x^* \left(t - \frac{\tau}{2} \right) x \left(t + \frac{\tau}{2} \right) w(\tau) e^{-j\omega\tau} d\tau \tag{4.56}$$

In order to suppress the undesired cross-terms the two-dimensional WV distribution may be convolved with a TF-domain window. The window is a two-dimensional lowpass filter, which satisfies the time and frequency marginal (uncertainty) conditions, as described earlier. This can be performed as:

$$C_x(t, \omega) = \frac{1}{2\pi} \int_{-\infty}^{\infty} \int_{-\infty}^{\infty} X_{WV}(t', \omega') \Phi(t - t', \omega - \omega') dt' d\omega' \tag{4.57}$$

where

$$\Phi(t, \omega) = \frac{1}{2\pi} \int_{-\infty}^{\infty} \int_{-\infty}^{\infty} \varphi(\tau, \nu) e^{-j\nu t} e^{-j\omega\tau} d\nu d\tau \tag{4.58}$$

and $\varphi(., .)$ is often selected from a set of well-known signals, the so-called *Cohen's Class*. The most popular member of the Cohen's class of functions is the bell-shaped function defined as:

$$\varphi(\tau, \nu) = e^{-\nu^2\tau^2/(4\pi^2\sigma)}, \qquad \sigma > 0 \tag{4.59}$$

A graphical illustration of such a function can be seen in Figure 4.5. In this case the distribution is referred to as a *Choi–Williams distribution*.

To improve the distribution a signal-dependent kernel may also be used [12].

4.5   Hermite Transform

The problem of the Hermite transform was initially addressed in image coding [13], where the feature used to play a central role in the Hermite analysis. It was argued that the use of the Hermite transform is in close agreement with the action of the human visual system. In this work it was demonstrated that the Hermite transform is in better agreement with human visual modelling than Gabor expansions [13, 14].

The Hermite transform of an EEG signal is obtained by computing the inner product between the signal and the Gaussian functions. These functions are the eigenvectors of a centred or shifted Fourier matrix. The discrete Hermite transform (DHT) of a digital signal is obtained using discrete basis functions, $\mathbf{h}_k$, which are generated as a set of eigenvectors of a centred or

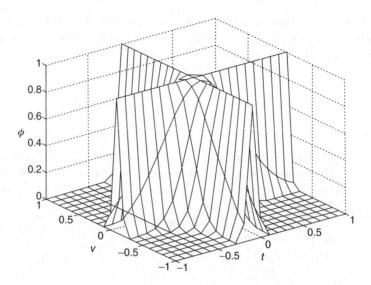

Figure 4.5   Illustration of $\varphi(\tau, \nu)$ for the Choi–Williams distribution

shifted Fourier transform [15] of that signal. The eigenvectors of the shifted Fourier transform, $\mathbf{F}_c$, comply with the following equation:

$$\mathbf{F}_c \mathbf{h}_k = j^k \mathbf{h}_k \qquad (4.60)$$

where $j = \sqrt{-1}$ and $k = 1, \ldots, m$ and m is the number of basis functions $\mathbf{h}_1, \mathbf{h}_2, \ldots, \mathbf{h}_m$ that are the basis functions for a matrix with dimension m in DHT. In order to generate the basis functions a tridiagonal matrix, $\mathbf{T}$, which has the same eigenvectors as $\mathbf{F}_c$ is used [16]. The eigenvectors of matrix $\mathbf{T}$ are orthogonal to each other as $\mathbf{T}$ is a symmetric $m \times m$ matrix. The elements of $\mathbf{T}$ can be computed using the following equations:

The kth element on the main diagonal is computed by:

$$T(k, k) = -2 \cos\left(\frac{\pi}{\sigma^2}\right) \sin\left(\frac{\pi k}{m\sigma^2}\right) \sin\left(\frac{\pi}{m\sigma^2}((m-1)-k)\right) \qquad (4.61)$$

where $1 < k < m$ and the kth off-diagonal element is computed by:

$$T(k, k-1) = T(k-1, k) = \sin\left(\frac{\pi k}{m\sigma^2}\right) \sin\left(\frac{\pi}{m\sigma^2}(m-k)\right) \qquad (4.62)$$

where $2 < k < m-1$. The remaining elements of T are set to zero.

Parameter $\sigma \geq 1$ in the above equations, known as the dilation parameter, controls the width of the digital basis functions. Figure 4.6 shows BCG models using different values of the dilation parameter.

Appropriately selecting the value of this parameter is important. A proper value for this parameter allows the Hermite transform to model the signal with a minimum number of terms.

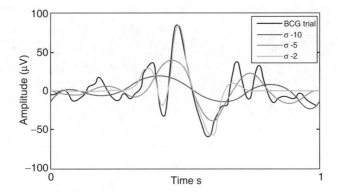

Figure 4.6   Modelling a cycle of the BCG with different values of the dilation parameter

The Hermite transform of a digital signal $\mathbf{x}$ of size $1 \times m$, can be computed by the inner product between the input signal and the basis functions. The result of this product is a set of transform parameters denoted as:

$$\mathbf{C}_k = \langle \mathbf{x}, \mathbf{h}_k \rangle k = 1, \ldots, m \qquad (4.63)$$

In fact, the coefficients of $\mathbf{C}_k$ represent the shape contents of the signal. Therefore, to achieve the result of applying this transform to $\mathbf{x}$ it is required that the signal artefact can be modelled by a number (e.g. p) of elements of $\mathbf{C}_k$. This can be achieved by inverse discrete Hermite transform:

$$\hat{\mathbf{x}} = \sum_{k=1}^{p} \mathbf{C}_k \mathbf{h}_k \qquad (4.64)$$

In [17] an algorithm using DHT has been proposed for the removal of ballistocardiogram (BCG) artefact from the EEG signals jointly recorded with fMRI. The main objective in this method is modelling the BCG artefact using discrete Hermite transform. The shape of BCG is modelled using Gaussian functions which are the primitive Hermite functions. Later, this method has been combined with ICA for a better performance in BCG artefact removal [16]. In this work the Hermite transform has been applied to the independent components of the multichannel EEG signals. These components represent the BCG artefact. In Figure 4.7 a comparison between ICA and ICA-DHT for BCG removal from EEG has been illustrated.

DHT based algorithms are sensitive to the value of the dilation parameter σ. Given the electrocardiogram (ECG) signal is available, one way to decrease the sensitivity of the algorithm is by using an adaptive strategy such as what follows:

$$\begin{cases} \sigma_{i+1} \leftarrow \sigma_i & \rho \leq T_r \\ \sigma_{i+1} \leftarrow \sigma_i - \beta \dfrac{\rho}{\rho_{\max}} & \rho > T_r \end{cases} \qquad (4.65)$$

where ρ is the normalized correlation between the ECG and BCG signals after applying the proposed method, i is the current iteration of the algorithm, and β is the step size which is

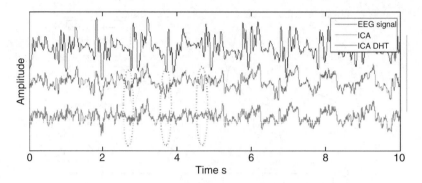

Figure 4.7   Comparison between ICA and ICA-DHT for BCG removal from EEG

manually selected by the user. The initial value of σ is manually selected between 10 and 15. The value of this parameter is updated until it reaches a predefined threshold, T_r.

4.6   Conclusions

The field of signal processing was revolutionized by the invention of signal transform techniques. In the majority of signal processing applications signal transformation plays a crucial role. Sinusoidal signals have narrow frequency bands and, therefore, appear sparse in Fourier or cosine transform domains. On the other hand flat spectrum data, such as wideband noise, turns up as a narrow impulse in time. Hence, to more effectively and efficiently analyse the data transformation becomes the main aspect of signal processing. Time–frequency domain information exploits the data characteristics and properties in both time and frequency. Therefore, variations of the signals in both domains can be seen and tracked in this domain.

Time–frequency approaches decompose the data into its dominant components more effectively. In processing the brain signals time–frequency transforms such as the wavelet can better show the events in the brain compared with time domain data. Besides the conventional transform methods introduced in this chapter there are other suboptimal ones and some special purpose transforms such as Hilbert and Gabor transforms. These transforms may be referred to in other chapters where the corresponding applications are discussed. The transforms introduced in this chapter are used throughout this book.

References

[1] Bro, R. (1998) Multi-way Analysis in the Food Industry: Models, Algorithms, and Applications. PhD thesis, University of Amsterdam (NL) and Royal Veterinary and Agricultural University; MATLAB toolbox available on line at: http://www.models.kvl.dk/users/rasmus/.

[2] Harris, F.J. (2004) *Multirate Signal Processing for Communication Systems*, Prentice Hall.

[3] Franaszczuk, P.J., Bergey, G.K., Durka, P.J. and Eisenberg, H.M. (1998) Time-frequency analysis using the matching pursuit algorithm applied to seizures originating from the mesial temporal lobe. *J. Electroencephalogr. Clin. Neurophysiol.*, **106**(6), 513–521.

[4] Murenzi, R., Combes, J.M., Grossman, A. and Tchmitchian, P. (eds) (1988) *Wavelets*, Springer, Berlin, Heidelberg, New York.

[5] Vaidyanathan, P.P. (1993) *Multirate Systems and Filter Banks*, Prentice Hall.

[6] Holschneider, M., Kronland-Martinet, R., Morlet, R.J. and Tchamitchian, Ph. (1989) A real-time algorithm for signal analysis with the help of the wavelet transform, in *Wavelets: Time-Frequency Methods and Phase Space*, eds J. M. Combes, A. Grossman and Ph. Tchamitchian, Springer-Verlag, Berlin, pp. 286–297.

[7] Chui, C.K. (1992) *An Introduction to Wavelets*, Academic Press.

[8] Stein, E.M. (1958) On the functions of Littlewood-Paley, Lusin and Marcinkiewicz. *Trans. Am. Math. Soc.*, **88**, 430–466.

[9] Vetterli, M. and Kovačevic, J. (1995) *Wavelets and Subband Coding*, Prentice-Hall.

[10] Glassman, E.L. (2005) A wavelet-like filter based on neuron action potentials for analysis of human scalp electroencephalographs. *IEEE Trans. Biomed. Eng.*, **52**(11), 1851–1862.

[11] Achard, S. (2006) A resilient, low-frequency, small-world human brain functional network with highly connected association cortical hub. *J. Neurosci.*, **26**(1), 63–72.

[12] Richiardi, J. (2011) Decoding brain states from fMRI connectivity graphs. *Neuroimage*, **56**(2), 616–626.

[13] Martens, J.-B. (1990) The Hermite transform-applications. **38**(9), 1607–1618.

[14] Martens, J.-B. (2006) The Hermite transform-theory, *EURASIP J. Adv. Signal Process.* doi:10.1155/ASP/2006/26145.

[15] Clary, S. and Mugler, D.H. (2003) Shifted Fourier matrices and their tridiagonal commutors. *SIAM J. Matrix Anal. Appl.*, **24**(3), 809.

[16] Ferdowsi, S., Sanei, S., Nottage, J. *et al.* (2012) A hybrid ICA-Hermite transform for removal of ballistocardiogram from EEG. Proceedings of the European Signal Processing Conference, EUSIPCO, Romania, 2012.

[17] Mahadevan, A., Acharya, A., Sheffer, S. and Mugler, D.H. (2008) Ballistocardiogram artifact removal in EEG-fMRI Signals Using Discrete Hermite Transforms. *Sel. Top. Signal Process.*, **2**(6), 839–853.

5

Chaos and Dynamical Analysis

The brain is a complex system whose characteristics change dynamically through time. Borrowing the concepts from time series analysis and forecasting, often used by mathematicians, astronomers, weather broadcasters, and researchers in economy, signal trends are not limited only to random or well defined trends (with correlated samples, stationary, cyclo-stationary, etc.). The trends and signals can follow particular dynamics when they are generated by well-defined linear or nonlinear systems. In physiological systems there is no clear model for the system and, therefore, only some approximations for the system model can be reached using the generated signals. Hence, characterisation and evaluation of chaotic behaviour of such signals and systems become an important problem.

The change in the brain as a dynamic system can be clearly seen in many cases; when a human goes to sleep the brain signal trends change significantly during the four stages of sleep. In the case of epileptic seizure, the signals from the neurogenerators around the epileptic loci gradually change from chaotic to ordered before the seizure onset. The same may happen with many biological signals. As another example, an intention to move a body part results in desynchronizing the alpha rhythm and, as the movement starts, synchronizing beta rhythm in the brain.

As an effective tool for prediction and characterization of the signals, *deterministic chaos* plays an important role. Although the EEG signals are considered chaotic, there are rules, which do not in themselves involve any element of change, which can be used in their characterisation [1]. Mathematical research on chaos started before 1890 when certain people, such as Andrey Kolmogorov and Henri Poincarétried to establish whether planets would remain indefinitely in their orbits. In the 1960s Stephan Smale formulated a plan to classify all the typical kinds of dynamic behaviour. Many chaos-generating mechanisms have been created and used to identify the behaviour of the dynamics of the system. The Rossler system was designed to model a strange attractor using a simple stretch and fold mechanism. This was, however, inspired by the Lorenz attractor introduced more than a decade earlier [1].

To evaluate how chaotic a dynamical system is different measures can be taken into account. A straightforward parameter is the attractor dimension. Different multidimensional attractors have been defined by a number of mathematicians. In many cases it is difficult to find the attractor dimension unless the parameters of the system can be approximated. However, later

Adaptive Processing of Brain Signals, First Edition. Saeid Sanei.
© 2013 John Wiley & Sons, Ltd. Published 2013 by John Wiley & Sons, Ltd.

in this chapter we will show that the attraction dimension [2] can be simply achieved using the Lyapunov exponents.

5.1 Entropy

Entropy is a measure of uncertainty. The level of chaos may also be measured using the entropy of the system. Higher entropy represents higher uncertainty and a more chaotic system. Entropy is given as

$$\text{Entropy of the Signal } x(n) = \int_{\min(x)}^{\max(x)} p_x \log(1/p_x)\,\mathrm{d}x \tag{5.1}$$

where p_x is the probability density function (pdf) of signal $x(n)$. Although this measure is defined for a single-channel time series it can be easily extended to multichannel or even multidimensional signals. As an example, multichannel EEG signals can be processed together. Therefore, generally, the distribution can be a joint pdf when the multichannel signals such as EEG are jointly processed. On the other hand, the pdf can be replaced by a conditional pdf in places where the occurrence of an event is subject to another event. In this case, the entropy is called conditional entropy. Entropy is very sensitive to noise. Noise increases the uncertainty and noisy signals have higher entropy, even if the original signal is ordered.

Entropy is used in the calculation of many other useful parameters, such as mutual information, negentropy, nongaussianity and Kulback–Leibler divergence. These variables are widely used in estimation of the degree of nonlinearity of the systems and correspondence of signals.

5.2 Kolmogorov Entropy

Kolmogorov entropy or *metric entropy*, is an effective measure of the complexity of a system. This metric is often used as a measure of chaos for dynamic systems. To estimate Kolmogorov entropy the phase space is divided into multidimensional hypercubes. The phase space is the space in which all possible states of a system are represented, with each corresponding to one unique point in the phase space. In this space, every degree of freedom or parameter of the system is represented as an axis of a multidimensional space. A phase space may contain many dimensions. The hypercube is a generalization of a 3-cube to n-dimensions, also called an n-cube or measure polytope. It is a regular polytope with mutually perpendicular sides and is, therefore, an orthotope. Now, let $P_{i_0,\ldots,i_n}$ be the probability that a trajectory falls inside the hypercube; i_0 at $t = 0$, i_1 at $t = T$, i_2 at $t = 2T$, .. Then define

$$K_n = - \sum_{i_0,\ldots i_n} P_{i_0,\ldots i_n} \ln P_{i_0,\ldots i_n} \tag{5.2}$$

where $K_{n+1} - K_n$ is the information needed to predict which hypercube the trajectory will be in at $(n + 1)T$, given trajectories up to nT. The Kolmogorov entropy is then defined as

$$K = \lim_{N \to \infty} \frac{1}{NT} \sum_{n=0}^{N-1} (K_{n+1} - K_n) \tag{5.3}$$

Estimation of the above joint probabilities for large dimensional data is computationally costly. On the other hand, in practice, long data sequences are normally required to perform a precise estimation of the Kolmogorov entropy.

5.3 Lyapunov Exponents

A chaotic model can be generated by a simple feedback system. Consider a quadratic iterator of the form $x(n) \rightarrow \alpha x(n)(1 - x(n))$ with an initial value of x_0. This generates a time series such as that in Figure 5.1 (for $\alpha = 3.8$).

Although in the first 20 samples the time series seems to be random noise, its semi-ordered alterations (cyclic behaviour) later show that some rules govern its chaotic behaviour. This time series is subject to two major parameters α and x_0.

Now to adopt this model within a chaotic system a different initial value may be selected. Perturbation of an initial value generates an error E_0, which propagates during the signal evolution. After n samples the error changes to E_n. E_n/E_0 is a measure of how fast the error grows. The average growth of infinitesimally small errors in the initial point x_0 is quantified by Ljapunov (Lyapunov) exponents $\lambda(x_0)$. The total error amplification factor $|E_n/E_0|$, can be written in terms of sample error amplifications as:

$$\left| \frac{E_n}{E_0} \right| = \left| \frac{E_n}{E_{n-1}} \right| \cdot \left| \frac{E_{n-1}}{E_{n-2}} \right| \cdots \left| \frac{E_1}{E_0} \right| \tag{5.4}$$

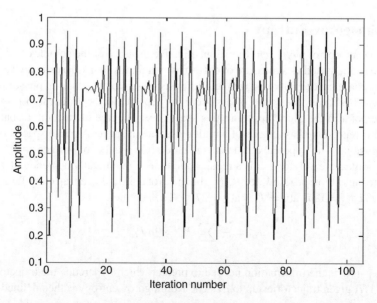

Figure 5.1 Generated chaotic signal using the model $x(n) \rightarrow \alpha x(n)(1 - x(n))$ using $\alpha = 3.8$ and $x_0 = 0.2$

The average logarithm of this becomes

$$\frac{1}{n}\ln\left|\frac{E_n}{E_0}\right| = \frac{1}{n}\sum_{k=1}^{n}\ln\left|\frac{E_k}{E_{k-1}}\right| \tag{5.5}$$

Obviously, the problem is how to measure $|E_k/E_{k-1}|$. For the iterator $f(x(n))$, where $f(x(n)) = \alpha x(n)(1 - x(n))$ in the above example, having a small perturbation ε at the initial point, the term in the above equation may be approximated as:

$$\frac{1}{n}\ln\left|\frac{E_n}{E_0}\right| = \frac{1}{n}\sum_{k=1}^{n}\ln\left|\frac{\tilde{E}_k}{\varepsilon}\right| \tag{5.6}$$

where $\tilde{E}_k = f(x_{k-1} + \varepsilon) - f(x_{k-1})$. By replacing this in the above equation the Lyapunov exponent is approximated as:

$$\lambda(x_0) = \lim_{n\to\infty}\frac{1}{n}\sum_{k=1}^{n}\ln\left|f'(x_{k-1})\right| \tag{5.7}$$

This measure is very significant in separating unstable, unpredictable, or chaotic behaviour from predictable, stable, or ordered ones. If λ is positive the system is chaotic whereas it is negative for ordered systems.

Kaplan and Yorke [3] empirically concluded that it is possible to predict the dimension of a strange attractor from knowledge of the Lyapunov exponents of the corresponding transformation. This is termed the Kaplan–Yorke conjecture, and has been investigated by many other researchers [4]. This is a very important conclusion since in many dynamical systems the various dimensions of the attractors are hard to compute, while the Lyapunov exponents are relatively easy to compute. This conjecture also claims that generally the information dimension D_I and Lyapunov dimension D_L, respectively, are defined as [1]

$$D_I = \lim_{s\to0}\frac{I(s)}{\log_2 1/s} \tag{5.8}$$

where s is the size of a segment of the attractor and $I(s)$ is the entropy of s, and

$$D_L = m + \frac{1}{|\lambda_{m+1}|}\sum_{k=1}^{m}\lambda_k \tag{5.9}$$

where m is the maximum integer with $\gamma(m) = \lambda_1 + \cdots + \lambda_m \geq 0$ given that $\lambda_1 > \lambda_2 > \cdots > \lambda_m$ (for $\lambda_1 < 0$ we set $D_L = 0$) are the same in both D_I and D_L.

5.4 Plotting the Attractor Dimensions from Time Series

Very often it is necessary to visualize a phase space attractor and decide about the stability, chaosity, or randomness of a signal (time series). The attractors can be multidimensional. For

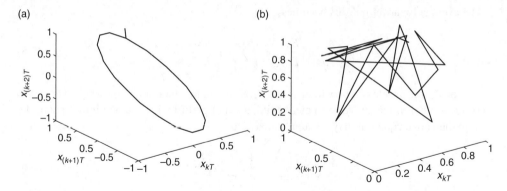

Figure 5.2   The attractors for (a) a sinusoid and (b) the above chaotic time sequence, both starting from the same initial point

a three-dimensional attractor we can choose a time delay T (a multiple of τ) and construct the following sequence of vectors:

$$[x(0) \qquad x(T) \qquad x(2T)]$$
$$[x(\tau) \quad x(\tau + T) \quad x(\tau + 2T)]$$
$$[x(2\tau) \quad x(2\tau + T) \quad x(2\tau + 2T)]$$
$$\cdot \qquad \cdot \qquad \cdot$$
$$\cdot \qquad \cdot \qquad \cdot$$
$$\cdot \qquad \cdot \qquad \cdot$$
$$[x(k\tau) \quad x(k\tau + T) \quad x(k\tau + 2T)]$$

By plotting these points in a three-dimensional coordinate space and linking the points together successively we can observe the attractor. Figure 5.2 shows the attractors for a sinusoidal and the above chaotic time series.

Although the attractors can be defined for a higher dimensional space, visualization of the attractors is not possible when the number of dimensions goes beyond three.

5.5   Estimation of Lyapunov Exponents from Time Series

Estimation of the Lyapunov exponents from time series was first proposed by Wolf *et al.* [5]. In their method, initially a finite embedding sequence is constructed from the finite time series of $2N + 1$ components as:

$$x(0), x(\tau), x(2\tau), \ldots$$

This is the basic data (often called reference trajectory or reference orbit) upon which the model builds. This can be shown in Figure 5.3.

However, we do not know the start point since there is no explicit governing equation that would generate the trajectory. From this sequence we may choose a point $x(k_0\tau)$, which

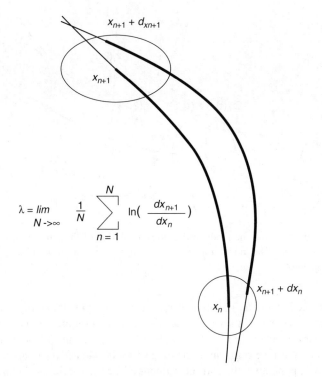

Figure 5.3   General definition of Lyapunov exponent from time series

approximates the desired initial point $z(0)$. Considering Figure 5.4, these approximations should satisfy

$$|x(k_0\tau) - x(0)| < \delta \qquad (5.10)$$

where δ is an a priori chosen tolerance. We may rename this point as

$$z_0(0) = x(k_0\tau) \qquad (5.11)$$

The successors of this point are known as:

$$z_0(r\tau) = x((k_0 + r)\tau), \quad r = 1, 2, 3, \ldots \qquad (5.12)$$

Now there are two trajectories to compare. The logarithmic error amplification factor for the first time interval becomes

$$l_0 = \frac{1}{\tau} \log \frac{|z_0(\tau) - z_0(0)|}{|x(\tau) - x(0)|} \qquad (5.13)$$

This procedure is repeated for the next point $x(\tau)$ of the reference trajectory. For that point we need to find another point $z_1(\tau)$ from the trajectory, which represents an error with a direction

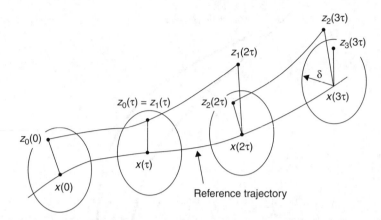

Figure 5.4 The reference and the model trajectories, evolution of the error, and start and end of the model trajectory segments. The model trajectory ends when its deviation from the reference trajectory is more than a threshold

close to that obtained from $z_0(\tau)$ relative to $x(\tau)$. If the previous trajectory is still close to the reference trajectory we may simply continue with that, thus setting $z_1(\tau) = z_0(\tau)$. This yields an error amplification factor l_1. Other factors, $l_1, l_2, \ldots, l_{m-1}$, can also be found by following the same procedure until the segment of the time series is exhausted. An approximation to the largest Lyapunov exponent for the current segment of the time series is obtained by averaging the logarithmic amplification factors over the whole reference trajectory.

$$\lambda = \frac{1}{m} \sum_{j=0}^{m-1} l_j \tag{5.14}$$

Instead of the above average, the maximum value of the error amplification factor may also be considered as the largest Lyapunov exponent. It is necessary to investigate the effect of noise here. The data usually stem from a physical measurement and, therefore, contain noise. Hence, the perturbed points, $z_k(k\tau)$ should not be taken very close to each other, since the noise would dominate the stretching effect on the chaotic attractor. On the other hand, we should not allow the error to become too large in order to avoid nonlinear effects. Thus, in practice, we prescribe some minimal error, δ_1, and a maximal error, δ_2, and require

$$\delta_1 < |x(k\tau) - z_k(k\tau)| < \delta_2 \tag{5.15}$$

5.5.1 Optimum Time Delay

In the above calculation it is important to find the *optimum time delay* τ. Very small time delays may result in near-linear reconstructions with high correlations between consecutive phase space points and very large delays might ignore any deterministic structure of the sequence. In an early research work [6] the autocorrelation function was used to estimate the time delay. In this method τ is equivalent to the duration after which the autocorrelation reaches a minimum or drops to a small fraction of its initial value. In another attempt [7, 8] it has been verified that the values of τ at which the mutual information has a local minimum are equivalent to the values of τ at which the logarithm of the correlation sum has a local minimum.

5.5.2 Optimum Embedding Dimension

To further optimise the measurement of Lyapunov exponents we need to specify the optimum value for m, named the *embedding dimension*. Before doing that some definitions have to be given as follows.

Fractal dimension is a statistic related to the dynamical measurement. The strange attractors are fractals and their fractal dimension D_f is simply related to the minimum number of dynamical variables needed to model the dynamics of the attractor. Conceptually, a simple way to measure D_f is to measure the *Kolomogorov capacity*. In this measurement a set is covered with small cells, depending on the dimensionality (i.e. squares for sets embedded in two dimensions, cubes for sets embedded in three dimensions, and so on), of size ε. If $M(\varepsilon)$ denotes the number of such cells within a set, the fractal dimension is defined as

$$D_f = \lim_{\varepsilon \to 0} \frac{\log (M(\varepsilon))}{\log \left(\frac{1}{\varepsilon}\right)} \tag{5.16}$$

for a set of single points $D_f = 0$, for a straight line $D_f = 1$, and for a flat plane $D_f = 2$. The fractal dimension, however, may not be an integer.

Correlation dimension is defined as

$$D_r = \lim_{r \to 0} \frac{\log C(r)}{\log r} \tag{5.17}$$

where

$$C(r) = \sum_{i=1}^{M(r)} p_i^2 \tag{5.18}$$

is the correlation sum.

Optimal embedding dimension, m, as required for accurate estimation of the Lyapunov exponents, has to satisfy $m \geq 2D_f + 1$. D_f is, however, not often known a priori. The Grassberger–Procaccia algorithm can nonetheless be employed to measure the correlation dimension, C_r. The minimum embedding dimension of the attractor is $m + 1$, where m is the embedding dimension above which the measured value of the correlation dimension C_r remains constant.

As another very important conclusion:

$$D_f = D_L = 1 + \frac{\lambda_1}{|\lambda_2|} \tag{5.19}$$

that is, the fractal dimension D_f is equivalent to the Lyapunov dimension [1].

Chaos has been used as a measure in analysis of many types of signals and time series. Its application to epileptic seizure prediction will be shown in Chapter 15.

5.6   Approximate Entropy

Another popular metric for chaos is *approximate entropy* (AE). AE can be estimated from the discrete time sequences, especially for real-time applications [9, 10]. This measure can quantify the complexity or irregularity of the system. AE is less sensitive to noise and can be used for short-length data. In addition, it is resistant to short strong transient interferences (outliers) such as spikes [10].

Given the embedding dimension m, the m-vector $\mathbf{x}(i)$ is defined as:

$$\mathbf{x}(i) = [x(i), x(i+1), \ldots, x(i+m-1)], \quad i = 1, \ldots, N-m+1 \tag{5.20}$$

where N is the number of data points. The distance between any two of the above vectors, $\mathbf{x}(i)$ and $\mathbf{x}(j)$, is defined as:

$$d[\mathbf{x}(i), \mathbf{x}(j)] = \max_k |x(i+k) - x(j+k)| \tag{5.21}$$

where |.| denotes the absolute value. Considering a threshold level of β, we find the number of times, $M^m(i)$, that the above distance satisfies $d[\mathbf{x}(i), \mathbf{x}(j)] \leq \beta$. This is performed for all i. For the embedding dimension m we form

$$\xi_\beta^m(i) = \frac{M^m(i)}{N-m+1}, \quad \text{for } i = 1, \ldots, N-m+1 \tag{5.22}$$

Then, the average natural logarithm of $\xi_\beta^m(i)$ is found as:

$$\psi_\beta^m = \frac{1}{N-m+1} \sum_{i=1}^{N-m+1} \ln \xi_\beta^m(i) \tag{5.23}$$

By repeating the same approach for an embedding dimension of $m+1$, the AE is approximated as:

$$AE(m, \beta) = \lim_{N \to \infty} \left[\psi_\beta^m - \psi_\beta^{m+1} \right] \tag{5.24}$$

In practice, however, N is limited and, therefore, the *AE* is calculated for N data samples. In this case the AE depends on m, β, and N, that is,

$$AE(m, \beta, N) = \psi_\beta^m - \psi_\beta^{m+1} \tag{5.25}$$

The embedding dimension can be found as previously mentioned. However, the threshold value has to be set correctly. In some applications the threshold value is taken as a value between 0.1 and 0.25 times the data standard deviation [9].

5.7   Using Prediction Order

Correct estimation of the prediction order is useful for recovery of the desired signal component from its noisy version. It is apparent that for signals with highly correlated time samples the prediction order of an AR or ARMA model is low, and for noise type signals where the correlation amongst the samples is low the order is high. This means for the latter case a

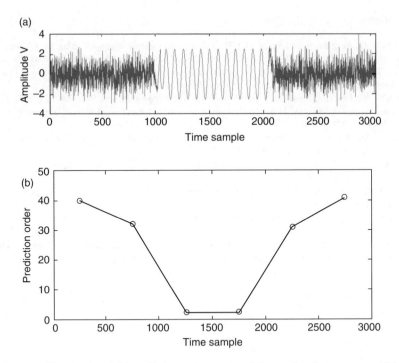

Figure 5.5 (a) The signal and (b) prediction order measured for overlapping segments of the signal

large number of previous samples are required for prediction of the current sample. A different criterion such as Akaike AIC may be employed to find the prediction order from the time series. Although this approach is mainly to differentiate between noise and correlated signals it clearly shows a high value for chaotic signals. Figure 5.5 shows the prediction order automatically computed for overlapping segments of three sections of a time series in which the middle section is sinusoidal and the first and third sections are random noise signals.

5.8   Conclusions

The brain is a complex system with nonlinear behaviour. It is generally very difficult to accurately characterize and express the dynamics of a complex system and the behaviour of a signal trend generated by a nonlinear system unless the nonlinearity is well defined and mathematically expressed. In order to more accurately define chaos in a system other parameters such as optimum time delay and attractor dimension often have to be estimated. These parameters inherently change in time and, therefore, a more complex evaluation has to be in place. Without correctly estimating these parameters the above measures are only rough estimators of chaos. On the other hand most of the measures introduced in this chapter, particularly different entropy estimators, cannot distinguish between chaos and noise in a system.

Another interesting conclusion of this chapter is that in a multichannel case, an ensemble measure of entropy corresponds to dependency between the channels and, therefore, can represent the mutual information.

References

[1] Peitgen, H.-O., Lurgens, H. and Saupe, D. (1992) *Chaos and Fractals*, Springer-Verlag, New York.

[2] Grassberger, P. and Procaccia, I. (1983) Characterization of strange attractors. *Phys. Rev. Lett.*, **50**, 346–349.

[3] Kaplan, L. and Yotke, J.A. (1979) Chaotic behaviour of multidimensional difference equations, in *Functional Differential Equations and Approximation of Fixed Points* (ed. H.-O. Walther), Springer-Verlag.

[4] Russell, D.A., Hanson, J.D. and Ott, E. (1980) Dimension of strange attractors. *Phys. Rev. Lett.*, (**45**), 1175–1179.

[5] Wolf, A., Swift, J.B., Swinny, H.L. and Vastano, J.A. (1985) Determining Lyapunov exponents from a time series. *Physica*, **16D**, 285–317.

[6] Albano, A.M., Muench, J., Schwartz, C. *et al.* (1988) Singular value decomposition and the Grassberger-Procaccia algorithm. *Phys. Rev. A*, **38**, 3017.

[7] King, G.P., Jones, R. and Broomhead, D.S. (1987) Phase portraits from a time series: A singular system approach. *Nucl. Phys. B*, **2**, 379.

[8] Fraser, A.M. and Swinney, H. (1986) Independent coordinates for strange attractors from mutual information. *Phys. Rev. A*, **33**, 1134–1139.

[9] Pincus, S.M. (1991) Approximate entropy as a measure of system complexity. *Proc. Natl. Acad. Sci., USA*, **88**, 2297–2301.

[10] Fusheng, Y., Bo, H. and Qingyu, T. (2001) Approximate entropy and its application in biosignal analysis, in *Nonlinear Biomedical Signal Processing*, vol. **II** (ed. M. Akay), IEEE Press, pp. 72–91.

6

Classification and Clustering of Brain Signals

6.1   Introduction

Separating or dividing the data, objects, samples, and so on into a number of classes is called clustering. Clustering methods are unsupervised and only the number of clusters may be identified and fed into the clustering algorithm by the user. Classification of data is similar to clustering except the classifier is trained using a set of labelled data before hand. Therefore, classification of the test data is supervised since the main criterion for classification is somehow similarity of the test data to a category of labelled data. In practice, the objective of classification is to draw a boundary between two or more classes and to label them based on their measured features. In a multidimensional feature space this boundary takes the form of a separating hyperplane. The art of the work here is to find the best hyperplane, which has maximum distance from all the classes while the members of each class are as close to each other as possible.

There is always an ambiguity in clustering or classification with regard to what features to use and how to either extract or enhance those features. PCA and ICA have been two very common approaches. This is an open question but in identification of ERPs, common spatial patterns (CSP) have been more common and indeed successful. Therefore, CSP is briefly reviewed in this chapter.

In the context of biomedical signal processing, especially with application to EEG signals, the classification of the data in feature spaces is often required. For example, the strength, locations, and latencies of P300 subcomponents may be classified not only to detect if the subject has Alzheimer's disease but also to determine the stage of the disease. As another example, to detect whether there is a left or right finger movement in the BCI systems one needs to classify the time, frequency, and spatial features.

There have been several clustering and classification techniques developed within the last forty years. Amongst them artificial neural networks (ANNs), linear discriminant analysis (LDA), hidden Markov modelling (HMM), k-mean clustering, fuzzy logic, and support vector machines (SVMs) have been very popular. These techniques have been developed and well explained in the literature [1]. The explanation of all these methods is beyond the objective

Adaptive Processing of Brain Signals, First Edition. Saeid Sanei.
© 2013 John Wiley & Sons, Ltd. Published 2013 by John Wiley & Sons, Ltd.

of this chapter. However, here we provide a summary of a SVM since it has been applied to EEG signals for the removal of the eye-blinking artefact [2], detection of epileptic seizures [3], detection of evoked potentials (EPs), classification of left and right finger movement in BCI [4], and many other issues related to EEGs [5].

Unlike many mathematical problems in which some form of explicit formula based on a number of inputs results in an output; in classification of data there is no model or formula of this kind. In such cases the system should be trained to be able to recognise the inputs. Many classification algorithms do not perform efficiently when

1. The number of features is high
2. There is a limited time for performing the classification
3. There is a non-uniform weighting amongst the features
4. There is a nonlinear map between the inputs and the outputs
5. The distribution of the data is not known
6. The convergence is not convex (monotonic), so it may fall into a local minimum.

There are two types of machine learning algorithms for classification of data; supervised learning and unsupervised learning. In the former case the target is known and the classifier is trained to minimise a difference between the actual output and the target values. Good examples of such classifiers are support vector machines (SVM) and the multilayered perceptrons (MLPs). In unsupervised learning, however, the classifier clusters the data into the groups having farthest distances from each other. A popular example for these classifiers is the k-means algorithm.

On the other hand, to estimate the features many algorithms can be used. These algorithms are often capable of changing the dimensionality of the signals to enhance their separability. Independent component analysis (ICA) as discussed in Chapter 10, principal component analysis (PCA), and also tensor factorization can be used to generate/estimate the necessary features. These features can then be clustered or classified.

CSP on the other hand, exploits spatial filters in order to discriminate between two classes. This approach and its extensions have become very popular in the EEG feature detection, with particular application to BCI. This method has been very successful mainly because it exploits the variations in the electrode space. In the following sections we briefly explain the most popular approaches that is, LDA, SVM and CSP for classification and the k-mean algorithm for clustering the data.

6.2   Linear Discriminant Analysis

Linear discriminant analysis (LDA) is a method used to find a linear combination of features which characterizes or separates two or more classes of objects or events. The resulting combination may be used as a linear classifier. In LDA it is assumed that the classes have normal distributions. Like PCA, LDA is used for both dimensionality reduction and data classification.

In a two-class dataset, given the a priori probabilities for class 1 and class 2 are respectively p_1 and p_2, and class means and overall mean as μ_1, μ_2, and μ, and the class variances as cov_1 and cov_2.

$$\mu = p_1 \times \mu_1 + p_2 \times \mu_2 \qquad (6.1)$$

Then, within-class and between-class scatters are used to formulate the necessary criteria for class separability. Within-class scatter is the expected covariance of each of the classes. The scatter measures for multiclass case are computed as:

$$S_w = \sum_{j=1}^{C} p_j \times cov_j \qquad (6.2)$$

where C refers to the number of classes and

$$cov_j = (\mathbf{x}_j - \mu_j)(\mathbf{x}_j - \mu_j)^T \qquad (6.3)$$

Slightly differently, the between-class scatter is estimated as:

$$S_b = \frac{1}{C} \sum_{j=1}^{C} (\mu_j - \mu)(\mu_j - \mu)^T \qquad (6.4)$$

Then, the objective is to find a discriminant plane such as $\mathbf{w}$ to maximize the ratio of between-class to within-class scatters (variances);

$$J_{LDA} = \frac{\mathbf{w} S_b \mathbf{w}^T}{\mathbf{w} S_w \mathbf{w}^T} \qquad (6.5)$$

In practice, the class means and covariances are not known. They can, however, be estimated from the training set. Either the maximum likelihood estimate or the maximum a posteriori estimate may be used in place of the exact value in the above equations.

6.3 Support Vector Machines

Amongst all supervised classifiers, SVM is the one which performs well in the above situations [6–11]. The concept of SVM was initiated in 1979 by Vapnik [11]. To understand the concept of SVM consider a binary classification for the simple case of a two-dimensional feature space of linearly separable training samples (Figure 6.1) $S = \{(\mathbf{x}_1, y_1), (\mathbf{x}_2, y_2), \ldots, (\mathbf{x}_m, y_m)\}$ where $\mathbf{X} \in R^d$ is the input vector and $y \in \{-1, 1\}$ is the class label. A discriminating function could be defined as:

$$f(\mathbf{x}) = sgn(\langle \mathbf{w}, \mathbf{x} \rangle + b) = \begin{cases} +1 & \text{if } \mathbf{x} \text{ belongs to the first class } \bullet \\ -1 & \text{if } \mathbf{x} \text{ belongs to the second class } \circ \end{cases} \qquad (6.6)$$

In this formulation $\mathbf{w}$ determines the orientation of a discriminant plane (or hyperplane). Clearly, there are an infinite number of possible planes that could correctly classify the training data. An optimal classifier finds the hyperplane for which the best generalising hyperplane is equidistant or farthest from each set of points. Optimal separation is achieved when there is no separation error and the distance between the closest vector and the hyperplane is maximal.

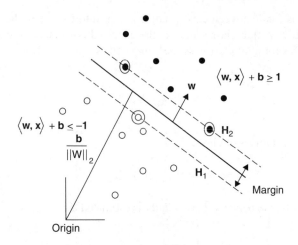

Figure 6.1   The SVM separating hyperplane and support vectors for the separable case

One way to find the separating hyperplane in a separable case is by constructing the so-called *convex hulls* of each data set. The encompassed regions are the convex hulls for the data sets. By examining the hulls one can then determine the closest two points lying on the hulls of each class (note that these do not necessarily coincide with actual data points). By constructing a plane that is perpendicular and equivalent to these two points an optimal hyperplane should result and the classifier should be robust in some sense.

In the design of an optimal separating hyperplane often a few points, referred to as the *support vectors* (SVs) are utilised (e.g. the three circled data points in Figure 6.1).

In places where the data are multidimensional and the number of points is high a mathematical solution rather than graphical solution will be necessary. To formulate an SVM, start with the simplest case: linear machines trained on separable data (as we shall see, the analysis for the general case; nonlinear machines trained on non-separable data results in a very similar quadratic programming problem). Again label the training data $\{\mathbf{x}_i, y_i\}$, $i = 1, \ldots, m$, $y_i \in \{-1, 1\}$, $\mathbf{x}_i \in R^d$

Suppose we have some hyperplane which separates the positive from the negative examples. The points $\mathbf{x}$ which lie on the hyperplane satisfy $\langle \mathbf{w}, \mathbf{x} \rangle + b = 0$, where $\mathbf{w}$ is normal to the hyperplane, $|b| / \|\mathbf{w}\|_2$ is the perpendicular distance from the hyperplane to the origin, and $\|\mathbf{w}\|_2$ is the Euclidean norm of $\mathbf{w}$. Define the "margin" of a separating hyperplane as in Figure 6.1 and, for the linearly separable case, the algorithm simply looks for the separating hyperplane with largest margin. The approach here is to reduce the problem to a convex optimisation problem by minimising a quadratic function under linear inequality constraints. To find the plane farthest from both classes of data, the margin between the supporting canonical hyperplanes for each class is maximised. The support planes are pushed apart until they meet the closest data points, which are then deemed to be the support vectors (circled in Figure 6.1). Therefore, the SVM problem to find $\mathbf{w}$ is stated as:

$$\langle \mathbf{x}_i, \mathbf{w} \rangle + b \geq +1 \quad \text{for } y_i = +1$$
$$\langle \mathbf{x}_i, \mathbf{w} \rangle + b \leq -1 \quad \text{for } y_i = -1$$

(6.7)

which can be combined into one set of inequalities as $y_i (\langle \mathbf{x}_i . \mathbf{w} \rangle + b) - 1 \geq 0 \quad \forall i$, the margin between these supporting planes (H$_1$ and H$_2$) can be shown to be $\gamma = 2/\|\mathbf{w}\|_2$. To maximise this margin we therefore need to

$$
\text{minimise } \langle \mathbf{w}, \mathbf{w} \rangle
$$
$$
\text{subject to } y_i (\langle \mathbf{x}_i . \mathbf{w} \rangle + b) - 1 \geq 0 \quad i = 1, \ldots, m.
$$
(6.8)

This constrained optimisation problem can be changed into an unconstrained problem by using Lagrange multipliers. This leads to minimisation of an unconstrained empirical risk function (Lagrangian) which consequently results in a set of conditions called Kuhn–Tucker (TK) conditions. The new optimisation problem leads to the so-called *primal form* as:

$$
L(\mathbf{w}, b, \alpha) = \frac{1}{2} \langle \mathbf{w}, \mathbf{w} \rangle - \sum_{i=1}^{m} \alpha_i \left[y_i (\langle \mathbf{x}_i, \mathbf{w} \rangle + b) - 1 \right]
$$
(6.9)

where $\alpha_i, i = 1, \ldots, m$ are the Lagrangian multipliers. Thus, the Lagrangian primal has to be minimised with respect to $\mathbf{w}, b$ and maximised with respect to $\alpha_i \geq 0$. Constructing the classical Lagrangian dual form facilitates this solution. This is achieved by setting the derivatives of the primal to zero and re-substituting them back into the primal. Hence,

$$
\frac{\partial L(\mathbf{w}, b, \alpha)}{\partial \mathbf{w}} = \mathbf{w} - \sum_{i=1}^{m} y_i \alpha_i \mathbf{x}_i = 0
$$
(6.10)

thus

$$
\mathbf{w} = \sum_{i=1}^{m} y_i \alpha_i \mathbf{x}_i
$$
(6.11)

and

$$
\frac{\partial L(\mathbf{w}, b, \alpha)}{\partial b} = \sum_{i=1}^{m} y_i \alpha_i = 0
$$
(6.12)

By replacing these into the primal form we get the dual form as

$$
L(\mathbf{w}, b, \alpha) = \frac{1}{2} \sum_{j=1}^{m} \sum_{i=1}^{m} y_i y_j \alpha_i \alpha_j \langle \mathbf{x}_i, \mathbf{x}_j \rangle - \sum_{i=1}^{m} y_i y_j \alpha_i \alpha_j \langle \mathbf{x}_i, \mathbf{x}_j \rangle + \sum_{i=1}^{m} \alpha_i
$$
(6.13)

which is reduced to

$$
L(\mathbf{w}, b, \alpha) = \sum_{i=1}^{m} \alpha_i - \frac{1}{2} \sum_{i=1}^{m} y_i y_j \alpha_i \alpha_j \langle \mathbf{x}_i, \mathbf{x}_j \rangle
$$
(6.14)

considering that $\sum_{i=1}^{m} y_i \alpha_i = 0$ and $\alpha_i \geq 0$. These equations can be solved mathematically (with the aid of a computer) using quadratic programming (QP) algorithms. There are many algorithms available within numerous publicly viewable websites [12, 13].

In non-separable cases (i.e. the classes have overlaps in the feature space) the maximum margin classifier described above is no longer applicable. Obviously, we may be able to define a complicated nonlinear hyperplane to perfectly separate the datasets but, as we see later, this causes an over-fitting problem which reduces the robustness of the classifier. The ideal solution where no points are misclassified and no points lie within the margin is no longer feasible. This means that we need to relax the constraints to allow for the minimum amount of misclassification. In this case, the points that subsequently fall on the wrong side of the margin are considered to be errors. They are, however, apportioned a lower influence (according to a preset *slack* variable) on the location of the hyperplane and as such are considered to be support vectors. The classifier obtained in this way is called a *soft margin classifier*.

In order to optimise the soft margin classifier, we must allow the margin constraints to be violated according to a preset *slack* variable ξ_i in the constraints, which then become:

$$\langle \mathbf{x}_i.\mathbf{w} \rangle + b \geq +1 - \xi_i \quad \text{for } y_i = +1$$
$$\langle \mathbf{x}_i.\mathbf{w} \rangle + b \leq -1 + \xi_i \quad \text{for } y_i = -1 \tag{6.15}$$
$$\text{and } \xi_i \geq 0 \,\, \forall i$$

Thus, for an error to occur, the corresponding ξ_i must exceed unity, so $\sum_i \xi_i$ is an upper bound on the number of training errors. Hence, a natural way to assign an extra cost for errors is to change the objective function to

$$\text{minimise } \langle \mathbf{w}, \mathbf{w} \rangle + C \sum_{i=1}^{m} \xi_i \tag{6.16}$$
$$\text{subject to } y_i \left(\langle \mathbf{x}_i, \mathbf{w} \rangle + b \right) \geq 1 - \xi_i \quad \text{and } \xi_i \geq 0 \quad i = 1, \ldots, m.$$

The primal form will then be:

$$L(\mathbf{w}, b, \xi, \alpha, \mathbf{r}) = \frac{1}{2} \langle \mathbf{w}, \mathbf{w} \rangle + C \sum_{i=1}^{m} \xi_i - \sum_{i=1}^{m} \alpha_i \left[y_i \left(\langle \mathbf{w}, \mathbf{x}_i \rangle + b \right) - 1 + \xi_i \right] - \sum_{i=1}^{m} r_i \xi_i$$
$$\tag{6.17}$$

Hence, by differentiating with respect to $\mathbf{w}$, ξ, and b we can achieve

$$\mathbf{w} = \sum_{i=1}^{m} y_i \alpha_i \mathbf{x}_i \tag{6.18}$$

and

$$\alpha_i + r_i = C \tag{6.19}$$

By replacing these into the primal form we obtain the dual form as

$$L(\mathbf{w}, b, \xi_i, \alpha, \mathbf{r}) = \sum_{i=1}^{m} \alpha_i - \frac{1}{2} \sum_{i=1}^{m} y_i y_j \alpha_i \alpha_j \langle \mathbf{x}_i, \mathbf{x}_j \rangle \tag{6.20}$$

again by considering that $\sum_{i=1}^{m} y_i \alpha_i = 0$ and $\alpha_i \geq 0$. This is similar to the maximal marginal classifier. The only difference is the new constraints of $\alpha_i + r_i = C$, where $r_i \geq 0$, hence $0 \leq \alpha_i \leq C$. This implies that the value C, sets an upper limit on the Lagrangian optimisation variables α_i. This is sometimes referred to as the box constraint. The value of C offers a trade-off between accuracy of data fit and regularization. A small value of C (i.e. < 1) significantly limits the influence of error points (or outliers), whereas if C is chosen to be very large (or infinite) then the soft margin (as in Figure 6.2) approach becomes identical to the maximal margin classifier. Therefore, in the use of the soft margin classifier, the choice of the value of C will depend heavily on the data. Appropriate selection of C is of great importance and is an area of research. One way to set C is to gradually increase C from max (α_i) for $\forall i$, and find the value for which the error (outliers, cross validation, or number of misclassified points) is a minimum. Finally, C can be found empirically [14].

There will be no change in formulation of the SVM for the multidimensional cases. Only the dimension of the hyperplane changes depending on the number of feature types.

In many non-separable cases use of a nonlinear function may help to make the datasets separable. As can be seen in Figure 6.3, the datasets are separable if a nonlinear hyperplane is used. *Kernel mapping* offers an alternative solution by non-linearly projecting the data into a (usually) higher dimensional feature space to allow the separation of such cases.

The key success of Kernel mapping is that special types of mapping that obey Mercer's theorem, sometimes called reproducing Kernel Hilbert spaces (RKHSs) [11], offer an implicit mapping into feature space.

$$K(\mathbf{x}, \mathbf{z}) = \langle \phi(\mathbf{x}), \phi(\mathbf{z}) \rangle \tag{6.21}$$

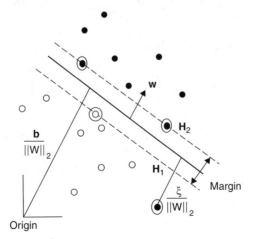

Figure 6.2   Soft margin and the concept of the slack parameter

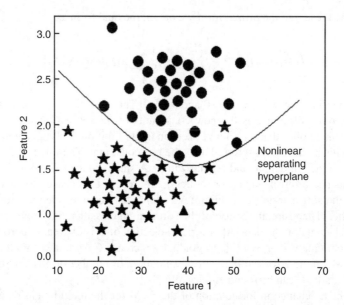

Figure 6.3 Nonlinear discriminant hyperplane

This means that the explicit mapping need not be known or calculated, rather the inner-product itself is sufficient to provide the mapping. This simplifies the computational burden dramatically and, in combination with the inherent generality of SVMs, largely mitigates the dimensionality problem. Further, this means that the input feature inner-product can simply be substituted with the appropriate Kernel function to obtain the mapping whilst having no effect on the Lagrangian optimisation theory. Hence,

$$L(\mathbf{w}, b, \xi_i, \alpha, \mathbf{r}) = \sum_{i=1}^{m} \alpha_i - \frac{1}{2} \sum_{i=1}^{m} y_i y_j \alpha_i \alpha_j K\left(\mathbf{x}_i, \mathbf{x}_j\right) \tag{6.22}$$

The relevant classifier function then becomes

$$f(\mathbf{x}) = \text{sgn}\left(\sum_{i=1}^{nSVs} y_i \alpha_i K(\mathbf{x}_i, \mathbf{x}_j) + b\right) \tag{6.23}$$

In this way all the benefits of the original linear SVM method are maintained. We can train a highly nonlinear classification function, such as a polynomial or a radial basis function or even a sigmoidal neural network, using a robust and efficient algorithm that does not suffer from local minima. The use of Kernel functions transforms a simple linear classifier into a powerful and general nonlinear classifier [14]. Radial basis function, defined as

$$K(u, v) = \exp\left(-\frac{\|u - v\|_2^2}{2\sigma^2}\right) \tag{6.24}$$

is the most popular nonlinear kernel used in SVM for the non-separable classes. It is possible to fit a hyperplane, using an appropriate Kernel, to the data in order to avoid overlapping the sets (or non-separable cases) and, therefore, produce a classifier with no error on the training set. This, however, is unlikely to generalize well. More specifically, the main problem with this is that the system may no longer be robust since a testing or new input can be easily misclassified.

Another issue related to the application of SVMs is the cross-validation problem. The distribution of the output of the classifier (without the hard limiter "sign" in equation (6.21)) for a number of inputs of the same class may be measured. The probability distributions of the results (which are centred at −1 for class "−1" and at +1 for class "+1") are plotted in the same figure. Less overlap between the distributions represents a better performance of the classifier. The choice of the kernel influences the performance of the classifier with respect to the cross-validation concept.

SVMs may be slightly modified to enable classification of multiclass data [15]. Moreover, some research has been undertaken to speed up the training step of the SVMs [16].

6.4 k-Means Algorithm

The k-means algorithm [17] is an effective and generally a simple clustering tool that has been widely used for many applications, such as in [18]. This algorithm divides a set of features (such as points in Figure 6.4) into k clusters.

The algorithm is initialised by setting 'k' to be the assumed number of clusters. Then, the centre for each cluster, k, is identified by selecting k representative data points. The next step in the k-means clustering algorithm after initialization is to assign the remaining data points to the closest cluster centre. Mathematically, this means that each data point needs to

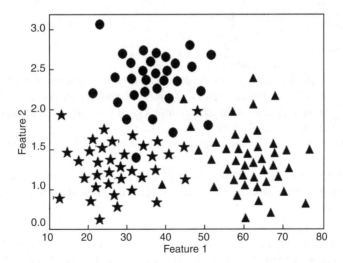

Figure 6.4   A two-dimensional feature space with three clusters, each with members of different shapes (circle, triangle, and asterisk)

be compared with every existing cluster centre and the minimum distance must be found. This is performed most often in the form of error checking (which will be discussed shortly). Before this though, new cluster centres are calculated. This is essentially the remaining step in k-means clustering: once clusters have been established (i.e. each data point is assigned to its closest cluster centre), the geometric centre of each cluster is recalculated.

The Euclidian distance of each data point within a cluster from its centre can be calculated. It can be repeated for all other clusters, whose resulting sums can themselves be summed together. The final sum is known as the *sum of within-cluster sum-of-squares*. Consider the within-cluster variation (*sum of squares* for cluster c) error as ε_c:

$$\varepsilon_c = \sum_{i=1}^{n_c} d_i^2 = \sum_{i=1}^{n_c} \left\| x_i^c - \bar{x}_c \right\|_2^2 \quad \forall c \tag{6.25}$$

Where d_i^2 is the squared Euclidean distance between data point i and its designated cluster centre $\bar{x}_c$, n_c is the total number of data points (features) in cluster c, and x_i^c is an individual data point in cluster c. The cluster centre (mean of data points in cluster c) can be defined as:

$$\bar{x}_c = \frac{1}{n_c} \sum_{i=1}^{n_c} x_i^c \tag{6.26}$$

and the total error is

$$E_k = \sum_{c=1}^{k} \varepsilon_c \tag{6.27}$$

The overall k-means algorithm may be summarized as:

1. Initialization
    i. Define the number of clusters (k).
    ii. Designate a cluster centre (a vector quantity that is of the same dimensionality as the data) for each cluster, typically chosen from the available data points.
2. Assign each remaining data point to the closest cluster centre. That data point is now a member of that cluster.
3. Calculate the new cluster centre (the geometric average of all the members of a certain cluster).
4. Calculate the sum of within-cluster sum-of-squares. If this value has not significantly changed over a certain number of iterations, stop the iterations. Otherwise, go back to Step 2.

Therefore, an optimum clustering depends on an accurate estimation of the number of clusters. A common problem in k-means partitioning is that if the initial partitions are not chosen carefully enough the computation will run the chance of converging to a *local* minimum rather than the *global* minimum solution. The initialization step is therefore very important.

To combat this problem one way is to run the algorithm several times with different initializations. If the results converge to the same partition then it is likely that a global minimum has been reached. This, however, has the drawback of being very time consuming and computationally expensive. Another solution is to dynamically change the number of partitions (i.e. number of clusters) as the iterations progress. The ISODATA (iterative self-organising data analysis technique algorithm) is an improvement on the original k-means algorithm that does exactly this. ISODATA introduces a number of additional parameters that allow it to progressively check within- and between-cluster similarities so that the clusters can dynamically split and merge.

Another approach for solving this problem is so-called gap statistics [19]. In this approach the number of clusters is iteratively estimated. The steps of this algorithm are:

1. For a varying number of clusters $k = 1, 2, \ldots, K$, compute the error measurement E_k using equation (6.25).
2. Generate a number B of reference datasets. Cluster each one with the k-mean algorithm and compute the dispersion measures, $\breve{E}_{kb}$, $b = 1, 2, \ldots, B$. The gap statistics are then estimated using

$$G_k = \frac{1}{B} \sum_{b=1}^{B} \log(\breve{E}_{kb}) - \log(E_k) \qquad (6.28)$$

where the dispersion measure $\breve{E}_{kb}$ is the E_k of the reference dataset B.
3. To account for the sample error in approximating an ensemble average with B reference distributions, the standard deviation is computed as

$$S_k = \left[\frac{1}{B} \sum_{b=1}^{B} (\log(\breve{E}_{kb}) - \bar{E}_b)^2 \right]^{1/2} \qquad (6.29)$$

where

$$\bar{E}_b = \frac{1}{B} \sum_{b=1}^{B} \log(\breve{E}_{kb}) \qquad (6.30)$$

4. By defining $\breve{S}_k = S_k \left(1 + \frac{1}{B}\right)^{1/2}$, the number of clusters is estimated as the smallest k such that $G_k \geq G_{k+1} - \breve{S}_{k+1}$,
5. With the number of clusters identified, use the k-mean algorithm to partition the feature space into k sub-sets (clusters).

The above clustering method has several advantages since it can estimate the number of clusters within the feature space. It is also a multiclass clustering system and, unlike SVM, can provide the boundary between the clusters.

6.5 Common Spatial Patterns

CSP is a popular feature extraction method for EEG classification. CSP aims at estimating spatial filters which discriminate between two classes based on their variances. The CSP ($\mathbf{w}$) minimises the Rayleigh quotient of the spatial covariance matrices to achieve the variance imbalance between the two classes of data $\mathbf{X}_1$ and $\mathbf{X}_2$. Before applying the CSP the signals are bandpass filtered and centred. The CSP approach aims to find a spatial filter $\mathbf{w} \in \mathfrak{R}^c$ such that the variance of the projected samples of one class is maximized while the other's is minimized. The CSP are estimated by the following maximization [20, 21]:

$$\mathbf{w}^{(CSP)} = \arg\max_{\mathbf{w}} \frac{Tr(\mathbf{w}^T\mathbf{C}_1\mathbf{w})}{Tr(\mathbf{w}^T\mathbf{C}_2\mathbf{w})} \tag{6.31}$$

where $\mathbf{C}_1$ and $\mathbf{C}_2$ are the covariance matrices of the two clusters $\mathbf{X}_1$ and $\mathbf{X}_2$. This optimization problem can be solved (though this is not the only way) by first observing that the function $J(\mathbf{w})$ remains unchanged if the filter $\mathbf{w}$ is rescaled, that is, $J(k\mathbf{w}) = J(\mathbf{w})$, with k as any real constant. Hence, extremizing $J(\mathbf{w})$ is equivalent to extremizing $\mathbf{w}^T\mathbf{C}_1\mathbf{w}$ subject to the constraint $\mathbf{w}^T\mathbf{C}_2\mathbf{w} = 1$ as it is always possible to find a rescaling of $\mathbf{w}$ such that $\mathbf{w}^T\mathbf{C}_2\mathbf{w} = 1$. Using the Lagrange multiplier method, this constrained problem changes to an unconstrained problem as [22]:

$$L(\lambda, \mathbf{w}) = \mathbf{w}^T\mathbf{C}_1\mathbf{w} - \lambda(\mathbf{w}^T\mathbf{C}_2\mathbf{w} - 1) \tag{6.32}$$

For a simple way to derive the optimum $\mathbf{w}$ the derivatives of L can be taken and zeroed as follows:

$$\frac{\partial L}{\partial \mathbf{w}} = 0 \Rightarrow \mathbf{C}_2^{-1}\mathbf{C}_1\mathbf{w} = \lambda\mathbf{w} \tag{6.33}$$

This is a standard eigenvalue problem; the spatial filters extremizing equation (6.32) are then the eigenvectors of $\mathbf{M} = \mathbf{C}_2^{-1}\mathbf{C}_1$ which correspond to its largest and lowest eigenvalues. When using CSP, the extracted features are the logarithm of the EEG signal variance after projection onto the filters $\mathbf{w}$.

The eigenvalue λ measures the ratio of variances of the two classes. CSP is suitable for classification of both spectral [21] and spatial data. It is based on the fact that the power of the latent signal is larger for the first cluster than for the second cluster.

Applying this approach to separation of event-related potentials (ERPs) enhances one of the classes against the rest. This allows a better discrimination between the two classes and can be separated more easily.

In early 2000 in a 2-class BCI set-up, Ramoser et al. [23] proposed application of CSP that learned to maximise the variance of bandpass filtered EEG signals from one class while minimizing their variance from the other class. Currently, CSP is widely used in BCI where the signals are changed by evoked potentials or movement.

The model uses the four most important CSP filters. The variance is then calculated from the CSP time series. Then, a log operation is applied and the weight vector obtained with the LDA is used to discriminate between left- and right-hand movement imaginations. Finally, the signals are classified and based on whether an output signal is produced to control the cursor

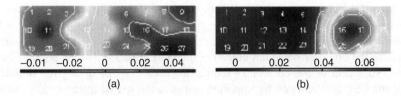

(a)   (b)

Figure 6.5 The CSP patterns related to (a) right-hand movement and (b) left-hand movement. The EEG channels are indicated by numbers which correspond to three rows of channels (electrodes) within the central and centro-parietal regions (see Plate 3 for the coloured version)

on the screen. Figure 6.5 shows the patterns related to right-hand movement in Figure 6.5a and left-hand movement in Figure 6.5b. These patterns are the strongest patterns.

As stated in [24] most of the existing CSP-based methods exploit covariance matrices on a subject-by-subject basis so that inter-subject information is neglected. In that paper, the CSP has been modified for subject-to-subject transfer, where a linear combination of covariance matrices of subjects under consideration has been exploited. We develop two methods to determine a composite covariance matrix that is a weighted sum of covariance matrices involving subjects, leading to composite CSP.

Accurate estimation of the CSPs under weak assumptions (mainly noisy cases) requires derivation of an asymptotically optimal solution. This then needs calculation of the loss in signal to noise-plus-interference ratio due to the finite sample effect in a closed form. This is often important in detection/classification of ERPs since the EEG or MEG signals do not only contain the spatio-temporal patterns bounded to the events but also ongoing brain activity as well as other artifacts, such as eye-blink and muscular artifacts. Therefore, to improve the estimation results the CSPs are applied to the recorded multichannel signals.

In equation (6.32) the trace values are equivalent to L2-norm. Due to the sensitivity of L2-norm to outliers [25], L2-norm has been replaced by L1-norm. This changes equation (6.32) to

$$\mathbf{w}_i^{(CSP)} = \arg\max_{\mathbf{w}} \frac{\left\|\mathbf{w}^T\mathbf{X}_1\right\|_1}{\left\|\mathbf{w}^T\mathbf{X}_2\right\|_1} = \arg\max_{\mathbf{w}} \frac{\sum_{k=1}^{m}\left|\mathbf{w}^T\mathbf{x}_{1k}\right|}{\sum_{l=1}^{n}\left|\mathbf{w}^T\mathbf{x}_{2l}\right|} \tag{6.34}$$

where m and n are the total sampled points of the two classes. It has been shown that the effect of outliers is alleviated using this approach.

Most of the existing CSP-based methods exploit covariance matrices on a subject-by-subject basis so that inter-subject information is neglected. CSP and its variants have received much attention and have been one of the most efficient extraction methods for BCI. However, despite its straightforward mathematics CSP overfits the data and is highly sensitive to noise.

To enhance the CSP performance and address these shortcomings, recently it has been proposed to improve the CSP learning process with prior information in terms of regularization terms or constraints. In a review by Lotte *et al.* [22], eleven different regularization methods for CSP have been categorised and compared. The methods included regularization in the estimation of the EEG covariance matrix [26], composite CSP [24], regularized CSP with generic learning [27], regularized CSP with diagonal loading [21] and invariant CSP [28].

They applied all the algorithms to a number of trials from 17 subjects and finally suggested, as the best method, the CSP with Tikhonov regularization in which the optimization of $J(\mathbf{w})$ is penalized by minimizing $\|\mathbf{w}\|^2$, hence, minimising the influence of artifacts and outliers.

To alleviate the effect of noise the objective function in (6.33) may be regularized either during the estimation of the covariance matrix for each class or during the minimization process of the CSP cost function by imposing priors to the spatial filters $\mathbf{w}$ [22]. A straight-forward approach to regularize the estimation of the covariance matrix for class i that is, $\hat{\mathbf{C}}_i$ is expressed as:

$$\hat{\mathbf{C}}_i = (1 - \gamma)\mathbf{P} + \gamma\mathbf{I} \tag{6.35}$$

where $\mathbf{P}$ is estimated as

$$\mathbf{P} = (1 - \beta)\mathbf{C}_i + \beta\mathbf{G}_i \tag{6.36}$$

In these equations $\mathbf{C}_i$ is the initial spatial covariance matrix for class i, $\mathbf{G}_i$, the so-called generic covariance matrix, is computed by averaging the covariance matrices of a number of trials for class i, although it may be defined based on neurophysiological priors only, and γ and $\beta \in [0,1]$ are the regularizing parameters. The generic matrix represents a given prior on how the covariance matrix for the mental state considered should be.

In regularizing the CSP objective function on the other hand, a regularization term is added to the CSP objective function in order to penalize the resulting spatial filters that do not satisfy a given prior. This results in a slightly different objective function as [22]:

$$J(\mathbf{w})\frac{\mathbf{w}^T\mathbf{C}_1\mathbf{w}}{\mathbf{w}^T\mathbf{C}_2\mathbf{w} + \alpha\mathbf{Q}\mathbf{w}} \tag{6.37}$$

The penalty function $\mathbf{Q}(\mathbf{w})$, weighted by the penalty (or Lagrange) parameter $\alpha \geq 0$, indicates how much the spatial filter $\mathbf{w}$ satisfies a given prior. Similar to the first term in the denominator of (6.37), this term needs to be minimized in order to maximize the cost function $J(\mathbf{w})$. Different quadratic and non-quadratic penalty functions have been defined in the literature, such as in [22, 28, 29]. A quadratic constraint such as $\mathbf{Q}(\mathbf{w}) = \mathbf{w}^T\mathbf{K}\mathbf{w}$ results in solving a new set of eigenvalue problems as [22]:

$$(\mathbf{C}_2 + \alpha\mathbf{K})^{-1}\mathbf{C}_1\mathbf{w} = \lambda\mathbf{w} \tag{6.38}$$

This has to be performed for the second pattern, resulting in $(\mathbf{C}_1 + \alpha\mathbf{K})^{-1}\mathbf{C}_2\mathbf{w} = \lambda\mathbf{w}$. The composite CSP algorithm proposed in [24] performs subject-to-subject transfer by regularizing the covariance matrices using other subjects' data. In this approach α and γ are zero and only β is nonzero. The generic covariance matrices $\mathbf{G}_i$ are defined according to covariance matrices of other subjects.

The approach based on regularized CSP with generic learning [27] uses both β and γ regularization terms, which means it intends to shrink the covariance matrix towards both the identity matrix and a generic covariance matrix $\mathbf{G}_i$. Similarly to composite CSP, $\mathbf{G}_i$ here is computed from the covariance matrices of other subjects.

Another form of covariance matrix regularization used in the BCI literature is diagonal loading (DL) [21], which consists of shrinking the covariance matrix towards the identity matrix. Hence, in this approach only γ is used ($\alpha = \beta = 0$) and its value can be automatically identified using Ledoit and Wolf's method [27] or by cross-validation [22].

In the approach based on Tikhonov regularization [22] the penalty term is defined as $\mathbf{P}(\mathbf{w}) = \|\mathbf{w}\|^2 = \mathbf{w}^T\mathbf{w} = \mathbf{w}^T\mathbf{I}\mathbf{w}$ which is obtained by using $\mathbf{K} = \mathbf{I}$ in (6.38). Such regularization is expected to constrain the solution to filters with a small norm, hence mitigating the influence of artifacts and outliers.

Finally, as expected, in the weighted Tikhonov regularization method, it is considered that some EEG channels are more important than others. Therefore, the signals from these channels are weighted more than those of others. This consequently causes different penalizations for different channels. To achieve this regularization define $\mathbf{P}(\mathbf{w}) = \mathbf{w}^T\mathbf{D_w}\mathbf{w}$, where $\mathbf{D_w}$ is a diagonal matrix whose entries are defined based on the measures of CSPs from other subjects [22].

The results achieved in implementing 11 different regularization methods in [22] shows that, in places where the data is noisy, the regularization improves the results by approximately 3% on average. However, the best algorithm outperforms CSP by about 3 to 4% in mean classification accuracy and by almost 10% in median classification accuracy. The regularized methods are also more robust, that is, they show lower variance across both classes and subjects. Amongst various approaches the weighted Tikhonov regularized CSP proposed in [22] has been reported to have the best performance.

Regularisation of $\mathbf{w}$ can also be used to reduce the number of EEG channels without compromising the classification score. Farquhar *et al.* [28] converted CSP into a quadratically constrained quadratic optimization problem with l_1-norm penalty and Arvaneh *et al.* [30] used the l_1/l_2-norm constraint. Recently, in [31] a computationally-expensive quasi l_0-norm based principle has been applied to achieve a sparse solution for $\mathbf{w}$. In [24] the CSP has been modified for subject-to-subject transfer, where a linear combination of covariance matrices of subjects under consideration has been exploited. In this approach a composite covariance matrix has been used that is a weighted sum of covariance matrices involving subjects, leading to composite CSP.

Although two-class classifiers have been very common for classification of cortical activities related to body movements, multiclass classifiers have also been developed and used. As an example, neural networks (NNs) may better cope with both multiclass cases and nonlinearity of the data. The Hidden Markov model (HMM) is able to best estimate the states of a system (such as the states representing various stages of sleep). Fuzzy clustering, particularly by local approximation of memberships, (FLAME) [32] is another option. FLAME captures the nonlinear relationships between objects and can automatically determine the number of clusters for the input data set. This clustering algorithm is a combination of k-nearest neighbour (KNN) and fuzzy classification algorithms. In the KNN algorithm an object is classified by a majority vote of its neighbors, with the object being assigned to the class most common amongst its k nearest neighbors (k is a positive integer, typically small). If $k = 1$, then the object is simply assigned to the class of its nearest neighbour.

6.6   Conclusions

It is difficult to explain all classification/clustering methods used for various pattern recognition applications. However, in recent analysis of EEG and MEG CSP, SVM, and K-mean have been used more often. Therefore, the above sections are mainly devoted to these approaches. CSP is generally used for feature detection of two classes. The features can then be classified or clustered using LDA, SVM or other classifiers. Neural networks have their own attractions since they can cope with nonlinearity and multiclass cases. They are more flexible in selection of the dimensionality and also the number of hidden neurons. For detection of the movement-related EEG/MEG patterns, however, the CSP method and its variants have shown a very good performance in feature selection and classification for BCI purposes.

References

[1] Vapnik, V. (1998) *Statistical Learning Theory*, Wiley Press.

[2] Shoker, L., Sanei, S. and Chambers, J. (2005) Artifact removal from electroencephalograms using a hybrid BSS-SVM algorithm. *IEEE Signal Process. Lett.*, **12**(10), 721–724.

[3] Shoker, L., Sanei, S., Wang, W. and Chambers, J. (2004) Removal of eye blinking artifact from EEG incorporating a new constrained BSS algorithm. *IEE J. Med. Biol. Eng. Comput.*, **43**(2), 290–295.

[4] Gonzalez, B., Sanei, S. and Chambers, J. (2003) Support vector machines for seizure detection. Proceedings of the IEEE, ISSPIT2003, Germany, December 2003, pp. 126–129.

[5] Shoker, L., Sanei, S. and Sumich, A. (2005) Distinguishing between left and right finger movement from EEG using SVM. *Proceedings of the IEEE, EMBS, Shanghai, China*, 2005, pp. 5420–5423.

[6] Bennet, K.P. and Campbell, C. (2000) Support vector machines: Hype or Hallelujah? *SIGKDD Explorations*, **2**(2), 1–13 (http://www.rpi.edu/~bennek).

[7] Christianini, N. and Shawe-Taylor, J. (2000) *An Introduction to Support Vector Machines*, Cambridge University Press.

[8] DeCoste, D. and Scholkopf, B. (2001) *Training Invariant Support Vector Machines*, Machine Learning, Kluwer Press.

[9] Burges, C. (1998) A tutorial on support vector machines for pattern recognition. *Data Min. Knowl. Disc.*, **2**, 121–167.

[10] Gunn, S. (1998) Support vector machines for classification and regression. Technical Reports, Dept. of Electronics and Computer Science, Southampton University.

[11] Vapnik, V. (1995) *The Nature of Statistical Learning Theory*, Springer, New York.

[12] http://www.support-vector.net

[13] http://www.kernel-machines.org

[14] O. Chapelle, and V. Vapnik, Choosing multiple parameters for support vector machines. *Mach. Learn.*, **46**, 131–159, 2002.

[15] Weston, J. and Watkins, C. (1999) Support vector machines for multi-class pattern recognition. Proceedings of the 7th European Symposium on Artificial Neural Networks, 1999.

[16] Platt, J. (1998) Sequential minimal optimisation: A fast algorithm for training support vector machines. Technical Report, MSR-TR-98-14, Microsoft Research.

[17] Hartigan, J. and Wong, M. (1979) A k-mean clustering algorithm. *Appl. Stat.*, **28**, 100–108.

[18] Fellous, J.-M., Tiesinga, P.H.E., Thomas, P.J. and Sejnowski, T.J. (2004) Discovering spike patterns in neural responses. *J. Neurosci.*, **24**(12), 2989–3001.

[19] Hastie, T., Tibshirani, R. and Walter, G. (2000) Estimating the number of clusters in a dataset via the gap statistic. Tech. Rep. 208, Stanford University, Stanford, CA, USA.

[20] Koles, Z. (1991) The quantitative extraction and topographic mapping of the abnormal components in the clinical EEG. *Electroencephalogr. Clin. Neurophysiol.*, **79**(6), 440–447.

[21] Blankertz, B., Tomioka, R., Lemm, S. *et al.* (2008) Optimizing spatial filters for robust EEG single-trial analysis. *IEEE Signal Proc. Magazine*, **25**(1), 41–56.

[22] Lotte, F. and Guan, C. (2011) Regularizing common spatial patterns to improve BCI designs: unified theory and new algorithms. *IEEE Trans. Biomed. Eng.*, **58**(2), 355–362.

[23] Ramoser, H., Muller-Gerking, J. and Pfurtscheller, G. (2000) Optimal spatial filtering of single trial EEG during imagined hand movement. *IEEE Trans. Rehab. Eng.*, **8**(4), 441–446.

[24] Kang, H., Nam, Y. and Choi, S. (2009) Composite common spatial pattern for subject-to-subject transfer. *IEEE Signal Proc. Lett.*, **16**(8), 683–686.

[25] Wang, H., Tang, Q. and Zheng, W. (2012) L1-norm-based common spatial patterns. *IEEE Trans. Biomed. Eng.*, **59**(3), 653–662.

[26] Lu, H., Plataniotis, K. and Venetsanopoulos, A. (2009) Regularized common spatial patterns with generic learning for EEG signal classification. Proceedings of the IEEE Engineering in Medicine and Biology Conference, EMBC, 2009, Minnesota, USA, pp. 6599–6602.

[27] Ledoit, O. and Wolf, M. (2004) A well-conditioned estimator for large dimensional covariance matrices. *J. Multivariate Anal.*, **88**(2), 365–411.

[28] Farquhar, J., Hill, N., Lal, T. and Schölkopf, B. (2006) Regularised CSP for sensor selection in BCI. Proceedings of the 3rd International BCI Workshop, 2006.

[29] Yong, X., Ward, R. and Birch, G. (2008) Sparse spatial filter optimization for EEG channel reduction in brain-computer interface. Proceedings of the IEEE International Conference on Acoustics, Speech, and Signal Processing, ICASSP, Las Vegas, USA, 2008, pp. 417–420.

[30] Arvaneh, M., Guan, C.T., Kai, A.K. and Chai, Q. (2011) Optimizing the Channel Selection and Classification Accuracy in EEG-Based BCI. *IEEE Trans. Biomed. Eng.*, **58**(6), 1865–1873.

[31] Goksu, F., Ince, N.F. and Tewfik, A.H. (2011) Sparse common spatial patterns in brain computer interface applications. Proceedings of the IEEE International Conference on Acoustics, Speech, and Signal Processing, ICASSP, Prague, Czech Republic, 2011, pp. 533–536.

[32] Fu, L. and Medico, E. (2007) FLAME: a novel fuzzy clustering method for the analysis of DNA microarray data. *BMC Bioinformatics*, **8**(3). doi: 10.1186/1471-2105-8-3

7

Blind and Semi-Blind Source Separation

7.1 Introduction

Most physiological information, signals, and images are found as mixtures of more than one component. Often, if the number of recording channels is large enough and the constituent sources are uncorrelated, independent, or disjoint in some domain, an accurate solution to the blind source separation (BSS) problem can be found and sources accurately separated from each other. Separation, extraction or identification of source signals from their recorded mixtures have been under development during the past two decades. Although, some a priori, assumptions such as independence of the sources, are considered to solve this problem, the term *blind* in BSS refers to the case of having both the sources and the mixing system (or medium) unknown. Different approaches based on some fundamental mathematical concepts, such as adaptive filtering, principal component analysis (PCA), matrix factorization, independent component analysis (ICA), and recently tensor factorization, have been proposed for separation of the sources. ICA, however, has been more popular than the other methods.

Most of the BSS approaches require the data to remain stationary within the segment under process. This is generally needed to enable definition of some invariant statistics within each segment of the data. However, this may not always be true, particularly for many physiological signals and image sequences, and often the mixtures involve nonstationarity of the sources (mainly due to the existing transient events), nonlinearity of the system (due to the involvement of a time-varying medium or moving sources,) and also other effects such as time-delays in the source signals reaching the sensors and the way the sources are spread (e.g. in the case of distributed sources where a number of synchronous sources are spread in space).

Section 7.10 of this chapter is dedicated to separation of nonstationary sources from their mixtures. It is necessary to emphasise how such nonstationarity can be ignored, mitigated, or exploited during the separation process.

Unfortunately, the data cannot necessarily be be found in multichannels. Therefore, single-channel signal decomposition or subspace data analysis has its own merit. One of the most effective single-channel subspace analysis methods is singular spectrum analysis (SSA) which

Adaptive Processing of Brain Signals, First Edition. Saeid Sanei.
© 2013 John Wiley & Sons, Ltd. Published 2013 by John Wiley & Sons, Ltd.

is an established approach within mathematics and financial forecasting communities. Here, before moving to multichannels we have a section on SSA.

7.2 Singular Spectrum Analysis

The basic SSA method consists of two complementary stages: decomposition and reconstruction [1]; each stage includes two separate steps. In the first step the series is decomposed and in the second the original series is reconstructed and used for further analysis. The main concept in studying the properties of SSA is *separability*, which characterizes how well different components can be separated from each other. The absence of approximate separability is often observed in series with complex structure. For these series, and series with special structure, there are different ways of modifying SSA leading to different versions, such as SSA with single and double centring, Toeplitz SSA, and sequential SSA [1].

SSA is becoming an effective and powerful tool for time series analysis in meteorology, hydrology, geophysics, climatology, economics, biology, physics, medicine and other sciences where short and long, one-dimensional and multidimensional, stationary and nonstationary, almost deterministic and noisy time series are to be analysed. The SSA is carried out in two stages; decomposition and reconstruction. A brief description of the SSA stages is given in the following subsections.

7.2.1 Decomposition

This stage includes an embedding operation followed by singular value decomposition (SVD). Embedding operation maps a one-dimensional time series $\mathbf{f}$ into a $k \times l$ matrix with rows of length l as [1]:

$$\mathbf{X} = \{x_{ij}\} = [\mathbf{x}_1, \mathbf{x}_2, \ldots, \mathbf{x}_k]$$

$$= \begin{bmatrix} f_0 & f_1 & f_2 & \cdots & f_{k-1} \\ f_1 & f_2 & f_3 & \cdots & f_k \\ f_2 & f_3 & f_4 & \cdots & f_{k+1} \\ \vdots & \vdots & \vdots & \ddots & \vdots \\ f_{l-1} & f_l & f_{l+1} & \cdots & f_{r-1} \end{bmatrix} \quad (7.1)$$

with vectors $\mathbf{x}_i = [f_{i-1}, f_i, \ldots, f_{i+l-2}]^\mathrm{T} \in R^L$, where $k = r - l + 1$ is the window length ($1 \leq l \leq r$), and superscript T denotes the transpose of a vector. Vectors $\mathbf{x}_i$ are called *l-lagged vectors* (or, simply, *lagged vectors*). The window length l should be sufficiently large. Note that the trajectory matrix $\mathbf{X}$ is a Hankel matrix, which means that all the elements along the diagonal $i + j = const.$ are equal.

In the SVD stage the SVD of the trajectory matrix is computed and represented as a sum of rank-one bi-orthogonal elementary matrices. Consider the eigenvalues of the covariance matrix $\mathbf{C}_x = \mathbf{X}\mathbf{X}^\mathrm{T}$ are denoted by $\lambda_1, \ldots, \lambda_l$ in decreasing order of magnitude ($\lambda_1 \geq \ldots \geq \lambda_l \geq 0$) and the corresponding orthogonal eigenvectors by $\mathbf{u}_1, \ldots, \mathbf{u}_l$. Set $d = \max(i;$ such that

$\lambda_i > 0) = rank(\mathbf{X})$: If we denote $\mathbf{v}_i = \mathbf{X}^T \mathbf{u}_i / \sqrt{\lambda_i}$, then the SVD of the trajectory matrix can be written as:

$$\mathbf{X} = \sum_{i=1}^{d} \mathbf{X}_i = \sum_{i=1}^{d} \sqrt{\lambda_i} \mathbf{u}_i \mathbf{v}_i^T \tag{7.2}$$

The constituting components are separated through a constrained procedure of configuring $\mathbf{X}$ and the SVD operation. Since the desired signal might have a different energy level, in general there is no assumption about the relationship between the order of the eigenvalues and the physiologically dominant component (note that if $\lambda_1 \gg \lambda_2$, the information energy will be mostly concentrated in the most dominant matrix $\mathbf{X}_1 = \sqrt{\lambda_1} \mathbf{u}_1 \mathbf{v}_1$ associated with the first singular values).

In general, ICA can be used instead of SVD. This changes the subspace decomposition problem into an ICA problem which can be solved by employing a suitable ICA technique. ICA is discussed later in this chapter.

7.2.2   Reconstruction

During reconstruction the elementary matrices are first split into several groups (depending on the number of components in the time series – in this application EMG and ECG signals, the matrices within each group are added together). The size of the group or, in other words, the length of each subspace may be specified based on some a priori information. In the case of ECG removal from recorded EMGs this feature and the desired component are jointly (and automatically) identified based on a constraint on the statistical properties of the signal to be extracted. Let $I = \{i_1, \ldots, i_p\}$ be the indices corresponding to the p eigenvalues of the desired component. Then the matrix $\hat{\mathbf{X}}_I$ corresponding to the group I is defined as $\hat{\mathbf{X}}_I = \sum_{j=i_1}^{i_p} \mathbf{X}_j$. In splitting the set of indices $J = \{1, \ldots, d\}$ into disjoint subsets I_1 to I_m we always have

$$\mathbf{Z} = \sum_{j=I_1}^{I_m} \hat{\mathbf{X}}_j \tag{7.3}$$

The procedure of choosing the sets $I_1, \ldots I_m$ is called eigentriple grouping. For a given group I the contribution of the component $\mathbf{X}_I$ in expansion (7.3) is measured by the contribution of the corresponding eigenvalues: $\sum_{i \in I} \lambda_i / \sum_{i=1}^{d} \lambda_i$. In the next step the obtained matrix is transformed to the form of a Hankel matrix which can be subsequently converted to a time series. If z_{ij} stands for an element of a matrix $\mathbf{Z}$, then the k-th term of the resulting series is obtained by averaging z_{ij} over all i,j such that $i + j = k + 1$. This procedure is called diagonal averaging or Hankelization of matrix Z. The hankelized $m \times n$ matrix $\mathbf{X}_h$ is defined as

$$\mathbf{X}_h = \begin{bmatrix} \hat{x}_1 & \hat{x}_2 & \cdots & \hat{x}_i \\ \hat{x}_2 & \hat{x}_3 & \cdots & \hat{x}_{i+1} \\ \vdots & \vdots & \ddots & \vdots \\ \hat{x}_j & \hat{x}_{j+1} & \cdots & \hat{x}_{i+j-1} \end{bmatrix} \tag{7.4}$$

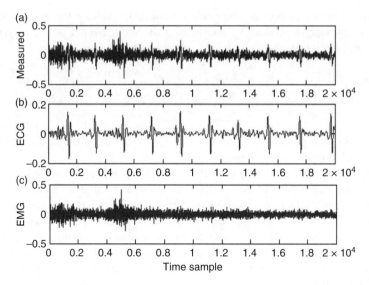

Figure 7.1 Separation of EMG and ECG using the SSA technique: (a) the measurement signal, (b) the separated ECG and (c) the EMG signal

where

$$\hat{x}_k = \frac{1}{num(D_k)} \sum_{i,j \in D_k} z_{i,j} \text{ for } D_k = \{(p,q) : 1 \leq p \leq i, \quad 1 \leq q \leq j, \quad p+q = k+1\}$$

(7.5)

where *num* abbreviates number.

A recursive SSA algorithm has been developed recently [2]. Such a method is advantageous for real-time processing of the data. The main problems in solving an SSA problem are selection of the appropriate embedding dimension (i.e. the window lengths) and establishing an effective criterion for selection of the desired subspace of eigentriples.

As an example, Figure 7.1 shows the separation of ECG signals from an EMG measurement by carefully selecting the above parameters. This figure clearly demonstrates the high potential of SSA in single channel source decomposition.

7.3 Independent Component Analysis

The concept of ICA lies in the fact that the signals may be decomposed into their constituent independent components. In places where the combined source signals can be assumed independent from each other this concept plays a crucial role in separation and de-noising the signals [3].

A measure of independence may be easily described to evaluate the independence of the decomposed components. Generally, considering the multichannel signal as $\mathbf{y}(n)$ and the constituent signal components as $y_i(n)$, the $y_i(n)$ are independent if

$$p_{\mathbf{Y}}(\mathbf{y}(n)) = \prod_{i=1}^{m} p_y(y_i(n)) \quad \forall n \tag{7.6}$$

where $p(\mathbf{Y})$ is the joint probability distribution, $p_y(y_i(n))$ are the marginal distributions and m is the number of independent components.

An important application of ICA is in BSS. BSS is an approach to estimate and recover the independent source signals using only the information of their mixtures observed at the recording channels. Due to its variety of applications BSS has attracted much attention recently. BSS of acoustic signals is often referred to as the Cocktail Party Problem [4], which means separation of individual sounds from a number of recordings in an uncontrolled environment such as a cocktail party. Figure 7.2 illustrates the BSS concept for separation of brain sources. As expected, ICA can be useful if the original sources are independent that is, $p(\mathbf{s}(n)) = \prod_{i=1}^{m} p_i(\mathbf{s}_i(n))$.

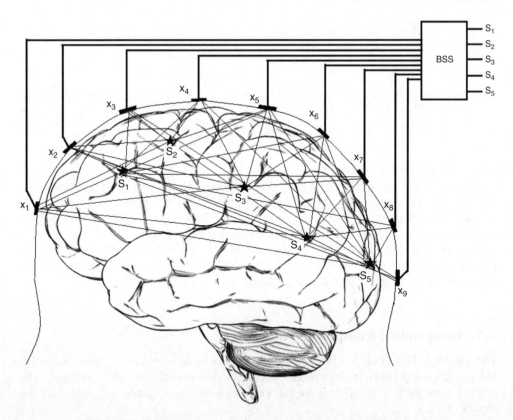

Figure 7.2 BSS concept; mixing and blind separation of the EEG signals

A perfect separation of the signals requires taking into account the structure of the mixing process. In a real-life application, however, this process is unknown, but some assumptions may be made about the source statistics.

Generally, the BSS algorithms do not make realistic assumptions about the environment in order to make the problem more tractable. There are typically three assumptions about the mixing medium. The most simple but widely used case is the instantaneous case, where the source signals arrive at the sensors at the same time. This has been considered for separation of biological signals such as the EEG where the signals have narrow bandwidths and the sampling frequency is normally low. The BSS model in this case can be easily formulated as:

$$\mathbf{x}(n) = \mathbf{H}\mathbf{s}(n) + \mathbf{v}(n) \tag{7.7}$$

where $m \times 1$ $\mathbf{s}(n)$, $n_e \times 1$ $\mathbf{x}(n)$, and $n_e \times 1$ $\mathbf{v}(n)$ denote respectively the vector of source signals, observed signals, and noise at discrete time n. $\mathbf{H}$ is the mixing matrix of size $n_e \times m$. The separation is performed by means of a separating $m \times n_e$ matrix, $\mathbf{W}$, which uses only the information about $\mathbf{x}(n)$ to reconstruct the original source signals (or the independent components) as:

$$\mathbf{y}(n) = \mathbf{W}\mathbf{x}(n) \tag{7.8}$$

In the context of EEG signal processing n_e denotes the number of electrodes. The early approaches in instantaneous BSS started from the work by Herault and Jutten [3] in 1986. In their approach, they considered non-Gaussian sources with similar number of independent sources and mixtures. They proposed a solution based on a recurrent artificial neural network for separation of the sources.

In acoustic applications, however, there are usually time lags between the arrival of the signals at the sensors. The signals also may arrive through multiple paths. This type of mixing model is called a convolutive model. Based on these assumptions the convolutive mixing model can be classified into two more types: anechoic and echoic. In both cases the vector representations of mixing and separating processes are changed to $\mathbf{x}(n) = \mathbf{H}(n)^*\mathbf{s}(n) + \mathbf{v}(n)$ and $\mathbf{y}(n) = \mathbf{W}(n)^*\mathbf{x}(n)$ respectively, where $*$ denotes the convolution operation.

In an anechoic model, however, the expansion of the mixing process may be given as:

$$x_i(n) = \sum_{j=1}^{m} h_{ij} s_j(n - \delta_{ij}) + v_i(n), \quad \text{for } i = 1, \ldots, n_e \tag{7.9}$$

where the attenuation, h_{ij}, and delay, δ_{ij}, of source j to sensor i would be determined by the physical position of the source relative to the sensors. Then the unmixing process will be given as:

$$y_j(m) = \sum_{i=1}^{n_e} w_{ji} x_i(m - \delta_{ji}), \quad \text{for } j = 1, \ldots, m \tag{7.10}$$

where w_{ji} are the elements of W. In an echoic mixing environment it is expected that the signals from the same sources reach the sensors through multiple paths. Therefore, the expansion of the mixing and separating models is changed to

$$x_i(n) = \sum_{j=1}^{m} \sum_{k=1}^{K} h_{ij}^k s_j(n - \delta_{ij}^k) + v_i(n), \quad \text{for } i = 1, \ldots, N \tag{7.11}$$

where K denotes the number of paths and $v_i(n)$ is the accumulated noise at sensor i. The unmixing process will be formulated similarly to the anechoic one. Obviously, for a known number of sources an accurate result may be expected if the number of paths is known.

The aim of BSS using ICA is to estimate an unmixing matrix **W** such that $\mathbf{Y} = \mathbf{WX}$ best approximates the independent sources S, where **Y** and **X** are, respectively, matrices with columns $\mathbf{y}(n) = [y_1(n), y_2(n), \cdots y_m(n)]^T$ and $\mathbf{x}(n) = [x_1(n), x_2(n), \cdots x_{ne}(n)]^T$.

In any case, the unmixing matrix for the instantaneous case is expected to be equal to the inverse of the mixing matrix that is, $\mathbf{W} = \mathbf{H}^{-1}$. However, since all ICA algorithms are based upon restoring independence, the separation is subject to permutation and scaling ambiguities in the output-independent components, that is, $\mathbf{W} = \mathbf{PDH}^{-1}$, where **P** and **D** are, respectively, the permutation and scaling matrices.

There are three major approaches in using ICA for BSS:

1. Factorising the joint pdf of the reconstructed signals into its marginal pdfs. Under the assumption that the source signals are stationary and non-Gaussian, the independence of the reconstructed signals can be measured by a statistical distance between the joint distribution and the product of its marginal pdfs. Kullback–Leibler (KL) divergence (distance) is an example. For non-stationary cases and for the short-length data, there will be poor estimation of the pdfs. Therefore, in such cases, this approach may not lead to favourable results. On the other hand, such methods are not robust for noisy data since in this situation the pdf of the signal will be distorted.

2. Decorrelating the reconstructed signals through time, that is, diagonalizing the covariance matrices at every time instant. If the signals are mutually independent, the off-diagonal elements of the covariance matrix vanish, although the reverse of this statement is not always true. If the signals are non-stationary we can utilise the time-varying covariance structure to estimate the unmixing matrix. An advantage of this method is that it only uses second-order statistics, which implies that it is likely to perform better in noisy and short data length conditions than higher order statistics.

3. Eliminating the temporal cross-correlation functions of the reconstructed signals as much as possible. In order to perform this, the correlation matrix of observations can be diagonalized at different time lags simultaneously. Here, second-order statistics are also normally used. As another advantage, it can be applied in the presence of white noise, since such noise can be avoided by using the cross-correlation only for $\tau \neq 0$. Such a method is appropriate for stationary and weakly stationary sources (i.e when the stationarity condition holds within a short segment of data).

It has been shown [5] that mutual information (MI) is a measure of independence and that maximizing the non-Gaussianity of the source signals is equivalent to minimizing the mutual information between them.

In the majority of cases the number of sources is known. This assumption avoids any ambiguity caused by false estimation of the number of sources. In exactly-determined cases the number of sources is equal to the number of mixtures. In over-determined situations, however, the number of mixtures is more than the number of sources.

BSS have been employed for decomposing the EEG signals [6, 7] for separation of normal brain rhythms, event-related signals, or mental or physical movement-related sources.

If the number of sources is unknown, a criterion has to be established to estimate the number of sources beforehand. This process is a difficult task especially when noise is involved.

In those cases where the number of sources is more than the number of mixtures (known as underdetermined systems), the above BSS schemes cannot be applied simply because the unmixing matrix will not be invertible, and generally the original sources cannot be extracted. However, when the signals are sparse other methods based on clustering may be utilized.

A signal is said to be sparse when it has many zeros or at least approximately zero samples. Separation of the mixtures of such signals is potentially possible in the situation where at each sample instant the number of nonzero sources is not more than the number of sensors. The mixtures of sparse signals can also be instantaneous or convolutive. However, as we will briefly describe later, the solution for only the simple case of a small number of idealized sources has been given in the literature.

In the context of EEG analysis, although the number of signals mixed at the electrodes seems to be limited, the number of sources corresponding to the neurons firing at a time can be enormous. However, if the objective is to study a certain rhythm in the brain the problem can be transformed to the time-frequency domain or even to the space-time-frequency domain. In such domains the sources may be considered disjoint and generally sparse. Also it is said that the brain neurons encode data in a sparse way if their firing pattern is characterised by a long period of inactivity [8, 9].

7.4   Instantaneous BSS

This is the most commonly used scheme for processing the EEGs. The early work by Herault and Jutten led to a simple but fundamental adaptive algorithm [10]. Linsker [11] proposed unsupervised learning rules based on information theory that maximise the average mutual information between the inputs and outputs of an artificial neural network. Comon [5] performed minimisation of mutual information to make the outputs independent. The Infomax algorithm [12] was developed by Bell and Sejnowski, and in spirit is similar to the Linsker method. Infomax uses an elegant stochastic gradient learning rule that was proposed by Amari *et al.* [13]. Non-Gaussianity of the sources was first exploited by Hyvarinen and Oja [14] in developing their fast ICA (fICA) algorithm. fICA is actually a blind source extraction (BSE) algorithm, which extracts the sources one by one based on their kurtosis; the signals with transient peaks have high kurtosis. Later it was demonstrated that the Infomax algorithm and maximum likelihood estimation are in fact equivalent [15, 16].

Based on the Infomax algorithm [12] for signals with positive kurtosis, such as simultaneous EEG-fMRI and speech signals, minimising the mutual information between the source estimates and maximising the entropy of the source estimates are equivalent. Therefore, a stochastic gradient ascent algorithm can be used to iteratively find the unmixing matrix by maximization of the entropy. The infomax finds a $\mathbf{W}$ which minimises the following cost function:

$$J(\mathbf{W}) = I(\mathbf{z}, \mathbf{x}) = H(\mathbf{z}) - H(\mathbf{z}|\mathbf{x}) \tag{7.12}$$

where $H(\mathbf{z})$ is the entropy of the output, $H(\mathbf{z}|\mathbf{x})$ is the entropy of the output subject to a known input, and $\mathbf{z} = f(\mathbf{y})$ is a non-linear activation function applied element-wise to $\mathbf{y}$, the estimated sources. $I(\mathbf{z},\mathbf{x})$ is the mutual information between the input and output of the constructed adaptive neural network (ANN). $H(\mathbf{z}|\mathbf{x})$ is independent of $\mathbf{W}$, therefore, the gradient of J is only proportional to the gradient of $H(\mathbf{z})$. Correspondingly, the natural gradient [13] of J denoted as $\nabla_{\mathbf{W}} J$ will be:

$$\nabla_{\mathbf{W}} J = \nabla_{\mathbf{W}} I(\mathbf{z}, \mathbf{x})\mathbf{W}^{\mathrm{T}}\mathbf{W} = \nabla_{\mathbf{W}} I(\mathbf{z}, \mathbf{x})\mathbf{W}^{\mathrm{T}}\mathbf{W} \tag{7.13}$$

in which the time index n is dropped for convenience of presentation. Then, the sequential adaptation rule for the unmixing matrix $\mathbf{W}$ becomes:

$$\mathbf{W}(n + 1) = \mathbf{W}(n) + \mu \left(\mathbf{I} + (1 - 2f(\mathbf{y}(n)))\mathbf{y}^{\mathrm{T}}(n)\right) \mathbf{W}(n) \tag{7.14}$$

where $f(\mathbf{y}(n)) = (1 + \exp(-\mathbf{y}(n)))^{-1}$, assuming the outputs are super-Gaussian, and μ is the learning rate, which is either a small constant or gradually changes following the speed of convergence.

Joint diagonalization of eigen matrices (JADE) is another well known BSS algorithm [17] based on higher order statistics (HOS). The JADE algorithm effectively diagonalizes the fourth-order cumulant of the estimated sources. This procedure uses certain matrices $Q_z(M)$ formed by the inner product of the fourth-order cumulant tensor of the outputs with an arbitrary matrix M, that is,

$$\{\mathbf{Q}_z(\mathbf{M})\}_{ij} = \sum_{k=1}^{n_e} \sum_{l=1}^{n_e} Cum(\mathbf{z}_i, \mathbf{z}_j^*, \mathbf{z}_k, \mathbf{z}_l^*)m_{lk} \tag{7.15}$$

where the (l,k)th component of the matrix M is written as m_{lk}, $\mathbf{Z} = \mathbf{CY}$, and * denotes the complex conjugate. The matrix $\mathbf{Q}_z(M)$ has the important property that it is diagonalized by the correct rotation matrix $\mathbf{U}$, that is, $\mathbf{U}^H\mathbf{QU} = \mathbf{\Lambda}_M$, and $\mathbf{\Lambda}_M$ is a diagonal matrix whose diagonal elements depend on the particular matrix M as well as $\mathbf{Z}$. By using equation (7.15), for a set of different matrices $\mathbf{M}$, a set of cumulant matrices $Q_z(\mathbf{M})$ can be calculated. The desired rotation matrix $\mathbf{U}$ then, jointly diagonalizes these matrices. In practice, only approximate joint diagonalization is possible [17] that is, the problem can be stated as minimization of $J(\mathbf{u}) = \sum_{j=1}^{n_e} \sum_{i=1}^{n_e} off\left\{\mathbf{u}^H Q_{ij}\mathbf{u}\right\}$ where

$$off(\mathbf{M}) = \sum_{i \neq j} |m_{ij}|^2 \tag{7.16}$$

EEG signals are, however, nonstationary. Nonstationarity of the signals has been exploited in developing an effective BSS algorithm based on second-order statistics called SOBI (second-order blind identification) [18]. In this algorithm separation is performed at a number of discrete time lags simultaneously. At each lag the algorithm unitarily diagonalizes the whitened data covariance matrix. It also mitigates the effect of noise on the observation by using a whitening matrix calculation, which can improve robustness to noise. Unitary diagonalization can be explained as follows: If $\mathbf{V}$ is a whitening matrix and $\mathbf{X}$ is the observation matrix, the covariance matrix of the whitened observation is $\mathbf{C}_X = E[\mathbf{VXX}^H\mathbf{V}^H] = \mathbf{VR}_x\mathbf{V}^H = \mathbf{VHR}_s\mathbf{H}^H\mathbf{V}^H = \mathbf{I}$, where $\mathbf{R}_x$ and $\mathbf{R}_s$ denote, respectively, the covariance matrices of the observed data and the original sources. It is assumed that $\mathbf{R}_s = \mathbf{I}$, that is, the sources have unit variance and they are uncorrelated, so $\mathbf{VH}$ is a unitary matrix. Therefore $\mathbf{H}$ can be factored as $\mathbf{H} = \mathbf{V}^{-1}\mathbf{U}$, where $\mathbf{U} = \mathbf{VH}$. The joint approximate diagonalization for a number of time lags can be obtained efficiently using a generalization of the Jacobi technique for the exact diagonalization of a single Hermitian matrix. The SOBI algorithm is implemented through the following steps, as given in [18]:

1. The sample covariance matrix $\hat{\mathbf{R}}(0)$ is estimated from T data samples. The m largest eigenvalues and their corresponding eigenvectors of $\hat{\mathbf{R}}(0)$ are denoted as $\lambda_1, \lambda_2, \ldots, \lambda_m$ and $\mathbf{h}_1, \mathbf{h}_2, \ldots, \mathbf{h}_m$ respectively.
2. Under the white noise assumption, an estimate $\hat{\sigma}^2$ of the noise variance is the average of the $n_e - m$ smallest eigenvalues of $\hat{\mathbf{R}}(0)$. The whitened signals are $\mathbf{z}(n) = [z_1(n), z_2(n), \ldots, z_{ne}(n)]^T$, computed by $z_i(n) = (\lambda_i - \hat{\sigma}^2)^{-\frac{1}{2}}\mathbf{h}_i^H\mathbf{x}(n)$ for $1 \le i \le n_e$. This is equivalent to forming a whitening matrix as $\hat{W} = [(\lambda_1 - \hat{\sigma}^2)^{-\frac{1}{2}}\mathbf{h}_1, \ldots, (\lambda_{n_e} - \hat{\sigma}^2)^{-\frac{1}{2}}\mathbf{h}_{n_e}]^H$.
3. Form sample estimates $\hat{\mathbf{R}}(\tau)$ by computing the sample covariance matrices of $\mathbf{z}(t)$ for a fixed set of time lags $\tau \in \{\tau_j | j = 1, \ldots, K\}$.
4. A unitary matrix $\hat{\mathbf{U}}$ is then obtained as a joint diagonalizer of the set $\{\hat{R}(\tau_j) | j = 1, \ldots, K\}$.
5. The source signals are estimated as $\hat{\mathbf{s}}(t) = \hat{\mathbf{U}}^H\hat{\mathbf{W}}\mathbf{x}(t)$ or the mixing matrix $\mathbf{A}$ is estimated as $\hat{\mathbf{A}} = \hat{\mathbf{W}}^{\#}\hat{\mathbf{U}}$, where superscript $^{\#}$ denotes the Moore–Penrose pseudoinverse.

The fICA algorithm [14] is another very popular BSS technique, which can either extract the signals one by one (by deflation) or all together based on their kurtosis. In fact, the algorithm uses an independence criterion that exploits non-Gaussianity of the estimated sources. In some places where the objective is to remove the spiky artefacts, such as the removal of the fMRI artefact from the EEG signals in simultaneous EEG–fMRI recordings, application of an iterative fICA followed by deflation of the artefact component gives excellent results [19]. A typical example of this type is given in Figure 7.3.

Practically, fICA maximizes the negentropy, which represents the distance between a distribution and a Gaussian distribution having the same mean and variance, that is,

$$Neg(y) \propto \{E\left[f(y)\right] - E\left[f(y_{\text{Gaussian}})\right]\}^2 \qquad (7.17)$$

where, f is a score function [20] and Neg stands for negentropy. This, as mentioned previously, is equivalent to maximizing the kurtosis. Therefore, the cost function can be simply defined as

$$J(\mathbf{W}) = -\frac{1}{4}|k_4(\mathbf{y})| = -\frac{\beta}{4}k_4(\mathbf{y}) \qquad (7.18)$$

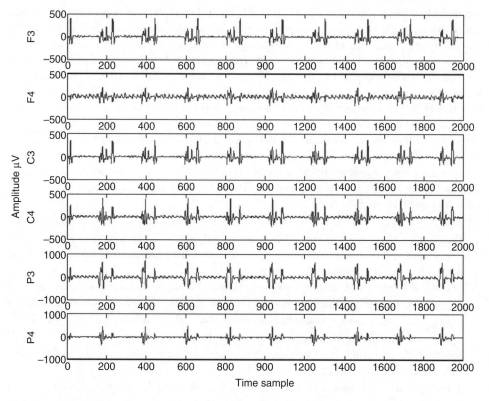

Figure 7.3 A sample of an EEG signal simultaneously recorded with fMRI

where $k_4(y)$ is the kurtosis, β is the sign of the kurtosis. Applying the standard gradient decent approach to minimise the cost function one can obtain

$$\mathbf{W}(n+1) = \mathbf{W}(n) - \mu \frac{\partial J(\mathbf{W})}{\partial \mathbf{W}}\bigg|_{\mathbf{w}=\mathbf{w}(n)} \tag{7.19}$$

where

$$-\mu \frac{\partial J(\mathbf{W})}{\partial \mathbf{W}}\bigg|_{\mathbf{W}=\mathbf{W}(n)} = \mu(n)\phi\left(\mathbf{y}(n)\right)\mathbf{x}(n) \tag{7.20}$$

Here $\mu(n)$ is a learning rate,

$$\phi(y_i) = \beta \frac{\hat{m}_4(y_i)}{\hat{m}_2^3(y_i)}\left[\frac{\hat{m}_2(y_i)}{\hat{m}_4(y_i)} y_i^3 - y_i\right] \tag{7.21}$$

and $\hat{m}_q(y_i) = E\left[y_i^q(n)\right]$, which is an estimate of the qth-order moment of the actual sources. Since fICA extracts the sources one by one a deflation process is followed to exclude the extracted source from the mixtures. The process reconstructs the mixtures iteratively by

$$\mathbf{x}_{j+1} = \mathbf{x}_j - \tilde{\mathbf{w}}_j \mathbf{y}_j, \quad j = 1, 2, \ldots \tag{7.22}$$

where $\tilde{\mathbf{w}}_j$ is estimated by minimization of the following cost function:

$$J\left(\tilde{\mathbf{w}}_j\right) = \frac{1}{2} E\left[\sum_{p=1}^{n_r} \mathbf{x}_{j+1,p}^2\right] \tag{7.23}$$

where n_r is the number of remaining mixtures.

Figure 7.4 shows the results after application of fICA to remove the scanner artefact from the EEGs.

In a time–frequency (TF) approach which assumes that the sources are approximately cyclostationary (generally nonstationary), the auto-terms and cross-terms of the covariance matrix of the mixtures are first separated and BSS is applied to both terms [21, 22]. In this approach, the spatial TF distribution (STFD) of the mixed signals is defined as

$$\mathbf{D}_{\mathbf{xx}}(n, \omega) = \frac{1}{2\pi} \sum_{u=\tau/2}^{N-\tau/2} \sum_{\tau=0}^{N/2-1} \varphi(n-u, \tau)e^{-i\omega\tau} E\left[\mathbf{x}\left(u+\frac{\tau}{2}\right)\mathbf{x}\left(u-\frac{\tau}{2}\right)\right] \tag{7.24}$$

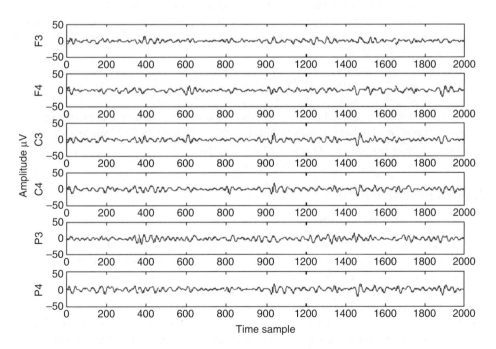

Figure 7.4   The EEG signals after removal of the scanner artefact

where $\phi(.)$ is the discretized kernel function defining a distribution from Cohen's class of TF distributions [23] and $\mathbf{x}(.)$ is an N sample observation of the signals, which is normally contaminated by noise. Assuming $\mathbf{x}(t) = \mathbf{As}(t) + \mathbf{v}(t)$, using the above equation we find

$$\mathbf{D}_{\mathbf{xx}}(n, \omega) = \mathbf{A}\mathbf{D}_{\mathbf{ss}}(n, \omega)\mathbf{A}^{H} + \sigma^2\mathbf{I} \tag{7.25}$$

where $\mathbf{D}_{\mathbf{ss}}(.)$ is the STFD of the source signals, σ^2 is the noise variance and depends on both noise power and the kernel function. From this equation it is clear that both $\mathbf{D}_{\mathbf{xx}}$ and $\mathbf{D}_{\mathbf{ss}}$ exhibit the same eigen-structure. The covariance matrix of the source signals is then replaced by the source STFD matrix composed of auto- and cross-source TFDs respectively, on the diagonal and off-diagonal entries.

Defining a whitening matrix $\mathbf{W}$ such that $\mathbf{U} = \mathbf{WA}$ is unitary, a whitened and noise compensated STFD matrix is defined as:

$$\begin{aligned}\tilde{\mathbf{D}}_{\mathbf{xx}}(n, \omega) &= \mathbf{W}(\mathbf{D}_{\mathbf{xx}}(n, \omega) - \sigma^2\mathbf{I})\mathbf{W}^{H} \\ &= \mathbf{U}\mathbf{D}_{\mathbf{ss}}(n, \omega)\mathbf{U}^{H}\end{aligned} \tag{7.26}$$

$\mathbf{W}$ and σ^2 can be estimated from the sample covariance matrix and $\mathbf{D}_{\mathbf{xx}}$ is estimated based on the discrete time formulation of the TFDs. From equation (7.21) it is known that the sensor STFD matrix exhibits the same eigen-structure as the data covariance matrix commonly used for cyclic data [21]. The covariance matrix of the source signals is replaced by a source STFD matrix composed of the auto- and cross-source TFDs on the diagonal and off-diagonal entries, respectively. The peaks occur in mutually exclusive locations on the TF plane. The kernel function can be defined in such a way as to maximise disjointedness of the points in the TF plane. By estimation of the STFD at appropriate TFD points, one may recover the source signals by estimating a unitary transformation $\hat{\mathbf{U}}$, via optimisation of a joint diagonal- and off-diagonal criterion, to have

$$\hat{\mathbf{s}}(n) = \hat{\mathbf{U}}^{H}\mathbf{W}\mathbf{x}(n) \text{ for } n = 1, \ldots, N - 1 \tag{7.27}$$

In order to define and extract the peaks of the $\mathbf{D}_{\mathbf{xx}}$ a suitable clustering approach has to be followed. This algorithm has potential application for estimating the EEG sources since in the most normal cases the sources are cyclic or quasi-cyclic.

7.5 Convolutive BSS

7.5.1 General Applications

In many practical situations the signals reach the sensors with different time delays. The corresponding delay between source j and sensor i, in terms of number of samples, is directly proportional to the sampling frequency and conversely to the speed of sound, that is, $\delta_{ij} \propto d_{ij} \times f_{s}/c$, where d_{ij}, f_{s}, and c are, respectively, the distance between source j and sensor i, the sampling frequency, and the speed of sound. For speech and music in the air as an example we may have d_{ij} in terms of metres, f_{s} between 8 and 44 KHz, and $c = 330$ m s^{-1}. Also, in an acoustic environment the sound signals can reach the sensors through multipaths after reflections by obstacles (such as walls). The above cases have been addressed, respectively, as anechoic and echoic BSS models and formulated at the beginning of this section. The solution to echoic cases is obviously more difficult and it normally involves some approximations to the actual system. As an example, in the previously mentioned cocktail party problem the source signals propagate through a dynamic medium with many parasitic effects, such as multiple echoes and reverberation. So, the received signals are a weighted sum of mixed and delayed components. In other words, the received signals at each microphone are the convolutive mixtures of speech signals.

In the case of spatial ICA, convolutive BSS should be applied where the source locations change within a segment. A good example is separation of video signals. Spatial ICA can also be applied to EEGs in order to separate the movement-related cortical sources when distributed sources are considered.

Unfortunately, most of the proposed BSS approaches to instantaneous mixtures fail or are limited in the separation of convolutive mixtures, mainly because either there are delays involved in propagation of the sources or a number of copies of the sources, delayed and weighted differently, reach the sensors.

Convolutive BSS has been a focus of research in the acoustic signal processing community. Two major approaches have been followed for both anechoic and echoic cases; the first approach is to solve the problem in the time domain. In such methods, in order to have accurate results, both the weights of the unmixing matrix and the delays have to be estimated. However, in the second approach, the problem is transformed into the frequency domain as $h(n)*s(n) \xrightarrow{F} H(\omega) \cdot S(\omega)$ and instantaneous BSS applied to each frequency bin mixed signal. The separated signals at different frequency bins are then combined and transformed to the time domain to reconstruct the estimated sources. The short-term Fourier transform is often used for this purpose. However, the inherent permutation problem of BSS severely deteriorates the results since the order of the separated sources in different frequency bins can vary.

An early work in convolutive BSS was by Platt and Faggin [24] who applied the adaptive noise cancellation network to the BSS model of Herault and Jutten [3], which has delays in the feedback path, was based on the minimum output power principle. This scheme exploits the fact that the signal corrupted by noise has more power than the clean signal. The feedback path cancels out the interferences as the result of delayed versions of the other sources. This circuit was also used later to extend the Infomax BSS to convolutive cases [25]. The combined network maximises the entropy at the output of the network with respect to the weights and delays. Torkkola [26] extended this algorithm to the echoic cases too. In order to achieve a reasonable convergence, some prior knowledge of the recording situation is necessary.

In another work an extension of SOBI has been used for anechoic BSS [27]; the problem has been transformed to the frequency domain and joint diagonalization of spectral matrices has been utilised to estimate the mixing coefficients as well as the delays [10]. In attempts by Parra [28], Ikram [29], and Cherkani [30] second-order statistics have been used to ensure that the estimated sources, $\mathbf{Y}(\omega,m)$, are uncorrelated at each frequency bin. $\mathbf{W}(\omega)$ is estimated in such a way that it diagonalizes the covariance matrices $\mathbf{R}_Y(\omega,k)$ simultaneously for all time blocks $k; k = 0,1, \ldots, K\text{-}1$, that is,

$$
\begin{aligned}
\mathbf{R}_Y(\omega, k) &= \mathbf{W}(\omega)\mathbf{R}_X(\omega, k)\mathbf{W}^{\mathrm{H}}(\omega) \\
&= \mathbf{W}(\omega)\mathbf{H}(\omega)\mathbf{\Lambda}_S(\omega, k)\mathbf{H}^{\mathrm{H}}(H\omega)\mathbf{W}^{\mathrm{H}}(\omega) \\
&= \mathbf{\Lambda}_c(\omega, k)
\end{aligned}
\tag{7.28}
$$

where $\mathbf{\Lambda}_S(\omega, k)$ is the covariance matrix of the source signals, which changes with k, $\mathbf{\Lambda}_c(\omega, k)$ is an arbitrary diagonal matrix, and $\mathbf{R}_X(\omega, k)$ is the covariance matrix of $\mathbf{X}(\omega)$.

The unmixing filter $\mathbf{W}(\omega)$ for each frequency bin ω that simultaneously satisfies the K decorrelation equations can be obtained using an over-determined least-squares solution. Since the output covariance matrix $\mathbf{R}_Y(\omega,k)$ has to be diagonalized the update equation for estimation of the unmixing matrix $\mathbf{W}$ can be found by minimizing the off-diagonal elements of $\mathbf{R}_Y(\omega,k)$, which leads to

$$\mathbf{W}_{\rho+1}(\omega) = \mathbf{W}_\rho(\omega) - \mu(\omega) \cdot \frac{\partial}{\partial \mathbf{W}_\rho^H(\omega)} \left\{ \left\| \mathbf{V}_\rho(\omega, k) \right\|^2 \right\} \tag{7.29}$$

where ρ is the iteration index, $\| . \|^2$ is the squared Frobenius norm,

$$\mu(\omega) = \frac{\alpha}{\sum_k \| \mathbf{R}_X(\omega, k) \|^2} \tag{7.30}$$

and

$$\mathbf{V}(\omega, k) = \mathbf{W}(\omega)\mathbf{R}_X(\omega, k)\mathbf{W}^H(\omega) - diag\left[\mathbf{W}(\omega)\mathbf{R}_X(\omega, k)\mathbf{W}^H(\omega)\right] \tag{7.31}$$

where α is a constant which is adjusted empirically.

In these methods a number of solutions for mitigating the permutation ambiguity have been suggested. Smaragdis [31] reformulated the Infomax algorithm for the complex domain and used it to solve the BSS in the frequency domain. Murata [32] also formulated the problem of BSS in each frequency bin using a simultaneous diagonalization method similar to the SOBI method. To mitigate the permutation problem a method based on the temporal structure of signals, which exploits the nonstationarity of speech was introduced. The method exploits the correlations between the frequency bins of the spectrum of the signals. Detecting and exploiting the silence periods either to localise the speakers individually or to detect the source statistics, such as pitch period, for each speaker, are other options to alleviate the permutation problem.

7.5.2   Application of Convolutive BSS to EEG

For the EEG mixing model the f_s is normally low (since the bandwidth <100 Hz) and the propagation velocity is equivalent to that of electromagnetic waves (300 000 km s^{-1}). Therefore, the delay is almost zero and we can always consider the mixing model for temporal BSS as instantaneous. Spatial BSS of EEGs, where the mixed signals are in sample per channel (i.e. alternating samples in the electrode space), however, can be convolutive. Spatial shift, analogous to time delay, is mainly due to source movement within one spatial signal segment. This idea has been developed in [33] in a spatio-temporal analysis of EEG using maximum likelihood convolutive ICA.

The main drawbacks for the application of BSS to separation of EEG signals is due to:

1. Unknown physiological and measurement noise statistics.
2. Unknown and varying number of sources.
3. Nonstationarity of the sources mainly due to evoked and movement related brain responses.
4. Existence of distributed moving sources related to synaptic currents.

Although many attempts have been made to solve the above problems more efforts are required to provide robust solutions for different applications.

7.6 Sparse Component Analysis

In nature, there are many events and sources with sparse nature. Some examples include interictal discharges within the hippocampus, QRS signals within ECGs, eye blinks within EEG, mutations within genomic sequence, and so on. In places where the sources are sparse that is, at each time instant only one of the sources has significant nonzero value, the columns of the mixing matrix may be calculated individually, which makes the solution to the underdetermined case possible. The problem can be stated as a clustering problem since the lines in the scatter plot can be separated based on their directionalities by means of clustering [34, 35]. The same idea has been followed more comprehensively by Li *et al.* [36]. In their method, however, the separation has been performed in two different stages. First, the unknown mixing matrix is estimated using the K-mean clustering method. Then, the source matrix is estimated using a standard linear programming algorithm. The line orientation of a data set may be thought of as the direction of its greatest variance. One way is to perform eigenvector decomposition on the covariance matrix of the data, the resultant principal eigenvector, that is, the eigenvector with the largest eigenvalue, indicates the direction of the data. There are many cases for which the sources are disjoint in other domains rather than the time-domain. In these cases the sparse component analysis can be performed in those domains more efficiently. One such approach called DUET [37] transforms the anechoic convolutive observations into the time–frequency domain using a short-time Fourier transform and the relative attenuation and delay values between the two observations are calculated from the ratio of corresponding time–frequency points. The regions of significant amplitudes (atoms) are then considered to be the source components in the time–frequency domain.

For instantaneous cases, in separation of sparse sources the common approach used by most researchers is to attempt to maximize the sparsity of the extracted signals in the output of the separator. The columns of the mixing matrix $\mathbf{A}$ assign each observed data point to only one source based on some measure of proximity to those columns [38] that is, at each instant only one source is considered active. Therefore the mixing system can be presented as:

$$x_i(n) = \sum_{j=1}^{M} a_{ji} s_j(n) \quad i = 1, \ldots, N \tag{7.32}$$

where, in an ideal case, $a_{ji} = 0$ for $i \neq j$. Minimization of the L_1-norm is one of the most logical methods for estimation of the sources. L_1-norm minimisation is a piecewise linear operation that partially assigns the energy of $x(n)$ to the M columns of $\mathbf{A}$ that form a cone around $x(n)$ in $\Re^M$ space. The remaining N-M columns are assigned zero coefficients, therefore the L_1-norm minimisation can be manifested as:

$$\min \|\mathbf{s}(n)\|_1 \text{ subject to } \mathbf{A}\mathbf{s}(n) = \mathbf{x}(n) \tag{7.33}$$

A detailed discussion of signal recovery using L_1-norm minimisation is presented by Takigawa *et al.* [39].

Most SCA-based methods for source separation and underdetermined blind identification (UBI) problems use the information about the signal intervals which include only one dominant source active in particular instants. These methods fail when there are not enough intervals including single dominant source to enable estimation of all columns of **A**.

Some more recent approaches have extended the SCA to the cases where more than one source, say k_a, where $k_a < N_x$, are active in an instant or interval. There, methods are able to use the information about these intervals to estimate the columns of mixing matrix **A** [40–42]. These methods are called k-SCA methods [43] and under some mild conditions (i.e. (i) having enough time samples to have enough planes/subspaces to cover all columns of **A** and (ii) having $k_a < N_x$ [43,44] are able to estimate **A** and the sources using some clustering methods.

In [40] it is assumed that at each instant of time there are on average $k_a < N_x/2$ sources active (non-zero). The main idea in this paper is to convert the problem of multiple dominant sources to a series of single dominant source problems, which may be solved by the well-known methods. Each column of **A** is then separately estimated by solving these single dominant problems. In this approach the number of sources has to be known. This latter problem has been addressed in [41] and a solution has been proposed. The adaptivity to changing environment and the robustness against noise and outliers (i.e. the intervals which do not satisfy sparseness conditions,) have been improved by the algorithm proposed in [42]. On the other hand, in this method k_a can be $N_x - 1 > k_a > N_x/2$.

7.7 Nonlinear BSS

Consider the cases where the parameters of the mixing system change because of the changes in the mixing environment or the changes in the source statistics. For example, if the images of both sides of a semi-transparent paper are photocopied the results will be two mixtures of the original sources. However, since the minimum observable grey level is black (or zero) and the maximum is white (say 1), the sum of the grey levels cannot go beyond these limits. This represents a nonlinear mixing system. As another example, think of the joint sounds heard from surface electrodes from over the skin. The mixing medium involves acoustic parameters of the body tissues. Since the tissues are not rigid, in such cases if the tissues vibrate due to the sound energy then the mixing system will be nonlinear. The mixing and unmixing can be generally modelled respectively as:

$$\mathbf{x}(n) = f(\mathbf{As}(n) + \mathbf{n}(n)) \tag{7.34}$$

$$\mathbf{y}(t) = g(\mathbf{Wx}(n)) \tag{7.35}$$

where $f(.)$ and $g(.)$ represent, respectively, the nonlinearities in the mixing and unmixing processes. There have been some attempts to solve nonlinear BSS problems, especially for the separation of image mixtures [45,46]. In the attempt in [46] the mixing system has been modelled as an RBF neural network. The parameters of this network are then computed iteratively. However, in these methods often an assumption about the mixing model is made.

An interesting example of a nonlinear mixing system is the combination of heart and lung sounds mixing environment. Both heart and lung vary in shape during both heart beating and inhaling and exhaling processes. In order to solve the BSS for such a system, both heart and lung acoustics should be accurately modelled. Such a model has not been revealed so far and, unfortunately, none of the nonlinear methods currently give satisfactory results.

7.8 Constrained BSS

The optimisation problem underlying the solution to the BSS problem may be subject to fulfilment of a number of conditions. These may be based on a priori knowledge of the sources or the mixing system. Any constraint on the estimated sources or the mixing system (or unmixing system) can lead to a more accurate estimation of the sources. Statistical [16] as well as geometrical constraints [17] have been very recently used in developing new BSS algorithms. In most of the cases the constrained problem is converted to an unconstrained one by means of a regularization parameter, such as a Lagrange multiplier or more generally a nonlinear penalty function, as used in [16].

Incorporating nonlinear penalty functions [30] into a joint diagonalization problem not only exploits nonstationarity of the signals but also ensures fast convergence of the update equation. A general formulation for the cost function of such a system can be in the form of

$$J(\mathbf{W}) = J_{\mathrm{m}}(\mathbf{W}) + \kappa\phi\left(J_{\mathrm{c}}(\mathbf{W})\right) \tag{7.36}$$

where $J_{\mathrm{m}}(\mathbf{W})$ and $J_{\mathrm{c}}(\mathbf{W})$ are, respectively, the main and the constraint cost functions, $\phi(\cdot)$ is the nonlinear penalty function, and κ is the penalty parameter.

Constrained BSS has a very high potential in incorporating clinical and physiological information into the main optimisation formulation.

As a new application of constrained BSS, an effective algorithm has been developed for removing the eye-blinking artefacts from EEGs. A similar method to the joint diagonalization of correlation matrices by using gradient methods [47] has been developed [48], which exploits the temporal structure of the underlying EEG sources. The algorithm is an extension of SOBI with the aim of iteratively performing the joint diagonalization of multiple time-lagged covariance matrices of the estimated sources and exploiting the statistical structure of the eye-blink signal as a constraint. The estimated source covariance matrix is given by

$$\mathbf{R}_{\mathbf{Y}}(k) = \mathbf{W}\mathbf{R}_{\mathbf{X}}(k)\mathbf{W}^{\mathrm{T}} \tag{7.37}$$

where $\mathbf{R}_{\mathbf{X}}(k) = E\left\{\mathbf{x}(n)\mathbf{x}^{\mathrm{T}}(n-k)\right\}$ is the covariance matrix of the electrode data. Following the same procedure as in [49], the least-squares (LS) estimate of $\mathbf{W}$ is found from

$$J_{\mathrm{m}}(\mathbf{W}) = \arg\min_{W} \sum_{k=1}^{T_{\mathrm{B}}} \|E(k)\|_{\mathrm{F}}^{2} \tag{7.38}$$

where $\|.\|_{\mathrm{F}}^{2}$ is the squared Frobenius norm and $E(k)$ is the error to be minimised between the covariances of the source signals, $\mathbf{R}_{\mathbf{S}}(k)$ and the estimated sources, $\mathbf{R}_{\mathbf{Y}}(k)$. The corresponding cost function has been defined based on minimizing the off-diagonal elements for each time block, that is

$$J(\mathbf{W}) = J_{\mathrm{m}}(\mathbf{W}) + \Lambda J_{\mathrm{c}}(\mathbf{W}) \tag{7.39}$$

where

$$J_{\mathrm{m}}(\mathbf{W}) = \sum_{k=1}^{T_{\mathrm{B}}} \|\mathbf{R}_{\mathbf{Y}}(k) - \mathrm{diag}(\mathbf{R}_{\mathbf{Y}}(k)\|_{\mathrm{F}}^2 \qquad (7.40)$$

and

$$J_{\mathrm{c}}(\mathbf{W}) = F\left(E\left\{\mathbf{g}(n)\mathbf{y}^{\mathrm{T}}(n)\right\}\right) \qquad (7.41)$$

is a second-order constraint term. $F(.)$ is a nonlinear function approximating the cumulative density function (CDF) of the data and $\Lambda = \{\lambda_{ij}\}$ ($i = 1, \ldots, N$) is the weighted factor which is governed by the correlation (matrix) between the EOG and EEG signals ($\mathbf{R}_{\mathrm{GY}}$), defined as $\Lambda = \kappa \mathrm{diag}(\mathbf{R}_{\mathrm{GY}})$, where κ is an adjustable constant. Then a gradient approach [23] is followed to minimise the cost function. The incremental update equation is

$$W(n+1) = W(n) - \mu \frac{\partial J(W)}{\partial W} \qquad (7.42)$$

which concludes the algorithm.

BSS has been widely used for the processing of EEG signals [50–53]. Although, the main assumptions about the source signals, such as uncorrelatedness or independence of such signals have not been verified yet, the empirical results illustrate the effectiveness of such methods. EEG signals are noisy and nonstationary and are normally affected by one or more types of internal artefacts. The most efficient approaches are those which consider all different domain statistics of the signals and take the nature of the artefacts into account. In addition, a major challenge is in how to incorporate and exploit the physiological properties of the signals and characteristics of the actual sources into the BSS algorithm. Some examples have been given here; some more will be presented in the other chapters of this book.

In the case of brain signals the independence or uncorrelatedness conditions for the sources may not be satisfied. This, however, may be acceptable for abnormal sources, movement-related sources, or ERP signals. Transforming the problem into a domain such as the space–time–frequency domain, where the sources can be considered disjoint, may be a good solution.

7.9   Application of Constrained BSS; Example

In practice, natural signals such as EEG source signals are not always independent. A topographic ICA proposed in [54] incorporates the dependence amongst the nearby sources in not only grouping the independent components related to nearby sources but also separating the sources originating from different regions of the brain. In this ICA model it is proposed that the residual dependence structure of the ICs, defined as dependences that cannot be cancelled by ICA, could be used to establish a topographic order between the components. Based on this model, if the topography is defined by a lattice or grid, the dependence of the components is a function of the distance of the components on that grid. Therefore, the generative model that implies correlation of energies for components that are close in the topographic grid is defined. The main assumption is then, the nearby sources are correlated and those far from each other are independent.

To develop such an algorithm a neighbourhood relation is initially defined as:

$$h(i, j) = \begin{cases} 1, & \text{if}\,|i - j| \leq m \\ 0, & \text{otherwise} \end{cases} \tag{7.43}$$

where the constant m specifies the width of the neighbourhood. Such a function is therefore, a matrix of hyperparameters. This function can be incorporated into the main cost function of BSS. The update rule is then given as [54]:

$$\mathbf{w}_i \propto E \left\{ \mathbf{z} \left(\mathbf{w}_i^\mathrm{T} \mathbf{z} \right) r_i \right\} \tag{7.44}$$

where $\mathbf{z}_i$ is the whitened mixed signals and

$$r_i = \sum_{k=1}^{n} h(i, k) g \left(\sum_{j=1}^{n} h(k, j) \left(\mathbf{w}_j^\mathrm{T} \mathbf{z} \right)^2 \right) \tag{7.45}$$

The function g is the derivative of a nonlinear function, such as those defined in [54]. It is seen that the vectors $\mathbf{w}_i$ are constrained to some topographic boundary defined by $h(i,j)$. Finally, the orthogonalization and normalization can be accomplished, for example, by the classical method involving matrix square roots;

$$\mathbf{W} \leftarrow (\mathbf{W}\mathbf{W}^\mathrm{T})^{-\frac{1}{2}} \mathbf{W} \tag{7.46}$$

where $\mathbf{W}$ is the matrix of the vectors $\mathbf{w}_i$, that is, $\mathbf{W} = [\mathbf{w}_1, \mathbf{w}_2, \ldots, \mathbf{w}_n]^\mathrm{T}$. The original mixing matrix A can be computed by inverting the whitening process as $\mathbf{A} = (\mathbf{W}\mathbf{V})^{-1}$, where $\mathbf{V}$ is the whitening matrix.

This algorithm has been modified for separation of seizure signals, by (i) iteratively finding the best neighbourhood m, and (ii) constraining the desired estimated source to be within a specific frequency band and originating from certain brain zones (confirmed clinically). Figure 7.5 illustrates the independent components of a set of EEG signals from an epileptic patient, using the above constrained topographic ICA method. In Figure 7.6 the corresponding topographic maps for all independent components (i.e. backprojection of each IC to the scalp using the inverse of estimated unmixing matrix,) are shown. From these figures the sixth IC from the top clearly shows the seizure component. Consequently, the corresponding topograph shows the location of a seizure over the left temporal electrodes.

7.10 Nonstationary BSS

As emphasised at the beginning of this chapter, most physiological signals have nonstationary behaviour, the existing artefacts are nonstationary or the mixing system may be nonlinear. On the other hand, the additive noise may not be Gaussian. As an example, consider the state of the brain when transferring from relax to attention, from preictal to ictal for epileptic patients,

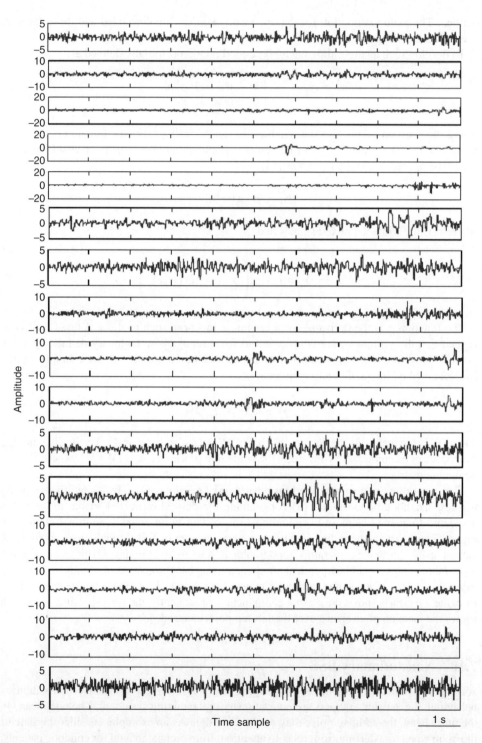

Figure 7.5 Estimated independent components of a set of EEG signals, acquired from 16 electrodes, using constrained topographic ICA

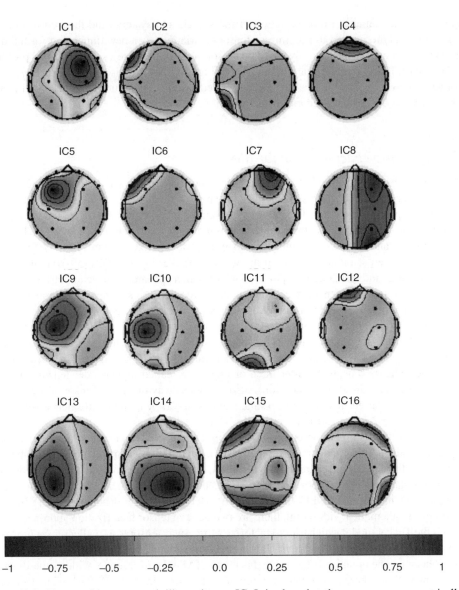

Figure 7.6 Topographic maps, each illustrating an IC. It is clear that the sources are geometrically localized

or from before to after hand movement (when the motor area is engaged). In these and many other cases the nonstationarity stems from the changes in the signal trends or statistics. These shortcomings have been taken into consideration and in some cases nicely exploited in the development of a new class of BSS systems called nonstationary BSS. Such algorithms not only exploit the mutual statistical behaviour of the source signals but also exploit the variations in the statistical properties of those sources across smaller signal segments. To accommodate this, a new mathematical representation of the signals and systems and their corresponding mixing model are needed to accommodate these variations as a new dimension for the data. The multiway approaches, mainly based on tensor factorization, are a solution to this problem.

In what follows, therefore, a number of these approaches are discussed. Before we move forward a new concept which has been recently used in source separation, channel identification and source localization, called tensor factorization, is described.

7.10.1   Tensor Factorization for BSS

Without doubt it is favourable to represent the data in a domain where the variation of such data in all aspects can be assessed and tracked. The so-called *multiway* signal processing is an effective step in this direction. The multiway technique incorporates the concept of tensors and tensor factorization to achieve that. Tensor decomposition was first introduced by Hitchcock in 1927 [55,56], and the idea of a multiway model is attributed to Cattell in 1944 [57,58]. These concepts received scant attention until the work of Tucker in the 1960s [59–61] and Carroll and Chang [62] and Harshman [63] in 1970, all of which appeared in psychometrics. Appellof and Davidson [64] are generally credited as being the first to use tensor decompositions (in 1981) in chemometrics, and tensors have since become extremely popular in that field [65].

A tensor is a multidimensional array. More formally, an *N*-way or *N*th-order tensor is an element of the tensor product of *N* vector spaces, each of which has its own coordinate system. A first-order tensor is a vector, a second-order tensor is a matrix, and a third-order tensor has three indices. Tensors of order three or higher are called higher-order tensors. Therder of a tensor is the number of dimensions, also known as way or mode. Tensor decomposition has been widely used in signal processing, such as in [66–75], and neuroscience, such as in [76–86]. Figure 7.7 illustrates the concept.

Decomposition of higher-order tensors has applications in chemometrics, signal processing, communications, linear algebra, computer vision, data mining, neuroscience, graph analysis, and so on, for detection, separation, localization, noise removal, and tracking. Traditionally, two particular tensor decompositions can be considered to be higher-order extensions of the matrix SVD, namely CANDECOMP/PARAFAC (CP) and Tucker decomposition [87].

From a mathematical view point, there are two general tensor factorization approaches. The first is the Tucker model which tries to factorize a three-way tensor to a smaller tensor (core tensor) and three other factors/matrices. This model is not unique since there are various ways the core tensor can be organized. The second is the PARAFAC tensor model with a superdiagonal core tensor. Since there is no other variation of such a core tensor the PARAFAC factorization is unique. Figure 7.8 represents the above models. Also PARAFAC decomposes a tensor to the sum of rank one tensors and the Tucker decomposition is a higher-order form of PCA.

There are many other tensor decompositions, including INDSCAL, PARAFAC2, CANDELINC, DEDICOM, and PARATUCK2, as well as nonnegative variants of all of the above. The N-way Toolbox, Tensor Toolbox, and Multilinear Engine are examples of software packages for working with tensors.

Recently, tensor factorization has been widely used for BSS of both instantaneous and convolutive mixtures. To be able to benefit from tensor modelling, we need to build up a tensor from our two-dimensional multichannel measurements. As an example, a time–space–frequency tensor can be built by initially transforming the data into the time–frequency domain

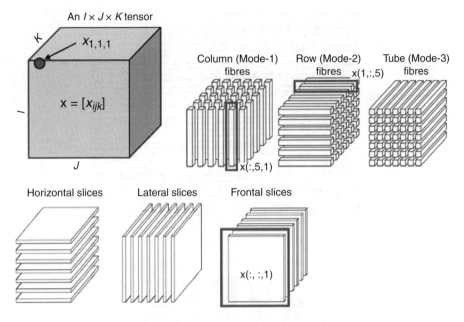

Figure 7.7   A tensor and its various modes

and then incorporating the space (geometrical locations of the sensors) dimension. For analysis of EEG or MEG this can be illustrated as in Figure 7.9.

In general, any other dimension in which the data variation is significant and desired can be selected. For example, to exploit the nonstationarity, one of the dimensions can be represented by the signal segment number. This allows the changes in statistics of the data to be tracked effectively within the tensor model.

In [85] an approach for the removal of eye-blink artefacts from EEG signals based on a novel space–time–frequency (STF) model of EEGs and the robust minimum variance beamformer (RMVB) has been proposed. In this method, in order to remove the artefact, the RMVB has been provided with a priori information, namely, an estimation of the steering vector corresponding to the point source of the eye-blink. The artefact-removed EEGs are subsequently reconstructed

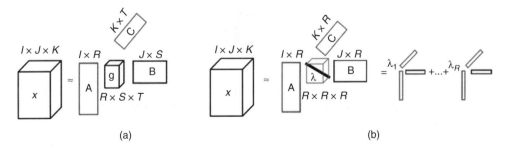

Figure 7.8   Tensor factorization using (a) Tuker and (b) PARAFAC models

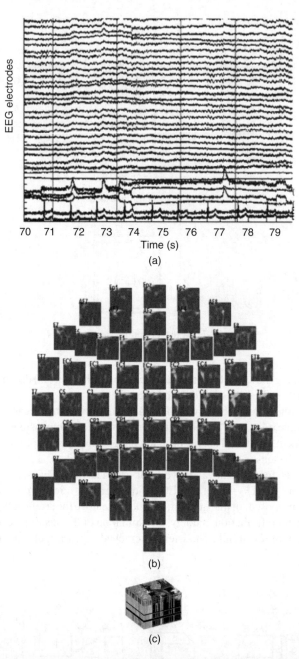

Figure 7.9 A tensor representation of a set of multichannel EEG signals; (a) time domain representation, (b) time–frequency representation of each electrode signal, and (c) the constructed tensor (see Plate 4 for the coloured version)

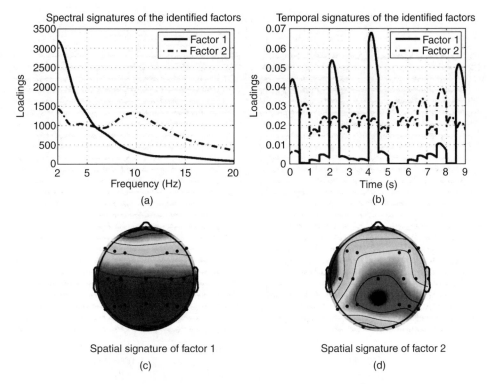

Figure 7.10 The extracted factors by using STF-TS modelling; (a) and (b) illustrate respectively the spectral and temporal signatures of the extracted factors and (c) and (d) represent the spatial distributions of those extracted factors [85]

by deflation. The a priori knowledge, the vector corresponding to the spatial distribution of the EB factor, is identified using the STF model of EEGs, provided by the PARAFAC method. In order to reduce the computational complexity present in the estimation of the STF model using the three-way PARAFAC, the time domain is sub-divided into a number of segments and a four-way array is then set to estimate the space–time–frequency-time/segment (STF-TS) model of the data using the four-way PARAFAC. The correct number of the factors of the STF model has been effectively estimated by using a novel core consistency diagnostic-(CORCONDIA-) based measure.

Subsequently, the STF-TS model is shown to approximate closely the classic STF model, with significantly lower computational cost. The results demonstrate that such an algorithm effectively identifies and removes the EB artefact from raw EEG measurements. Using this approach, Figure 7.10 shows the space, time, and frequency signatures of an eye-blink as the result of applying PARAFAC and Figure 7.11 shows the restored EEGs after applying the STF-TS method. In this method the nonstationarity of the data is also exploited through segmentation, as discussed in the following section.

A similar result has also been achieved when the PARAFAC was combined with a BSS in [86] to make a semi-blind scheme for removal of eye-blink artefact. In this paper, a novel iterative blind signal extraction (BSE) scheme for the removal of the eye-blink artefact

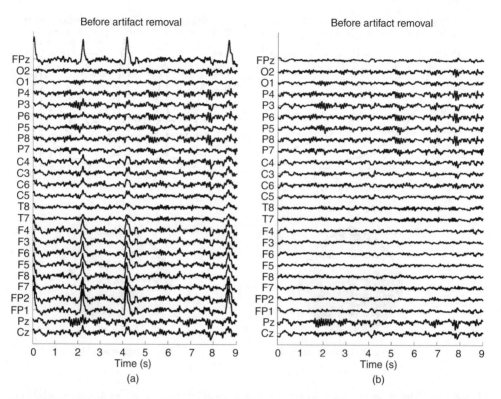

Figure 7.11 Restoration of the EEG signals in (a) from multiple eye-blinks after applying the STF-TS method [85]; the results are shown in (b)

from electroencephalogram (EEG) signals has been proposed. In this method, in order to remove the artefact, the extraction algorithm is subject to an estimation of the column of the mixing matrix corresponding to the point source eye-blink artefact. The restored EEGs are subsequently reconstructed in a deflation framework. The *a priori* knowledge, namely the vector corresponding to the spatial distribution of the eye-blink factor, is identified using PARAFAC. Hence, the BSE approach is called semi-blind signal extraction (SBSE). The results demonstrate that the proposed algorithm effectively identifies and removes the eye-blink artefact from raw EEG measurements.

These two approaches verify that PARAFAC can be combined with another modality in source detection or localization to not only guarantee the uniqueness of the solution but also improve the results that can be achieved by any one of those modalities.

More applications of PARAFAC such as in [84] related to brain–computer interfacing (BCI) will be explained in other chapters of this book.

7.10.2 Solving BSS of Nonstationary Sources Using Tensor Factorization

Using a simple PARAFAC model may provide some insight about various components in the signal mixtures. As an example, Figure 7.12 demonstrates the results of PARAFAC operation

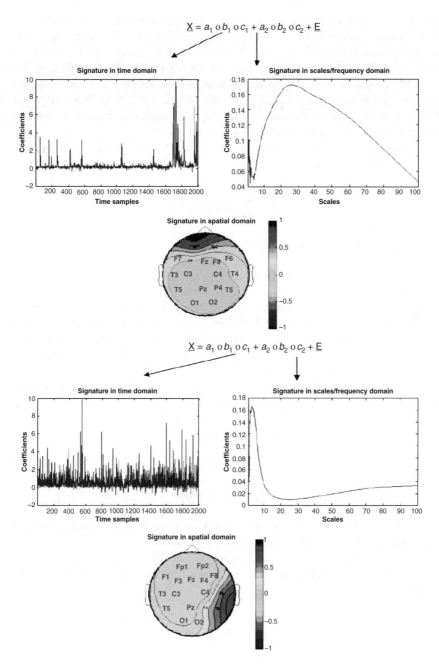

Figure 7.12 Representation of the first two components in the time–space–frequency domain for a segment of EEG signal during left finger movement

for detecting two components (atoms) of an EEG segment recording during left finger movement. The main component in Figure 7.12(b) is in the right-hand side (contralateral) and the
second component in Figure 7.12(a) is in the same side (ipsilateral) of the brain motor cortex.

In this approach, in order to localize seizure signals (within the scalp through an analysis of EEG), a tensor with three modes, that is, time samples, scales and electrodes, has
been constructed through wavelet analysis of multichannel EEG. Next, they demonstrate that
PARAFAC provides promising results in modelling the complex structure of epileptic seizure,
localizing a seizure origin and extracting the artefacts.

The fundamental expression of the PARAFAC model, which is used to describe decomposition of trilinear data sets, is given as:

$$x_{ijk} = \sum_{r=1}^{R} a_{ir} b_{jr} c_{kr} + e_{ijk} \tag{7.47}$$

In matrix form this can be described as:

$$\mathbf{X}_k = \mathbf{B} \mathbf{D}_k \mathbf{A}^{\mathrm{T}} + \mathbf{E}_k \; k = 1, 2, \ldots, K \tag{7.48}$$

where $(.)^{\mathrm{T}}$ refers to the transpose operation, $\mathbf{X}_k$ represents the kth frontal slice of tensor $\underline{\mathbf{X}}$ built
from multichannel data $\mathbf{X}$, and $\mathbf{A}$ and $\mathbf{B}$ are the component matrices in the first and second
modes, respectively. $\mathbf{D}_k$ is a diagonal matrix, whose diagonal elements correspond to the kth
row of the third component matrix $\mathbf{C}$ denoted as $\mathbf{D}_k = \mathrm{diag}(\mathbf{C}_{k,:})$. Finally, $\mathbf{E}_k$ contains the error
terms corresponding to the entries in the kth frontal slice.

In order to estimate the factors an alternating least-square (ALS) is applied. Ignoring the
noise term, this leads to performing the following alternating operations till convergence:

$$\mathbf{A} = \mathbf{X}_{(1)} \left((\mathbf{C} \circ \mathbf{B})^{\mathrm{T}} \right)^{\dagger} \tag{7.49}$$

$$\mathbf{B} = \mathbf{X}_{(2)} \left((\mathbf{C} \circ \mathbf{A})^{\mathrm{T}} \right)^{\dagger} \tag{7.50}$$

$$\mathbf{C} = \mathbf{X}_{(3)} \left((\mathbf{B} \circ \mathbf{A})^{\mathrm{T}} \right)^{\dagger} \tag{7.51}$$

Where $(.)^{\dagger}$ denotes the pseudo-inverse operation, $\mathbf{X}_{(i)}$ represents the unfolded version of tensor
$\mathbf{X}$ in the ith dimension and $\circ$ denotes the Khatri–Rao product.

The results of PARAFAC decomposition are unique if $k_A + k_B + k_C \geq 2R + 2$, where
k_A refers to the Kruskal rank of $\mathbf{A}$, and so on. For an N rank tensor, however, this has been
modified to [88]

$$\sum_{n=1}^{N} k_n \geq 2R + (N-1) \tag{7.52}$$

Although most of the properties of the sources can be evaluated using these so-called
"signatures" of the data [89] and they are useful for diagnostic applications, unfortunately, the
original sources (in the time domain) cannot be recovered. Also, it is seen that the representing
signatures do not vary for different time segments. Hence, in this model the signals are
considered stationary. Therefore, such a model cannot be utilised for BSS. So, the main

question here is how can tensor factorization be used for source separation. Moreover, how can this be extended to separation of nonstationary data.

As an extension to PARAFAC, PARAFAC2 supports variation in one mode of the tensor. Mathematically, PARAFAC can be expressed in the following form:

$$\mathbf{X}_k = \mathbf{A}_k \mathbf{D}_k \mathbf{B}^\mathrm{T} + \mathbf{E}_k \qquad (7.53)$$

Compared with PARAFAC, PARAFAC2 is designed to deal with non-trilinear data sets, while keeping the uniqueness of the solutions as for PARAFAC. To do so, PARAFAC2 allows some freedom in the shape of the k slabs ($\mathbf{X}_k$) in the variable mode. To maintain the uniqueness of the solutions $\mathbf{X}_k \mathbf{X}_k^\mathrm{T}$ are forced to have the same structure for all k, that is, $\mathbf{X}_k \mathbf{X}_k^\mathrm{T} = \mathbf{A}\mathbf{D}_k \mathbf{\Phi} \mathbf{D}_k \mathbf{A}^\mathrm{T}$. In this case the above equation is subject to $\mathbf{A}_k^\mathrm{T} \mathbf{A}_k = \mathbf{\Phi}$. Consequently, $\mathbf{A}_k$ can be modelled as $\mathbf{A}_k = \mathbf{P}_k \mathbf{A}$, where $\mathbf{P}_k \mathbf{P}_k^\mathrm{T} = \mathbf{I}$. Therefore, the *direct* form for PARAFAC2 can be in the following form:

$$\mathbf{X}_k = \mathbf{P}_k \mathbf{H} \mathbf{D}_k \mathbf{A}^\mathrm{T} + \mathbf{E}_k \quad \text{subject to} \quad \mathbf{P}_k \mathbf{P}_k^\mathrm{T} = \mathbf{I} \qquad (7.54)$$

ALS optimization is proposed to estimate the direct model parameters. Unlike for indirect PARAFAC2 [90], for the direct PARAFAC it is possible to apply priories, impose constraints, and generalize it to n-way models. The steps in estimating the factors are as follows [91]: Considering $\mathbf{X}_k$ as an $R \times J \times K$ three-way array with frontal planes $\mathbf{P}_k^\mathrm{T} \mathbf{X}_k$, $k = 1, \ldots, K$.

Step 0. If $J < n_k$, where n_k is the number of columns in the kth data matrix, replace $\mathbf{X}_k$ by $\mathbf{H}_k$, for example, from the Cholesky decomposition $\mathbf{X}_k \mathbf{X}_k^\mathrm{T} = \mathbf{H}_k \mathbf{H}_k^\mathrm{T}$.

Step 1. Initialize $\mathbf{B}$ as the loading matrix by applying PCA to $\sum_{k=1}^{K} \mathbf{X}_k \mathbf{X}_k^\mathrm{T}$ and initialize $\mathbf{A}$ and $\mathbf{D}_1, \ldots, \mathbf{D}_K$ as $\mathbf{I}_R$.

Step 1a. Compute the SVD $\mathbf{A}\mathbf{D}_k \mathbf{B}^\mathrm{T} \mathbf{X}_k^\mathrm{T} = \mathbf{U}_k \mathbf{\Lambda}_k \mathbf{V}_k^\mathrm{T}$ and update $\mathbf{P}_k$ as $\mathbf{U}_k \mathbf{V}_k^\mathrm{T}$, $k = 1$, $\ldots, K$.

Step 1b. Update $\mathbf{A}$, $\mathbf{B}$ and $\mathbf{D}_1, \ldots, \mathbf{D}_K$ by one cycle of a PARAFAC1 algorithm applied to the $R \times J \times K$ three-way array with frontal planes $\mathbf{P}_k^\mathrm{T} \mathbf{X}_k$, $k = 1, \ldots, K$.

Step 1c. Evaluate $\sigma(\mathbf{P}_1, \ldots, \mathbf{P}_k, \mathbf{A}, \mathbf{D}_1, \ldots, \mathbf{D}_k) = \sum_{k=1}^{K} \left\| \mathbf{X}_k - \mathbf{P}_k \mathbf{A} \mathbf{D}_k \mathbf{B}^\mathrm{T} \right\|^2$.

If $\sigma_{\mathrm{old}} - \sigma_{\mathrm{new}} > \varepsilon \sigma_{\mathrm{old}}$ for some small value ε, repeat Step 1; else go to Step 2.

Step 2. Given that $\mathbf{X}_k$ has been replaced by $\mathbf{H}_k$ in Step 0, now replace $\mathbf{H}_k$ by $\mathbf{X}_k$ again and compute $\mathbf{P}_k$ according to Step 1a, $k = 1, \ldots, K$.

Since all the factors are estimated through the above steps, the sources can be easily estimated as $\mathbf{S}_k = \mathbf{P}_k \mathbf{H} \mathbf{D}_k$.

In the case of nonstationary sources it is logical to have k as the number of segments. This allows one to track and exploit the variation in signal statistics across the segments and, therefore, the nonstationarities can be tolerated.

A number of nonstationary BSS approaches using tensor factorization have been developed recently. SOBI [20] was probably the first tried to solve the problem of nonstationarity in BSS by applying diagonalization of covariance matrices in multiple lags. Although for nonstationary signal sources the performance of the SOBI algorithm is better than those of many other BSS approaches, for highly nonstationary cases not only is the performance not acceptable but also the statistical fluctuation within the source signals is ignored. The fundamental approach followed by the researchers recently involves dividing the input into small segments and converting the entire input signal matrices into tensors which can be processed using the corresponding mathematical techniques, as described above. A number of approaches have been followed recently, such as those in [92–94].

In the work by Hyvarinen [92] inspired by Matsukas *et al.*'s research [95], the nonstationarity is interpreted by smooth variation in variance and it is shown that the cross-cumulants for nonstationary sources are positive. Consequently, a method similar to that for maximizing the kurtosis for BSS has been developed to maximize the cumulants which measure the nonstationarity of the separated source. Since the defined positive cross-cumulants do not require the marginal distribution of the source signals to be Gaussian, it can be used for nonstationary Gaussian sources too.

In another attempt [93] the algorithm utilizes the principle of maximum likelihood and minimum mutual information. In both cases the optimization amounts to approximate joint diagonalization of a set of L matrices. Intuitively, nonstationarity allows blind identification in the Gaussian case. This is because a single covariance matrix does not provide enough constraints to uniquely determine $\mathbf{A}$, A collection of several covariance matrices estimated over different time periods does determine $\mathbf{A}$, provided the source distributions change enough over the whole observation period. In his algorithm it is considered that the profile of changing variance of each estimated source is simply a smoothed version of the square of that source. Therefore, the variances are calculated over smaller segments of the data.

A similar concept has also been exploited in [94]. Similarly, this method can be used for both Gaussian and non-Gaussian sources. Also, in this approach we have assumed that the additive noise signals are temporally uncorrelated and mutually uncorrelated with the source signals. Unlike general BSS methods using second-order statistics (SOS) or higher order statistics (HOS) concepts, a first-order blind source separation (FOBSS) has been proposed by defining a first-order model for the mixtures. This system was then applied to separation of seizure components of a set of EEG signals. The use of the FOBSS method is based on the following assumptions:

- A1: The columns of all $\mathbf{S}_k$ are mutually uncorrelated, particularly, $\mathbf{S}_k$ matrices are all orthogonal.
- A2: The source columns of all $\mathbf{S}_k$ can be temporally correlated/uncorrelated.
- A3: The source envelopes (as in [94]) change independently, that is, each source has varying variance over its segments k and the variation for different sources is independent.
- A4: The mixing channel A is full rank and remains unchanged for all time segments k.
- A5: The additive noise signals are temporally uncorrelated and mutually uncorrelated with the sources.

Based on A1 all the source covariance matrices, that is, $\mathbf{S}_k \mathbf{S}_k^{\mathrm{T}}$, are diagonal with non-negative diagonal values for all $k = 1, \ldots, K$. So, the basic model after temporal segmentation is simply stated as

$$\mathbf{X}_k = \mathbf{S}_k \mathbf{A} + \mathbf{E}_k \text{ subject to } \mathbf{S}_k^{\mathrm{T}} \mathbf{S}_k = \mathbf{D}_k^2 \ k = 1, \ldots, K \qquad (7.55)$$

where $\mathbf{D}_k^2$ is a diagonal matrix. $\mathbf{A}$ is fixed across the segments. Each $\mathbf{S}_k$ is decomposed into one row-wise orthonormal matrix $\mathbf{P}_k$ and one diagonal matrix $\mathbf{D}_k$ which absorbs the norm of different columns of $\mathbf{S}_k$ at each segment k as $\mathbf{S}_k = \mathbf{P}_k\mathbf{D}_k$. This satisfies orthogonality of the segmented sources, that is, $\mathbf{S}_k\mathbf{S}_k^{\mathrm{T}} = \mathbf{D}_k^2$. Similar to the previous model, this can change the above model to

$$\mathbf{X}_k = \mathbf{P}_k\mathbf{D}_k\mathbf{A}^{\mathrm{T}} + \mathbf{E}_k \text{ subject to } \mathbf{P}_k^{\mathrm{T}}\mathbf{P}_k = \mathbf{I}_{N_s}, k = 1, \ldots, K \qquad (7.56)$$

The parameters (factors) of the above model using the FOBSS approach in [95] can be estimated by going through the following steps:

Step 1: Initialize all the model parameters randomly.

Step 2: Estimation of $\mathbf{P}_k$ using $\mathbf{P}_k = \mathbf{U}_k\mathbf{V}_k^{\mathrm{T}}$ for all k $= 1, \ldots, K$, where $\mathbf{U}_k$ and $\mathbf{V}_k$ are the eigenvectors.

Step 3: $\mathbf{A}$ is estimated using $\mathbf{A} = \sum_{k=1}^{K} \mathbf{D}_k\mathbf{P}_k^{\mathrm{T}}\mathbf{X}_k$

Step 4: To calculate $\mathbf{D}_k$ the following relations (as the result of solving the above constrained optimization problem) are used:

$$\sum_{k=1}^{K} \mathbf{A}\mathbf{P}_k^{\mathrm{T}}\mathbf{X}_k = \sum_{k=1}^{K} \mathbf{D}_k\mathbf{\Lambda} \qquad (7.57)$$

where Λ is a diagonal matrix defined as $\mathbf{\Lambda} = D\mathrm{diag}(\mathbf{A}^{\mathrm{T}}\mathbf{A})$ which sets all the non-diagonal elements of $\mathbf{A}^{\mathrm{T}}\mathbf{A}$ to zero. Then, for all k we use

$$\sum_{m=1}^{K} \mathbf{Q}_{km}\mathbf{D}_m = \mathbf{D}_k\mathbf{\Lambda} \qquad (7.58)$$

where $\mathbf{Q}_{km} = D\mathrm{diag}(\mathbf{P}_k^{\mathrm{T}}\mathbf{X}_k\mathbf{X}_m^{\mathrm{T}}\mathbf{P}_m)$ are diagonal matrices for $k,m = 1, \ldots, K$. Finally, for the rth source using $\mathbf{G}_r\mathbf{d}_r = \lambda_r\mathbf{d}_r$, where $\mathbf{d}_r$ is the result of stacking all rth diagonal elements of all $\mathbf{D}_k$. Solving the above problem is equal to finding the nearest eigenvalue of $\mathbf{G}_r$ to $\mathbf{a}_r^{\mathrm{T}}\mathbf{a}_r$ where $\mathbf{a}_r$ is the rth column of mixing matrix $\mathbf{A}$, and, respectively, its corresponding eigenvectors $\mathbf{d}_r$.

Step 5: If $\|\mathbf{X}_k - \mathbf{P}_k\mathbf{D}_k\mathbf{A}^{\mathrm{T}}\| \geq \sigma$, where σ is the expected noise variance, go to **Step 2**, otherwise, stop.

Figure 7.13 shows the results of the FOBSS algorithm in the separation of the seizure component from the scalp EEG. The data are just the last segment of a large amount of data recorded from the scalp of an epileptic patient in the Clinical Neuroscience Department, King's College London simultaneously with a set of intracranial signals using electrodes implanted within the brain. In Figure 7.14 the signals from three intracranial source signals have been illustrated. From this figure the instant of ictal onset is very clear.

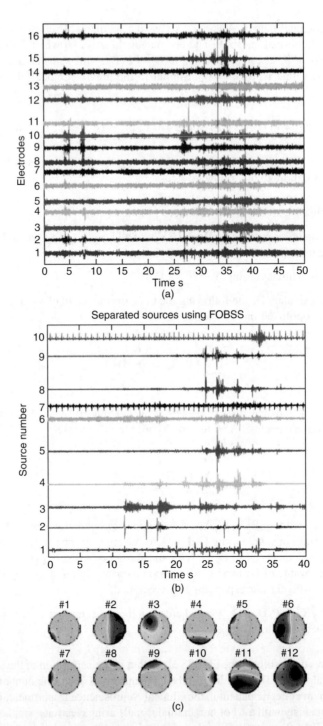

Figure 7.13 The results of application of the FOBSS algorithm to a set of scalp EEGs in (a); (b) the separated sources, and (c) the corresponding topoplots (see Plate 5 for the coloured version of (c))

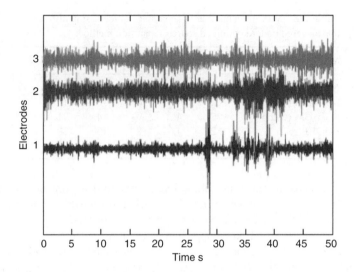

Figure 7.14   The intracranial records from three electrodes. These signals were simultaneously recorded together with the scalp EEGs in Figure 7.13 (a)

7.11   Tensor Factorization for Underdetermined Source Separation

The SCA methods in Section 7.4 exploit the sparsity of the constituent sources in the grouping and separation of the signals. Moreover, the main assumption in these methods is that only one signal has non-zero amplitude at each instant of time. Although in some applications with sparse sources identification of the mixing channels is achieved successfully, in many cases reconstruction of the original source signals is either difficult or less accurate.

In tensor factorization approaches recently developed for underdetermined or sparse source separation, such as in [96–99], two main concepts have been exploited. The first property is sparsity of the sources, as for the SCA approaches, and the second, which represents one of the main advantages of tensor factorization over other separation methods, is the fact that tensors can have ranks higher than those of the related matrices. Therefore, it is expected that the maximum number of separable sources increases with increase in the rank of the tensor. Moreover, in some applications, such as in [100], the idea of sparse events has been presented. This means that at each period of time up to a certain number of sources can be active. Such an approach can also be considered as a generalization of the SCA methods described in Section 7.4 to the situation where we can have more than one source active and, also, the duration of activity of a source can be more than a sample period.

Therefore, tensor factorization techniques, especially PARAFAC [69, 98], have been used to solve the UBI problem, even when the sources are non-sparse or fully active (e.g. $k_a \leq N_s$) [94]. Later, second- and higher-order statistics based methods have been developed to solve the UBI problem using joint/simultaneous diagonalization of a series of symmetric square matrices, such as second-order blind identification of underdetermined mixtures (SOBIUM) [96] and fourth-order cumulant-based blind identification of under-determined mixtures (so-called FOOBI) [97]. The SOBIUM method deals with the covariance matrices of different segments and tries to convert the UBI problem to a PARAFAC problem, with two identical

Table 7.1 Maximum number of separable sources using different tensor-based underdetermined source separation methods

N_x						
N_s	2	3	4	5	6	7
SOBIUM (real)	2	4	6	10	15	20
SOBIUM (complex)	2	4	9	14	21	30
FOOBI (real)	2	4	8	13	20	29
FOOBI (complex)	2	5	12	22	36	55

factors called individual differences in scaling (INDSCAL) [68], to joint block diagonalization of some exact-determined hermitian matrices. Table 7.1 shows the upper bounds for SOBIUM and FOOBI in terms of the maximum number of separable sources.

In [100] a tensor factorization-based approach to k_a-SCA (called k-SCA in [49]) has been developed to solve the underdetermined blind source separation (UBSS) and especially underdetermined blind identification (UBI) problems. k_a sources have been considered active in each signal segment. Similar to k_a-SCA methods, introduced in Section 7.4, k_a can be $k_a \leq N_x - 1$. However, in this approach instead of sparse signals (with limited signals active in each instant) sparse events (for which limited sources are active within some signal segments of particular duration) have been introduced and considered. Therefore, initially, the number of active sources in each signal segment has been estimated. A necessary condition for such a method to work is to have a sufficient number of segments to have each source active in some of the segments. The UBI is then performed using the segments of sparsely active sources.

The algorithm, so-called UOM-BSS (underdetermined orthogonal model BSS) for solving a simiar problem as in (7.51) has the following steps [100]:

Step 1: Initialize all the model parameters randomly; $iter = 1$;

Step 2: Estimate $\mathbf{P}_k$ using $\mathbf{P}_k = \mathbf{U}_k \mathbf{V}_k^T$ for all $k = 1, \ldots, K$.

Step 3: Estimate $\mathbf{A}$ using

$$\mathbf{A} = \left(\sum_{k=1}^{K} \mathbf{X}_k^T \mathbf{P}_k \mathbf{D}_k \right) \left(\sum_{k=1}^{K} \mathbf{D}_k^2 \right)^{\dagger}$$

Step 4: Estimate $\mathbf{D}_k$ for all $k = 1, \ldots, K$ using

$$\mathbf{D}_k \leftarrow \lambda \breve{\mathbf{D}}_k + (1 - \lambda)\mathbf{D}_k$$

where
$\lambda = \lambda_{\max}(1 - e^{-0.01(iter-1)})$; $\lambda_{\max} = 0.2$ as an empirical choice, and $\breve{\mathbf{D}}_k = \text{diag}(\breve{\mathbf{d}}_k)$, a diagonal matrix with $\breve{\mathbf{d}}_k$ as the diagonal elements, computed separately as given later.

$$iter = iter + 1;$$

Step 5: If $\left\| \mathbf{X}_k - \mathbf{P}_k \mathbf{D}_k \mathbf{A}^{\mathrm{T}} \right\| \geq \sigma$, where σ is the expected noise variance, go to **Step 2**, otherwise, stop.

$\breve{\mathbf{d}}_k$ is also estimated for each segment k separately using the following steps [100]:

```
Step 1:  n = 0, y=vec (PᵀₖXₖ) =vec (DₖA),  Φ=A∘I_{Ns},  s = []
Step 2: Find the index number of the maximum element of |ΦᵀE| as m.
Update the selected active list s by s ← [sm]
Step 3: Build Φ̆ by selecting columns of Φ indexed in s.
e=y − Φ̆ (Φ̆†y)
n ← n + 1
Step 4: if ‖e‖ > ε OR n < Nₓ - 1
        go to Step 2
  else
        converged: d̆ₖ(s)=Φ̆†y
  end if.
```

This approach improves the general upper bound in terms of the maximum possible number of sources that can be achieved by the SOBIUM method [96]. Using some synthetically mixed signals it has been shown that the separation error is less for the above UOM-BSS method than for both SOBIUM and FOOBI methods.

7.12 Tensor Factorization for Separation of Convolutive Mixtures in the Time Domain

Due to flexibility in the definition of multiway systems, convolutive BSS in the time domain can be modelled using tensors. In [73, 101–103] the anechoic convolutive BSS problem has been formulated as a tensor factorization problem. In a noiseless environment the problem has been stated as:

$$\mathbf{X}_k = \sum_{\tau=1}^{M} \mathbf{\Xi}_\tau \mathbf{P}_k \mathbf{D}_k \mathbf{A}_\tau^{\mathrm{T}} \text{ subject to } \mathbf{P}_k^{\mathrm{T}} \mathbf{P}_k = \mathbf{I}_{N_s} \qquad (7.59)$$

where $\mathbf{\Xi}_\tau$ accounts for the delays and $\mathbf{A}_\tau$ accommodates all the weights for the delayed terms. An effective method for estimation of the parameters/factors has been proposed, particularly for estimation of $\mathbf{A}_\tau$ [103].

In this approach it has been demonstrated that the impulse response of the channels can be more accurately estimated if the locations of the sources and the sensors are known. This geometrical information can be nicely fused into the separation process.

7.13 Separation of Correlated Sources via Tensor Factorization

Unfortunately, many physiological and biological signal sources are somehow correlated mainly because of having the same origin or cause for their generation. Consider, for example, the actual components of a QRS signal within an ECG signal. Such components cannot be separated using conventional BSS which relies on independence of the sources. Another example is separation of P3a and P3b, subcomponents of a P300 signal, from EEGs. These components are related to the same event or stimulus and partially overlap in time and space (and albeit in frequency).

In [104] this problem has been addressed using tensor factorization. Promising outcomes have been obtained using synthetically mixed components. However, assessment of the application of this method to real signals is still under research.

7.14   Conclusions

During the past two decades a large number of BSS methods have been proposed. A BSS problem becomes more difficult to solve when the signals are convolutively mixed, the number of sources is more than the number of sensors, the signals are nonstationary, or the system is nonlinear. In this chapter the BSS problem for different scenarios has been explained. Recently introduced tensor factorization approaches can highly alleviate these problems by relying on their three fundamental properties; first, they can map the signals into a multidimensional space where the sources can become disjoint; second, their ranks are often higher than any of the corresponding two-dimensional representations of the data; and third, any related constraint can be easily incorporated to ensure the desired unique solution. The concepts in this chapter can be a good foundation for understanding many other approaches in BSS.

References

[1] Golyandina, N., Nekrutkin, V. and Zhigljavsky, A. (2001) *Analysis of Time Series Structure: SSA and Related Techniques*, Chapman & Hall/CRC, New York.

[2] Haavisto, O. (2010) Detection and analysis of oscillations in a mineral flotation circuit. *J. Control Eng. Pract.*, **18**(1), 23–30.

[3] Herault, J. and Jutten, C. (1986) Space or time adaptive signal processing by neural models, *Proceedings of AIP Conference, Neural Network for Computing*, American Institute of Physics, pp. 206–211.

[4] Cherry, C.E. (1953) Some experiments in the recognition of speech, with one and two ears. *J. Acoust. Soc. Am.*, **25**, 975–979.

[5] Comon, P. (1994) Independent component analysis: A new concept. *Signal Process.*, **36**, 287–314.

[6] Jung, T.P., Makeig, S., Westerfield, M. *et al.* (1999) Analyzing and visualizing single-trial event-related potentials, in *Advances in Neural Information Processing Systems*, **11**, MIT Press, pp. 118–124.

[7] Makeig, S., Jung, T.P., Bell, A.J. *et al.* (1997) Blind separation of auditory event-related brain responses into independent components. *Proc. Nat. Acad. Sci. U. S. A.*, **94**, 10979–10984.

[8] Földiák, P. and Young, M. (1995) Sparse coding in the primate cortex, in *The Handbook of Brain Theory and Neural Networks*, MIT Press, pp. 895–898.

[9] Einhauser, W., Kayser, C., Konig, P. and Kording, K.P. (2002) Learning the invariance properties of complex cells from their responses to natural stimuli. *Eur. J. Neurosci.*, (**15**), 475–486.

[10] Jutten, C. and Herault, J. (1991) Blind separation of sources, part I: an adaptive algorithm based on neuromimetic architecture. *Signal Process.*, **24**, 1–10.

[11] Linsker, R. (1989) An application of the principle of maximum information preservation to linear systems, in *Advances in Neural Information Processing Systems*, Morgan Kaufmann, pp. 186–194.

[12] Bell, A.J. and Sejnowski, T.J. (1995) An information-maximization approach to blind separation, and blind deconvolution. *Neural Comput.*, **7**(6), 1129–1159.

[13] Amari, S., Cichocki, A. and Yang, H.H. (1996) A new learning algorithm for blind signal separation, in *Advances in Neural Information Processing Systems*, **8**, MIT Press, pp. 757–763.

[14] Hyvarinen, A. and Oja, E. (1997) A fast-fixed point algorithm for independent component analysis. *Neural Comput.*, **9**(7), 1483–1492.

[15] Parra, L. and Spence, C. (2000) Convolutive blind separation of non-stationary sources. *IEEE Trans. Speech Audio Process*, **8**(3), 320–327.

[16] Cardoso, J.-F. (1997) Infomax and maximum likelihood for blind source separation. *IEEE Signal Process. Lett.*, **4**(4), 112–114.

[17] Cardoso, J. (1989) Source separation using higher order moments. Proceedings of International Conference on Acoustics, Speech and Signal Processing (ICASSP), pp. 2109–2112.

[18] Belouchrani, A., Abed-Meraim, K., Cardoso, J-F. and Moulines, E. (1997) A blind source separation technique using second order statistics. *IEEE Trans. Signal Process*, **45**(2), 434–444.

[19] Jing, M., and Sanei, S. (2006) Scanner artifact removal in simultaneous EEG-fMRI for epileptic seizure prediction, Proceedings of the IEEE 18th International Conference on Pattern Recognition, ICPR, vol. **3**, pp. 722 – 725.

[20] Mathis, H. and Douglas, S.C. (May 2004) On the existence of universal nonlinearities for blind source separation. *IEEE Trans. Signal Process.*, **50**, 1007–1016.

[21] Belouchrani, A., Abed-Mariam, K., Amin, M.G. and Zoubir, A.M. (2004) Blind source separation of nonstationary signals. *IEEE Signal Process. Lett.*, **11**(7), 605–608.

[22] Cirillo, L. and Zoubir, A. (2005) On blind separation of nonstationary signals. Proceedings of the 8th Symposium on Signal Processing and its Applications (ISSPA), Sydney, Australia.

[23] Cohen, L. (1995) *Time-Frequency Analysis*, Prentice Hall.

[24] Platt, C. and Fagin, F. (1992) Networks for the separation of sources that are superimposed and delayed, in *Advances in Neural Information Processing 4*, Morgan Kaufmann, pp. 730–737.

[25] Torkkola, K. (1996) Blind separation of delayed sources based on information maximization. Proceedings of ICASSP, Atlanta, Georgia, pp. 3509–3512.

[26] Torkkola, K. (1996) Blind separation of convolved sources based on information maximization. Proceedings of IEEE workshop on Neural Networks and Signal Processing (NNSP), pp. 315–323.

[27] Wang, W., Sanei, S. and Chambers, J.A. (2005) Penalty function based joint diagonalization approach for convolutive blind separation of nonstationary sources. *IEEE Trans. Signal Process.*, **53**(5), 1654–1669.

[28] Parra, L., Spence, C., Sajda, P. *et al.* (2000) Unmixing hyperspectral data, in *Advances in Neural Information Processing*, **13**, MIT Press, pp. 942–948.

[29] Ikram, M. and Morgan, D. (2000) Exploring permutation inconstancy in blind separation of speech signals in a reverberant environment. Proceedings of ICASSP, Turkey.

[30] Cherkani, N. and Deville, Y. (1999) Self adaptive separation of convolutively mixed signals with a recursive structure, Part 1: stability analysis and optimisation of asymptotic behaviour. *Signal Process.*, **73**(3), 225–254.

[31] Smaragdis, P. (1998) Blind separation of convolved mixtures in the frequency domain. *Neurocomputing*, **22**, 21–34.

[32] Murata, N., Ikeda, S. and Ziehe, A. (2001) An approach to blind source separation based on temporal structure of speech signals. *Neurocomputing*, **41**, 1–4.

[33] Dyrholm, M., Makeig, S. and Hansen, L.K. (2007) Model selection for convolutive ICA with an application to spatio-temporal Analysis of EEG. *Neural Comput.*, **19**(4), 934–955.

[34] Zibulevsky, M. (2002) Relative Newton Method for Quasi-ML Blind Source Separation. http://ie.technion.ac.il/mcib.

[35] Luo, Y., Chambers, J., Lambotharan, S. and Proudler, I. (2006) Exploitation of source non-stationarity in underdetermined blind source separation with advanced clustering techniques. *IEEE Trans. Signal Process.*, **54**(6), 2198–2212.

[36] Li, Y., Amari, S., Cichocki, A. *et al.* (2006) Underdetermined blind source separation based on sparse representation. *IEEE Trans Signal Process.*, **54**(2), 423–437.

[37] Jurjine, A., Rickard, S. and Yilmaz, O. (2000) Blind separation of disjoint orthogonal signals: demixing N sources from 2 mixtures. Proceedings of IEEE Conference Acoustic, Speech, and Signal Processing (ICASSP 2000), vol. 5, pp. 2985–2988.

[38] Vielva, L., Erdogmus, D., Pantaleon, C. *et al.* (2002) Underdetermined blind source separation in a time-varying environment. Proceedings of IEEE Conference on Acoustic, Speech, and Signal Processing (ICASSP 2000), vol. 3, pp. 3049–3052.

[39] Takigawa, I., Kudo, M., Nakamura, A. and Toyama, J. (2004) On the minimum 11-norm signal recovery in underdetermined source separation. Proceedings of 5th International Conference Independent Component Analysis, pp. 22–24.

[40] Noorshams, N., Babaie-Zadeh, M. and Jutten, C. (2007) Estimating the mixing matrix in sparse component analysis based on converting a multiple dominant to a single dominant problem. ICA'07 Proceedings of the 7th International Conference on Independent Component Analysis and Signal Separation, pp. 397–405

[41] Movahedi Naini, F., Mohimani, G.H., Babaie-Zadeh, M. and Jutten, C. (2008) Estimating the mixing matrix in sparse component analysis (SCA) based on partial k-dimensional subspace clustering. *Neurocomput.*, **71**(10–12), 2330–2343.

[42] Washizawa, Y. and Cichocki, A. (2006) On-line k-plane clustering learning algorithm for sparse component analysis. in Proceedings of the International Conference on Acoustics, Speech and Signal Processing, ICASSP2006, Toulouse, France, pp. 681–684.

[43] Gao, F., Sun, G., Xiao, M. and Lv, J. (2010) Matrix estimation based on normal vector of hyperplane in sparse component analysis. ICSI (2)'10, pp. 173–179.

[44] Georgiev, P., Theis, F. and Cichocki, A. (2005) Sparse component analysis and blind source separation of underdetermined mixtures. *IEEE Trans. Neural Net.*, **16**(4), 992–966.

[45] Almeida, L. (2005) Nonlinear source separation, in. *Synthesis Lectures on Signal Processing*, Morgan & Claypool Publishers.

[46] Jutten, C. and Karhunen, J. (2004) Advances in blind source separation (BSS) and independent component analysis (ICA) for nonlinear mixtures. *Int. J. Neural Syst.*, **14**(5), 267–292.

[47] Joho, M. and Mathis, H. (2002) Joint diagonalization of correlation matrices by using gradient methods with application to blind signal processing. Proceedings of 1st Annual Conference in Sensor Array and Multichannel Signal Processing, SAM2002, pp. 273–277.

[48] Shoker, L., Sanei, S. and Chambers, J. (Oct. 2005) Artefact removal from electroencephalograms using a hybrid BSS-SVM algorithm. *IEEE Signal Process. Lett.*, **12**(10), 721–724.

[49] Parra, L. and Spence, C. (2000) Convolutive blind separation of non-stationary sources. *IEEE Trans. Speech Audio Process.*, **8**, 320–327.

[50] Jing, M., Sanei, S. (2007) A novel constrained topographic ICA for separation of epileptic seizure signals, *EURASIP J. Comput. Intell. Neurosci.*, doi: 10.1155/2007/21315.

[51] Jung, T.-P., Humphries, C., Lee, T. W., McKeown, M. J., Iragui, V., Makeig, S. and Sejnowski, T. J. (2000) Removing electroencephalographic artifacts by blind source separation, *Psychophysiology*, **37**, 163–178.

[52] Ghaderi, F., Sanei, S., Nazarpour, K. and McWhirter, J. (2010) Removal of ballistocardiogram artifacts from EEG/fMRI data using cyclostationary source extraction method, *IEEE Trans. Biomed. Eng.*, **57**(11), 2667–2676.

[53] Spyrou, L., Jing, M., Sanei, S. and Sumich, A. (2007) Separation and localisation of P300 sources and the subcomponents using constrained blind source separation, *EURASIP J. Adv. Signal Process*, paper ID: 82912, pp. 1–10.

[54] Hyvärinen, A., Hoyer, P.O. and Inkl, M. (2001) Topographic independent component analysis. *Neural Comput.*, **13**, 1527–1558.

[55] Hitchcock, F.L. (1927) The expression of a tensor or a polyadic as a sum of products. *J. Math. Phys.*, **6**, 164–189.

[56] Hitchcock, F.L. (1927) Multiple invariants and generalized rank of a p-way matrix or tensor. *J. Math. Phys.*, **7**, 39–79.

[57] Cattell, R.B. (1944) Parallel proportional profiles and other principles for determining the choice of factors by rotation. *Psychometrika*, **9**, 267–283.

[58] Cattell, R.B. (1952) The three basic factor-analytic research designs I their interrelations and derivatives. *Psychol. Bull.*, **49**, 499–452.

[59] Tucker, L.R. (1963) Implications of factor analysis of three-way matrices for measurement of change, in *Problems in Measuring Change* (ed. C.W. Harris), University of Wisconsin Press, pp. 122–137.

[60] Tucker, L.R. (1964) The extension of factor analysis to three-dimensional matrices, in *Contributions to Mathematical Psychology* (eds H. Gulliksen and N. Frederiksen), Holt, Rinehardt, & Winston, New York.

[61] Tucker, L.R. (1966) Some mathematical notes on three-mode factor analysis. *Psychometrika*, **31**, 279–311.

[62] Carroll, J.D. and Chang, J.J. (1970) Analysis of individual differences in multidimensional scaling via an N-way generalization of 'Eckart-Young' decomposition, *Psychometrika*, **35**, 283–319.

[63] Harshman, R.A. (1970) Foundations of the PARAFAC procedure: Models and conditions for multi-modal factor analysis. UCLA working papers in phonetics, vol. 16, pp. 1–84.

[64] Appellof, C.J. and Davidson, E.R. (1981) Strategies for analyzing data from video fluorometric monitoring of liquid chromatographic effuents. *Anal. Chem.*, **53**, 2053–2056.

[65] Smilde, A., Bro, R. and Geladi, P. (2004) *Multi-Way Analysis: Applications in the Chemical Sciences*, Wiley, West Sussex, England.

[66] De Lathauwer, L., Castaing, J. and Cardoso, J.-F. (2007) Fourth-order cumulant based blind identification of underdetermined mixtures. *IEEE Trans. Signal Process.*, **55**, 2965–2973.

[67] De Lathauwer, L. and de Baynast, A. (2008) Blind deconvolution of DS-CDMA signals by means of decomposition in rank-(1; l; l) terms. *IEEE Trans. Signal Process.*, **56**(4), 1562–1571.

[68] De Lathauwer, L. and De Moor, B. (1998) From matrix to tensor: Multilinear algebra and signal processing, in *Mathematics in Signal Processing IV* (eds J. McWhirter and I. Proudler), Clarendon Press, Oxford, pp. 1–15.

[69] De Lathauwer, L. and Vandewalle, J. (2004) Dimensionality reduction in higher-order signal processing and rank-(R1;R2; : : :;RN) reduction in multilinear algebra. *Linear Algebra Appl.*, **391**, 31–55.

[70] Chen, B., Petropulu, A. and De Lathauwer, L. (2002) Blind identification of convolutive MIMO systems with 3 sources and 2 sensors. *EURASIP J. Appl. Signal Process.* (Special Issue on Space-Time Coding and Its Applications, Part II), pp. 487–496, doi:10.1155/S1110865702000823.

[71] Comon, P. (2001) Tensor decompositions: State of the art and applications, in *Mathematics in Signal Processing V* (eds J.G. McWhirter and I.K. Proudler), Oxford University Press, pp. 1–24.

[72] Muti, D. and Bourennane, S. (2005) Multidimensional filtering based on a tensor approach. *Signal Process.*, **85**, 2338–2353.

[73] Sanei, S. and Makkiabadi, B. (2009) Tensor factorization with application to convolutive blind source separation of speech, in *Machine Audition: Principles, Algorithms and Systems* (ed. W. Wang), IGI-Global Pub.

[74] Makkiabadi, B. and Sanei, S. (2012) A new time domain convolutive BSS of heart and lung sounds. Proceedings of IEEE International Conference on Acoustics, Speech and Signal Processing, ICASSP 2012, Kyoto, Japan.

[75] Makkiabadi, B., Sarrafzadeh, A., Jarchi, D. *et al.* (2009) Semi-blind signal separation and channel estimation in MIMO communication systems by tensor factorization. Proceedings of IEEE Workshop on Statistical Signal Processing, SSP2009, Cardiff, UK.

[76] Beckmann, C. and Smith, S. (2005) Tensorial extensions of independent component analysis for multisubject FMRI analysis. *NeuroImage*, **25**, 294–311.

[77] Acar, E., Aykut-Bingol, C., Bingol, H. *et al.* (2007) Multiway analysis of epilepsy tensors. *Bioinformatics*, **23**, i10–i18.

[78] De Vos, M., De Lathauwer, L., Vanrumste, B. *et al.* (2007) Canonical decomposition of ictal scalp EEG and accurate source localisation: principles and simulation study. *Comput. Intell. Neurosci.*, **2007**, 1–8.

[79] De Vos, M., Vergult, A., De Lathauwer, L. *et al.* (2007) Canonical decomposition of ictal scalp EEG reliably detects the seizure onset zone. *NeuroImage*, **37**, 844–854

[80] Martinez-Montes, E., Valdés-Sosa, P.A., Miwakeichi, F. *et al.* (2004) Concurrent EEG/fMRI analysis by multiway partial least squares. *NeuroImage*, **22**, 1023–1034.

[81] Miwakeichi, F., Martinez-Montes, E., Valds-Sosa, P.A. *et al.* (2004) Decomposing EEG data into space-time-frequency components using parallel factor analysis. *NeuroImage*, **22**, 1035–1045.

[82] Ocks, J.M. (1988) Topographic components model for event-related potentials and some biophysical considerations. *IEEE Trans. Biomed. Eng.*, **35**, 482–484.

[83] Morup, M., Hansen, L.K. and Arnfred, S.M. (2007) ERPWAVELAB a toolbox for multi-channel analysis of time-frequency transformed event related potentials. *J. Neurosci. Method.*, **161**, 361–368.

[84] Nazarpour, K., Praamstra, P., Miall, R.C. and Sanei, S. (2009) Steady-state movement related potentials for brain computer interfacing. *IEEE Trans. Biomed. Eng.*, **56**(8), 2104–2113.

[85] Nazarpour, K., Wangsawat, Y., Sanei, S. *et al.* (2008) Removal of the eye-blink artefacts from EEGs via STF-TS modeling and robust minimum variance beamforming. *IEEE Trans. Biomed. Eng.*, **55**(9), 2221–2231.

[86] Nazarpour, K., Mohseni, H.R., Hesse, C. *et al.* (2008) A novel semi-blind signal extraction approach incorporating PARAFAC for the removal of eye-blink artefact from EEGs. *EURASIP J. Adv. Signal Process.*, 2008, Article ID 857459, 12. doi: 10.1155/2008/857459

[87] Kolda, T.G. and Bader, B.W. (2008) Tensor decompositions and applications. *SIAM Rev.*, **51**(3), 455–500

[88] Stegeman, A. and Sidiropoulos, N.D. (2007) On Kruskal's uniqueness condition for the candecomp/parafac decomposition. *Linear Algebra Appl.*, **420**(2–3), 540–552.

[89] Rong, Y., Vorobyov, S.A., Gershman, A.B. and Sidiropoulos, N.D. (2005) Blind spatial signature estimation via time-varying user power loading and parallel factor analysis. *IEEE Trans. Signal Process.*, **53**(5), 1697–1710.

[90] Harshman, R.A. and Lundy, M.E. (1994) PARAFAC: parallel factor analysis. *Comput. Stat. Data Anal.*, **8**, 32–79.

[91] Kiers, H.A.L., Ten Berge, J.M.F. and Bro, R. (1999) PARAFAC2 - Part I. A direct fitting algorithm for the PARAFAC2 model. *J. Chemometrics*, **13**, 275–294.

[92] Hyvarinen, A. (2001) Blind source separation by nonstationarity of variance: a cumulant-based approach. *IEEE Trans. Neural Networks*, **12**(6), 1471–1474.

[93] Matsuoka, K. and Kawamoto, M. (1994) A neural net for blind separation of nonstationary signal sources. *IEEE World Congr. Comput. Intell.*, **1**, 221–232.

[94] Pham, D.T. and Cardoso, J.F. (2001) Blind separation of instantaneous mixtures of nonstationary sources. *IEEE Trans Signal Process.*, **49**(9), 1837–1848.

[95] Makkiabadi, B. (2011) Advances in factorization based blind source separation. PhD Thesis, University of Surrey.

[96] De Lathauwer, L. and Castaing, J. (2008) Blind identification of underdetermined mixtures by simultaneous matrix diagonalization. *IEEE Trans. Signal Process*, **56**(3), 1096–1105.

[97] De Lathauwer, L., Castaing, J. and Cardoso, J.-F. (2007) Fourth-order cumulant-based blind identification of underdetermined mixtures. *IEEE Trans. Signal Process.*, **55**(6), 2965–2973.

[98] Bro, R. (1997) PARAFAC. tutorial and applications. *Chemometr. Intell. Lab. Syst.*, **38**(2), 149–171.

[99] Shoker, L., Sanei, S., Wang, W. and Chambers, J. (2004) Removal of eye blinking artefact from EEG incorporating a new constrained BSS algorithm. *IEEE J. Med. Biol. Eng. Comput.*, **43**(2), 290–295.

[100] Makkiabadi, B., Sanei, S. and Marshall, D. (2010) A k-subspace based tensor factorization approach for under-determined blind identification. Proceedings of IEEE Asilomar Conference on Signals, Systems, and Computers, CA, USA, pp. 18–22.

[101] Makkiabadi, B., Jarchi, D. and Sanei, S. (2011) A new time domain approach for convolutive blind source separation. Proceedings of IEEE Asilomar Conference on Signals, Systems, and Computers, CA, USA.

[102] Makkiabadi, B. and Sanei, S. (2012) A new time domain convolutive BSS of heart and lung sounds. Proceedings of IEEE International Conference on Acoustics, Speech and Signal Processing, ICASSP 2012, Kyoto, Japan.

[103] Makkiabadi, B., Jarchi, D., Abolghasemi, V. and Sanei, S. (2011) A time domain geometrically constrained multimodal approach for convolutive blind source separation. Proceedings of EUSIPCO 2011 (19th European Signal Processing Conference 2011), Barcelona, Spain.

[104] Makkiabadi, B., Jarchi, D., and Sanei, S. (2011) Blind separation and localization of correlated P300 subcomponents from single trial recordings using extended PARAFAC2 tensor model. Proceedings of IEEE EMBC 2011, Boston, USA.

8

Connectivity of Brain Regions

8.1 Introduction

The notion of brain connectivity has been well recognized by researchers in neuroimaging through EEG, fMRI, and MEG information collected noninvasively. Amongst different neuroimaging methods, EEG and MEG directly reflect neuronal firing, exhibiting a good temporal resolution (in milliseconds) despite a poor spatial resolution (of the order of a few square centimetres). This data can therefore be used in the estimation of the brain regions connectivity. Most brain functions rely on interactions between neuronal assemblies distributed within and across different cerebral regions.

According to an unpublished work called EEG Connectivity: A Tutorial, by Thatcher *et al.* the nature of brain sources has to be investigated first. In this work EEG sources are divided into two different types; one includes the electrical fields of brain operating at the speed of light where dipoles distributed in space turn on and off while oscillating at different amplitudes, and the second type is the source of the electrical activity, which is an excitable medium of locally connected networks. This has been nicely modelled by Hodgkin and Huxley who wrote the fundamental excitable medium equations of the brain in 1952 (as discussed in Chapter 3). The brain network comprises axons, synapses, dendritic membranes, and ionic channels that behave like "kindling" at the leading edge of a confluence of different fuels and excitations. Approximately 80% of the cortex is excitatory, with recurrent loop connections. This obviously calls for stability of the whole network, stemming from the fact that the refractory periods are relatively long and this allows for self-organizing and stability of the cortical activities.

The connected network within the brain results in EEG connectivity. This connectivity is defined by the magnitude of coupling between neurons. Consequently, connectivity is evaluated by measuring the magnitude strength, duration and time delays from the electrical recording of electrical fields of the brain produced by the excitable medium.

Unlike dipole propagation, connectivity does not occur at the speed of light and is best measured when there are time delays; in fact, electrical volume conduction is not a property of an EEG excitable medium and occurs at zero time delay. This very important property of the excitable medium sources versus electrical properties means that time delays determine whether or not and to what extent an excitable medium is responsible for the electrical potentials measured at the scalp surface.

Adaptive Processing of Brain Signals, First Edition. Saeid Sanei.
© 2013 John Wiley & Sons, Ltd. Published 2013 by John Wiley & Sons, Ltd.

So, volume conduction defined at zero phase lag is related to one type of EEG sources and the second type is lagged correlations related to an excitable medium.

The network of the latter sources is often considered as coupled oscillators. This is mainly because electrical potentials are ionic fluxes across polarized membranes of neurons with intrinsic rhythms and driven rhythms (self-sustained oscillations) [1–3].

Volume conduction involves near-zero phase delays between any two points within the electrical field as collections of dipoles oscillate in time [2]. Zero phase delay is one of the important properties of volume conduction. Therefore, measures such as the cross-spectrum, coherence, bi-coherence and coherence of phase delays become crucial and significant in evaluating brain connectivity independent of volume conduction.

For this purpose, correlation coefficient methods, such as Pearson product correlation (e.g. "co-modulation" and "Lexicor correlation") do not compute phase and thus are incapable of controlling volume conduction. The use of the above approaches (bispectrum, etc.) for the study of brain connectivity is not only due to the ability to control volume conduction but also the need to measure the fine temporal details and temporal history of coupling or "connectivity" within and between different regions of the brain. From the physiology of the brain, both the thalamus and septo-hippocampal systems are approximately beneath and in the centre of the brain and contain "pacemaker" neurons and neural circuits that regularly synchronize widely disparate groups of cortical neurons [4].

The cross-spectrum is the sum of both the in-phase (i.e. cospectrum) and out-of-phase (i.e. quadspectrum) potentials. The in-phase component contains volume conduction and the synchronous activation of local neural generators. The out-of-phase component contains network or connectivity contributions from locations distant to a given source. In other words, cospectrum reflects volume conduction and quadspectrum represents non-volume conduction.

The cross-spectrum of coherence and the phase difference can distinguish between volume conduction and network zero phase differences that is, produced by the thalamus or the septal hippocampus-entorhinal cortex, and so on.

The Pearson product correlation ("comodulation" and Lexicor "spectral correlation coefficient") is one of the earliest measures of brain connectivity. This coefficient is often used to estimate the degree of association between EEG amplitudes or magnitudes over intervals of time and frequency [4]. The Pearson product correlation coefficient is not used to calculate a cross-spectrum and, therefore, it is neither used to calculate the phase nor involved in the measurement of phase relationship consistency such as with coherence and bispectrum.

Coherence and the Pearson product correlation coefficient, however, are statistical measures with similar accuracy and statistical significance. The Pearson coefficient is a valid and important normalised measure of coupling which has been used over 40 years. In recent works, application of the Pearson product correlation coefficient (PCC) for magnitude has been called "comodulation" [5].

"Spectral correlation" or "spectral amplitude correlation" are more popular terms while comodulation is a limited term since it fails to refer to the condition of a third source affecting the two other sources without these two latter sources being directly connected. In addition comodulation cannot correct for volume conduction. "Comodulation" has a different meaning than "synchronization" [5] and to reduce confusion the term correlation or PCC is used.

Further research by Blinowska et al. in the 1990s [6–8] demonstrated how the directionality of cortical signal patterns changes within the brain using multivariate auto-regressive (MVAR) modelling followed by directed transfer functions (DTF). In some applications, such as detection and classification of finger movement, it is very useful to discover how the associated movement signals propagate within the neural network of the brain. There is a consistent movement of the source signals from occipital to temporal regions. It is also apparent that during mental tasks different regions within the brain communicate with each other. The interaction and cross-talk amongst the EEG channels may be the only clue to understanding this process. This requires recognition of transient periods of synchrony between various regions in the brain. These phenomena are not easy to observe by visual EEG inspection. Therefore, some signal processing techniques have to be used in order to infer such causal relationships. One time series is said to be causal to another if its inherent information enables the prediction of other time series. In some approaches to the evaluation of connectivity, coherency, or synchronization of brain regions the spatial statistics of scalp EEG are presented as coherence in individual frequency bands, these coherences result from both correlations amongst neocortical sources and volume conduction through tissues of the head, that is, brain, cerebrospinal fluid, skull and scalp.

One way to select the cortical areas is based on a list of ROIs included in the so-called parieto-frontal integration theory (P-FIT [9]), which describes the regions of the human brain involved in intelligence and reasoning tasks. This P-FIT model includes the dorsolateral prefrontal cortex (i.e. Brodmann areas 10), the superior (Brodmann area 7) parietal lobe, and the anterior cingulate (Brodmann area 32). In addition, there is evidence that the most caudal region of the medial frontal cortex, containing cingulate motor areas (CMA) is involved in movements of the hands and other body parts [10]. In particular, the activity in the regions including CMA has also been related directly to behavioural response rate [11].

8.2 Connectivity Through Coherency

Spectral coherence [12] is a common method for determination of synchrony in EEG activity. Coherency is a normalized form of cross-spectrum and is given as:

$$Coh_{ij}^2(\omega) = \frac{E\left\{\left|C_{ij}(\omega)\right|^2\right\}}{E\left\{C_{ii}(\omega)\right\} E\left\{C_{jj}(\omega)\right\}} \tag{8.1}$$

where $C_{ij}(\omega) = X_i(\omega)X_j^*(\omega)$ is the Fourier transform of cross-correlation coefficients between channel i and channel j of the EEGs. Figure 8.1 shows an example of the cross-spectral coherence around 1 s prior to finger movement. A measure of this coherency, such as an average over a frequency band, is capable of detecting zero time lag synchronization and fixed time non-zero time lag synchronization, possibly occurring when a significant delay between the two neuronal population sites exists [13]. Nonetheless, it does not provide any information on the directionality of the coupling between the two recording sites.

Granger causality (also called as Wiener–Granger causality) [14] is another measure which attempts to extract and quantify directionality from EEGs. Granger causality is based on bivariate AR estimates of the data. In a multichannel environment this causality is calculated from pair-wise combinations of electrodes. This method has been used to evaluate the directionality of the source movement from the local field potential in the cat's visual system [15].

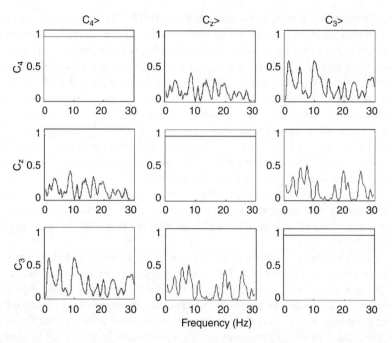

Figure 8.1 Cross-spectral coherence for a set of three electrode EEGs, 1 s prior to right-finger movement. Each block refers to one electrode. By careful inspection of the figure, one can observe same waveform transferred from C_z to C_3

Based on Granger causality, if past samples of a time series $y(t)$ can be used in prediction of another series $x(t)$ then, $y(t)$ is said to cause $x(t)$. In order to determine an estimate of this causality consider

$$e(t) = x(t) - \sum_{i=1}^{p} \breve{a}(i)x(t-i) \tag{8.2}$$

and

$$e_1(t) = x(t) - \sum_{i=1}^{p} a_{11}(i)x(t-i) - \sum_{i=1}^{p} a_{12}(i)y(t-i) \tag{8.3}$$

which also implies

$$e_2(t) = y(t) - \sum_{i=1}^{p} a_{22}(i)y(t-i) - \sum_{i=1}^{p} a_{21}(i)x(t-i) \tag{8.4}$$

The Granger causality index is therefore defined as:

$$GCI_{1-2} = \ln(e/e_1) \tag{8.5}$$

Such a definition can easily be extended to the multichannel case when the included channels change the residual variance ratios. This is a time domain measure. The Granger causality measure, however, is sensitive to arbitrary mixtures of independent noise.

8.3   Phase-Slope Index

To overcome the effect of noise and establish a more robust measure of connectivity the phase-slope index (PSI) was introduced [16]. The main idea behind PSI is that the cause precedes the effect in time and hence the slope of phase of the cross-spectrum between two time series reflects the directionality [17]. The cross-spectrum of two signals $z_i[n]$ and $z_j[n]$ is defined as:

$$S_{ij}(f) = E[Z_i(f)Z_j^*(f)] \tag{8.6}$$

Using this definition the complex coherence is defined as:

$$C_{ij}(f) = \frac{S_{ij}(f)}{\sqrt{S_{ii}(f)S_{jj}(f)}} \tag{8.7}$$

The unnormalized PSI is then measured in terms of the complex cross-spectrum as [16]:

$$\tilde{P}_{ij} = \mathrm{Im}\left(\sum_{f \in B} C_{ij}^*(f)C_{ij}(f + \delta f) \right) \tag{8.8}$$

where B is the desired frequency band. Often PSI is normalized with respect to its variance as [17]:

$$P_{ij} = \tilde{P}_{ij}/\mathrm{var}(\tilde{P}_{ij}) \tag{8.9}$$

It has been shown that the absolute values of $P_{ij} > 2$ are significant. This measure, however, has been applied to ECoG signals rather than scalp EEG signals and it gives higher connectivity between the brain regions which are more synchronised.

8.4   Multivariate Directionality Estimation

However, application of the Granger causality for multivariate data in a multichannel recording is not computationally efficient [15, 18]. DTF [19], as an extension of Granger causality, is obtained from multichannel data and can be used to detect and quantify the coupling directions. The advantage of DTF over spectral coherence is that it can determine the directionality in coupling when the frequency spectra of the two brain regions overlap. The DTF has been adopted by some researchers to determine the coupling directionality [20, 21], since a directed flow of information or cross-talk between the sensors around the sensory motor area before finger movement has been demonstrated [22]. DTF is based on fitting the EEGs to an

MVAR model. Assuming $\mathbf{x}(n)$ is an M-channel EEG signal, then in vector form, it can be modelled as

$$\mathbf{x}(n) = -\sum_{k=1}^{p} \mathbf{L}_k \mathbf{x}(n-k) + \mathbf{v}(n) \tag{8.10}$$

where n is the discrete time index, p the prediction order, $\mathbf{v}(n)$ zero-mean noise and $\mathbf{L}_k$ is generally an $M \times p$ matrix of prediction coefficients. A similar method to the Durbin algorithm for single channel signals, namely the Levinson–Wiggins–Robinson (LWR) algorithm is used to calculate MVAR coefficients [23]. The Akaike AIC criterion [24] is also used for estimation of the prediction order p. By multiplying both sides of the above equation by $\mathbf{x}^T(n-k)$ and performing statistical expectation the following Yule–Walker equation is obtained [25].

$$\sum_{k=0}^{p} \mathbf{L}_k \mathbf{R}(-k+p) = 0; \quad \mathbf{L}_0 = 1 \tag{8.11}$$

where $\mathbf{R}(q) = E[\mathbf{x}(n)\mathbf{x}^T(n+q)]$ is the covariance matrix of $\mathbf{x}(n)$. Cross-correlation of the signal and noise is zero since they are assumed to be uncorrelated. Similarly, the noise autocorrelation is zero for nonzero shift since the noise samples are uncorrelated. The data segment is considered short enough for the signal to remain statistically stationary within that interval and long enough to enable accurate measurement of the prediction coefficients.

8.4.1   Directed Transfer Function

Given the MVAR model coefficients, a multivariate spectrum can be obtained. Here it is assumed that the residual signal, $\mathbf{v}(n)$, is white noise. Therefore,

$$\mathbf{L}_f(\omega)\mathbf{X}(\omega) = \mathbf{V}(\omega) \tag{8.12}$$

where

$$\mathbf{L}_f(\omega) = \sum_{m=0}^{p} \mathbf{L}_m e^{-j\omega m} \tag{8.13}$$

and $\mathbf{L}(0) = \mathbf{I}$. Rearranging the above equation and replacing noise by $\sigma_v{}^2\mathbf{I}$ yields

$$\mathbf{X}(\omega) = \mathbf{L}_f^{-1}(\omega) \times \sigma_v^2\mathbf{I} = \mathbf{H}(\omega) \tag{8.14}$$

which represents the model spectrum of the signals or transfer matrix of the MVAR system. The DTF or the causal relationship between channel i and channel j can be defined directly from the transform coefficients [18] given by

$$\Theta_{ij}^2(\omega) = \left| H_{ij}(\omega) \right|^2 \tag{8.15}$$

Electrode i is causal to j at frequency f if

$$\Theta_{ij}^2(\omega) > 0 \tag{8.16}$$

A time-varying DTF can also be generated (mainly to track the source signals) by calculating the DTF over short windows to achieve the short-time DTF (SDTF) [18].

As an important feature in classification of left and right finger movements, or tracking the mental task related sources, SDTF plays an important role. Some results of using SDTF for detection and classification of finger movement have been given in Chapter 13.

In [26], and consequently in [27] another measure called direct DTF (dDTF) has been defined as a multiplication of a modified DTF by partial coherence:

$$\xi_{ij}^2(f) = \frac{C_{ij}^2(f)|H_{ij}(f)|^2}{\sum\limits_{f}\sum\limits_{m=1}^{k}|H_{im}(f)|^2} \tag{8.17}$$

where $C_{ij}(f)$ is the partial coherence defined as

$$C_{ij}(f) = \frac{M_{ij}(f)}{\sqrt{M_{ii}(f)M_{jj}(f)}} \tag{8.18}$$

and $M_{ii}(f)$ and $M_{ij}(f)$ are matrices of spectra and cross-spectra, respectively.

Moreover, distinction of direct from indirect transmission, in the case of signals from implanted electrodes, is essential. Both DTF and dDTF show propagation when there is a phase difference between the signals. Based on the phase coherence the delay between the onsets of similar frequency components is estimated and it reveals the direction of propagation. The phase, however, should vary within 2π (or modulo 2π), otherwise the directionality may be misjudged.

Partial coherence analysis has been used to determine graphical models for brain functional connectivity [28]. It has been investigated that the outcome of such analysis may be considerably influenced by factors such as the degree of spectral smoothing, line and interference removal, matrix inversion stabilization and the suppression of effects caused by side-lobe leakage, combination of results from different epochs and people, and multiple hypothesis testing. Another similar measure of coherency, called partial directed coherence (PDC), has also been defined as [27, 29]

$$P_{ij}(f) = \frac{A_{ij}(f)}{\sqrt{a_j^*(f)a_j(f)}} \tag{8.19}$$

Where $A_{ij}(f)$ is an element of $A(f)$, a Fourier transform of MVAR model coefficients $\mathbf{A}(f)$, where $\mathbf{a}_j(f)$ is the jth column of $\mathbf{A}(f)$ and the asterisk denotes a vector hermitian operator. PDC represents only direct flows between channels [27]. Finally, a generalised PDC approach has been defined as [27]

$$GP_{ij}(f) = \frac{A_{ij}(f)}{\sum\limits_{i=1}^{k}\left|A_{ij}(f)\right|^2} \tag{8.20}$$

which is used for connectivity estimation. These methods can be applied to consecutive segments of the signals for approximation of the dynamics of the signal propagation. Alternatively SDTF can be used instead of DFT.

8.5   Modelling the Connectivity by Structural Equation Modelling

As described in [30, 31], connectivity can be modelled and covariance between the node signals can be compared with real measurements.

Structural equation modelling (SEM) is meant to set up an a priori connectivity model and aims to answer (i) what the influence of a variable signal-to-noise ratio (SNR) level is on the accuracy of the pattern connectivity estimation obtained by SEM, (ii) what the amount of data necessary to get good accuracy for connectivity estimation between cortical areas should be, and (iii) how the SEM performance is degraded by an imprecise anatomical model formulation; is it able to perform a good estimation of connectivity pattern when connections between the cortical areas are not correctly assumed, and (iv) which kind of errors should be avoided. SEM has also been used to model such activities from high resolution (both spatial and temporal) EEG data. Anatomical and physiological constraints have been exploited to change an underdetermined set of equations to a determined one.

SEM consists of a set of linear structural equations containing observed variables and parameters defining causal relationships amongst the variables [32]. The variables can be endogenous (i.e. independent from the other variables in the model) or exogenous (independent from the model itself). Considering a set of variables (expressed as deviations from their means) with N observations, estimation of SEM for these variables may be expressed as:

$$\mathbf{y} = \mathbf{B}\mathbf{y} + \mathbf{\Gamma}\mathbf{x} + \mathbf{\xi} \tag{8.21}$$

where, $\mathbf{y}$ is an $m \times 1$ vector of dependent (endogenous) variables, $\mathbf{x}$ is an $n \times 1$ vector of independent (exogenous) variables, $\mathbf{\xi}$ is the $m \times 1$ vector of equation errors (random disturbances), and $\mathbf{B}$ ($m \times m$) and $\mathbf{\Gamma}$ ($m \times n$) are respectively the coefficient matrices of the endogenous and exogenous variables. $\mathbf{\xi}$ is assumed to be uncorrelated with the data and $\mathbf{B}$ to be a zero-diagonal matrix. If $\mathbf{z}$ is a vector containing all the $p = m + n$ exogenous and endogenous variables in the following order:

$$\mathbf{z}^{\mathrm{T}} = [x_1 \cdots x_n \; y_1 \cdots y_m] \tag{8.22}$$

The observed covariances can be expressed as

$$\Sigma_{\mathrm{obs}} = \frac{1}{N-1}\mathbf{Z}.\mathbf{Z}^{\mathrm{T}} \tag{8.23}$$

where $\mathbf{Z}$ is a $p \times N$ matrix of p observed variables for N observations. The covariance matrix implied by the model can be obtained as follows:

$$\Sigma_{\mathrm{mod}} = \begin{bmatrix} E[\mathbf{x}\mathbf{x}^{\mathrm{T}}] & E[\mathbf{x}\mathbf{y}^{\mathrm{T}}] \\ E[\mathbf{y}\mathbf{x}^{\mathrm{T}}] & E[\mathbf{y}\mathbf{y}^{\mathrm{T}}] \end{bmatrix} \tag{8.24}$$

where if $E[\mathbf{x}\mathbf{x}^{\mathrm{T}}] = \mathbf{\Phi}$, then

$$E[\mathbf{x}\mathbf{y}^{\mathrm{T}}] = ((\mathbf{I} - \mathbf{B})^{-1} \mathbf{\Gamma}\mathbf{\Phi})^{\mathrm{T}} \tag{8.25}$$

$$E[\mathbf{y}\mathbf{x}^{\mathrm{T}}] = (\mathbf{I} - B)^{-1} \mathbf{\Gamma}\mathbf{\Phi} \tag{8.26}$$

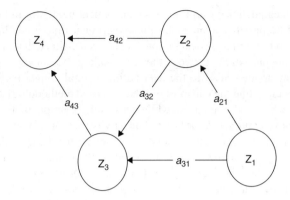

Figure 8.2 Connectivity pattern imposed in the generation of simulated signals. Values on the links represent the connection strength

and

$$E[\mathbf{y}\mathbf{y}^{\mathrm{T}}] = (\mathbf{I} - \mathbf{B})^{-1} (\mathbf{\Gamma}\mathbf{\Phi}\mathbf{\Gamma}^{\mathrm{T}} + \mathbf{\Psi})((\mathbf{I} - \mathbf{B})^{-1})^{\mathrm{T}} \tag{8.27}$$

where $\mathbf{\Psi} = \mathrm{E}[\boldsymbol{\xi}\boldsymbol{\xi}^{\mathrm{T}}]$. With no constraints, the problem of minimisation of the differences between the observed covariances and those implied by the model is underdetermined, mainly because the number of variables ($\mathbf{B}$, $\mathbf{\Gamma}$, $\mathbf{\Phi}$ and $\mathbf{\Psi}$) is greater than the number of equations ($m + n$)($m + n + 1$)/2. The significance of SEM is that it eliminates some of the connections in the connectivity map based on some a priori anatomical and physiological (and possibly functional) information. For example, in Figure 8.2 if the connection a_{42} is not in the hypothesised model it may be set to zero.

So, if r is the number of parameters to be estimated, we will need to have $r \leq (m + n)(m + n + 1)/2$. The parameters are estimated by minimising a function of the observed and implied covariances. The most widely used objective function for SEM is the maximum-likelihood (ML) function [30];

$$F_{\mathrm{ML}} = \log|\Sigma_{\mathrm{mod}}| + \mathrm{tr}\left(\Sigma_{\mathrm{obs}}.\Sigma_{\mathrm{mod}}^{-1}\right) - \log|\Sigma_{\mathrm{obs}}| - p \tag{8.28}$$

where p is the number of observed variables (endogenous + exogenous).

For multivariate normally distributed variables, the minimum of the ML function multiplied by $N - 1$, approximates a χ^2 distribution with ($p(p + 1)/2) - t$ degrees of freedom, where t is the number of parameters to be estimated and p the total number of observed endogenous and exogenous variables. The χ^2 statistics test can then be used to infer the statistical significance of the structural equation model obtained. LISREL [33], a publicly available software, may be used to implement the SEM for brain connectivity.

In [30] SEM has been applied to EEG, possibly for the first time, and the effect of noise and data length on the accuracy of the results and the modelling error have been examined.

In this experiment the authors investigated the influence of SNR on the accuracy of the pattern connectivity estimation obtained by SEM, the amount of EEG data necessary to achieve an acceptable accuracy of the estimation of connectivity between cortical areas,

SEM performance degradation by an imprecise anatomical model formulation, and the error type which should be possibly avoided. In this experiment they used simulated models of connectivity between four brain cortical regions. It has been claimed that the proposed method retrieved the cortical connections between the areas under different experimental conditions. In the next trial they applied SEM to the data during a simple finger tapping experiment in humans, in order to underline the capability of the proposed methodology to draw patterns of cortical connectivity between brain areas while performing a simple motor task.

In most of the above approaches for brain connectivity measure involving distributed brain sources, application of a realistic head model is favourable. A true head model helps in accurate localization (and tracking) of the sources involved in the connectivity measure. Although solving the inverse problem, such as in [3, 34], and using fMRI [35] have been used for this purpose, each has its own drawbacks. The inverse problem is solved for localization of dipole sources and cannot track the distributed sources adequately. On the other hand, fMRI reveals information about the oxygenation of blood vessels as a consequence of event-related activity and cortical activations. It also has a long time lag of 2–5 s with respect to the onset of the event or the activity (such as movement-related) of the cortex.

In [36] both SEM and DTF have been implemented for estimation of the connectivity from EEG movement-related potentials. The results of the application of the SEM method for estimation of the connectivity show the statistically significant cortical connectivity patterns obtained for the period preceding the movement onset in the alpha frequency band. The connectivity pattern during the period preceding the movement in the alpha band involves mainly the left parietal region, functionally connected with the left and right premotor cortical regions, the left sensorimotor area, and both the prefrontal brain regions.

The stronger functional connections correspond to the link between the left parietal and the premotor areas of both cerebral hemispheres. After the preparation and the beginning of finger movement, the changes in connectivity pattern can be noted. In particular, the origin of the functional connectivity links is positioned in the sensorimotor left cortical areas. From there, functional links are established with left prefrontal and both premotor areas. A functional link in this condition connects the right parietal area with the right sensorimotor area. The left parietal area, much active in the previous condition, was linked instead with the left sensorimotor and right premotor cortical areas.

As stated previously, the origin of functional connectivity links is positioned in the sensorimotor left cortical areas (SMl). From there, functional links are established between left prefrontal and both premotor areas (left and right) [36]. Model order, number and locations of the regions of interest are determined before applying the algorithm.

8.6 EEG Hyper-Scanning and Inter-Subject Connectivity

8.6.1 Objectives

The aim here is to design a robust cooperative BCI system which estimates the states and connectivity of the brains of two or more subjects while performing cooperative or competitive tasks. Simultaneous EEG recording from more than one subject's brain is referred to as *EEG hyper-scanning*. Interpersonal body movement synchronization has been observed widely. For example, often it is experienced that one's footsteps unconsciously synchronized with those of his friend while walking together. On the other hand, in a cooperative or competitive performance one's action is the response to the action of other(s). This can be examined in the case of team working or one competing against another player in a match. Such mechanisms of body movement coherency and their relation to implicit social interaction remain obscure.

8.6.2   Technological Relevance

Very recently, a study and measurement of fingertip movement between two participants while recording their EEG simultaneously have been introduced [37]. In particular, the aim has been to evaluate body movement synchrony and implicit social interactions between two or more participants, and assess the underlying dynamics and the effective connectivity between and within their brain regions. The concurrent activity in multiple brains of the group may be estimated and the causal connections between regions of different brains (hyper-connectivity) indicated.

The study of concurrent and simultaneous brain activities of different subjects while they perform cooperative or competitive tasks is very important in investigating paradigms in many cases where the knowledge of simultaneous interactions between individuals has a value. A person's ability to concentrate, predict, cooperate, follow, compete, learn, and engage in long and tedious tasks, for both normal subjects and patients with various brain abnormalities, such as Alzheimer's, mental fatigue, and during drug infusion, may effectively be studied from simultaneous EEG records of subjects cooperating with or competing against each other.

Most recently, methods established for brain connectivity measures, from single subject EEG recordings, have been extended and applied to simultaneous multiple brain recordings [38–43]. The recorded data were processed in different ways. In a method based on the time–frequency domain approach, changes in amplitude as well as synchronization of the EEG rhythms were exploited in the conventional frequency bands of delta, theta, alpha, and beta. These approaches aim at estimation of coherency or synchronisation between the recorded EEGs from multiple brains.

The second approach, which is based on connectivity or synchronization measures, allows an estimation of concurrent activity in multiple brains and connections between regions of different brains (so called hyper-connectivity). Granger causality was suggested to be used in this approach. A variety of techniques, such as MVAR followed by DTF, could also be used. The aforementioned approaches may also be combined to increase the effectiveness of the system.

As a good example it has been shown that Alzheimer's Disease (AD) is closely related to alteration in the functional brain network, that is, functional connectivity between different brain regions [44]. Both within-lobe connectivity and between-lobe connectivity as well as between hemisphere connectivity have been estimated for AD, normal control (NC), and mild cognitive impairment (MCI). The temporal lobe of AD has a significantly lesser amount of direct within-lobe connectivity than NC. This direct link within the temporal lobe may be attenuated by AD. It has also been shown that the hippocampus and parahippocampal are much more separated from other regions in AD than in NC. The temporal lobe of MCI, however, does not show a significant decrease in the amount of direct connections, compared with NC. The frontal lobe of AD shows considerably more connectivity than NC. This has been interpreted as compensatory reallocation or recruitment of cognitive resources, as denoted in

[44] and the references herein. There is no significant difference between AD, MCI, and NC in terms of connectivity within the parietal lobe and within the occipital lobe.

In terms of between-lobe connectivity, in general, human brains tend to have fewer between-lobe connections than within-lobe connections. In addition, AD has significantly more parietal-occipital direct connections than NC. An increase in the amount of connections between the parietal and occipital lobes of AD has previously been reported [45]. It may also be interpreted as a compensatory effect. Furthermore, MCI also shows an increase in the amount of direct connections between parietal and occipital lobes compared with NC, but the increase is not as significant as AD. While the amount of direct connections between frontal and occipital lobes shows little difference between AD and NC, for MCI it shows a significant decrease. Also, AD causes less temporal–occipital and frontal–parietal connections, but a higher parietal-temporal connectivity than NC.

Finally, between hemisphere connectivity estimation results indicate that AD disrupts the strong connection between the same regions in both the left and right hemispheres, whereas this disruption is not significant in MCI [44].

For the same objectives another research was carried out by Escodero *et al.* [46] for differentiation between NC, AD and MCI from MEG signals. In this work the scalp electrode space has been divided into nine regions, as illustrated in Figure 8.3. Instead of measuring the connectivity directly, MEG signals have been processed using empirical mode decomposition and blind source separation first, for NC, AD and MCI recorded signals.

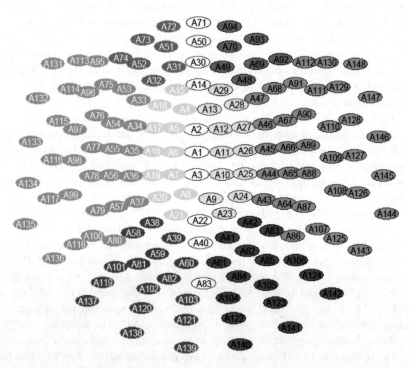

Figure 8.3 Nine regions of the brain used for measuring connectivity using spectral coherency (see Plate 6 for the coloured version)

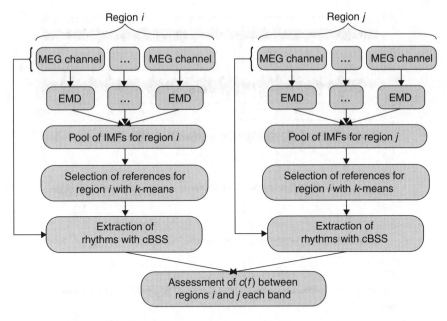

Figure 8.4 Block diagram of the spectral coherency, $c(f)$, measure

In the proposed algorithm empirical mode decomposition (EMD) decomposed each MEG channel into a set of intrinsic mode functions (IMFs). Then, a constraint blind source separation block extracts the representative delta, theta, alpha and beta rhythms of each region In addition to the corresponding spectral coherence, magnitude square coherence (MSC) is used here to measure the brain synchrony. It quantifies linear correlations as a function of frequency. MSC is bounded between 0 and 1. This measure can detect the linear synchronisation between two signals, but it does not discriminate the directionality of the coupling [46].

A simplified block diagram of the algorithm has been presented in Figure 8.4. In Figure 8.5 a channel of an MEG signal can be viewed. Figure 8.6 illustrates the corresponding IMFs computed using the EMD algorithm.

To select the right IMFs in terms of centre frequency from each region, the IMFs have been clustered based on their amplitudes, frequency and proximity to any one of the four centre

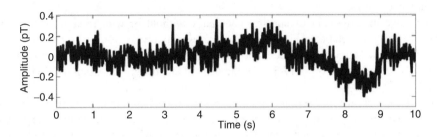

Figure 8.5 One channel of MEG signals

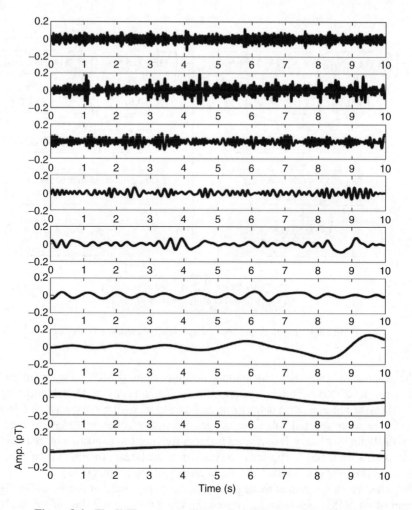

Figure 8.6 The IMFs corresponding to the MEG channel in Figure 8.4

frequencies. As depicted in Figure 8.7 the k-mean algorithm has been used to select a reference signal frequency for each band. This is necessary since the frequency band centres change from subject to subject.

Quantitative measurement of spectral coherency without or with regional extraction of rhythms has led tothe charts in Figure 8.8a and b, respectively.

As a result, raw MEG measurements of subject group pairs produced $c(f)$ accuracy rates of 51.6–61.1%, compared to 66.7–77.3% for the pairs of subject groups when the rhythms were extracted.

None of these approaches (including all previously used for brain connectivity estimation) takes into account any model of such processes where the brain attempts to achieve certain objectives, in a concise manner, considering evolutions in both time and (electrode) space, particularly when one brain sets the goals and the other brain follows.

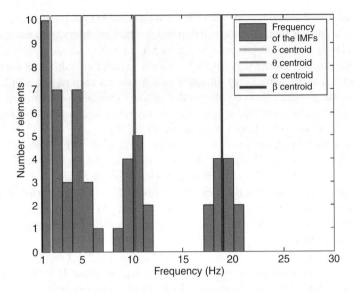

Figure 8.7 Reference selection using the k-mean algorithm

8.7   State-Space Model for Estimation of Cortical Interactions

Most of the above approaches for the estimation of brain cortical connectivity from EEG or MEG suffer from noise involved in the measurements. Therefore, it would be useful if a suitable technique could be developed to improve the estimation of the prediction/connectivity parameters and be more robust against noise [47].

Nalatore *et al.* [48] developed a state-space approach to MVAR model estimation for invasive electrophysiological recordings and show that explicitly modelling noise results in improved connectivity estimates. Shumway and Stoffer [49] introduced an expectation maximization approach to ML parameter estimation in state-space MVAR models.

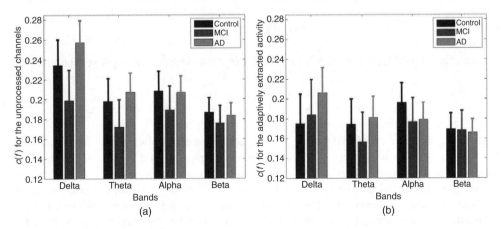

Figure 8.8 Spectral coherency levels without (left) and with (right) regional rhythms extraction

In [50] a state-space model has been developed to estimate MVAR parameters. Such a model can represent the MVAR model of cortical dynamics, while an observation equation describes the physics relating the cortical signals to the measured EEG and the presence of spatially correlated noise. In this model it is assumed that the cortical signals originate from known regions of cortex, but the spatial distribution of activity within each region is unknown. Then, the MVAR parameters have been computed through an expectation-maximization approach. The algorithm also calculates the spatial activity distribution components, and the spatial covariance matrix of the noise from the measured EEG [50].

The state-space approach does not circumvent the limitations of EEG and MEG regarding noise, but it uses a maximum likelihood (ML) criterion to solve for MVAR parameters from the signals. ML estimates are known to be asymptotically unbiased with variance approaching the Cramer–Rao lower bound as data length increases. Therefore, in a two-stage approach first, the cortical sources are estimated by solving a variant of the inverse problem and next, an MVAR model is fitted to the estimated cortical signals [50]. Mathematically, the process can be expressed as follows:

Consider $\mathbf{x}_{n,j} = [x_{n,j}^1, \ldots, x_{n,j}^M]^T$, $n = 1, \ldots, N$, $j = 1, \ldots, J$ being the jth trial of an $M \times 1$ state vector representing samples of cortical signals from M regions at time n. An MVAR of order P for representing the cortical signals $\mathbf{x}_{n,j}$ can be described as

$$\mathbf{x}_{n,j} = \sum_{p=1}^{P} \mathbf{A}_p \mathbf{x}_{n-p,j} + \mathbf{w}_{n,j} \tag{8.29}$$

where $\mathbf{A}_p$ is the matrix of prediction coefficients. This can be rewritten as:

$$\mathbf{x}_{n,j} = \mathbf{A}\mathbf{z}_{n-1,j} + \mathbf{w}_{n,j} \tag{8.30}$$

where $\mathbf{A} = [\mathbf{A}_1, \ldots, \mathbf{A}_P]$ is an $M \times MP$ matrix of prediction coefficients and $\mathbf{z}_{n-1,j} = [x_{n-1,j}^T, \ldots, x_{n-P,j}^T]^T$ as an $MP \times 1$ vector containing the past P state vectors. Here it is assumed that the initial vector $\mathbf{z}_{0,j}$ is Gaussian distributed with unknown mean μ_0 and unknown covariance matrix $\mathbf{\Sigma}^0$. The dynamical state-space model for an observation $\mathbf{y}_{n,j}$ is presented as:

$$\mathbf{y}_{n,j} = \mathbf{C}\mathbf{\Lambda}\mathbf{x}_{n,j} + \mathbf{v}_{n,j} \tag{8.31}$$

This can be combined with the MVAR model in (8.27) to have the state-space equations [50]

$$\begin{aligned} \mathbf{x}_{n,j} &= \mathbf{A}\mathbf{z}_{n-1,j} + \mathbf{w}_{n,j} \\ \mathbf{y}_{n,j} &= \mathbf{C}\mathbf{\Lambda}\mathbf{x}_{n,j} + \mathbf{v}_{n,j} \end{aligned} \tag{8.32}$$

where $\mathbf{w}_{n,j}$ and $\mathbf{v}_{n,j}$ are $M \times 1$ and $L \times 1$ vectors of state and observation, respectively, $\mathbf{C}$ is the $L \times MF$ forward matrix of source to electrode mapping, and $\mathbf{\Lambda}$ is an $MF \times M$ block diagonal matrix with $F \times 1$ vectors $\mathbf{\Lambda}_m$ on the diagonal, representing any unknown parameters describing the spatial activity of the mth region [47]. Although such a presentation is not unique, the non-uniqueness problem does not affect the model parameter estimations [50]. An ML approach including expectation maximization can then be employed to estimate the model (8.31) parameters, including $\mathbf{A}$ which represents the connectivity parameters.

A similar state-space method has been applied to a radial basis function (RBF)-based connectivity model. An MVAR of order P for representing the cortical signals $\mathbf{x}_{n,j}$ can be described as in [51]. This model is important since it complies with the invariance property required to evaluate Granger causality. Also, the universal approximation theorem [52] states that an RBF network is capable of approximating any smooth function to an arbitrary degree of accuracy [47]. Therefore, RBF is able to model nonlinear dynamics of cortical signals. RBF has been used to estimate nonlinear Granger causality for intracranial EEG signal analysis [51]. To use RBF for estimation of the cortical connectivity between M cortical regions we have [47]

$$\mathbf{x}_{n,j} = \mathbf{\Psi} \Phi(\mathbf{Z}_{n-1,j}) + \mathbf{w}_{n,j} \tag{8.33}$$

Where $\mathbf{\Psi}$ is an $M \times MI$ matrix of RBF weights, and $\Phi(\mathbf{Z}_{n-1,j}) = [\varphi^1(\mathbf{z}_{n-1,j}^1)^{\mathrm{T}}, \ldots, \varphi^M(\mathbf{z}_{n-1,j}^M)^{\mathrm{T}}]^{\mathrm{T}}$ is an $MI \times 1$ vector of nonlinear functions of P variables, where $\varphi^m = [\varphi_1^m, \ldots, \varphi_I^m]$ is an $I \times 1$ vector of RBF kernels [47]. The elements of $\mathbf{\Psi}$ reflect the coupling between brain cortical regions, which can be estimated from the cortical signals by solving a system of linear equations.

8.8 Application of Adaptive Filters

Distributed sources assumption of movement-related cortical sources and the dynamics of synaptic currents motivates a new solution for investigation of brain connectivity. Recent evidences from a finger movement experiment show that the neurons in both contralateral and ipsilateral brain lobes are engaged in the movement [53]. However, there is no further investigation about the dynamics of the movement in this literature. In this section a different and novel approach based on cooperative adaptive filtering theory, incorporating *diffusion adaptation* [54–56], is introduced. The initial research on this system was established by Saeid Sanei and his research team at the University of Surrey (UoS), UK in collaboration with Ali Sayed in the University of California in Los Angeles (UCLA), USA. The fundamental aim of this research is to model the brain dynamics (i.e. varying connectivity in both time and space) for each particular body movement. There are enormous applications for this design, such as brain behaviour during prolonged movement for healthy and Parkinson subjects and EEG hyper-scanning.

In the original work on diffusion adaptation [57–59] a simple strategy in sharing information was introduced. In this interesting work both time and space evolutions have been exploited in the design of an adaptive filter. Therefore, a response from a node in a network corresponds not only to the input (or stimulus) of the network but also to the network model, including cooperation of nearby nodes, which evolves in time.

Adaptive networks in mobile communications consist of collection of nodes with learning and motion abilities that interact with each other locally in order to solve distributed processing and distributed inference problems in real-time. The objective of such a network is to estimate some filter coefficients in a fully distributed manner and in real-time, where each node is allowed to interact only with its neighbours.

In [54] a similar adaptation algorithm, inspired by bacteria mobility, has been developed for adaptation over networks with mobile nodes. The nodes have limited functionalities and are allowed to cooperate with their neighbours to optimise a common objective function. In this elegant approach the nodes do not know the form of the cost function beforehand. They can only sense variations in values of the objective function as they diffuse through the space. A good nature-inspired application example is in sensing the variation in the concentration of nutrients in the environment. In this model, however, sharing the information has been limited to binary choices between run and tumble for the movement of mobile nodes.

Following the above work, a diffusion adaptation algorithm that exhibited self-organization properties was developed and applied to the model of cooperative hunting (of prey fish) amongst predators (sharks) while the fish herd and move toward a nutrition point [56]. In this new application the nodes of the network wish to track the location of the food source and the location of the predator. The modelling equations are therefore applied to both targets.

To better estimate the parameters, such as weights of the links between nodes, other previously described methods, such as SEM, can be used to initialise the nodes and their mutual connectivity (presented by the link weights) levels. Time evolution of the changes in these levels can be estimated while undertaking cooperative or competitive tasks (such as hand movement in a synchronised dance). Then, a constrained diffusion adaptation algorithm is developed to model the brain responses to these tasks performed by two or more human brains to achieve or avoid a particular objective (stimulus).

This approach requires the definition of a number of bio-inspired and physiological constraints for each type of brain activity, and also definition of regions of connectivity and the spatial connectivity likelihoods within each region. Eventually, depending on what the brain responses are, which regions of the brain are involved, how synchronised the brains are and how they follow (or avoid) each other, the diffusion adaptation algorithm can be designed. Here, each measurement point (electrode) will be considered as a node and the expected time–space response of the brain to each stimulus as the target. Hence, from an algorithmic perspective a relatively complex constrained optimization problem is solved for which the optimum filter parameters are subject to decisions by the other cooperating or competing parties.

Subject to accurate definition of the constraints and pre-determination of time–space parameters, such a filter is able to model the time varying interactions between two or more phenomena more accurately.

8.8.1   Use of Kalman Filter

Most techniques in connectivity evaluation consider the signals to be stationary. They also consider the sources separately. Application of adaptive filtering in time and space in so called diffusion adaptation approach tracks a time-space varying process which results in particular function. This involves sources from a number of neurons which communicate to each other through space and evolve in time. An important initial conclusion from this approach is that the signals are considered stationary, but often the system can cope with nonstationary data.

In [5] and [60] a diffusion Kalman filtering approach to cooperative multi-agent sensor networking has been introduced. In most of these applications (excluding decentralised network) it is assumed that the correlation between estimates of two neighbouring agents is known to both agents. In addition, communication between the agents is required to be fast enough such that consensus can be reached between two consecutive Kalman filter updates. In [61] the filtering is based on covariance intersection (CI) method [62]. This approach fuses multiple consistent estimates with unknown correlations with a convex combination of the estimates and chooses the corresponding combination weights by minimizing the trace or determinant of an upper bound of the error covariance matrix [61].

In the diffusion Kalman filtering approach it is assumed that $N_{k,i}$ nodes spatially distributed over a region, centred (so called current node) at k at time i, communicate with each other (within a limited range).

Being at state x_i the measurement at node k and time i is denoted by $y_{k,i}$, the corresponding state-space model is of the form

$$
\begin{aligned}
x_{i+1} &= F_i x_i + w_i \\
y_{k,i} &= H_{k,i} x_i + v_{k,i}
\end{aligned}
\tag{8.34}
$$

where w_i and $v_{k,i}$ are the state and measurement noises at time i and node k, respectively. F_i and $H_{k,i}$ are generally time-varying (but bounded) matrices and the noises are zero-mean, white and have [61]

$$
E\left[\begin{bmatrix} w_i \\ v_{k,i} \end{bmatrix}\begin{bmatrix} w_i \\ v_{k,i} \end{bmatrix}^{\mathrm{T}}\right] = \begin{bmatrix} Q_i \delta_{ij} & 0 \\ 0 & R_{k,i}\delta_{kl}\delta ij \end{bmatrix}
\tag{8.35}
$$

where δ_{ij} is the Kronecker delta, Q_i and $R_{k,i}$ are assumed to be positive definite and bounded.

The diffusion Kalman filtering approach in [60], shown schematically in Figure 8.9, aims at estimating the stable state x_i, while sharing data with its neighbours only. In this algorithm at every time instant i, node k transmits the quantities $H_{k,i}^{\mathrm{T}} R_{k,i}^{-1} H_{k,i}$ and $H_{k,i}^{\mathrm{T}} R_{k,i}^{-1} H_{k,i}$ to its neighbours in updating each intermediate estimate $\psi_{k,i}$. Information from the neighbouring nodes is combined using a $p \times p$ diffusion matrix $C_{k,l,i}$ subject to

$$
\sum_{l \in N_{k,i}} C_{k,l,i} = \mathbf{I}_p, \quad C_{k,l,i} = 0 \text{ for } l \notin N_{k,i}
\tag{8.36}
$$

where $\mathbf{I}_p$ is a $p \times p$ identity matrix. $C_{k,l,i}$ matrices play an important role in the diffusion update. Therefore, identification or estimation of the link weights, either based on physical or

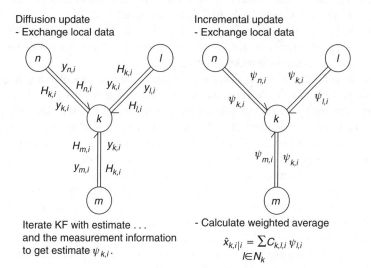

Figure 8.9 Diffusion Kalman filtering algorithm. Taken from [61], © IEEE

physiological constraints or, as in [61], based on the CI algorithm, enhances the accuracy of the overall diffusion filtering.

As mentioned earlier, connectivity estimation methods such as SEM or MVAR may be used to define the link weights. This is currently under research. CI estimation on the other hand, assumes that node k is to fuse the estimates $\hat{x}_{l,i|i-1}(l \in N_{k,i})$ using its neighbours and through the estimates of their error covariance matrices $P_{l,i|i-1} > 0$. Based on the CI algorithm by Julier and Uhlman as reported in [61],

$$\psi_{k,i} = C_{k,l,i}\hat{x}_{l,i|i-1} \tag{8.37}$$

where

$$
\begin{aligned}
C_{k,l,i} &= \beta_{k,l,i}\Lambda_{k,i}P_{l,i|i-1}^{-1} \\
\Lambda_{k,i} &= \left(\sum_{l \in N_{k,i}} \beta_{k,l,i}P_{l,i|i-1}^{-1} \right)^{-1}
\end{aligned}
\tag{8.38}
$$

$0 \le \beta_{k,l,i} \le 1$ where $\sum_{l \in N_{k,i}} \beta_{k,l,i} = 1$ is computed such that the trace or determinant of $\Lambda_{k,i}$ is minimised. The above optimisation is often nonlinear. Therefore, some CI algorithms, such as the one presented in [61], have been proposed to estimate these parameters.

8.8.2   Task-Related Adaptive Connectivity

From the above connectivity estimation methods, it may be concluded that particular connectivity patterns are related to particular mental, cognitive or movement activity. Such patterns are those which vary in both time and space (i.e. the engaged electrodes or actual brain source locations) and originate from distributed synaptic current sources. Modelling of such connectivity patterns can be of great importance in places where the neural activity corresponding to a movement or mental task, such as in BCI, is to be learned.

Diffusion adaptation filters are similar to conventional adaptive filters where the inputs are the sum of the contributions (or cooperation) between the input nodes (sources). The cooperation pattern between the nodes itself varies in time.

As a motivation for this study, very recently, a study and measurement of fingertip movement between two participants, while recording their EEG simultaneously, has been carried out [37]. In particular, the aim was evaluation of body movement synchrony during implicit social interactions between two or more participants, assessing the underlying dynamics and the effective connectivity between and within their brain, mainly cortical regions. Concurrent activity in multiple brains of a group of people may be estimated in order to indicate the causal connections between regions of different brains (hyper-connectivity).

Diffusion adaptation can be a very good candidate to model the evolution of neuronal network connectivity during a task performance. This method is more attractive when one subject performs a task as a model target and others follow. At each stage of adaptation, however, the tradition measures for connectivity, such as DTF, can be used to adjust the instantaneous cooperation weights amongst the nodes (EEG electrodes or brain sources). To understand the concept the method is briefly reviewed here.

8.8.3   Diffusion Adaptation

In diffusion adaptation, the main goal is to estimate an $M \times 1$ unknown vector $\mathbf{w}$ from measurements collected at N nodes spread over a network. Each node k has access to time realisations $\{d_k(i), u_{k,i}\}$ of the data $\{\mathbf{d}_k, \mathbf{u}_k\}$ where $\mathbf{d}_k$ is a vector of scalar measurements and $\mathbf{u}_k$ is a $1 \times M$ row regression vector [55]. Considering these vectors from all the nodes, the data form two global matrices $\mathbf{U}$ and $\mathbf{D}$, respectively. The objective is to estimate vector $\mathbf{w}$ that solves the minimisation problem:

$$\min_{w} E \, |\mathbf{d} - \mathbf{U}\mathbf{w}| \qquad (8.39)$$

Assuming that by applying $\mathbf{w}$ the signal from node k, at time instance i, generates a signal $\psi_k^{(i)}$. Each node combines the information from its neighbouring nodes to produce a new input to the filter as:

$$\varphi_k^{(i-1)} = f_k(\psi_l^{(i-1)}; l \in N_{k,i=1}) \qquad (8.40)$$

where f_k is a combining function and $N_{k,i}$ is the number of nodes included in the neighbourhood of node k at time instance i. An option for selection of the above function is to consider it as a fixed linear averaging process. Another alternative is to consider it as a linear sum of weighted inputs from neighbouring nodes which can be adapted through time to minimise the global error [27]. In both cases adaptation can be presented in the following set of alternating equations:

$$\varphi_k^{(i-1)} = \sum_{l \in N_{k,i=1}} C_{kl} \psi_l^{(i-1)} \qquad (8.41a)$$

$$\psi_k^{(i)} = \varphi_k^{(i-1)} + \mu_k u_{k,i}^* (d_k(i) - u_{k,i}\varphi_k^{(i-1)}) \qquad (8.41b)$$

In equation (8.41a) the new input to the model is estimated using received signals from the neighbouring nodes. In (8.41b) the output of the filter is updated for iteration/time sample i. By implementing the above equations using alternating adaptation we can track the signal component in both space, using k, and time, using i. Alternatively, by replacing $\phi_k^{(i)}$ by $w_{k,i}$ equation (8.41) can be written as:

$$\psi_k^{(i)} = w_{k,i-1} + \mu_k u_{k,i}^* \left(d_k(i) - u_{k,i}w_{k,i-1}\right) \qquad (8.42a)$$

$$w_{k,i} = \sum_{l \in N_{k,i=1}} C_{kl} \psi_l^{(i)} \qquad (8.42b)$$

8.8.4   Application of Diffusion Adaptation to Brain Connectivity

In order to apply this algorithm to brain connectivity we can model the time–space evolution of brain distributed sources for each particular brain activity based on the above concept. One important factor in equation (8.42) is the weights in the cooperative network, C_{kl}. This factor may be estimated using the aforementioned CI method or estimated using other connectivity measurement techniques and training. DTF is perhaps the most straightforward deterministic

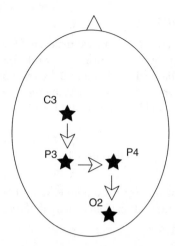

Figure 8.10 Communication path between the nodes (EEG electrodes)

method for estimation of C_{kl}. Consider the following simulation: assume that quick lifting of the right hand causes a cortical signal generated around the C3 electrode and evolves into P3, P4, and O2 as in Figure 8.10. Contribution and communication by other nodes also exist but have been considered negligible.

The simulated signals may look like those in Figure 8.11.

Any node can be considered to be the central node. The target signal can also be considered as delayed and a weighted noiseless version of any given input or, in practice, any output which can be used to move a robotic arm. In any case the convergence of the algorithm is convex and the solution tends to the global minima. As an example, when the target is defined

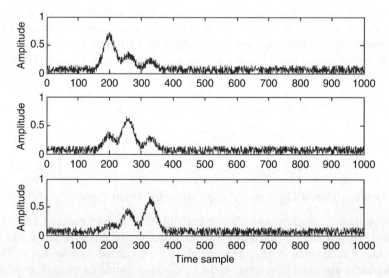

Figure 8.11 Signals from C3, P4, and O2 electrodes, respectively

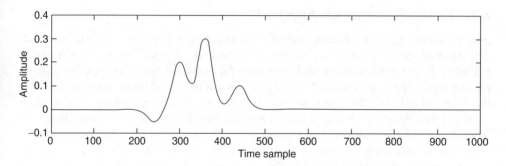

Figure 8.12   The target signal considered in the simulation

as the signal in Figure 8.12 and the node is considered to be P3 (using temporally correlated noise), it was noticed that the output resembled much better the target signal.

The main objective of using diffusion adaptation is to better process inter-subject connectivity while performing cooperative or competitive tasks. The error between actual and model outputs is expected to be different for any abnormality which affects the connectivity (such as Parkinson and Alzheimer diseases). A block diagram of the combined DTF and diffusion adaptation adaptive filter is shown in Figure 8.13. This is ongoing research and some preliminary results of the implementation of a diffusion algorithm for modelling the brain dynamics during body movement have been published in [67].

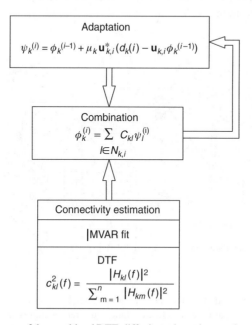

Figure 8.13   Block diagram of the combined DTF-diffusion adaptation used to estimate the cooperation link weights and the filter parameters through adaptation and combination

8.9   Tensor Factorization Approach

Another interesting approach in the analysis of connectivity is *link prediction*, which not only provides a measure of connectivity but also tracks the changes of the links between the nodes of a network. The problem of link prediction arises for groups of objects connected by multiple relation types. The solution should not only allow identification of the correlations (amongst the linked patterns) but also reveal the impact of various relations on prediction performance.

In [63], following the work in [64], it is assumed that the objects tend to form links with others having similar characteristics and related objects share a similar behaviour pattern. To formulate the link pattern prediction consider a set of N objects $X = \{x_1, x_2, \ldots, x_N\}$ and assume there are T different types of relations amongst object pairs defined by a set of T trajectory matrices [63]. An $N \times N \times T$ tensor $\underline{\mathbf{Y}}$ is then defined with entry $y_{i,j,t}$ representing the type t relation value between object pair x_i and x_j as [63]:

For $\forall i,j \in [1, \ldots N]^2$ and $\forall t \in [1, \ldots, T]$

$$y_{i,j,t} = \begin{cases} 1 & \text{if } x_i \text{ links to } x_j \text{ by relation type } t \\ 0 & \text{otherwise} \end{cases} \tag{8.43}$$

Based on this $\underline{\mathbf{Y}}_{(i,j,:)}$ is defined as the *link pattern* involving T different types of relations between each pair of objects x_i and x_j. The problem to solve here is: given some observed link patterns information for a subset of related objects in the multirelational network, how can the unobserved link patterns for the remaining object pairs be found? In Figure 8.14 an example of the problem and its tensor model (taken from an unpublished work by Andrzej Cichocki) has been illustrated.

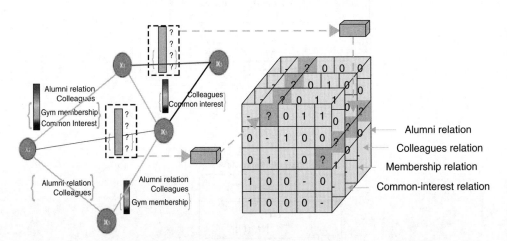

Figure 8.14 Example of modelling the multirelational social network as a tensor for predicting the unobserved link patterns in the network. To the right is the tensor representation of a network where each slice matrix represents one relation type and each tube fibre represents the link pattern between two nodes. Unknown link patterns are represented as "?" fibres

To mathematically express the problem consider a non-negative $N \times N$ matrix $\mathbf{W}$ such that its entries $w_{i,j}$ for $\forall i,j \in [1, \ldots, N]^2$ are defined as [63]

$$w_{i,j} = \begin{cases} 1 & \text{if link pattern between } x_i \text{ and } x_j \text{ is known} \\ 0 & \text{if link pattern between } x_i \text{ and } x_j \text{ is missing} \end{cases} \quad (8.44)$$

For the decomposition the following loss function is defined:

$$L(\mathbf{X}) = \sum_{i=1}^{N} \sum_{j=1}^{N} w_{i,j} \left\| y_{i,j,:} - x_{i,j,:} \right\|^2 \quad (8.45)$$

where $\| . \|$ denotes a Frobenius norm for a tensor. Following CANDECOMP/PARAFAC (CP) tensor decomposition and (8.45) we can solve the following optimization problem [63] using a conjugate gradient approach:

$$\arg \min_{\mathbf{X}} L(\mathbf{X}) \quad \text{subject to} \quad x_{i,j,t} = \sum_{k=1}^{K} u_{i,k} v_{j,k} r_{t,k} \quad (8.46)$$

Where K is the rank of the tensor, and $\mathbf{U}$, $\mathbf{V}$ and $\mathbf{R}$ are the latent factor matrices for CP tensor decomposition. A popular approach to solve the problem is by ALS, as in [63]. This, however, requires estimation of a large number of parameters and, consequently, leads to a high cost of computation and may result in over-fitting. To overcome these problems the above constrained problem can be changed to an unconstrained problem using regularization parameters (penalty functions), which leads to the following cost function:

$$\arg \min_{\mathbf{U}, \mathbf{V}, \mathbf{R}, \gamma} L(\mathbf{U}, \mathbf{V}, \mathbf{R}) + \frac{\gamma_0}{2} \left(\|\mathbf{U}\|_F^2 + \|\mathbf{V}\|_F^2 \right) + \frac{\gamma_T}{2} \left(\|\mathbf{R}\|_F^2 \right) \quad (8.47)$$

where γ_0 and γ_T are the regularization parameters. The above approach has a high potential in brain connectivity estimation and estimating the unknown links between different brain regions.

In the linked multiway BSS, an approximate decomposition of a set of data tensors $\mathbf{X}^{(s)} \in R^{I_1 \times I_2 \times \ldots \times I_N}$ ($s = 1, 2, \ldots, S$) representing multiple subjects and/or multiple tasks (see Figure 8.15) is performed. The objective is to find a set of constrained factor matrices $\mathbf{U}^{(n,s)} = [\mathbf{U}_C^{(n)}, \mathbf{U}_I^{(n,s)}] \in \Re^{I_n \times J_n}$, $n = 1, 2, 3$) and core tensors $\underline{\mathbf{G}}(s) \in \Re^{J_1 \times J_2 \times J_3}$, which are partially linked or maximally correlated, that is, they have the same common components or highly correlated components. In the case of EEG signals a four-dimensional tensor $\underline{\mathbf{X}}$ can be defined in terms of its three-dimensional components as:

$$\underline{\mathbf{X}}^{(s)} = \underline{G}^{(s)} \times_1 \mathbf{U}^{(1,s)} \cdots \times_N \mathbf{U}^{(N,s)} + \underline{\mathbf{E}}^{(s)} \quad (8.48)$$

where $\times_k$ denotes tensor multiplication along slab (dimension) k. Each factor $\mathbf{U}^{(n,s)}$ consists of two parts; one, $\mathbf{U}_C^{(n,s)}$, which are common bases for all subjects in the group and correspond to the same or maximally correlated components and another one, $\mathbf{U}_I^{(n,s)}$, which corresponds

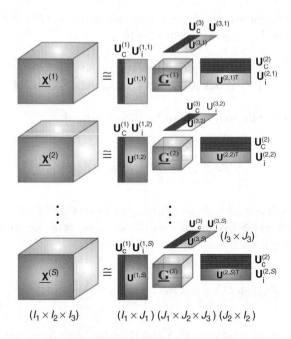

Figure 8.15 Conceptual model of tensors decomposition for linked multiway BSS. The objective is to find a set of constrained factor matrices $\mathbf{U}^{(n,s)} = [\mathbf{U}_C^{(n)}, \mathbf{U}_I^{(n,s)}] \in \Re^{I_n \times J_n}$, $n = 1, 2, 3$) and core tensors $\underline{\mathbf{G}}(s) \in \Re^{I_1 \times J_2 \times J_3}$, which are partially linked or maximally correlated, that is, they have the same common components or highly correlated components. Taken from [65], © IEEE

to stimuli/tasks with independent individual characteristics. $\underline{\mathbf{X}}^{(s)}$ represents space (channel) – time – frequency data for the sth subject.

The factors are usually estimated using an alternating optimization method which minimizes the distance between the terms in the left and right sides of equation (8.48). In order to achieve unique solutions for all the factors some constraints are required. For the EEG signals such constraints can be the approximate source locations, sparsity in space or frequency domain, or a known structure of the core matrix; for example, for independent factors the core matrix is diagonal.

8.10   Conclusions

Estimation of the connectivity of brain regions using EEG or MEG signals has its root in the work carried out approximately three decades ago. Although various methods, such as MVAR and DTF, SEM PDC and dDTF, have been popular for connectivity estimation, the new approaches, such as by using diffusion adaptation or tensor factorization, have opened a new front in both modelling and estimation of the connectivity. Application of these methods is in its early stages and, therefore, more research needs to be carried out to enable robust procedures for EEG or MEG connectivity measures. A major problem with most of these methods is that all the sources are considered cortical and thus nobody has attempted deep sources. This requires localization of multiple deep sources which is another challenging problem. In a recent work by Thatcher *et al.* [66] the LORETA source localization algorithm has been applied to determine the existing current sources within a large number of regions of interest in the brain. Temporal and spatial correlations are later computed as the measure of connectivity. It also makes sense to apply a source separation algorithm to the signals to reject noise and irrelevant information before estimation of connectivity and to better deal with non-cortical sources. Certainly, the correlated sources cannot be separated using conventional BSS approaches.

References

[1] Steriade, M. (1995) Cellular substrates of brain rhythms, in *Electroencephalography* (eds E. Niedermeyer and F. Lopes da Silva), Williams and Wilkins, Baltimore.

[2] Nunez, P. (1981) *Electrical Fields of the Brain*, Oxford University Press, New York.

[3] Nunez, P. (1994) *Neocortical Dynamics and Human EEG Rhythms*, Oxford University Press, New York.

[4] Adey, W.R., Walter, D.O. and Hendrix, C.E. (1961) Computer techniques in correlation and spectral analyses of cerebral slow waves during discriminative behavior. *Exp Neurol.*, **3**, 501–524.

[5] Sterman, M. and Kaiser, D. (2001) Comodulation: a new QEEG analywsis metric for assessment of structural and functional disorders of the central nervous system. *J. Neurother.*, **4**(3), 73–83.

[6] Kaminski, M. and Blinowska, K.J. (1991) A new method of the description of the information flow in brain structure. *Biol. Cybern.*, **65**, 203–210.

[7] Kaminski, M., Blinowska, K.J. and Szelenberger, W. (1997) Topographic analysis of coherence and propagation of EEG activity during sleep and wakefulness. *Electroencephalogr. Clin. Neurophysiol.*, **102**, 216–227.

[8] Korzeniewska, A., Kasicki, S., Kaminski, M. and Blinowska, K.J. (1997) Information flow between hippocampus and related structures during various types of rat's behaviour. *J. Neurosci. Methods*, **73**, 49–60.

[9] Jung, R.E. and Haier, R.J. (2007) The Parieto-Frontal Integration Theory (P-FIT) of intelligence: converging neuroimaging evidence. *Behav. Brain Sci.*, **30**(2), 135–54.

[10] Picard, N. and Strick, P.L. (1996) Motor areas of the medial wall: a review of their location and functional activation. *Cereb Cortex*, **6**, 342–353.

[11] Paus, T., Koski, L., Caramanos, Z. and Westbury, C. (1998) Regional differences in the effects of task difficulty and motor output on blood flow response in the human anterior cingulate cortex: a review of 107 PET activation studies. *Neuroreport*, **9**, R37–R47.

[12] Gerloff, G., Richard, J., Hadley, J. *et al.* (1998) Functional coupling and regional activation of human cortical motor areas during simple, internally paced and externally paced finger movements. *Brain*, **121**(8), 1513–1531.

[13] Sharott, A., Magill, P.J., Bolam, J.P. and Brown, P. (2005) Directional analysis of coherent oscillatory field potentials in cerebral cortex and basal ganglia of the rat. *J. Physiol.*, **562**(3), 951–963.

[14] Granger, C.W.J. (1969) Investigating causal relations in by econometric models and cross-spectral methods. *Econometrica*, **37**, 424–438.

[15] Bernosconi, C. and König, P. (1999) On the directionality of cortical interactions studied by spectral analysis of electrophysiological recordings. *Biol. Cybern.*, **81**(3), 199–210.

[16] Nolte, G., Ziehe, A., Nikulin, V.V. *et al.* (2008) Robustly estimating the flow direction of information of information in complex physical systems. *Phys. Rev. Lett.*, **100**(23), 1–4.

[17] Rana, P., Lipor, J., Lee, H. *et al.* (2012) Seizure detection using the phase-slope index and multichannel ECoG. *IEEE Trans. Biomed. Eng.*, **59**(4), 1125–1134.

[18] Kaminski, M., Ding, M., Truccolo, W. and Bressler, S. (2001) Evaluating causal relations in neural systems: Granger causality, directed transfer function, and statistical assessment of significance. *Biol. Cybern.*, **85**, 145–157.

[19] Kaminski, M. and Blinowska, K. (1991) A new method of the description of information flow in the brain structures. *Biol. Cybern.*, **65**, 203–210.

[20] Jing, H. and Takigawa, M. (2000) Observation of EEG coherence after repetitive transcranial magnetic stimulation. *Clin. Neurophysiol.*, **111**, 1620–1631.

[21] Kuś, R., Kaminski, M. and Blinowska, K. (2004) Determination of EEG activity propagation: Pair-wise versus multichannel estimate. *IEEE Trans. Biomed. Eng.*, **51**(9), 1501–1510.

[22] Ginter, J. Jr, Kaminski, M., Blinowska, K. and Durka, P. (2001) Phase and amplitude analysis in time-frequency-space; application to voluntary finger movement. *J. Neurosc. Meth.*, **110**, 113–124.

[23] Morf, M., Vieria, A., Lee, D. and Kailath, T. (1978) Recursive multichannel maximum entropy spectral estimation. *IEEE Trans. Geoscience Electronics*, **16**, 85–94.

[24] Akaike, H. (1974) A new look at statistical model order identification. *IEEE Trans. Automat. Contr.*, **19**, 716–723.

[25] Ding, M., Bressler, S.L., Yang, W. and Liang, H. (2000) Short-window spectral analysis of cortical event-related potentials by adaptive multivariate autoregressive modelling: data preprocessing, model validation, and variability assessment. *Biol. Cybern.*, **83**, 35–45.

[26] Korzeniewska, A., Manszak, A., Kaminski, M. *et al.* (2003) Determination of information flow direction among brain structures by a modified directed transfer function method (dDTF). *J. Neurosci. Methods*, **125**, 195–207.

[27] Blinowska, K.J. (2011) Review of the methods of determination of directed connectivity from multichannel data. *Med. Biol. Eng. Comput.*, **49**, 521–529

[28] Medkour, T., Walden, A.T. and Burgess, A. (2009) Graphical modelling for brain connectivity via partial coherence. *J. Neurosci. Methods*, **180**, 374–383.

[29] Baccala, L.A. and Sameshima, K. (2001) Partial directed coherence: a new conception in neural structure determination. *Biol. Cybern.*, **84**, 463–474.

[30] Astolfi, L., Cincotti, F., Babiloni, C. *et al.* (2005) Estimation of the cortical connectivity by high resolution EEG and structural equation modeling: simulations and application to finger tapping data. *IEEE Trans. Biomed.Eng.*, **52**(5), 757–768.

[31] David, O., Cosmelli, D. and Friston, K.J. (2004) Evaluation of different measures of functional connectivity using a neural mass model. *NeuroImage*, **21**, 659–673.

[32] Bollen, K.A. (1989) *Structural Equations with Latent Variable*, John Wiley & Sons, Inc., New York.

[33] McIntosh, A.R. and Gonzalez-Lima, F. (1994) Structural equation modeling and its application to network analysis in functional brain imaging. *Hum. Brain Mapp.*, **2**, 2–22.

[34] Gevins, A. (1989) Dynamic functional topography of cognitive task. *Brain Topography*, **2**, 37–56.

[35] Gevins, A., Brickett, P., Reutter, B. and Desmond, J. (1991) Seeing through the skull: advanced EEGs use MRIs to accurately measure cortical activity from the scalp. *Brain Topogr.*, **4**, 125–131.

[36] Astolfi, L. and Babiloni, F. (2008) *Estimation of Cortical Connectivity in Humans: Advanced Signal Processing Techniques*, Morgan & Claypool.

[37] Astolfi, L., Toppi, J., De Vico Fallani, F. *et al.* (2010) Neuroelectrical hyperscanning measures simultaneous brain activity in humans. *Brain Topogr.*, **23**(3), 243–56.

[38] Dumas, G., Nadel, J., Soussignan, R. *et al.* (2010) Inter-brain synchronization during social interaction. *PLoS One*, **5**(8), doi: 10.1371/journal.pone.0012166.

[39] Astolfi, L., Cincotti, F., Mattia, D. *et al.* (2009) Estimation of the cortical activity from simultaneous multi-subject recordings during the prisoner's dilemma. Proceedings of the Annual International Conference of the IEEE Engineering in Medicine and Biology Society, EMBC 2009, pp. 1937–1939.

[40] Lindenberger, U., Li, S.C., Gruber, W. and Müller, V. (2009) Brains swinging in concert: cortical phase synchronization while playing guitar. *BMC Neurosci.*, **17**, 10–22.

[41] Babiloni, F., Cincotti, F., Mattia, D. *et al.* (2007) High resolution EEG hyperscanning during a card game, Proceedings of the Annual International Conference of the IEEE Engineering in Medicine and Biology Society, 2007. EMBC 2007, pp. 4957–4960.

[42] Babiloni, F., Astolfi, L., Cincotti, F. *et al.* (2007) Cortical activity and connectivity of human brain during the prisoner's dilemma: an EEG hyperscanning study, Proceedings of the Annual International Conference of the IEEE Engineering in Medicine and Biology Society, 2007. EMBC 2007, 4953–4956.

[43] Babiloni, F., Cincotti, F., Mattia, D. *et al.* (2006) Hypermethods for EEG hyperscanning, Proceedings of the Annual International Conference of the IEEE Engineering in Medicine and Biology Society, 2006. EMBC 2006, pp. 3666–3669.

[44] Huang, S., Liang Sun, J.L., Ye, J. *et al.* (2010) Learning brain connectivity of Alzheimer's disease by sparse inverse covariance estimation. *NeuroImage*, **50**, 935–949.

[45] Supekar, K., Menon, V., Rubin, D., Musen, M. and Greicius, M.D. (2008) Network analysis of intrinsic functional brain connectivity in Alzheimer's disease. *PLoS Comput. Biol.*, **4**(6): e1000100. doi:10.1371/journal.pcbi.1000100.

[46] Escodero, J., Sanei, S., Jarchi, D. *et al.* (2011) Regional coherence evaluation in mild cognitive impairment and alzheimer's disease based on adaptively extracted magnetoencephalogram rhythms. *J. Physiol. Meas.*, **32**(8), 1163–1180.

[47] Cheong, B.L.P., Nowak, R., Lee, H.C. *et al.* (2012) Cross validation for selection of cortical interaction models from scalp EEG or MEG. *IEEE Trans. Biomed. Eng.*, **59**(2).

[48] Nalatore, H., Ding, M. and Rangarajan, G. (2009) Denoising neural data with state-space smoothing: Method and application. *J. Neurosci. Methods*, **179**, 131–141.

[49] Shumway, R. and Stoffer, D. (1982) An approach to time series smoothing and forecasting using the EM algorithm. *J. Time Series Anal.*, **3**(4), 253–264.

[50] Cheong, B.L., Riedner, B., Tononi, G. and van Veen, B. (2010) Estimation of cortical connectivity from EEG using sate-space models. *IEEE Trans. Biomed. Eng.*, **57**(9), 2122–2134.

[51] Ancona, N., Marrinazo, D. and Stramaglia, S. (2004) Radial basis function approach to nonlinear Granger causality of time series. *Phys. Rev. E.*, **70**, 56221–56227.

[52] Park, J. and Sandberg, I.W. (1991) Universal approximation using radial-basis function networks. *Neural Comput.*, **3**, 246–257.

[53] Diedrichsen, J., Wiestler, T. and Krakauer, W. (2012) Two distinct ipsilateral cortical representations for individuated finger movements. *Cerebral Cortex.* doi: 10.1093/cercor/bhs120

[54] Chen, J., Zhao, X. and Sayed, A.H. (2010) Bacterial motility via diffusion adaptation. Proceedings of the 44th Asilomar Conference on Signals, Systems and Computers, Pacific Grove, CA, Nov. 2010.

[55] Tu, S.-Y. and Sayed, A.H. (2010) Tracking behavior of mobile adaptive networks. Proceedings of the 44th Asilomar Conference on Signals, Systems and Computers, Pacific Grove, CA, Nov. 2010.

[56] Tu, S.-Y. and Sayed, A.H. (2011) Cooperative prey herding based on diffusion adaptation. Proceedings of the IEEE ICASSP 2011, Prague, Czech Republic, April 2011.

[57] Cattivelli, F.S. and Sayed, A.H. (2010) Diffusion LMS strategies for distributed estimation. *IEEE Trans. Signal Process.*, **58**(3), 1035–1048.

[58] Lopes, C. and Sayed, A.H. (2008) Diffusion least-mean squares over adaptive networks: formulation and performance analysis. *IEEE Trans. Signal Process.*, **56**(7), 3122–3136.

[59] Takahashi, N., Yamada, I. and Sayed, A.H. (2010) Diffusion least-mean squares with adaptive combiners: formulation and performance analysis. *IEEE Trans. Signal Process.*, **58**(9), 4795–4810.

[60] Cattivelli, F. and Sayed, A. (2010) Diffusion strategies for distributed Kalman filtering and smoothing. *IEEE Trans. Autom. Control.*, **55**(9), 2069–2084.

[61] Hu, J., Xie, L. and Zheng, C. (2012) Diffusion Kalman filtering based on covariance intersection. *IEEE Trans. Signal Process.*, **60**(2), 891–902.

[62] Julier, S. and Uhlmann, J. (2001) General decentralised data fusion with covariance intersection (CI), in *Handbook of Multiuser Data* Fusion.

[63] Gao, S., Denoyer, L. and Gallinari, P. (2011) Link pattern prediction with tensor decomposition in multi-relational networks. IEEE Symposium on Computational Intelligence and Data Mining (CIDM), Paris, France, April 2011.

[64] Acar, E., Dunlavy, D.M. and Kolda, T.G. (2009) Link prediction on evolving data using matrix and tensor factorization. ICDM Workshops, pp. 262–269.

[65] Sanei, S., Ferdowsi, S., Nazarpour, K. and Cichocki, A. (2013) Advances in EEG signal processing. *IEEE Signal Process. Mag.*, **30**(1), 170–176.

[66] Thatcher, R.W., Biver, C.J. and North, D. (2007) Spatial-temporal current source correlations and cortical connectivity. *Clin. EEG Neurosci.*, **38**(1), 35–48.

[67] Eftexias K., Sanei S., and Sayed A. (2013) A new approach to evaluation and modeling of brain connectivity using diffusion adaptation, Proc. of IEEE International Conference on Acoustics, Speech and Signal Processing, ICASSP, 2013.

9

Detection and Tracking of Event-Related Potentials

9.1 ERP Generation and Types

The brain generates different types of signals: (i) normal brain rhythms which constantly represent the state of the brain, (ii) event-related potentials (ERPs), which are the responses to brain stimulation, and (iii) movement-related potentials (MRPs) which appear due to intentional physical or imagery movements. ERPs were first explained in 1964 [1, 2], and have remained as useful diagnostic indicators in many applications in psychiatry and neurology as well as for brain–computer interfacing.

An ERP consists of a sequence of labelled positive and negative amplitude components. These components reflect various sensory, cognitive (e.g. stimulus evaluation) and motor processes that are classified on the basis of their scalp distribution and response to experimental stimulus. Therefore, ERPs (which can be detected from EEG or MEG signals) directly measure the electrical response of the cortex to sensory, affective or cognitive events. They are voltage fluctuations in the EEG induced within the brain, as a sum of a large number of action potentials (APs) that are time locked to sensory, motor, or cognitive events. They are typically generated in response to peripheral or external stimulations, and appear as somatosensory, visual, and auditory brain potentials, or as slowly evolving brain activity observed before voluntary movements or during anticipation of conditional stimulation.

Compared with the background EEG activity ERPs are quite small (1–30 µV). Therefore, traditionally, they often need the use of a signal-averaging procedure for their elucidation. In addition, although evaluation of the ERP peaks may not be considered as a reliable diagnostic procedure, the application of ERP in psychiatry has been very common and widely followed.

The ERP waveform can be quantitatively characterised across three main dimensions: amplitude, latency, and scalp distribution [3]. In addition, an ERP signal may also be analysed with respect to the relative latencies between its subcomponents. The amplitude provides an index of the extent of neural activity (and how it responds functionally to experimental variables), the latency (i.e. the time point at which peak amplitude occurs) reveals the timing of this activation, and the scalp distribution provides the pattern of the voltage gradient of a component over the scalp at any time instant.

Adaptive Processing of Brain Signals, First Edition. Saeid Sanei.
© 2013 John Wiley & Sons, Ltd. Published 2013 by John Wiley & Sons, Ltd.

ERP signals are either positive, represented by letter P, such as P300, or negative, represented by letter N, such as N100 and N400. The digits indicate the time in terms of milliseconds after the stimuli (audio, visual, or somatosensory). The amplitude and latency of the components occurring within 100 ms after stimulus onset are labelled oxogenous, and are influenced by physical attributes of stimuli. such as intensity, modality, and presentation rate. On the other hand, endogenous components, such as P300, are nonobligatory responses to stimuli, and vary in amplitude, latency, and scalp distribution with strategies, expectancies, and other mental activities triggered by the event eliciting the ERP. These components are not influenced by the physical attributes of the stimuli.

Analysis of ERPs, such as visual event potentials (VEP), within the EEG signals is important in the clinical diagnosis of many psychiatric diseases, such as dementia. Alzheimer's disease, the most common cause of dementia, is a degenerative disease of the cerebral cortex and subcortical structures. The relative severity of the pathological changes in the associated cortex accounts for the clinical finding of diminished visual interpretation skills with normal visual acuity. Impairment of visuo-cognitive skills often happens with this disease. This means that the patient may have difficulties with many complex visual tasks, such as tracing a target figure embedded in a more complex figure, or identifying single letters that are presented briefly and followed by a pattern-making stimulus. A specific indicator of dementia is the information obtained by using VEP. Several studies have confirmed that flash VEPs are abnormal while pattern VEPs are normal in patients with Alzheimer's disease.

The most consistent abnormality in the flash VEP is the increase in the latency of the P100 component and an increase in its amplitude [4, 5]. Other ERP components, such as N130, P165, N220, also have longer latencies in patients with Alzheimer's disease [6, 7]. Figure 9.1 illustrates the shape of two normal and two abnormal P100 components. It is seen that the peak of the normal P100s (normally shown in reversed polarity) has a latency of approximately 106 ms whereas, the abnormal P100 peak has a latency of approximately 135 ms and has lower amplitude.

Although determination of the locations of the ERP sources within the brain is a difficult task, the scalp distribution of an ERP component can often provide very useful and complementary information to that derived from amplitude and latency. Generally, two types of topographic maps can be generated: raw voltage (or surface potentials) and current source density (CSD), both derived from the electrode potentials.

The scalp-recorded ERP voltage activity reflects the summation of both cortical and subcortical neural activity within each time window. On the other hand, CSD maps reflect primary cortical surface activity [8, 9]. CSD is obtained by spatially filtering the subcortical areas as well as the cortical areas distal to the recording electrodes to remove the volume-conducted activity. CSD maps are useful for forming hypotheses about neural sources within the superficial cortex [10].

The fMRI technique has become another alternative to investigate brain ERPs since it can detect the hemodynamic of the brain. However, there are at least three shortcomings with this brain imaging modality: firstl, the temporal resolution is low; secondly, the activated areas based on hemodynamic techniques do not necessarily correspond to the neural activity identified by ERP measures; and thirdly, the fMRI is not sensitive to the type of stimulus (e.g. target, standard, or novel). It is considered that the state of the subject changes due to differences in the density of different stimulus types across blocks of trials. By target P300, we refer to the P300 component elicited by events about which the subject has been instructed and

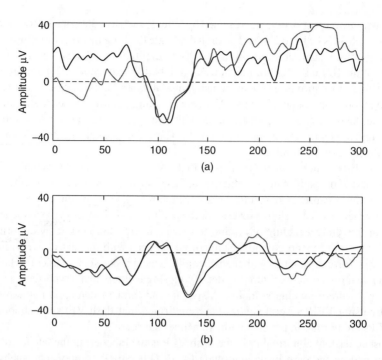

Figure 9.1 Four P100 components; (a) two normal P100 and (b) two abnormal P100 components. In (a) the P100 peak latency is at approximately 106 ms, whereas in (b) the P100 peak latency is at approximately 135 ms (waves are shown in reverse polarity)

to which the subject is required to generate some kind of response. A novel stimulus indicates a sole or irregular stimulus.

The ERP parameters, such as amplitude and latency, are the indicators of the function of the brain neurochemical systems and can potentially be used as predictors of the response of an individual to psychopharmacotherapy [11]. ERPs are also related to the circumscribed cognitive process. For example, there are interesting correlations between late evoked positivities and memory, N400 and semantic processes, or the latencies of ERPs and the timing of cognitive processes. Therefore, the ERP parameters can be used as indicators of cognitive processes and dysfunctions not accessible to behavioural testing. As an example Figure 9.2 shows how the average (over $N = 100$ subjects) P300 differs for normal and alcoholic subjects.

The fine-grained temporal resolution of ERPs has been traditionally limited. In addition, overlapping components within ERPs, which represent specific stages of information processing, are difficult to distinguish [13, 14]. An example is the composite P300 wave, a positive ERP component, which occurs with a latency of about 300 ms after novel stimuli, or task-relevant stimuli, which require an effortful response on the part of the individual under test [13–17].

The elicited ERPs are comprised of two main components; the mismatch negativity (MMN) and the novelty P300. By novelty P300, we refer to the P300 component elicited by events about which the subject has not been instructed prior to the experiment. The MMN is the earliest ERP activity (which occurs within the first 10 ms after the stimulus) indicating that the brain has detected a change in a background of brain homogeneous events. The MMN

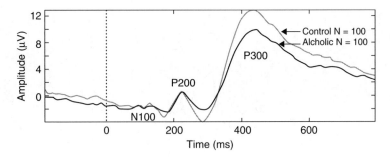

Figure 9.2 The average ERP signals for normal and alcoholic subjects. The curves show reduced P300 amplitude in alcoholics. Reprinted from [12]

is thought to be generated in and around the primary auditory cortex [18]. The amplitude of MMN is directly proportional, and its latency inversely related, to the degree of difference between the standard and the deviant stimuli. It is most clearly seen by subtraction of the ERP elicited by the standard stimulus from that elicited by the deviant stimulus during a passive oddball paradigm (OP) when both of those stimuli are unattended or ignored. Therefore, it is relatively automatic.

9.1.1   P300 and Its Subcomponents

The P300 wave represents cognitive functions involved in orientation of attention, contextual updating, response modulation, and response resolution [13, 15] and consists mainly of two overlapping subcomponents P3a and P3b [14, 17, 19]. P3a reflects an automatic orientation of attention to novel or salient stimuli independent of task relevance. Prefrontal, frontal, and anterior temporal brain regions play the main role in generating P3a giving it a frontocentral distribution [17]. In contrast, P3b has a greater centro-parietal distribution due to its reliance on posterior temporal, parietal and posterior cingulated cortex mechanisms [13, 14]. P3a is also characterised by a shorter latency and more rapid habituation than P3b [19].

A neural event is the frontal aspect of the novelty P300, that is, P3a. For example, if the event is sufficiently deviant, the MMN is followed by the P3a. The eliciting events in this case are highly deviant environmental sounds such as a dog barking. Non-identifiable sounds elicit larger P3a than identifiable sounds. Accordingly, a bigger MMN results in a dominant P3a. Figure 9.3 shows typical P3a and P3b subcomponents.

P3b is also elicited by infrequent events but, unlike P3a, it is task relevant, or involves a decision to evoke this component.

The ERPs are the responses to different stimuli, that is, novel or salient. It is important to distinguish between the ERPs when they are the response to novel or salient (i.e. what has already been experienced) stimuli, or when the degree of novelty changes.

The orienting response engendered by deviant or unexpected events consists of a characteristic ERP pattern, which is comprised sequentially of the MMN and the novelty P300 or P3a. Each of p3a and p3b has a different cognitive or possibly neurological base.

The orienting response [21, 22] is an involuntary shift of attention that appears to be a fundamental biological mechanism for survival. It is a rapid response to a new, unexpected or

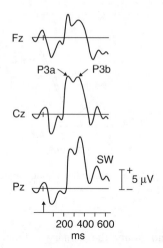

Figure 9.3 Typical P3a and P3b subcomponents of a P300 ERP signal viewed at Fz, Cz, and Pz electrodes [20]

unpredictable stimulus, which essentially functions as a what-is-it detector [23]. The plasticity of the orienting response has been demonstrated by showing that stimuli that initially evoked the response no longer did so with repeated presentation [22]. Habituation of the response is proposed to indicate that some type of memory for these prior events has been formed, which modifies the response to the repeated incidences.

Abnormalities in P300 are found in several psychiatric and neurological conditions [16]. However, the impact of the diseases on P3a and P3b may be different. Both audio and visual P300 (i.e. a P300 signal produced earlier due to an audio or a visual stimulation) are used. Audio and visual P300 appear to be differently affected by illnesses and respond differently to their treatment. This suggests differences in the underlying structures and neurotransmitter systems [14]. P300 has significant diagnostic and prognostic potential, especially when combined with other clinical symptoms and evidences.

In many applications, such as human–computer interaction (HCI), muscular fatigue, visual fatigue, and mental fatigue are induced as a result of physical and mental activities. In order for the ERP signals and their subcomponents to be reliably used for clinical diagnosis, assessment of mental activities, fatigue during physical and mental activities, and for provision of a human–computer interface, very effective and reliable methods for their detection and parameter evaluation have to be developed. In the following section a number of established methods for detection of ERP signals, especially P300 and its subcomponents P3a and P3b, are described in sufficient detail.

9.2 Detection, Separation, and Classification of P300 Signals

Although in many applications, such as assessment of memory, a number of ERPs are involved, detection and tracking of P300 and its subcomponents' variability have been under more intensive research. Traditionally, EPs are synchronously averaged to enhance the evoked signal and suppress the background brain activity [24].

Step-wise discriminant analysis (SWDA) followed by peak picking and evaluation of the covariance, was first introduced by Farwell and Dounchin [25]. Later, a discrete wavelet transform (DWT) was also added to the SWDA to better localize the ERP components in both time and frequency [26].

PCA has been employed to assess temporally-overlapping EP components [27]. However, the resultant orthogonal representation does not necessarily coincide with the true component structure since the actual physiological components need not be orthogonal, that is, the source signals may be correlated.

ICA has been applied to ERP analysis by many researchers including Makeig *et al.* [28]. Infomax ICA [29] was used by Xu *et al.* [30] to detect the ERPs for the P300-based speller. In their approach those independent components (ICs) with relatively larger amplitudes in the latency range of P300 were kept, while the others were set to zero. Also, they exploited a priori knowledge about the spatial information of the ERPs and decided whether a component should be retained or wiped out. To manipulate the spatial information denote the ith row and jth column element in the inverse of the unmixing matrix, $\mathbf{W}^{-1}$, by w'_{ij}, therefore, $\mathbf{x}(n) = \mathbf{W}^{-1}\mathbf{u}(n)$. Then, denote the jth column of $\mathbf{W}^{-1}$ by $\mathbf{w}'_j$, which reflects the intensity distribution at each electrode for the jth IC u_j [28]. Then, for convenience the spatial pattern $\mathbf{W}^{-1}$ is transformed into an intensity order matrix $\mathbf{M} = \{m_{ij}\}$ with the same dimension. The value of the element m_{ij} in $\mathbf{M}$ is set to be the order number of the value w'_{ij} in the column vector $\mathbf{w}'_j$. For example $m_{ij} = 1$ if w'_{ij} has the largest value, $m_{ij} = 2$, if it has the second largest value, and so on. Based on the spatial distribution of the brain activities, an electrode set $\mathbf{Q} = \{q_k\}$ of interest is selected, in which q_k is the electrode number and is equal to the row index of the multichannel EEG matrix $\mathbf{x}(n)$. For extraction of the P300 signals these electrodes are located around the vertex region (C_z, C_1, and C_2) since they are considered to have prominent P300. The spatial filtering of the ICs is simply performed as:

$$\tilde{\mathbf{u}}_j(n) = \begin{cases} \mathbf{u}_j(n) & \text{if } \exists q_k \in \mathbf{Q} \text{ and } m_{q_k j} \leq T_r \\ 0 & \text{else} \end{cases} \qquad (9.1)$$

where T_r is the threshold for the order numbers. T_r is introduced to retain the most prominent spatial information about the P300 signal. Therefore, $\tilde{\mathbf{u}}_j(n)$ holds most of the source information about P300 while other irrelevant parts are set to zero. Finally, after the temporal and spatial manipulation of the ICs, the $\tilde{\mathbf{u}}_j(n)$ s, $j = 1, 2, \ldots, M$, where M is the number of ICs, are back projected to the electrodes by using $\mathbf{W}^{-1}$ to obtain the scalp distribution of the P300 potential, that is,

$$\tilde{\mathbf{x}}(n) = \mathbf{W}^{-1}\tilde{\mathbf{u}}(n) \qquad (9.2)$$

where $\tilde{\mathbf{x}}(n)$ is the P300 enhanced EEG. The $\tilde{\mathbf{x}}(t)$ features can then be measured for classification purposes [30].

9.2.1   Using ICA

Assuming ERP as an IC from other brain generated and artefact signals, ICA has been used by many researchers for detection and analysis of the ERP components from EEG, EMG and fMRI (in the form of BOLD) data [31]. Four main assumptions underlie ICA decomposition of EEG or MEG time series: (i) signal conduction times are equal and summation of currents at the scalp sensors is linear; both are reasonable assumptions for currents carried to the scalp electrodes by volume conduction for EEG or for superposition of magnetic fields at SQUID sensors; (ii) spatial projections of components are fixed across time and conditions; (iii) source activations are temporally independent of one another across the input data; (iv) statistical distributions of the component activation values are not Gaussian [31].

After application of ICA to the signals, the brain activities of interest, accounted for by single or by multiple components, can be obtained by projecting the selected ICA component(s) back onto the scalp, $\tilde{\mathbf{X}} = \mathbf{W}^{-1}\mathbf{Y}^{(i)}$ where $\mathbf{W}$ is the estimated separating matrix using ICA and $\mathbf{Y}^{(i)}$ is the matrix of estimated/separated sources when all except source $\mathbf{y}_i$ have been set to zero. As the main conclusions here, ICA can separate stimulus-locked, response-locked, and non-event-related background EEG activities into separate components, allowing (i) removal of pervasive artefacts of all types from single trial EEG records, making possible analysis of highly contaminated EEG records from clinical populations; (ii) identification and segregation of stimulus- and response-locked event-related activity in single-trail EEG epochs; (iii) separation of spatially overlapping EEG activities over the entire scalp and frequency band that may show a variety of distinct relationships to task events, rather than focusing on activity at single frequencies in single scalp channels or channel pairs; (iv) investigation of the interaction between ERPs and ongoing EEG [31].

The work in [32] involved application of a matched filter together with averaging and applying a threshold for detecting the existence of the P300 signals. The block diagram in Figure 9.4 shows the method.

The IC corresponding to the P300 source is selected and segmented to form overlapping segments from 100 to 600 ms. Each segment is passed through a matched filter to give one feature that represents the maximum correlation between the segment and the average P300 template. However, a very obvious problem with this method is that the ICA system is very likely to be underdetermined (i.e. the number of sources is more than the number of sensors or observations) since only three mixtures are used. In this case the independent source signals are not separable.

Although ICA appears to be generally useful for EEG and fMRI analysis, it has some inherent limitations. First, ICA, in its general form, decomposes at most N sources from data collected at N scalp electrodes. Usually, the effective number of statistically independent source signals contributing to the scalp EEG is unknown and it is likely that the observed brain activity arises from more physically separable effective sources than the available number of EEG electrodes [31].

Second, ICA fails to separate correlated sources of ERP signals. The assumption of temporal independence used by ICA may not be satisfied when the training data set is small, or when separate topographically distinguishable phenomena nearly always co-occur in the data. In

Figure 9.4 Block diagram of the ICA-based algorithm proposed in [32]. Three recorded data channels C_z, P_z, and F_z are the inputs to the ICA algorithm

the latter case, simulations show that ICA may derive single components accounting for the co-occurring phenomena, along with additional components accounting for their brief periods of separate activation [31].

Third, ICA assumes that physical sources of artefacts and cerebral activity are spatially fixed over time. In general, this is not always true. Applying ICA to brain data shows the effectiveness of the independence assumption. ERP is generally the largest component (after eyeblink artefact,) within the scalp map emitted from one or two dipoles. This is unlikely to occur unless time courses of coherently synchronous neural activity in patches of cortical neurons generating the EEG are nearly independent of one another [33].

Fourth, application of ICA to fMRI does not necessarily provide the full information about the actual components of the fMRI sequences. However, by placing ICA in a regression framework, it is possible to combine some of the benefits of ICA with the hypothesis-testing approach of the general linear model (GLM) [34].

Current research on ICA algorithms has been focused on incorporating domain-specific constraints into the ICA framework. This would allow information maximization to be applied to the precise form and statistics of biomedical data [31].

9.2.2 Estimation of Single Trial Brain Responses by Modelling the ERP Waveforms

Detection or estimation of evoked potentials (EPs) from only a single trial EEG is favourable since (i) on-line processing of the signals can be performed, and (ii) variability of ERP parameters across trials can be tracked. Unlike the averaging (multiple trial) [35] scheme, in this approach the shape of the ERPs is first approximated and then used to recover the actual signals. A few good examples, including mental fatigue, are investigated in this book in which trial-to-trial variability of ERPs is of clinical importance.

A decomposition technique, which relies on the statistical nature of neural activity, is the one that efficiently separates the EEGs into their constituent components including the ERPs. A neural activity may be delayed when passing through a number of synaptic nodes each introducing a delay. Thus, the firing instants of many synchronized neurons may be assumed to be governed by Gaussian probability distributions. In a work by Lange $et\,al.$ [24] the evoked potential source waveform is assumed to consist of superposition of p components u_i delayed by τ_i:

$$s(n) = \sum_{i=1}^{p} k_i \cdot u_i(n - \tau_i) \tag{9.3}$$

The model equation for constructing the data from a number of delayed templates can be given in the z-domain as:

$$\mathbf{X}(z) = \sum_{i=1}^{p} \mathbf{B}_i(z) \cdot \mathbf{T}_i(z) + \mathbf{W}^{-1}(z)\mathbf{E}(z) \tag{9.4}$$

where $\mathbf{X}(z)$, $\mathbf{T}(z)$ and $\mathbf{E}(z)$ represent, respectively, the observed (or measured) signals, the average EP, and a Gaussian white noise. Assuming that the background EEG is statistically stationary, $\mathbf{W}(z)$ is identified from the pre-stimulus data via AR modelling of the pre-stimulus interval, and used for post-stimulus analysis [24]. The template and measured EEG are filtered through the identified $\mathbf{W}(z)$ to whiten the background EEG signal and thus, a closed-form least-squares (LS) solution of the model is formulated. Therefore, the only parameters to be identified are the matrices $\mathbf{B}_i(z)$. This avoids any iterative identification of the model parameters. This can be represented using a regression-type equation as:

$$\breve{x}(n) = \sum_{i=1}^{p} \sum_{j=-d}^{d} b_{i,j} \cdot \breve{T}_i(n-j) + e(n) \tag{9.5}$$

where d denotes the delay (latency), p is the number of templates, and $\breve{x}(n)$ and $\breve{T}(n)$ are, respectively, whitened versions of $x(n)$ and $T(n)$. Let $\mathbf{A}^{\mathrm{T}}$ be the matrix of input templates and $\mathbf{b}^{\mathrm{T}}$ the filter coefficients vector; then we construct:

$$\breve{\mathbf{x}} = [\breve{x}(d+1), \breve{x}(d+2), \ldots, \breve{x}(N-d)]^{\mathrm{T}} \tag{9.6}$$

$$\mathbf{A} = \begin{bmatrix} \breve{T}_1(2d+1)\breve{T}_1(2d+2)\cdots\breve{T}_1(N) \\ \breve{T}_1(2d)\breve{T}_1(2d+1)\cdots\breve{T}_1(N-1) \\ \vdots \\ \breve{T}_1(1)\breve{T}_1(2)\cdots\breve{T}_1(N-2d) \\ \breve{T}_2(2d+1)\breve{T}_2(2d+2)\breve{T}_2(N) \\ \vdots \\ \breve{T}_2(1)\breve{T}_2(2)\cdots\breve{T}_2(N-2d) \\ \vdots \\ \breve{T}_p(2d+1)\breve{T}_p(2d+2)\cdots\breve{T}_2(N) \\ \vdots \\ \breve{T}_p(1)\breve{T}_p(2)\cdots\breve{T}_p(N-2d) \end{bmatrix}^{\mathrm{T}} \tag{9.7}$$

and

$$b = [b_{1,-d}, b_{1,-d+1}, \ldots, b_{1,d}, b_{2,-d}, \ldots, b_{p,d}]^{\mathrm{T}} \tag{9.8}$$

The model can also be expressed in a vector form as

$$\breve{\mathbf{x}}(n) = \mathbf{A}\mathbf{b}(n) + \boldsymbol{\varepsilon}(n) \tag{9.9}$$

where $\boldsymbol{\varepsilon}(n)$ is the vector of prediction errors;

$$\boldsymbol{\varepsilon}(t) = [e(2d+1), e(2d+2), \ldots, e(N)]^{\mathrm{T}} \tag{9.10}$$

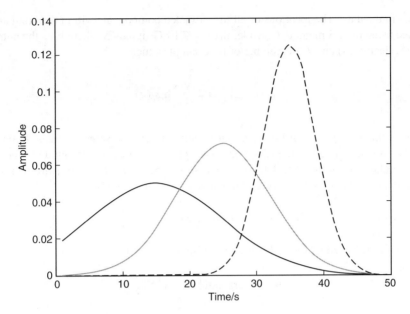

Figure 9.5 Synthetic ERP templates including a number of delayed Gaussian and exponential waveforms

To solve for the model parameters the sum squared error defined as

$$\xi(\mathbf{b}) = \|\boldsymbol{\varepsilon}(n)\|^2 \tag{9.11}$$

is minimised. Then, the optimal vector of parameters in the LS sense is obtained as:

$$\hat{\mathbf{b}} = (\mathbf{A}^\mathsf{T}\mathbf{A})^{-1}\mathbf{A}^\mathsf{T}\breve{\mathbf{x}} \tag{9.12}$$

The model provides an easier way of analysing the ERP components from single-trial EEG signals and facilitates tracking of amplitude and latency shifts of such components. The experimental results show that the main ERP components can be extracted with a high accuracy. The template signals may look like those in Figure 9.5.

A similar method has also been developed to detect and track the P300 subcomponents, P3a and P3b, of the ERP signals [36]. This is described in the following section.

9.2.3 ERP Source Tracking in Time

The major drawback of BSS and dipole methods is that the number of sources needs to be known a priori to achieve good results. In a recent work [36] P300 signals are modelled as spike-shape Gaussian signals. The latencies and variances of these signals are, however, subject to change. The spikes serve as reference signals onto which the EEG data are projected. The existing spatio-temporal information is then used to find the closest representation of the reference in the data. The locations of all the extracted components within the brain are later computed using the LS method. Consider the $q \times T$ EEG signals $\mathbf{X}$ where n_e is the number of channels (electrodes) and T the number of time samples, then

$$\mathbf{x}(n) = \mathbf{H}\mathbf{s}(n) = \sum_{i=1}^{m} \mathbf{h}_i s_i(n) \tag{9.13}$$

where $\mathbf{H}$ is the $n_e \times m$ forward mixing matrix and $s_i(n)$ are the source signals. Initially, the sources are all considered as the ERP components that are directly relevant and time-locked to the stimulus and assumed to have transient spiky shape. m filters $\{\mathbf{w}_i\}$ are then needed (although the number of sources is not known beforehand) to satisfy

$$s_i(n) = \mathbf{w}_i^T \mathbf{x}(n) \tag{9.14}$$

This can be achieved by the following minimisation criterion:

$$\mathbf{w}_{\text{opt}} = \arg \min_{\mathbf{w}} \left\| \mathbf{s}_i - \mathbf{w}_i^T \mathbf{X} \right\|_2^2 \tag{9.15}$$

where $\mathbf{X} = [\mathbf{x}_1, \mathbf{x}_2, \ldots, \mathbf{x}_{n_e}]^T$, which however requires some knowledge about the shape of the sources. A Gaussian type spike defined as:

$$s_i(n) = e^{-(n-\tau_i)^2/\sigma_i^2} \tag{9.16}$$

where τ_i, $i = 1, 2, \ldots T$, is the latency of the ith source and σ_i is its width, can be used as the model. The width is chosen as an approximation to the average width of the P3a and P3b subcomponents. Then, for each of the T filters ($T \gg m$) we have:

$$\mathbf{w}_{l\,\text{opt}}^T = \arg \min_{w_l} \left\| \mathbf{r}_l - \mathbf{w}_l^T \mathbf{X} \right\|_2^2, \quad \mathbf{y}_l = \mathbf{w}_l^T \mathbf{X} \tag{9.17}$$

where $(.)^T$ denotes the transpose operation. Therefore, each $\mathbf{y}_l$ has a similar latency to that of a source. Figure 9.6 shows the results of the ERP detection algorithm. It can be seen that all possible peaks are detected first. However, because the number of ERP signals is limited the outputs may be grouped into m clusters. To perform such classification we examine the following criteria for the latencies of the ith and $i-1$)th estimated ERP components according to the following rule;

$$\text{for } l = 1 \text{ to } T$$

if $l(i) - l(i-1) < \beta$, where β is an small empirical threshold, then $\mathbf{y}_i$ and $\mathbf{y}_{i-1}$ are assumed to belong to the same cluster,
if $l(i) - l(i-1) > \beta$ then $\mathbf{y}_i$ and $\mathbf{y}_{i-1}$ belong to different clusters.

Here $l(i)$ denotes the latency of the ith component. The signals within each cluster are then averaged to obtain the related ERP signal, $\mathbf{y}_c$. $c = 1, 2, \ldots, m$. Figure 9.7 presents the results

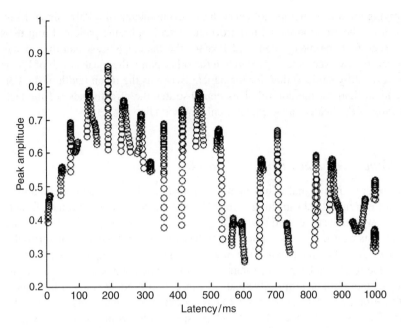

Figure 9.6 The results of the ERP detection algorithm [35]; the scatter plot which shows the peaks at different latencies. The concentration of samples around the peaks makes the ERP classification straightforward

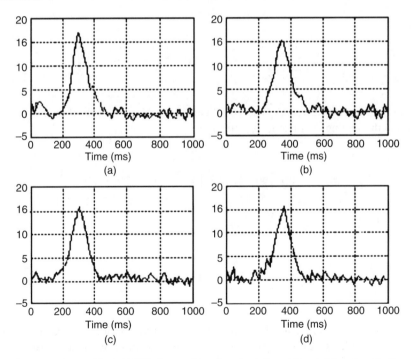

Figure 9.7 The average P3a and P3b for a schizophrenic patient (a) and (b), respectively, and for a healthy individual (c) and (d), respectively

after applying the above method to detect the subcomponents of a P300 signal from a single trial EEG for a healthy individual (control) and a schizophrenic patient. From these results it can be seen that the mean latency of P3a for the patient is less than that of the control. The difference, however, is less for P3b. On the other hand, the mean width of P3a is less for the control (healthy subject) than for the patient, whereas the mean width of the P3b is more for the patient than for the control. This demonstrates the effectiveness of the technique in classification of the healthy and schizophrenic individuals.

9.2.4 Time–Frequency Domain Analysis

In places where some signal components are transient, that is, a signal contains frequency components emerging and vanishing in time-limited intervals, localization of the active signal components within the time–frequency (TF) domain is very useful [37]. The traditional method proposed for such an analysis is application of the short-time frequency transform (STFT) [38]. The STFT enables the time localization of a certain sinusoidal frequency but with an inherent limitation due to Heisenberg's uncertainty principle which states that resolution in time and frequency cannot both be arbitrarily small, because their product is lower bounded by $\Delta t \Delta \omega \geq 1/2$. Therefore, wavelet transform (WT) has become very popular instead. An advantage of WT over STFT is that for the STFT the phase space is uniformly sampled, whereas in WT the sampling in frequency is logarithmic which enables one to analyse higher frequencies in shorter windows and lower frequencies in longer windows in time [39].

As explained in Chapter 4 a multi-resolution decomposition of signal $x(n)$ over I octaves is given as:

$$x(n) = \sum_{i=1}^{\infty} \sum_{k \in Z} a_{i,k} g_i(n - 2^i k) + \sum_{k \in Z} b_{I,k} h_I(n - 2^I k) \tag{9.18}$$

Where Z refers to integer values. The DWT computes the wavelet coefficients $a_{i,k}$ for $i = 1, \ldots, I$ and the scaling coefficients $b_{I,k}$ given by

$$a_{i,k} = DWT\left\{x(n); 2^I, k2^I\right\} = \sum_{n=0}^{N} x(n) g_i^*(n - 2^i k) \tag{9.19}$$

and

$$b_{I,k} = \sum_{n=0}^{N} x(n) h_I^*(n - 2^I k). \tag{9.20}$$

where $(.)^*$ denotes the conjugate. The functions $g(.)$ and $h(.)$ perform, respectively, as high-pass and lowpass filters. Amongst the wavelets discrete B-spline WTs have near-optimal TF localization [40] and also, generally have anti-symmetric properties, which make them more suitable for analysis of ERPs. The filters for an nth-order B-spline wavelet multiresolution decomposition are computed as;

$$h(k) = \frac{1}{2} \left[b^{2n+1} \right]^{-1} (k) \uparrow_2^* b^{2n+1}(k) * p^n(k) \tag{9.21}$$

and

$$g(k + 1) = \frac{1}{2} [b^{2n+1}]^{-1}(k) \uparrow_2 * (-1)^k p^n(k) \tag{9.22}$$

where $\uparrow_2$ indicates upsampling by 2 and n is an integer. For the quadratic spline wavelet ($n = 2$) employed by Ademoglu $et\ al.$ to analyse pattern reversal VEPs [39] substituting $n = 2$ in the above equations, the parameters can be derived mathematically as:

$$[(b^5)^{-1}](k) = Z^{-1} \left\{ \frac{120}{z^2 + 26z + 66 + 26z^{-1} + z^{-2}} \right\} \tag{9.23}$$

$$p^2(k) = \frac{1}{2^2} Z^{-1} \left\{ 1 + 3z^{-1} + 3z^{-2} + z^{-3} \right\} \tag{9.24}$$

where Z^{-1} denotes the inverse Z-transform. To find $[(b^5)^{-1}](k)$ the z-domain term may be factorized into:

$$[(b^5)^{-1}](k) = Z^{-1} \left[\frac{\alpha_1 \alpha_2}{(1 - \alpha_1 z^{-1})(1 - \alpha_1 z)(1 - \alpha_2 z^{-1})(1 - \alpha_2 z)} \right] \tag{9.25}$$

where $\alpha_1 = -0.04309$ and $\alpha_2 = -0.43057$. Therefore:

$$[(b^5)^{-1}](k) = \frac{\alpha_1 \alpha_2}{(1 - \alpha_1^2)(1 - \alpha_2^2)(1 - \alpha_1 \alpha_2)(\alpha_1 - \alpha_2)} \cdot \left[\alpha_1(1 - \alpha_2^2)\alpha_1^{|k|} - \alpha_2(1 - \alpha_1^2)\alpha_2^{|k|} \right] \tag{9.26}$$

On the other hand, $p(k)$ can also be computed easily by taking the inverse transform in equation (9.24). Therefore the sample values of $h(n)$ and $g(n)$ can be obtained. Finally, to perform a WT only lowpass and highpass filtering followed by down-sampling are needed. In such a multiresolution scheme the number of wavelet coefficients halves from one scale to the next. This requires and justifies longer time windows for lower frequencies and shorter time windows for higher frequencies.

In the above work the VEP waveform is recorded using a bipolar recording at electrode locations C_z and O_z of the conventional 10–20 EEG system. The sampling rate is 1 kHz. Therefore, each scale covers the following frequency bands: 250–500 Hz, 125–250 Hz, 62.5–125 Hz, 31.3–62.5 Hz, 15.5–31.5 Hz, 7.8–15.6 Hz (including N70, P100, and N130), and 0–7.8 Hz (residual scale). Although this method is not meant to discriminate between normal and pathological subjects, the ERPs can be highlighted and the differences between normal and abnormal cases are observed. By using the above spline wavelet analysis, it is observed that the main effect of a latency shift of the N70-P100-N130 complex is reflected in the sign and magnitude of the second, third, and fourth wavelet coefficients within the 7.8–15.6 Hz band. Many other versions of the WT approach have been introduced in the literature during the recent decades, such as those in [41–43].

As an example, single-trial ERPs have been analysed using wavelet networks (WNs) and the developed algorithm applied to the study of attention deficit hyperactivity disorder (ADHD) [41]. ADHD is a psychiatric and neurobehavioral disorder. It is characterised by significant occurrence of inattention or hyperactivity and impulsiveness. In this work it is assumed that the ERP and the background EEG are correlated, that is, the parameters, which describe the signals before the stimulation, do not hold for the post-stimulation period [44]. In a WN topology they presented an ERP, $\hat{s}(t)$ as a linear combination of K modified wavelet functions $h(n)$:

$$\hat{s}(n) = \sum_{k=1}^{K} w_k h\left(\frac{n - b_k}{a_k}\right) \tag{9.27}$$

where b_k, a_k, and w_k, are, respectively, the shift, scale, and weight parameters and $s(n)$ are called wavelet nodes. Figure 9.8 shows the topology of a WN for signal representation. In this figure $\psi_k = h((u - b_k)/a_k)$ in the case of a Morlet wavelet,

$$\hat{s}(n) = \sum_{k=1}^{K} w_{1,k} \cos\left(\omega_k \left(\frac{n - b_k}{a_k}\right)\right) + w_{2,k} \sin\left(\omega_k \left(\frac{n - b_k}{a_k}\right)\right) \exp\left[-\frac{1}{2}\left(\frac{n - b_k}{a_k}\right)^2\right] \tag{9.28}$$

In this equation each of the frequency parameters ω_k and the corresponding scale parameters a_k define a node frequency $f_k = \omega_k/2\pi a_k$, and are optimized during the learning process.

In the above approach a WN is considered as a single hidden layer perceptron whose neuron activation function [45] is defined by wavelet functions. Therefore, the well-known

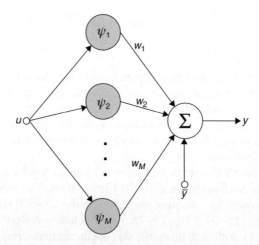

Figure 9.8 Construction of an ERP signal using a WN. The nodes in the hidden layer are represented by modified wavelet functions. The addition of the $\bar{y}$ value is to deal with functions whose means are nonzero

backpropagation neural network algorithm [46] is used to simultaneously update the WN parameters (i.e. $w_{\cos,k}$, $w_{\sin,k}$, ω_k, a_k, and b_k of node k). The minimisation is performed on the error between the node outputs and the bandpass filtered version of the residual error obtained using ARMA modelling. The bandpass filter and a tapering time-window is applied to ensure that the desired range is emphasized and the node sees its specific part of the TF plane [41]. It has been reported that for the study of ADHD this method enables detection of the N100 ERP components in places where the traditional time averaging approach fails due to the position variation of this component.

In a recent work there has been an attempt to create a wavelet-like filter from simulation of the neural activities which is then used in place of a specific wavelet to improve the performance of wavelet-based single-trial ERP detection. The main application of such a filter has been found to be in BCI [35].

9.2.5   Application of Kalman Filter

The main problem in linear time invariant filtering of ERP signals is that the signal and noise frequency components highly overlap. Another problem is that the ERP signals are transient and their frequency components may not fall into a certain range. Wiener filtering, which provides optimum solution in the mean square sense for the majority of noise removal cases, is not suitable for nonstationary signals such as ERPs.

With the assumption that the ERP signals are dynamically slowly varying processes the future realizations are predictable from the past realizations. These changes can be studied using a state-space model. Kalman filtering and generic observation models have been used to denoise the ERP signals [47]. Considering a single channel EEG observation vector at stimulus time t, as $\mathbf{x}_n = [x_n(1), x_n(2), \ldots, x_n(N)]^T$, the objective is to estimate a vector of parameters such as $\boldsymbol{\theta}_n$ of length k for which the map from $\mathbf{x}_n$ to $\boldsymbol{\theta}_n$ is (preferably) a linear map. An optimum estimator, $\hat{\boldsymbol{\theta}}_{n,\text{opt}}$, which minimises $E[||\hat{\boldsymbol{\theta}}_n - \boldsymbol{\theta}_n||^2]$, is $\hat{\boldsymbol{\theta}}_{n,\text{opt}} = E[\boldsymbol{\theta}_n|\mathbf{x}_n]$ and can be shown to be

$$\hat{\boldsymbol{\theta}}_{n,\text{opt}} = \mu_{\boldsymbol{\theta}_n} + C_{\boldsymbol{\theta}_n,\mathbf{x}_n} C_{\mathbf{x}_n}^{-1}(\mathbf{x}_n - \mu_{\mathbf{x}_n}) \tag{9.29}$$

where $\mu_{\boldsymbol{\theta}_n}$ and $\mu_{\mathbf{x}_n}$ are, respectively, the means of $\boldsymbol{\theta}_n$ and $\mathbf{x}_t$, and $C_{\boldsymbol{\theta}_n,\mathbf{x}_n}$ is the cross-covariance matrix of the observations and the parameters to be estimated, and $C_{\mathbf{x}_n}$ is the covariance matrix of the column vector $\mathbf{x}_n$. Such an estimator is independent of the relationship between $\boldsymbol{\theta}_n$ and $\mathbf{x}_n$. The covariance matrix of the estimated error $\boldsymbol{\varepsilon}_{\theta} = \boldsymbol{\theta}_n - \hat{\boldsymbol{\theta}}_n$ can also be found as:

$$C_{\boldsymbol{\varepsilon}} = C_{\boldsymbol{\theta}_n} - C_{\boldsymbol{\theta}_n,\mathbf{x}_n} C_{\mathbf{x}_n}^{-1} C_{\mathbf{x}_n,\boldsymbol{\theta}_n} \tag{9.30}$$

In order to evaluate $C_{\boldsymbol{\theta}_n,\mathbf{x}_n}$ some prior knowledge about the model is required. In this case the observations may be assumed to be of the form $\mathbf{x}_n = \mathbf{s}_n + \mathbf{v}_n$, where $\mathbf{s}_n$ and $\mathbf{v}_n$ are considered, respectively, as the response to the stimulus and the background EEG (irrelevant of the stimulus and ERP). Also, the ERP is modelled as:

$$\mathbf{x}_n = \mathbf{H}_n \boldsymbol{\theta}_n + \mathbf{v}_n \tag{9.31}$$

where $\mathbf{H}_n$ is a deterministic $N \times k$ observation model matrix. The estimated ERP $\hat{\mathbf{s}}_n$ can then be obtained as:

$$\hat{\mathbf{s}}_n = \mathbf{H}_n \hat{\boldsymbol{\theta}}_n \tag{9.32}$$

where the estimated parameters, $\hat{\boldsymbol{\theta}}_n$, of the observation model $\mathbf{H}_n$, using a linear mean square (MS) estimator and with the assumption of $\boldsymbol{\theta}_n$ and $\mathbf{v}_n$ being uncorrelated, is achieved as [48]:

$$\hat{\boldsymbol{\theta}}_n = \left(\mathbf{H}_n^{\mathrm{T}} C_{\mathbf{v}_n}^{-1} \mathbf{H}_n + C_{\boldsymbol{\theta}_n}^{-1} \right)^{-1} \left(\mathbf{H}_n^{\mathrm{T}} C_{\mathbf{v}_n}^{-1} \mathbf{x}_n + C_{\boldsymbol{\theta}_n}^{-1} \mu_{\boldsymbol{\theta}_n} \right) \tag{9.33}$$

This estimator is optimum if the joint distribution of $\boldsymbol{\theta}_n$ and $\mathbf{x}_n$ is Gaussian and the parameters and noise are uncorrelated.

However, in order to take into account the dynamics of the ERPs from trial to trial the evolution of $\boldsymbol{\theta}_n$ has to be taken into account. Such evolution may be denoted as:

$$\boldsymbol{\theta}_{n+1} = \mathbf{F}_n \boldsymbol{\theta}_n + \mathbf{G}_n \mathbf{w}_n \tag{9.34}$$

and some initial distribution for $\boldsymbol{\theta}_0$ assumed. Although the states are not observed directly the parameters are related to the observations through equation (9.31). In this model it is assumed that $\mathbf{F}_n$, $\mathbf{H}_n$, and $\mathbf{G}_n$ are known matrices, $(\boldsymbol{\theta}_0, \mathbf{w}_n, \mathbf{v}_n)$ is a sequence of mutually uncorrelated random vectors with finite variance, $E[\mathbf{w}_n] = E[\mathbf{v}_n] = 0 \; \forall n$, and the covariance matrices $C_{\mathbf{w}_n}$, $C_{\mathbf{v}_n}$, and $C_{\mathbf{w}_n, \mathbf{v}_n}$ are known.

With the above assumptions the Kalman filtering algorithm can be employed to estimate $\boldsymbol{\theta}$ as [49]:

$$\hat{\boldsymbol{\theta}}_n = \left(\mathbf{H}_n^{\mathrm{T}} C_{\mathbf{v}_n}^{-1} \mathbf{H}_n + C_{\boldsymbol{\theta}_{n|n-1}}^{-1} \right)^{-1} \left(\mathbf{H}_n^{\mathrm{T}} C_{\mathbf{v}_n}^{-1} \mathbf{x}_n + C_{\boldsymbol{\theta}_{n|n-1}}^{-1} \hat{\boldsymbol{\theta}}_{n|n-1} \right) \tag{9.35}$$

where $\hat{\boldsymbol{\theta}}_{n|n-1} = E[\boldsymbol{\theta}_n | \mathbf{x}_{n-1}, \ldots, \mathbf{x}_1]$ is the prediction of $\boldsymbol{\theta}_n$ subject to $\hat{\boldsymbol{\theta}}_{n-1} = E[\boldsymbol{\theta}_{n-1} | \mathbf{x}_{n-1}, \ldots, \mathbf{x}_1]$, which is the optimum mean-square estimate at $n-1$. Such an estimator is the best sequential estimator if the Gaussian assumption is valid and is the best linear estimator disregarding the distribution. The overall algorithm based on Kalman filtering for tracking the ERP signal may be summarized as follows:

$$C_{\hat{\boldsymbol{\theta}}_0} = C_{\boldsymbol{\theta}_0} \tag{9.36}$$

$$\hat{\boldsymbol{\theta}}_0 = E[\boldsymbol{\theta}_0] \tag{9.37}$$

$$\hat{\boldsymbol{\theta}}_{n|n-1} = \mathbf{F}_{n-1} \hat{\boldsymbol{\theta}}_{n-1} \tag{9.38}$$

$$C_{\hat{\boldsymbol{\theta}}_{n|n-1}} = \mathbf{F}_{n-1} C_{\hat{\boldsymbol{\theta}}_{n-1}} \mathbf{F}_{n-1}^{\mathrm{T}} + \mathbf{G}_{n-1} C_{\mathbf{w}_{n-1}} \mathbf{G}_{n-1}^{\mathrm{T}} \tag{9.39}$$

$$\mathbf{K}_t = C_{\hat{\boldsymbol{\theta}}_{n|n-1}} \mathbf{H}_n^{\mathrm{T}} \left(\mathbf{H}_n C_{\hat{\boldsymbol{\theta}}_{n|n-1}} \mathbf{H}_n^{\mathrm{T}} + C_{\mathbf{v}_n} \right)^{-1} \tag{9.40}$$

$$C_{\hat{\boldsymbol{\theta}}_n} = (\mathbf{I} - \mathbf{K}_n \mathbf{H}_n) C_{P \hat{\boldsymbol{\theta}}_{n|n-1}} \tag{9.41}$$

$$\hat{\boldsymbol{\theta}}_n = \boldsymbol{\theta}_{n|n-1} + \mathbf{K}_n \left(\mathbf{x}_n - \mathbf{H}_n \hat{\boldsymbol{\theta}}_{n|n-1} \right) \tag{9.42}$$

In the work by Georgiadis *et al.* [47], however, a simpler observation model has been proposed. In this model the state-space equations have the forms:

$$\boldsymbol{\theta}_{n+1} = \boldsymbol{\theta}_n + \mathbf{w}_n \tag{9.43}$$

$$\mathbf{x}_n = \mathbf{H}_n \boldsymbol{\theta}_n + \mathbf{v}_n \tag{9.44}$$

in which the state vector models a random walk process [50]. Now, assuming that the conditional covariance matrix of the parameter estimation error is $\mathbf{P}_t = C_{\boldsymbol{\theta}_t} + C_{\mathbf{w}_t}$, the Kalman filter equations will be simplified to

$$\mathbf{K}_n = \mathbf{P}_{n-1}\mathbf{H}_n^{\mathrm{T}} \left(\mathbf{H}_n \mathbf{P}_{n-1} \mathbf{H}_n^{\mathrm{T}} + C_{\mathbf{v}_n} \right)^{-1} \tag{9.45}$$

$$\mathbf{P}_n = (I - \mathbf{K}_n \mathbf{H}_n) \mathbf{P}_{n-1} + C_{\mathbf{w}_n} \tag{9.46}$$

$$\hat{\boldsymbol{\theta}}_n = \hat{\boldsymbol{\theta}}_{n-1} + \mathbf{K}_n \left(\mathbf{x}_n - \mathbf{H}_n \hat{\boldsymbol{\theta}}_{n-1} \right) \tag{9.47}$$

where $\mathbf{P}_n$ and $\mathbf{K}_n$ are called, respectively, the recursive covariance matrix estimate and the Kalman-gain matrix. At this stage it is interesting to compare these parameters for the four well-known recursive estimation algorithms, namely recursive least squares (RLS), least mean square (LMS), normalised least mean square (NLMS), and the above Kalman estimators for the above simple model. This is depicted in Table 9.1 [47].

In practice $C_{\mathbf{w}_n}$ is considered as a diagonal matrix, such as $C_{\mathbf{w}_n} = 0.1\mathbf{I}$. Also it is logical to consider $C_{\mathbf{v}_n} = \mathbf{I}$, and, by assuming the background EEG to have Gaussian distribution it is possible to use the Kalman filter to dynamically estimate the ERP signals. As an example, it is possible to consider $\mathbf{H}_n = \mathbf{I}$, so that the ERPs are recursively estimated as:

$$\hat{\mathbf{s}}_n = \hat{\boldsymbol{\theta}}_n = \hat{\boldsymbol{\theta}}_{n-1} + \mathbf{K}_n(\mathbf{x}_n - \hat{\boldsymbol{\theta}}_{n-1}) = \mathbf{K}_n \mathbf{x}_n + (\mathbf{I} - \mathbf{K}_n)\hat{\mathbf{s}}_{n-1} \tag{9.48}$$

Such assumptions make the application of the algorithms in Table 9.1 for ERP tracking and parameter estimation easier.

In a more advanced algorithm $\mathbf{H}_n$ may be adapted to the spatial topology of the electrodes. This means that the column vectors of $\mathbf{H}_n$ are weighted based on the expected locations of the

Table 9.1 $\mathbf{K}_n$ and $\mathbf{P}_n$ for different recursive algorithms

RLS	$\mathbf{K}_n = \mathbf{P}_{n-1}\mathbf{H}_n^{\mathrm{T}} \left(\mathbf{H}_n \mathbf{P}_{n-1} \mathbf{H}_n^{\mathrm{T}} + \lambda_n \right)^{-1}$ $\mathbf{P}_n = \lambda_n^{-1} (I - \mathbf{K}_n \mathbf{H}_n) \mathbf{P}_{n-1}$
LMS	$\mathbf{K}_n = \mu \mathbf{H}_n^{\mathrm{T}}$ $\mathbf{P}_n = \mu \left(\mathbf{I} - \mu \mathbf{H}_n^{\mathrm{T}} \mathbf{H}_n \right)^{-1}$
NLMS	$\mathbf{K}_n = \mu \mathbf{H}_n^{\mathrm{T}} (\mu \mathbf{H}_n \mathbf{H}_n^{\mathrm{T}} + 1)^{-1}$ $\mathbf{P}_n = \mu \mathbf{I}$
Kalman Filter	$\mathbf{K}_n = \mathbf{P}_{n-1}\mathbf{H}_n^{\mathrm{T}} \left(\mathbf{H}_n \mathbf{P}_{n-1} \mathbf{H}_n^{\mathrm{T}} + C_{\mathbf{v}_n} \right)^{-1}$ $\mathbf{P}_n = (I - \mathbf{K}_n \mathbf{H}_n) \mathbf{P}_{n-1} + C_{\mathbf{w}_n}$

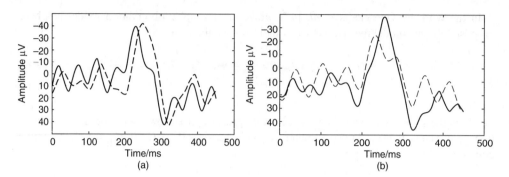

Figure 9.9 Dynamic variations of ERP signals; (a) first stimulus and (b) 20th stimulus, for the ERPs detected using LMS (dotted line) and Kalman filtering (thick line) approaches

ERP generators within the brain. Figure 9.9 compares the estimated ERPs for different stimuli for the LMS and recursive mean-square estimator (Kalman filter). From this figure it is clear that the estimates using the Kalman filtering appear to be more robust and show very similar patterns for different stimulation instants.

9.2.6 Particle Filtering and Its Application to ERP Tracking

Particle filtering (PF) is an attractive but complicated state-space approach widely used for object tracking from video sequences, target tracking in airborne radar or generally multi-dimensional data. PF is more suitable in places where the linearity condition required for KF is not satisfied.

In a wavelet-based PF approach [51] the time locked ERP in the wavelet domain in the kth trial is presented as $\mathbf{y}_k = [y_k(1), y_k(2), \ldots, y_k(M)]^{\mathrm{T}}$.

Following the fundamentals of PF, by modelling the wavelet domain signals (i.e. wavelet coefficients) in the state-space, in general, the evolution of the state $\{\mathbf{x}_k\}$ and the relation between the state and the estimated wavelet coefficients (the measurement equation) are given as follows.

$$\mathbf{x}_k = f_{k-1}\left(\mathbf{x}_{k-1}, \mathbf{w}_{k-1}\right) \tag{9.49}$$

$$\mathbf{y}_k = h_k(\mathbf{x}_k, \mathbf{v}_k) \tag{9.50}$$

where f_k and h_k are nonlinear functions of the state $\mathbf{x}_k$, and $\mathbf{w}_{k-1}$ and $\mathbf{v}_k$ are i.i.d. noise processes. In PF the filtered estimates of $\mathbf{x}_k$ are searched based on a set of all available wavelet coefficients. According to Bayes rule, an available measurement at time k, that is, $\mathbf{y}_k$ is iteratively used to update the posterior density

$$p\left(\mathbf{x}_k | \mathbf{y}_{1:k}\right) = \frac{p\left(\mathbf{y}_k | \mathbf{x}_k\right) p\left(\mathbf{x}_k | \mathbf{y}_{1:k-1}\right)}{p\left(\mathbf{x}_k | \mathbf{y}_{1:k-1}\right)} \tag{9.51}$$

where $p(\mathbf{x}_k|\mathbf{y}_{1:k-1})$ is computed in the prediction stage using the Chapman–Kolmogorov equation and $p(\mathbf{y}_k|\mathbf{y}_{1:k-1})$ is a normalizing constant.

In PF, the posterior distributions are approximated by discrete random measures defined by particles $\{\mathbf{x}^{(n)}; n = 1, \ldots, N\}$ and their associated weights $\{w^{(n)}; n = 1, \ldots, N\}$. The distribution based on these samples and weights at the kth trial is approximated as

$$p(\mathbf{x}) \approx \sum_{n=1}^{N} w^{(n)} \delta(\mathbf{x} - \mathbf{x}^{(n)}) \tag{9.52}$$

where δ is the Dirac delta function and $(.)^n$ refers to the nth weight. If the particles are generated according to distribution $p(\mathbf{x})$, the weights are identical and equal to $1/N$. When the distribution $p(\mathbf{x})$ is unknown, the particles are generated from a known distribution $\pi(\mathbf{x})$ called the importance density. Following the concepts of importance sampling and following the Bayes rule, the particle weights can be iteratively estimated as

$$w_k^{(n)} \propto w_{k-1}^{(n)} \frac{p\left(\mathbf{y}_k^{(n)}|\mathbf{x}_k^{(n)}\right) p\left(\mathbf{x}_k^{(n)}|\mathbf{x}_{k-1}^{(n)}\right)}{\pi\left(\mathbf{x}_k^{(n)}|\mathbf{x}_k^{(n)}, \mathbf{y}_{1:k}\right)} \tag{9.53}$$

The choice of importance density is crucial in the design of PF since it significantly affects the performance. This function must have the same support as the probability distribution to be approximated; the closer the importance function to the distribution, the better the weight approximations are. The most popular choice for the prior importance function, is given by

$$\pi\left(\mathbf{x}_k^{(n)}|\mathbf{x}_k^{(n)}, \mathbf{y}_{1:k}\right) \approx p\left(\mathbf{x}_k^{(n)}|\mathbf{x}_{k-1}^{(n)}\right) \tag{9.54}$$

This choice of importance density implies that we need to sample from $p(\mathbf{x}_k^{(n)}|\mathbf{x}_{k-1}^{(n)})$. A sample can be obtained by generating a noise sample for w_{k-1}^n, and setting

$$\mathbf{x}_k^{(n)} = f_{k-1}\left(\mathbf{x}_{k-1}^{(n)}, \mathbf{w}_{k-1}^{(n)}\right) \tag{9.55}$$

Therefore, the particle weights can be updated as [51]

$$w_k^{(n)} \propto w_{k-1}^{(n)} p\left(\mathbf{y}_k^{(n)}|\mathbf{x}_k^{(n)}\right) \tag{9.56}$$

The importance sampling weights indicate the level of importance of the corresponding particle. Relatively small weights imply that the sample is drawn far from the main body of the posterior distribution and has a small contribution in the final estimation. Such a particle is said to be ineffective. If the number of ineffective particles is increased, the number of particles contributing to the estimation of states is decreased, so the performance of the filtering procedure deteriorates. The degeneracy can be avoided by resampling. During the resampling process the particles with small weights are eliminated and those with large weights are replicated.

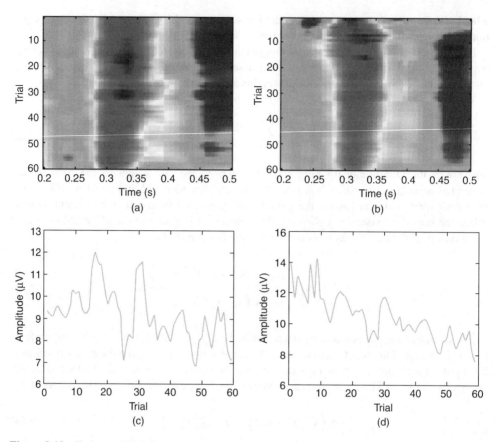

Figure 9.10 Estimated ERPs by applying KF and PF, (a) and (b) ERP latency over 60 trials using KF and PF, respectively, and (c) and (d) amplitudes of the P300 waves for consecutive trials extracted by KF and PF, respectively

An experiment using real EEG data over 60 trials, assuming Gaussian distributions for both state and observation noises and when w_{k-1}^n is sampled from a Gaussian distribution, using both KF and PF has been performed. The results are depicted in Figure 9.10 [52].

In another approach tracking variation of ERP parameters for both p3a and p3b subcomponents of P300 a Rao–Blackwellised PF (RBPF) approach has been developed [52]. In this approach the particles vary with respect to the changes in both p3a and p3b parameters simultaneously. RBPF combines KF and PF to track linear (such as amplitude in this work) and nonlinear (such as latency and width) parameters from a single EEG channel at the same time, that is,

$$\mathbf{x}_k = [\mathbf{a}_k(1)\mathbf{b}_k(1)\mathbf{s}_k(1) \ldots \mathbf{a}_k(p)\mathbf{b}_k(p)\mathbf{s}_k(p)]^T \qquad (9.57)$$

a, **b**, and **s** refer to the ERP parameters (amplitude, latency, and width) and p denotes the number of existing ERPs in a single trial k. If all the ERPs are masked out and only P300 is retained, then p3a and p3b can be considered as the ERP components in (9.49), that is, $p = 2$. Following RBPF, it is possible to take the amplitudes out and form a matrix and find a linear relation between the observation and amplitudes and a nonlinear relation with respect to other variables. Therefore, regarding the nonlinear portion of particles in equation (9.49) the state-space and observation can be formulated as:

$$\mathbf{x}_k^{(1)} = [\mathbf{b}_k(1)\mathbf{s}_k(1)\dots\mathbf{b}_k(p)\mathbf{s}_k(p)]^\mathrm{T} \tag{9.58}$$

$$\mathbf{x}_k^{(2)} = [\mathbf{a}_k(1).\dots\mathbf{a}_k(p)]^\mathrm{T} \tag{9.59}$$

$$\mathbf{x}_k = \mathbf{x}_{k-1} + \mathbf{v}_{k-1} \tag{9.60}$$

$$\mathbf{z}_k = \left[f_1(\mathbf{x}_k^{(1)})\dots f_p(\mathbf{x}_k^{(1)}) \right] \begin{bmatrix} \mathbf{a}_k(1) \\ \vdots \\ \mathbf{a}_k(p) \end{bmatrix} + \mathbf{n}_k \tag{9.61}$$

$$f_i(\mathbf{x}_k^{(1)}, t) = f_i\left(\mathbf{b}_k(i), \mathbf{s}_k(i), t\right) = \mathrm{e}^{-\frac{(t - \mathbf{b}_k(i))^2}{2s_k^2(i)}} \tag{9.62}$$

t denotes the time index and varies from the beginning to the end of the ERP component. In each iteration the relation between the two RBPFs has to be evaluated. As long as the shapes of P3a or P3b are assumed to remain the same in every channel, we can relate the pairs of particles with the same width for P3a and P3b. Using the relation between two RBPFs, a constrained RBPF is built up which not only links the widths but also relies on having p3a before p3b. This is useful for removing the invalid particles. The weights of these invalid particles are then set to zero disabling them in predicting the new particles. These invalid particles are those whose latency of P3b is shorter than the latency of P3a.

9.2.7 Variational Bayes Method

A very recent approach for ERP parameter estimation is by using variational Bayes. Variational Bayes methods, also called ensemble learning, can be seen as an extension of the expectation maximization (EM) algorithm from maximum a posteriori (MAP) estimation of the single most probable value of each parameter to full Bayesian estimation. Variational Bayes is employed to compute (an approximation to) the complete posterior distribution of the estimation parameters and latent variables [49]. It has been used effectively for spatiotemporal modelling of EEG and MEG systems for both distributed and dipole source models [52,53]. Variational Bayes has also been utilized for fast computation of source parameters [54]. The main shortcoming for this approach is that only specific priors (normally gamma distributions) are used for the model parameters. This constraint often limits the performance of the method in practical applications. A spatiotemporal variational Bayes using a dipole source model has also been formulated in [55]. In these methods, Markov chain Monte Carlo (MCMC) technique was used to sample the unknown high-dimensional parameters from the marginalized posterior distribution. It is, however, likely for the solutions of this approach to trap in local maxima [56].

In [56] a new approach to the detection and tracking of event-related potential (ERP) subcomponents based on variational Bayes has been suggested. In this work the ERP subcomponent sources have been taken as electric current dipoles (ECDs), and their locations and parameters (amplitude, latency, and width) are estimated and tracked from trial to trial. Variational Bayes allows the parameters to be estimated separately using the likelihood function of each parameter.

The locations of ECDs can vary from trial to trial in a realistic head model. ERPs are assumed to be the superposition of a small number of ECDs and their temporal bases are modelled by Gaussian-shaped signals. The amplitudes, means, and variances of the Gaussian waves can be interpreted as the amplitudes, latencies, and widths of ERP subcomponents, respectively.

Variational Bayes shows that, when the prior distribution is unknown, maximizing the likelihood of each parameter (via separate estimation of each) is equivalent to minimizing the Kullback–Leibler distance between the estimated and the true posterior distributions.

The locations are estimated using PF. Many studies have shown that PF is one of the best methods when the relation between the desired parameter (states) and the measurement is nonlinear [25]. A closed-form solution for the amplitude is also given by the ML approach.

Estimations of the amplitude and noise covariance matrix of the measurement are optimally estimated recursively by the maximum likelihood (ML) approach, while estimations of the latency and width are obtained by a recursive Newton–Raphson technique [56].

The Newton–Raphson algorithm has a rapid convergence. One challenge for this approach is that very low SNR in some trials can result in divergence of the filtering, which impacts negatively on the estimation of amplitudes, latencies, and widths. To compensate for this failure, recursive methods are suggested to improve the stability of the trial to trial filtering.

The main advantage of the method is the ability to track varying ECD locations.

To formulate the problem assume that there are q ECDs and their generated sources are combined over the L electrodes on the scalp, each consists of M samples, as [56]:

$$\mathbf{Y}_k = \sum_{i=1}^{q} \mathbf{H}(\rho_{k,i})\mathbf{a}_{k,i}\psi_{k,i} + \mathbf{N}_k \tag{9.63}$$

where $\mathbf{H}$ is the nonlinear forward matrix of size $L \times 3$ and is a nonlinear function of the ECD locations and $\mathbf{N}_k$ is an additive white zero-mean Gaussian noise of spatial variance $\mathbf{Q}_k$ which is unknown, and $\mathbf{I}$ the temporal variance (identity matrix). $\mathbf{a}_{k,i} \in \Re^{3\times1}$ is the strength (amplitude) of the ECD moment in three dimensions and $\psi_{k,i}(t) = [\psi_{k,i}(1), \ldots, \psi_{k,i}(M)] \in \Re^{1\times M}$ refers to the temporal basis of the ith ECD moment.

$\mathbf{H}$ can be calculated in a spherical head model with three layers: skull, scalp, and skin (the medium within each layer is assumed to be homogenous), or can be obtained using a realistic head model. In the realistic head model, after considering each small region to be a cell (or vertex) of a fine mesh, a pre-estimated forward matrix $\mathbf{H}$ is given for each cell (vertex). Also, the covariance of noise can be denoted as $\mathbf{I} \otimes \mathbf{Q}_k$, where $\otimes$ denotes a Kronecker product.

Noise is assumed to be independent of the source activities and distributed identically across time, but not necessarily across sensors. These assumptions provide a fast and simple estimation of the noise covariance matrix from trial to trial.

Each $\psi_{k,i}(t)$ is given by a Gaussian wave as [56]

$$\psi_{k,i}(t) = \frac{1}{\sigma_{k,i}\sqrt{2\pi}} \exp\left(-\frac{(t - \mu_{k,i})^2}{2\sigma_{k,i}^2}\right) \tag{9.64}$$

The primary objective of this work is to recursively estimate the model parameters $\theta_{k,i} = \{\rho_{k,i}, \mathbf{a}_{k,i}, \mathbf{Q}_k, \mu_k, \sigma_k\}$ based on their previous estimates $\theta_{k-1,i}$ and the available measurements $\mathbf{Y}_k$. Furthermore, it is assumed that the changes in $\theta_{k-1,i}$ are smooth and it can be often considered as a Markovian process [56]. In the recursive estimation the set of parameters is estimated first. Then, all the parameters are updated recursively.

The algorithm starts with estimating the posterior distribution $p(\theta|\mathbf{Y})$ instead of θ itself. The posterior distribution is often difficult to calculate, particularly in nonlinear cases. Therefore, the MAP is approximated by a simpler distribution called *variational distribution*, $r(\theta)$. This can be found using variational Bayes methods. To maximize the similarity between the MAP and $r(\theta)$ different criteria such as Kullback–Leibler or free energy can be utilized [56]. As the key restriction in variational Bayes, it is also assumed that the variational distributions are factorizable to the distributions of the constituent parameters in the set that is, $r(\theta) = \prod_i r(\theta_i)$.

In the above approach, the locations are placed in one group and the latencies and the widths in another group. The parameter θ is partitioned into ρ (location), $\mathbf{a}$ (amplitude), $\mathbf{Q}$ (noise variance), μ, and σ and different methods can be employed for estimation of the posterior distribution of each sub-parameter $r(\theta i)$ estimated as:

$$r(\theta_i) \propto p(\mathbf{Y}|\theta_i) \tag{9.65}$$

In the localization and tracking of multiple ECDs related to ERPs $\mathbf{R}$ can be considered as a matrix including the locations of all the dipoles, that is, $\mathbf{R}_k = [\rho_{k,1} \ldots, \rho_{k,q}] \in \mathbb{R}^{3 \times q}$. Then, following the process for estimation of particles in the previous section we have:

$$w_k^{(n)} \propto w_{k-1}^{(n)} p\left(\mathbf{y}_k^{(n)}|\mathbf{R}_k^{(n)}\right) \tag{9.66}$$

For tracking the ERP amplitude $\mathbf{a}_k$ and noise variance $\mathbf{Q}_k$ at trial k the log likelihood of (9.63) is maximized. Accordingly, the negative log likelihood to be minimized is defined as [56]:

$$f(\theta, \mathbf{Y}_k) = \text{tr}\left\{\left(\mathbf{Y}_k - \sum_{i=1}^{q} \mathbf{H}(\rho_{k,i})\mathbf{a}_{k,i}\psi_{k,i} + \mathbf{N}_k\right)^{\text{T}} \mathbf{Q}_k^{-1} \times \left(\mathbf{Y}_k - \sum_{i=1}^{q} \mathbf{H}(\rho_{k,i})\mathbf{a}_{k,i}\psi_{k,i} + \mathbf{N}_k\right)\right\} \tag{9.67}$$

An alternate minimization of the above equation results in stable values for $\mathbf{a}_k$ and $\mathbf{Q}_k$.

The same function has also been minimised using the Newton–Raphson algorithm to find the latency and width of the ERP components [56].

The results of the application of the method to a multi-channel EEG dataset including 60 time locked ERP trials for tracking p3a and p3b parameters can be seen in Figure 9.11. This figure shows the variations in latency and amplitude of both p3a and p3b.

9.2.8   Prony's Approach for Detection of P300 Signals

Based on Prony's approach the source signals are modelled as a series expansion of damped sinusoidal basis functions. Prony's method is often used for estimation of pole positions when the noise is white. Also, in [57] the basis functions are estimated using the least-squares

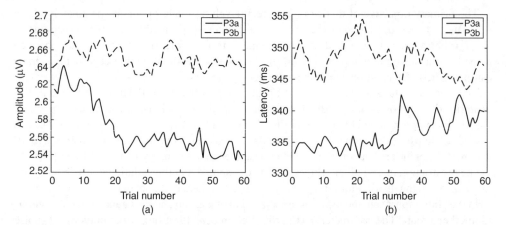

Figure 9.11 Estimated (a) amplitude and (b) latency of P3a (bold line) and P3b (dotted line) for real data

method, derived for coloured noise. This is mainly because after the set of basis functions are estimated using Prony's method, the optimal weights are calculated for each single trial ERP by minimizing the squared error.

The amplitude and latency of the P300 signal give information about the level of attention. In many psychiatric diseases this level changes. Therefore, a reliable algorithm for more accurate estimation of these parameters is very useful.

In the above attempt, the ERP signal is initially divided into the background EEG and ERP, respectively, before and after the stimulus time instant. The ERP is also divided into two segments, the early brain response, which is a low-level high-frequency signal, and the late response, which is a high-level low-frequency signal that is,

$$x(n) = \begin{cases} x_1(n) & -L \leq n < 0 \\ s_e(n - n_0) + x_2(n) & 0 \leq n \leq L_1 \\ s_l(n - n_0 - L_1) + x_3(n) & L_1 \leq n < N + L_1 \end{cases} \tag{9.68}$$

where $x_1(n)$, $x_2(n)$, and $x_3(n)$ are, respectively, the background EEGs before the stimulus, during the early brain response, and during the late brain response, and $s_e(n)$ and $s_l(n)$ are respectively the early and late brain responses to a stimulus. Therefore, to analyse the signals they are windowed (using a window $g(n)$) within $L_1 \leq n < N + L_1$ within the duration that the P300 signal exists. This gives

$$x_g(n) = x(n)g(n) \approx \begin{cases} s_l(n - L_1) + x_3(n) & L_1 \leq n < N + L_1 \\ 0 & \text{Otherwise} \end{cases} \tag{9.69}$$

We may consider $s(n) = s_l(n - L_1)$ for $L_1 \leq n < N + L_1$, for simplicity. Based on Prony's algorithm the ERP signals are represented through the following over-determined system:

$$s(n) = \sum_{i=1}^{M} a_i \rho_i^n e^{j\omega_i n} = \sum_{i=1}^{M} a_i z_i^n \tag{9.70}$$

where $z_i = \rho_i e^{j\omega_i}$ and $M < N$. Using vector notation we can write:

$$\mathbf{x}_g = \mathbf{Z}\mathbf{a} + \mathbf{x}_3 \tag{9.71}$$

where $\quad \mathbf{x}_g = [x_g(L_1) \ldots x_g(N + L_1 - 1)]^\mathrm{T}, \qquad \mathbf{x}_3 = [x_3(L_1) \ldots x_3(N + L_1 - 1)]^\mathrm{T},$
$\mathbf{a} = [a_1 \ldots a_M]^\mathrm{T}$ and

$$\mathbf{Z} = \begin{bmatrix} \rho_1 e^{j\omega_1} & \cdots & \rho_M e^{j\omega_M} \\ \vdots & & \\ \rho_1 e^{j\omega_1 N} & \cdots & \rho_M e^{j\omega_M N} \end{bmatrix} \tag{9.72}$$

Therefore, in order to estimate the waveform $\mathbf{s} = \mathbf{Z}\mathbf{a}$ both $\mathbf{Z}$ and $\mathbf{a}$ have to be estimated. The criterion to be minimised is the weighted least-squares criterion given by

$$\min_{\mathbf{Z},\mathbf{a}} \left\{ (\mathbf{x}_g - \mathbf{Z}\mathbf{a})^\mathrm{H} \mathbf{R}_{\mathbf{x}_3}^{-1} (\mathbf{x}_g - \mathbf{Z}\mathbf{a}) \right\} \tag{9.73}$$

where $\mathbf{R}_{\mathbf{x}_3}$ is the covariance matrix of the EEG (as a disturbing signal) after stimulation. For a fixed $\mathbf{Z}$ the optimum parameter values are obtained as:

$$\mathbf{a}_{\mathrm{opt}} = \left(\mathbf{Z}^\mathrm{H} \mathbf{R}_{\mathbf{x}_3}^{-1} \mathbf{Z} \right)^{-1} \mathbf{Z}^\mathrm{H} \mathbf{R}_{\mathbf{x}_3}^{-1} \mathbf{x}_g \tag{9.74}$$

Then, the corresponding estimate of the single trial ERP is achieved as:

$$\mathbf{s} = \mathbf{Z}\mathbf{a}_{\mathrm{opt}} \tag{9.75}$$

Prony's method is employed to minimise (9.73) for coloured noise. In order to do that a matrix $\mathbf{F}$ orthogonal to $\mathbf{Z}$ is considered, that is, $\mathbf{F}^\mathrm{H}\mathbf{Z} = 0$ [58]. The ith column of $\mathbf{F}$ is col $\{\mathbf{F}\}_i = [\underbrace{0 \ldots 0}_{i-1} \ f_0 \ldots f_M \ \underbrace{0 \ldots 0}_{N-M-i}]^T$. The roots of the polynomial $f_0 + f_1 z^{-1} + \cdots + f_M z^{-M}$ are the elements z_i in $\mathbf{Z}$ [58]. The minimization in (9.73) is then converted to

$$\min_{\mathbf{F}} \left\{ \mathbf{x}_g^\mathrm{H} \mathbf{F} (\mathbf{F}^\mathrm{H} \mathbf{R}_{\mathbf{x}_3} \mathbf{F})^{-1} \mathbf{F}^\mathrm{H} \mathbf{x}_g \right\} \tag{9.76}$$

In practice a linear model is constructed from the above nonlinear system using [58, 59]

$$\mathbf{F} \left(\mathbf{F}^\mathrm{H} \mathbf{R}_{\mathbf{x}_3} \mathbf{F} \right)^{-1} \mathbf{F}^\mathrm{H} \approx \tilde{\mathbf{F}} \mathbf{R}_{\mathbf{x}_3}^{-1} \tilde{\mathbf{F}}^\mathrm{H} \tag{9.77}$$

where the ith column of $\tilde{\mathbf{F}}$ is col $\{\tilde{\mathbf{F}}\}_i = [\underbrace{0 \ldots 0}_{i-1} \ f_0 \ldots f_p \ \underbrace{0 \ldots 0}_{N-p-i}]^T$ where $p \gg M$ and, therefore, it includes the roots of $\mathbf{F}$ plus p-M spurious poles. Then, the minimisation in (9.57) changes to

$$\min_{\tilde{\mathbf{F}}} \left\{ \mathbf{e}^\mathrm{H} \mathbf{R}_{\mathbf{x}_3}^{-1} \mathbf{e} \right\} = \min_{\tilde{\mathbf{F}}} \left\{ \mathbf{x}_g^\mathrm{H} \tilde{\mathbf{F}} \mathbf{R}_{\mathbf{x}_3}^{-1} \tilde{\mathbf{F}}^\mathrm{H} \mathbf{x}_g \right\} \tag{9.78}$$

where $\mathbf{e} = \tilde{\mathbf{F}}^H \mathbf{x}_g = \mathbf{Xf}$. In this equation

$$\mathbf{X} = \begin{bmatrix} x_g(p) & x_g(p-1) & \dots & x_g(0) \\ \vdots & \vdots & & \vdots \\ x_g(N-1) & x_g(N-2) & \dots & x_g(N-p-1) \end{bmatrix} = [\tilde{\mathbf{x}} \quad \tilde{\mathbf{X}}] \qquad (9.79)$$

where $\tilde{\mathbf{x}}$ is the first column of $\mathbf{X}$ and includes the spurious poles. The original Prony's method for solving (9.59) is found by replacing $\mathbf{R}_{x_3} = \mathbf{I}$ with $f_0 = 1$ or $f_1 = 1$. However, the background EEG cannot be white noise, therefore, in this approach a coloured noise, $\mathbf{R}_{x_3} \neq \mathbf{I}$, is considered. Hence the column vector $\mathbf{f}$ including the poles or the resonance frequencies of the basis functions in (9.51) is estimated as:

$$\mathbf{f} = [1; -[\tilde{\mathbf{X}}^H \mathbf{R}_{x_3}^{-1} \tilde{\mathbf{X}}]_M^{-1} [\tilde{\mathbf{X}}^H \mathbf{G}^{-H}]_M \mathbf{G}^{-1} \tilde{\mathbf{x}}] \qquad (9.80)$$

$\mathbf{GG}^H = \mathbf{R}_{x_3}$ and the notation $[.]_M$ denotes the rank M approximated matrix. The rank reduction is performed using a singular value decomposition (SVD). Although several realizations of the ERP signals may have similar frequency components, $\mathbf{a}$ may vary from trial to trial. Therefore, instead of averaging the data to find $\mathbf{a}$, the estimated covariance matrix is averaged over D realizations of the data. An analysis based on a forward–backward Prony solution is finally employed to separate the spurious poles from the actual poles. The parameter $\mathbf{a}$ can then be computed using (9.55).

Generally, the results achieved using Prony-based approaches for single-trial ERP extraction are rather reliable and consistent with the physiological and clinical expectations. The above algorithm has been examined for a number of cases, including pre-stimulus, after stimulation and during post-stimulus. Clear peaks corresponding to mainly P300 sources are evident in the extracted signals.

In another attempt, similar to the above approach, it is considered that for the ERP signals frequency, amplitude, and phase characteristics vary with time, therefore a piecewise Prony method (PPM) has been proposed [60]. The main advantages of using PPM are: first, this method enables modelling non-monotonically growing or decaying components with nonzero onset time. Therefore, by using PPM it is assumed that the sinusoidal components have growing or decaying envelopes, and abrupt changes in amplitude and phase. The PPM uses variable length windows (previously suggested by [60, 61]) and variable sampling frequencies (performed also by Kulp [62] for adjusting sampling rate to the frequencies to be modelled) to overcome the limitations of the original Prony's method. Secondly, the window size is determined based on the signal characteristics. Also, the signals with large bandwidths are modelled in several steps, focusing on smaller frequency bands per step, as proposed in [63]. Thirdly, the varying-length windows try to include the conventional Prony components. Therefore, there is more consistency in detecting the ERP features (frequency, amplitude, and phase). Finally, it is reasonable to assume that the signal components obtained from adjacent windows with identical frequency are part of the same component and they can be combined into one component.

9.2.9   Adaptive Time–Frequency Methods

Combination of an adaptive signal estimation technique and time–frequency signal representation can enhance the performance of the ERP and EP detections. This normally refers to modelling the signals with variable-parameter systems and application of adaptive estimators to estimate suitable parameters.

Using STFT the information about the duration of the activities is not exploited and, therefore, it cannot describe structures much shorter or much longer than the window length. WT can overcome this limitation by allowing for variable window length but there is still a reciprocal relation between the central frequency of the wavelet and its window length. Therefore, WT does not precisely estimate the low frequency components with short-time duration or narrow-band high frequency components.

In another work the adaptive chirplet transform (ACT) has been utilized to characterise the time-dependent behaviour of the VEPs from its transient to steady state section [64]. Generally, this approach uses a matching pursuit (MP) algorithm to estimate the chirplets and a maximum likelihood (ML) algorithm to refine the results and enable estimation of the signal with a low SNR. Using this method it is possible to visualize the early moments of a VEP response.

The ACT attempts to decompose the signals into Gaussian chirplet basis functions with four adjustable parameters of time-spread, chirp rate, time-centre and frequency centre. Moreover, it is shown that only three chirplets can be used to represent a VEP response.

In this approach the transient VEP which appears first following the onset of the visual stimulus [65] is analysed together with the steady-state VEP [66,67]. Identification of steady-state VEP has had many clinical applications and can help diagnose sensory dysfunction [68,69]. The model using only steady-state VEPs is, however, incomplete without considering the transient pattern. A true steady-state VEP is also difficult to achieve in the cases where the mental situation of the patient is not stable.

Chirplets (also called chirps) are windowed rapidly swept sinusoidal signals [70]. The bases for a Gaussian chirplet transform are derived from a simple Gaussian function $\pi^{-1/4} \exp(-t^2/2)$ through four operations, namely scaling, chirping, time- and frequency-shifting, which (as for wavelets) produces a family of wave packets with four adjustable parameters [71]. Such a continuous time chirplet may be represented as:

$$g(t) = \pi^{-1/4} \Delta_t^{-1/2} e^{-\frac{1}{2}\left(\frac{t-t_c}{\Delta_t}\right)^2} e^{j[c(t-t_c)+\omega_c](t-t_c)} \tag{9.81}$$

where t_c is the time centre, ω_c is the frequency centre, $\Delta_t > 0$ is the effective time-spread, and c is the chirp rate which refers to the speed of changing the frequency. The chirplet transform of a signal $f(t)$ is then defined as

$$a(t_c, \omega_c, c, \Delta_t) = \int_{-\infty}^{\infty} f(t)g^*(t)dt \tag{9.82}$$

Based on this approach, the signal $f(t)$ is reconstructed as a linear combination of Gaussian chirplets as:

$$f(t) = \sum_{n=1}^{p} a(t_c, \omega_c, c, \Delta_t)g(t) + e(t) \tag{9.83}$$

where p denotes the approximation order and $e(t)$ is the residual signal. Figure 9.12 illustrates the chirplets extracted from an EEG-type waveform. To estimate $g(t)$ at each round (iteration) six parameters have to be estimated (consider a complex). An optimal estimation of these

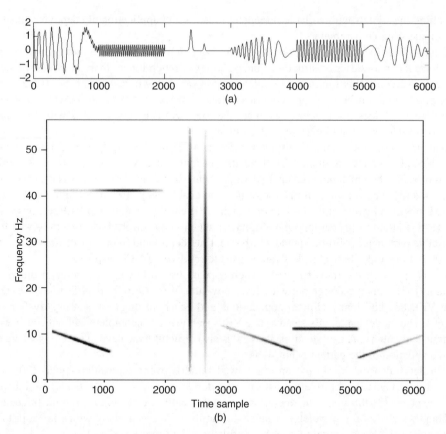

Figure 9.12 The chirplets extracted from a simulated EEG-type waveform [64]; (a) represents the combined waveforms and (b) shows the TF representation of the corresponding chirplets

parameters is generally not possible. However, there have been some suboptimal solutions such as in [71–74]. In the approach by Cui and Wong [64] a coarse estimation of the chirplets is obtained using the MP algorithm. Then, maximum likelihood estimation (MLE) is performed iteratively to refine the results.

To perform this the algorithm is initialised by setting $e(n) = f(n)$. A dictionary is constructed using a set of predefined chirplets to cover the entire time-frequency plane [71]. The MP algorithm projects the residual $e(n)$ to each chirplet in the dictionary and the optimal chirplet is decided based on the projection amplitude. In the next stage, Newton-Raphson iterative algorithm is used to refine the results and achieve the optimum match.

Under a low SNR situation a post processing step maybe needed to follow the MLE concept [75] followed by the EM algorithm proposed in [64].

9.3   Brain Activity Assessment Using ERP

Variations in P300 component parameters, such as amplitude, latency, and duration, from trial to trial indicate the depth of cognitive information processing. For example, it has been reported that the P300 amplitude elicited by the mental task loading decreases with the increase in the perceptual/cognitive difficulty of the task [76]. Assessment of mental fatigue using P300 signals by measuring the amplitude and latency of the signals and also the alpha-band power have shown that [77] the P300 amplitude tends to decrease immediately after the experimental task whereas the latency decreases at this time. The amplitude decrease is an indicator of decreasing the attention level of the subject, which can be due to habituation. The increase in the latency might be indicative of the prolonged temporal processing due to the difficulty of cognitive information processing [76–79]. Therefore, the mental and physical fatigues can be due to this problem. In this work, it is also pointed out that one aspect of the mentally and physically fatigued could be because of decreased activity of the central nervous system (CNS) that appears in both the temporal prolongation of cognitive information processing and the decrease in attention level. It has also been demonstrated that alpha-band power decreases due to attention to the task and beta-band power rises immediately after the task. This indicates that the activity of the CNS is decelerated with accumulation of mental and physical fatigue. However, the appearance of fatigue is reflected more strongly in the amplitude and latency of the P300 signal and its subcomponents p3a and p3b rather than the alpha-band power. This is analysed in detail in Chapter 10.

9.4   Application of P300 to BCI

Electrical cortical activities used in BCI may be divided into the five following categories. (i) beta (β) and mu (μ) rhythms: these activities range, respectively, within 8–12 Hz and 12–30 Hz. These signals are associated with those cortical areas most directly connected to the motor output of the brain and can be willingly modulated with an imagery movement [80]; (ii) P300 evoked potential (EP): it is a late appearing component of an auditory, visual, or somatosensory ERP as explained before; (iii) visual N100 and P200: the ERPs with short latency that represent the exogenous response of the brain to a rapid visual stimulus. These potentials are used as clues indicating the direction of user gaze [81, 82]; (iv) steady-state visual evoked potentials (SSVEP): these signals are natural responses to visual stimulations at specific frequencies. When the retina is excited by a visual stimulus ranging from 3.5 to 75 Hz, the brain generates an electrical activity at the same (or multiples of the) frequency of the visual stimulus. They are used for understanding which stimulus the subject is looking at in the case of stimuli with different flashing frequencies [83, 84]; (v) slow cortical potentials (SCP): they are slow potential variations generated in the cortex 0.5 to 10s after presenting the stimulus. Negative SCPs are generated by movement whereas positive SCPs are associated with reduced cortical activation. Adequately trained users can control these potentials and use them to control the movement of a cursor on the computer screen [85].

To enable application of these waves, especially P300 for a P300-based BCI, the data have to be pre-processed to reduce noise and enforce P300-related information. A pattern recognition algorithm has to be later developed to check the presence of the P300 wave in the recorded ERP epochs and label them. Also, a feedback mechanism has to be established to send the user a visible signal on the monitor correlated to the recorded epoch. Finally, the parameters of the pattern recognition algorithm have to be made adaptable to the subject's characteristics.

The P300 speller, as the first P300-BCI, described by Farwell and Dounchin, adapted the OP as the operating principle of BCI [25]. In this paradigm the participant is presented with a Bernoulli sequence of events, each belonging to one of two categories. The participant is assigned a task that cannot be performed without a correct classification of the events, each belonging to one of two categories. Using this speller, for example, a 6 × 6 matrix is displayed to the subject. The system is operated by briefly intensifying each row and column of the matrix and the attended row and column elicits a P300 response. In a later work [86] it was found that some people who suffer from amyotrophic lateral sclerosis (ALS) can better respond to the system with a smaller size matrix (less than 6 × 6).

In addition, measurement and evaluation of the movement-related events, such as event-related desynchronization (ERD), from the EEGs can improve the diagnosis of functional deficits in patients with cerebrovascular disorders and Parkinson's disease (PD).

There is a high correlation between morphological, such as computerised tomography (CT) and functional, such as EEG, findings in cerebrovascular disorders. For example, the ERD reduces over the affected hemisphere.

The pre-movement ERD in PD is less lateralized over the contralateral sensorimotor area and starts later than in control subjects. Also, post-movement beta ERS is of smaller magnitude and delayed in PD as compared to controls. It has been shown [87] that based on only two parameters, namely the central ERD within 6–10 Hz, and post-movement event-related synchronization (ERS) within 16–20 Hz, it is possible to discriminate PD patients with a Hoehn and Yahr scale of 1–3 from age-matched controls by a linear discriminant analysis. The Hoehn and Yahr scale is a commonly used system for describing how the symptoms of PD progress [54].

Also, following the above procedure in P300 detection and classification, two user adaptive BCI systems based on SSVEP and P300, have been proposed by Beverina *et al.* [88].

The P300 component has been detected (separated) using ICA for the BCI purpose in another attempt [30]. It has also been empirically confirmed that the visual spatial attention modulation of the SSVEP can be used as a control mechanism in a real-time independent BCI (i.e. when there is no dependence on peripheral muscles or nerves) [89].

P300 signals have also been used in the Wandsworth BCI development [90]. In this application a similar method to that proposed by Farwell and Donchin [25] is used for highlighting and detection of the P300 signals. It has been shown that this well established system can be used by severely disabled people in their homes with minimal ongoing technical oversight.

In another attempt, P300 signals have been used for the design of a speller (text-input application). Twenty five channels around C3, C4, Cz, CPz, and FCz electrodes are manually selected to have the best result. In addition P7 and P8 electrodes are also used. The number of channels is later reduced to 20 using PCA and selecting the largest eigenvalues. During this process the data are also whitened and the effect of the eye-blink is removed. Support vector machines (SVM) are then used to classify the principal components for each character, using a Gaussian kernel [55].

9.5   Conclusions

ERP signals indicate the types and states of many brain abnormalities and mental disorders. These signals are characterised by their spatial, temporal and spectrum locations. Also, they are often considered as independent sources within the brain. The ERP signals can be characterized by their amplitudes, latencies, source locations, and frequency contents. Automatic extraction and tracking of these signals from single EEG trials, however, requires sufficient knowledge and expertise in the development of mathematical and signal processing algorithms. Although so far there has not been any robust and well-established method to detect and characterize these signals, some recently developed single trial methods have high potential, supported by analysis of real EEG signals for evaluation and tracking of more popular ERP components such as P300. More recent methods, such as in [53], overcome the limiting conditions, for instance, lack of correlation or independence of various ERP components in their separation procedure.

References

[1] Walter, W.G. (1964) Contingent negative variation: an electrical sign of sensorimotor association and expectancy in the human brain. *Nature*, **203**, 380–384.

[2] Sutton, S., Braren, M., Zoubin, J. and John, E.R. (1965) Evoked potential correlates of stimulus uncertainty. *Science*, **150**, 1187–1188.

[3] Johnson, R. Jr. (1992) Event-related brain potentials, in *Progressive Supranuclear Palsy: Clinical and Research Approaches* (eds I. Litvan and Y. Agid), Oxford University Press, New York, pp. 122–154.

[4] Visser, S.L., Stam, F.C., Van Tilburg, W. *et al.* (1976) Visual evoked response in senile and presenile dementia. *Electrencephalogr. Clin. Neurophysiol.*, **40**(4), 385–392.

[5] Cosi, V., Vitelli, E., Gozzoli, E. *et al.* (1982) Visual evoked potentials in aging of the brain. *Adv. Neurol.*, **32**, 109–115.

[6] Coben, L.A., Danziger, W.L. and Hughes, C.P. (1981) Visual evoked potentials in mild senile dementia of Alzheimer type. Electroencephalogr. *Clin. Neurophysiol.*, **52**, 100.

[7] Visser, S.L., Van Tilburg, W., Hoojir, C. *et al.* (1985) Visual evoked potentials (VEP's) in senile dementia (Alzheimer type) and in nonorganic behavioural disorders in the elderly: Comparison with EEG parameters. Electroencephalogr. *Clin. Neurophysiol.*, **60**(2), 115–121.

[8] Nunez, P.L. (1981) *Electric Fields of the Brain*, Oxford University Press, New York.

[9] Picton, T.W., Lins, D.O. and Scherg, M. (1995) The recording and analysis of event-related potentials, in *Hand Book of Neurophysiology*, vol. **10** (eds F. Boller and J. Grafman), Elsevier, Amsterdam, pp. 3–73.

[10] Perrin, P., Pernier, J., Bertrand, O. and Echallier, J.F. (1989) Spherical splines for scalp potential and current density mapping. *Electroencephalogr. Clin. Neurophysiol.*, **72**, 184–187.

[11] Hegerl, U. (1999) Event-related potentials in psychiatry, Chapter 31, *Electroencephalography*, 4th edn (eds E. Niedermayer and F. Lopez Da Silva), Lippincott Williams & Wilkins, pp. 621–636.

[12] Rangaswamy, M., Jones, K.A., Porjesz, B. *et al.* (2007) Delta and theta oscillations as risk markers in adolescent offspring of alcoholics. *Int. J. Psychophysiol.*, **63**, 3–15.

[13] Diez, J., Spencer, K. and Donchin, E. (2003) Localization of the event-related potential novelty response as defined by principal component analysis. *Brain Res. Cogn.*, **17**(3), 637–650.

[14] Frodl-Bauch, T., Bottlender, R. and Hegerl, U. (1999) Neurochemical substrates and neuro-anatomical generators of the event-related P300. *Neuropsychobiology*, **40**, 86–94.

[15] Kok, A., Ramautar, J., De Ruiter, M. *et al.* (2004) ERP components associated with successful and unsuccessful stopping in a stop-signal task. *Psychophysiology*, **41**(1), 9–20.

[16] Polich, J. (2004) Clinical application of the P300 event-related brain potential. *Phys. Med. Rehab. Clin. N. Am.*, **15**(1), 133–161.

[17] Friedman, D., Cycowics, Y. and Gaeta, H. (2001) The novelty P3: an event-related brain potential (ERP) sign of the brains evaluation of novelty. *Neurosci. Biobehav. Rev.*, **25**(4), 355–373.

[18] Kropotov, J.D., Alho, K., Näätänen, R. *et al.* (2000) Human auditory-cortex mechanisms of preattentive sound discrimination. *Neurosci. Lett.*, **280**, 87–90.

[19] Comerchero, M. and Polich, J. (1999) P3a and P3b from typical auditory and visual stimuli. *Clin Neurophysiol.*, **110**(1), 24–30.

[20] Friedman, D. and Squires-Wheeler, E. (1994) Event-related potentials (ERPs) as indicators of risk for schizophrenia. *Schizophrenia Bull.*, **20**(1), 63–74.

[21] Luria, A.R. (1973) *The Working Brain*, Basic Books, New York.

[22] Sokolov, E.N. (1963) *Perception and the Conditioned Reflex*, Pergamon Press, Oxford, UK.

[23] Friedman, D., Cycowicz, Y.M. and Gaeta, H. (2001) The novelty P3: an event-related brain potential (ERP) sign of the brain's evaluation of novelty. *Neurosci. Biobehav. Rev.*, **25**, 355–373.

[24] Lange, D.H., Pratt, H. and Inbar, G.F. (1997) Modeling and estimation of single evoked brain potential components. *IEEE Trans. Biomed. Eng.*, **44**(9), 791–799.

[25] Farwell, L.A. and Dounchin, E. (1998) Talking off the top of your heard: Toward a mental prosthesis utilizing event-related brain potentials. *Electroenceph. Clin. Neurophysiol.*, **70**, 510–523.

[26] Donchin, E., Spencer, K.M. and Wijesingle, R. (2000) The mental prosthesis: assessing the speed of a P300-based brain-computer interface. *IEEE Trans. Rehab. Eng.*, **8**, 174–179.

[27] McGillem, C.D. and Aunon, J.I. (1997) Measurement of signal components in single visually evoked brain potentials. *IEEE Trans. Biomed. Eng.*, **24**, 232–241.

[28] Makeig, S., Jung, T.P., Bell, A.J. and Sejnowsky, T.J. (1997) Blind separation of auditory event-related brain responses into independent components. *Proc. Nat. Acad. Sci.*, **94**, 10979–10984.

[29] Bell, A.J. and Sejnowsky, T.J. (1995) An information maximisation approach to blind separation and blind deconvolution. *Neural Comput.*, **7**, 1129–1159.

[30] Xu, N., Gao, X., Hong, B. *et al.* (2004) BCI competition 2003-data set IIb: enhancing P300 wave detection using ICA-based subspace projections for BCI applications. *Trans. Biomed. Eng.*, **51**(6), 1067- 1072.

[31] Jung, T.-P., Makeig, S., Mckeown, M. *et al.* (2001) Imaging brain dynamics using independent component analysis. *Proc. IEEE*, **89**(7), 1107–1122.

[32] Serby, H., Yom-Tov, E. and Inbar, G.F. (2005) An improved P300-based brain-computer interface. *IEEE Trans. Neural Syst. Rehab. Eng.*, **13**(1), 89–98.

[33] Makeig, S., Enghoff, S., Jung, T.-P. and Sejnowski, T.J. (2000) Moving-window ICA decomposition of EEG data reveals event-related changes in oscillatory brain activity. 2nd International Workshop on Independent Component Analysis and Signal Separation, 2000, pp. 627–632.

[34] McKeown, M.J. (2000) Detection of consistently task-related activations in fMRI data with hybrid independent component analysis. *Neuroimage*, **11**, 24–35.

[35] Glassman, E.L. (2005) A wavelet-like filter based on neuron action potentials for analysis of human scalp electroencephalographs. *IEEE Trans. Biomed. Eng.*, **52**(11), 1851–1862.

[36] Spyrou, L., Sanei, S. and Cheong Took, C. (2007) Estimation and location tracking of the P300 subcomponents from single-trial EEG. Proceedings of IEEE, International Conference on Acoustic Signal Speech Processing, ICASSP, USA, 2007.

[37] Marple, S.L. Jr. (1987) *Digital Spectral Analysis with Applications*, Prentice Hall, Englewood Cliffs, NJ.

[38] Bertrand, O., Bohorquez, J. and Pernier, J. (1994) Time-frequency digital filtering based on an invertible wavelet transform: An application to evoked potentials. *IEEE Trans. Biomed. Eng.*, **41**(1), 77–88.

[39] Ademoglu, A., Micheli-Tzanakou, E. and Istefanopulos, Y. (1997) Analysis of pattern reversal visual evoked potentials (PRVEP's) by Spline wavelets. *IEEE Trans. Biomed. Eng.*, **44**(9), 881–890.

[40] Unser, M., Aldroubi, A. and Eden, M. (1992) On the asymptotic convergence of B-spline wavelets to Gabor functions. *IEEE Trans. Inform. Theory*, **38**(2), 864–872.

[41] Heinrich, H., Dickhaus, H., Rothenberger, A. *et al.* (1999) Single sweep analysis of event-related potentials by wavelet networks – Methodological basis and clinical application. *IEEE Trans. Biomed. Eng.*, **46**(7), 867–878.

[42] Quiroga, R.Q. and Garcia, H. (2003) Single-trial event-related potentials with wavelet denoising. *Clin. Neurophysiol.*, **114**, 376–390.

[43] Bartnik, E.A., Blinowska, K. and Durka, P.J. (1992) Single evoked potential reconstruction by means of wavelet transform. *Biol.Cybern.*, **67**, 175–181.

[44] Basar, E. (1988) EEG dynamics and evoked potentials in sensory and cognitive processing by brain, in *Dynamics of Sensory and Cognitive Processing By Brain* (ed. E. Basar), Springer.

[45] Zhang, Q. and Benveniste, A. (1992) Wavelet networks. *IEEE Trans. Neural Networks*, **3**, 889–898.

[46] Rumelhart, D., Hinton, G.E. and Williams, R.J. (1986) Learning internal representations by error propagation, in *Parallel Distributed Processing*, vol. **1** (eds D. Rumelhart and J.L. McClelland), MIT Press, Cambridge, MA, pp. 318–362.

[47] Georgiadis, S.D., Ranta-aho, P.O., Tarvainen, M.P. and Karjalainen, P.A. (2005) Single-trial dynamical estimation of event-related potentials: A Kalman filter-based approach. *IEEE Trans. Biomed. Eng.*, **52**(8), 1397–1406.

[48] Sorenson, H.W. (1980) *Parameter Estimation: Principles and Problems*, Marcel Dekker, NY.

[49] Melsa, J. and Cohn, D. (1978) *Decision and Estimation Theory*, McGraw-Hill, NY.

[50] Yates, R.D. and Goodman, D.J. (2005) *Probability and Stochastic Processes*, 2nd edn, John Wiley &Sons.

[51] Mohseni, H.R., Wilding, E.L. and Sanei, S. (2008) Single trial estimation of event-related potentials using particle filtering. Proceedings of IEEE International Conference on Acoustics, Speech, and Signal Processing, ICASSP, Taiwan, 2008.

[52] Jarchi, D., Makkiabadi, B. and Sanei, S. (2009) Estimation of trial to trial variability of p300 subcomponents by coupled rao-blackwellised particle filtering. Proceedings of Statistical Signal Processing Workshop, Cardiff, UK, 2009.

[53] Jarchi, D., Sanei, S., Principe, J.C. and Makkiabadi, B. (2011) A new spatiotemporal filtering method for single-trial ERP subcomponent estimation. *IEEE Trans. Biomed. Eng.*, **58**(1), 132–143.

[54] Hoehn, M. and Yahr, M. (1967) Parkinsonism: onset, progression and mortality. *Neurology*, **17**(5), 427–442.

[55] Thulasides, M., Guan, C. and Wu, J. (2006) Robust classification of EEG signal for brain-computer interface. *IEEE Trans. Neural Syst. Rehab. Eng.*, **14**(1), 24–29.

[56] Mohseni, H.R., Ghaderi, F., Wilding, E.L. and Sanei, S. (2010) Variational Bayes for spatiotemporal identification of event-related potential subcomponents. *IEEE Trans. Biomed. Eng.*, **57**(10), 2413–2428.

[57] Hansson, M., Gansler, T. and Salomonsson, C. (1996) Estimation of single event-related potentials utilizing the Prony method. *IEEE Trans. Biomed. Eng.*, **43**(10), 973–978.

[58] Scharf, L.L. (1991) *Statistical Signal Processing*, Addison-Wesley, Reading, MA.

[59] Garoosi, V. and Jansen, B.H. (2000) Development and evaluation of the piecewise Prony method for evoked potential analysis. *IEEE Trans. Biomed. Eng.*, **47**(12), 1549–1554.

[60] Barone, P., Massaro, E. and Polichetti, A. (1989) The segmented Prony method for the analysis of nonstationary time series. *Astron. Astrophys.*, **209**, 435–444.

[61] Meyer, J.U., Burkhard, P.M., Secomb, T.W. and Intaglietta, M. (1989) The Prony spectral line estimation (PSLE) method for the analysis of vascular oscillations. *IEEE Trans. Biomed. Eng.*, **36**, 968–971.

[62] Kulp, R.W. (1981) An optimum sampling procedure for use with the Prony method. *IEEE Trans. Electromagn. Comput.*, **23**, 67–71.

[63] Steedly, W.M., Ying, C.J. and Moses, R.L. (1994) A modified TLS-Prony method using data decimation. *IEEE Trans. Signal Process.*, **42**, 2292–2303.

[64] Cui, J. and Wong, W. (2006) The adaptive chirplet transform and visual evoked potentials. *IEEE Trans. Biomed. Eng.*, **53**(7), 1378–1384.

[65] Regan, D. (1989) *Human Brain Electrophysiology: Evoked Potentials and Evoked Magnetic Fields in Science and Medicine*, Elsevier, NY.

[66] Middendorf, M., McMillan, G., Galhoum, G. and Jones, K.S. (2000) Brain computer interfaces based on steady-state visual-evoked responses. *IEEE Trans. Rehab. Eng.*, **8**(2), 211–214.

[67] Cheng, M., Gao, X.R., Gao, S.G. and Xu, D.F. (2002) Design and implementation of a brain-computer interface with high transfer rates. *IEEE Trans. Biomed. Eng.*, **49**(10), 1181–1186.

[68] Holliday, A.M. (1992) *Evoked Potentials in Clinical Testing*, 2nd edn, Churchill Livingston, UK.

[69] Heckenlively, J.R. and Arden, J.B. (1991) *Principles and Practice of Clinical Electrophysiology of Vision*, Mosby Year Book, St Louis, MO.

[70] Mann, S. and Haykin, S. (1995) The chirplet transform – physical considerations. *IEEE Trans. Signal Process.*, **43**(11), 2745–2761.

[71] Bultan, A. (1999) A four-parameter atomic decomposition of chirplets. *IEEE Trans. Signal Process.*, **47**(3), 731–745.

[72] Mallat, S.G. and Zhang, Z. (1993) Matching-pursuit with time-frequency dictionaries. *IEEE Trans. Signal Process.*, **41**(12), 3397–3415.

[73] Gribonval, R. (2001) Fast matching pursuit with a multiscale dictionary of Gaussian chirps. *IEEE Trans. Signal Process.*, **49**(5), 994–1001.

[74] Qian, S., Chen, D.P. and Yin, Q.Y. (1998) Adaptive chirplet based signal approximation. Proceedings of IEEE International Conference on Acoustics, Speech, and Signal Processing, ICASSP, 1998, vol. 1–6, pp. 1781–1784.

[75] O'Neil, J.C. and Flandrin, P. (1998) Chirp hunting. Proceedings of IEEE-SP International Symposium on Time-Frequency and Time-Scale Analysis, 1998, pp. 425–428.

[76] Ullsperger, P., Metz, A.M. and Gille, H.G. (1998) The P300 component of the event-related brain potential and mental effort. *Ergonomics*, **31**, 1127–1137.

[77] Uetake, A. and Murata, A. (2000) Assessment of mental fatigue during VDT task using event-related potential (P300). Proceedings of the IEEE International Workshop on Root and Human Interactive Communication, Osaka, Japan, Sept. 27–29, 2000, pp. 235–240.

[78] Ulsperger, P., Neumann, U., Gille, H.G. and Pictschan, M. (1986) P300 component of the ERP as an index of processing difficulty, in *Human Memory and Cognitive Capabilities* (eds F. Flix and H. Hagendorf), Elsevier, Amsterdam, pp. 723–773.

[79] Neumann, U., Ulsperger, P. and Erdman, U. (1986) Effects of graduated processing difficulty on P300 component of the event-related potential. *Z. Psychol.*, **194**, 25–37.

[80] Mc Farland, D.J., Miner, L.A., Vaugan, T.M. and Wolpaw, J.R. (2000) Mu and beta rhythms topographies during motor imagery and actual movement. *Brain Topogr.*, **3**, 177–186.

[81] Vidal, J.J. (1977) Real-time detection of brain events in EEG. *IEEE Proc.*, **65**, 633–664. Special Issue on Biological Signal Processing and Analysis.

[82] Sutter, E.E. (1992) The brain response interface communication through visually induced electrical brain response. *J. Microcomp. Appl.*, **15**, 31–45.

[83] Morgan, S.T., Hansen, J.C. and Hillyard, S.A. (1996) Selective attention to the stimulus location modulates the steady state visual evoked potential. *Neurobiology*, **93**, 4770–4774.

[84] Muller, M.M. and Hillyard, S.A. (1997) Effect of spatial selective attention on the steady-state visual evoked potential in the 20-28 Hz range. *Cogn. Brain Res.*, **6**, 249–261.

[85] Birbaumer, N., Hinterberger, T., Kubler, A. and Neumann, N. (2003) The thought-translation device (ttd): neurobehavioral mechanisms and clinical outcome. *IEEE Trans. Neural Syst. Rehabil. Eng.*, **11**, 120–123.

[86] Sellers, E.W., Kubler, A., and Donchin, E. (2006) Brain-computer interface research at the University of South Florida Cognitive Psychology Laboratory: The P300 speller *IEEE Trans. Neural Syst. Rehabil. Eng.*, **14**, 221–224.

[87] Diez, J., Pfurtscheller, G., Reisecker, F. *et al.* (1997) Event-related desynchronization and synchronization in idiopathic Parkinson's disease. *Electroencephalogr. Clin. Neurophysiol.*, **103**(1), 155–155.

[88] Beverina, F., Palmas, G., Silvoni, S. *et al.* (2003) User adaptive BCIs: SSVEP and P300 based interfaces. *PsychNol. J.*, **1**(4), 331–354.

[89] Kelly, S.P., Lalor, E.C., Finucane, C. *et al.* (2005) Visual spatial attention control in an independent brain-computer interface. *IEEE Trans. Biomed. Eng.*, **52**(9), 1588–1596.

[90] Vaughan, T.M., McFarland, D.J., Schalk, G. *et al.* (2006) The Wandsworth BCI Research and Development Program: at home with BCI. *IEEE Trans. Neural Syst. Rehab. Eng.*, **14**(2), 229–233.

10

Mental Fatigue

10.1 Introduction

Fatigue is a state of body stability and awareness, also called exhaustion, lethargy, languidness, languor, lassitude, and listlessness, associated with physical or mental weakness. *Physical fatigue* is the inability to continue normal body functioning. On the other hand *mental fatigue* can be defined as the state of reduced brain activity and associated cognitive functions arising during continuous mental activity. Mental fatigue can manifest itself both as somnolence (decreased wakefulness), or only as diminished attention not necessarily including sleepiness [1]. Fatigued people often experience difficulties in concentration and appear more easily distracted. An objective cognitive testing may differentiate various states of mental fatigue. Unfortunately, neurocognitive mechanisms underlying the effects of mental fatigue are not yet well understood.

Mental fatigue has turned out to be a necessary and attractive research subject due to its effects on personnel who require extended attention in execution of their work, such as drivers, pilots, security guards, and because of its effects on many other human biometrics. It particularly affects normal brain activity and also the brain response to various stimuli. These effects can be recognised by examining and analysing the brain signals (such as EEG and MEG) and images (such as fMRI).

Mental fatigue can be the result of boredom, depression, disease, jet lag, lack of sleep [2,3], mental stress, over and under stimulation, thinking, or prolonged working. Most of these causes lead to temporary mental fatigue. However, some diseases such as uraemia, blood disorder, and chronic fatigue syndrome (CFS) may cause chronic fatigue which can last even for a few months. Therefore, rarely, there is need for a full clinical check up to find out the causes of chronic fatigue. The majority of people with chronic fatigue do not have an underlying cause discovered after a year with a condition. However, there is no need for clinical examinations when the subject is under temporary mental fatigue.

Temporary mental fatigue, however, becomes an issue when the subject is involved in sensitive and critical jobs, such as driving or guarding. Consequently and unlike physical fatigue, mental fatigue can seriously affect normal sleep and often causes various sleep disorders [2,3]. Therefore, an inclusive study of mental fatigue from the EEG signals requires analysis of both background EEGs and event-related brain responses.

Adaptive Processing of Brain Signals, First Edition. Saeid Sanei.
© 2013 John Wiley & Sons, Ltd. Published 2013 by John Wiley & Sons, Ltd.

In a study by Lorist *et al.* [4] they examined whether (i) error-related brain activity, (ii) indexing performance monitoring by the anterior cingulate cortex (ACC), and (iii) strategic behavioural adjustments were modulated by mental fatigue. In this experiment the hand response times to follow a visual stimulus and its variance were measured and evaluated. It was found that first, mental fatigue is associated with compromised performance monitoring and inadequate performance adjustments after errors; secondly, monitoring functions of ACC and striatum rely on dopaminergic inputs from the midbrain; and thirdly, patients with striatal dopamine deficiencies show symptomatic mental fatigue. Based on these collective results it was concluded that mental fatigue results from a failure to maintain adequate levels of dopaminergic transmission to the striatum and the ACC, resulting in impaired cognitive control.

It is common to measure various time and frequency domain features from the EEG data recorded over several hours of experiments for the assessment of mental fatigue. These features, which can be statistical measures [5], are carefully selected and then classified using popular classifiers such as Random Forest [6], SVM [7–9], and HMM [10].

From the EEG rhythms, synchronization, coherency, and connectivity of the brain regions may be examined. In the case of mental fatigue, these parameters for different conventional frequency bands can gradually change when the brain moves from alert to fatigue states.

On the other hand, based on ERP studies, the ability of subjects to focus their attention on a task decreases with time [1, 4]. Attention is an important feature of dynamic human behaviour; it allows one to (i) directly process the incoming information in order to focus on the relevant information for achieving certain goals, and (ii) actively ignore irrelevant information that may interfere with the goals [11]. Therefore, in the analysis of mental fatigue one key direction is to examine how mental fatigue affects these attentional processes. This can be achieved by analysis of different ERP components and subcomponents before and during the fatigue state. Some variations in attention-related ERP components variations induced by mental fatigue have provided strong evidence that attentional processes are indeed influenced by mental fatigue [1].

In this chapter some attempts at detection, quantification, and monitoring of mental fatigue, using either brain rhythms or ERPs and their subcomponents, are described in detail. Some mathematical and signal processing techniques with potential application to fatigue analysis are also explored. Finally, the above approaches based on EEG and ERP are linked to pave the path for a unified approach to the analysis of mental fatigue.

10.2   Measurement of Brain Synchronization and Coherency

10.2.1   Linear Measure of Synchronization

Incorporation of brain synchronization and coherency measures between the cerebral regions [12] as estimators of fatigue is another major approach which has been followed by a number of researchers, such as in [13, 14]. A functional relationship between different brain regions is generally associated with synchronous electrical activities in these regions. Recorded EEGs can be used to measure synchronization of different brain regions. In the approach by our group [15] linear and nonlinear synchronization of different brain regions is evaluated by exploiting an empirical mode decomposition (EMD) algorithm [16]. The synchronization measure can reveal useful information about the changes in the functional connectivity of brain regions during the fatigue state.

Therefore, first a brain region is selected and then the EMD algorithm, as a signal-dependent decomposition method, is applied to one channel of the EEG time series in the selected region to decompose it to waveforms modulated in amplitude and frequency. The iterative extraction of these components called intrinsic mode functions (IMFs), is based on local representation of the signal as the sum of a local oscillating component and a local trend. The IMFs can be considered as the reference signals for the brain rhythmic activities. Then, the linear and nonlinear synchronization measures can be estimated using the selected IMFs in different brain regions for the subject before and during the fatigue state. These measures are important for detecting and evaluating the mental fatigue in real world applications. One important issue is that the extracted IMFs might be noisy. This happens for the first few IMFs. In several research works conventional filtering was suggested to remove the noise from the IMFs. This, however, may result in loss of phase information.

As an effective option an adaptive line enhancer (ALE) [17] can be applied to the resulting (narrowband) IMF which may contain wide-band noise. Since the ALE is traditionally meant for detection of periodic signals with known period, it exploits the cyclic nature of the IMFs to restore the noise. The filtered IMFs are then used in the measurement of synchronization.

The concepts of both ALE and EMD have been reviewed in previous chapters. ALE is adaptive; it is able to track non-stationary signals. There are many applications, such as sonar, biomedical and speech signal processing, which make use of ALE. The ALE can be considered as a degenerate form of adaptive noise canceller. The reference signal in the ALE, instead of being derived separately, as in the adaptive noise canceller, is a delayed version of the input signal [18]. Therefore, the delay Δ is the prediction depth of the ALE, measured in units of the sampling period. The reference input $u(n - \Delta)$ is processed by a transversal filter, resulting in an error signal $e(n)$, defined as the difference between the actual input $u(n)$ and the ALEs output. The error signal is used to activate the adaptive algorithm by adjusting the weights of the transversal filter, mainly according to Widrow's LMS algorithm. The tap-weight adaptation can then be obtained through:

$$w(n + 1) = w(n) + \mu u(n)e(n) \tag{10.1}$$

where n is the iteration time. In this equation μ is the step size, $e(n)$ is the adaptation error and $u(n)$ is the input vector at time n. In the case of a fixed step-size LMS algorithm, μ is chosen as a constant. After applying the ALE to the extracted IMF by the EMD algorithm, it is possible to measure the synchronization measures of the different enhanced IMFs obtained from different parts of the brain. Given a clean IMF around the frequency of interest (such as alpha or beta brain rhythms), linear and nonlinear synchronization parameters can be evaluated as follows.

Suppose we have two IMFs simultaneously obtained from different channels. Every IMF is a real valued signal. The discrete Hilbert transform (HT) [16] is used to compute the analytic signal for an IMF. The discrete HT, denoted by $H_d[\cdot]$ of signal $x(t)$ is given by:

$$H_d[x(t)] = \sum_{\substack{\delta=-\infty \\ \delta\neq0}}^{+\infty} \frac{x(\delta)}{t - \delta} \tag{10.2}$$

The analytic function for the ith IMF $d_i(t)$ is defined as:

$$C_i(t) = d_i(t) + j H_d[d_i(t)] = a_i(t)e^{j\theta_i(t)} \tag{10.3}$$

where $a_i(t)$ and $\theta_i(t)$ are the instantaneous amplitude and phase of the ith IMF, respectively. The analytic signal is utilised in determining the instantaneous quantities, such as energy, phase, and frequency. The discrete time instantaneous frequency (IF) of the ith IMF is then given as the derivative of the phase $\theta_i(t)$ calculated at t:

$$\omega_i(t) = \frac{d\theta_i(t)}{dt} \tag{10.4}$$

On the other hand, the Hilbert spectrum represents the distribution of the signal energy as a function of time and frequency. If we partition the frequency range into k frequency bins, the instantaneous amplitude of the ith IMF at the kth frequency bin can be defined as:

$$H_i(k, t) = a_i(t)v_i^k(t) \tag{10.5}$$

where the $v_i^k(t)$ takes the value 1 if $\omega_i(t)$ falls within the kth band. The marginal spectrum corresponding to the Hilbert spectrum $H_i(k,t)$ of the ith IMF is defined as:

$$\psi_i(k) = \sum_{t=1}^{T} H_i(k, t) \tag{10.6}$$

where T is the data length. Using this information, the coherence function, as a measure of synchronization between the two IMFs, is defined as:

$$\zeta_{ij}(k) = \frac{|\chi_{ij}(k)|^2}{\psi_i(k)\psi_j(k)} k = 1, 2, \ldots, B \tag{10.7}$$

where B is the number of frequency bins, $\chi_{ij}(k)$ is the cross-spectrum of the ith and jth IMFs, $\psi_i(k)$ and $\psi_j(k)$ are the marginal power spectra of the ith and jth IMFs, respectively, and $\zeta_{ij}(k)$ is a quantity which shows how much the ith and jth IMFs are correlated.

10.2.2 Nonlinear Measure of Synchronization

It is likely to have two dynamic systems with synchronized phases while their amplitudes are uncorrelated [13]. Therefore, it is useful to consider phase synchronization as a nonlinear measure of synchronization. Two signals are phase synchronous if the difference in their phases remains constant across time [14]. After finding the instantaneous phase of the ith and jth IMFs, the phase synchronization measure can be defined as:

$$\gamma_{ij} = \left| \langle e^{j\theta_{ij}(t)} \rangle_t \right| = \sqrt{\langle \cos \theta_{ij}(t) \rangle_t^2 + \langle \sin \theta_{ij}(t) \rangle_t^2} \tag{10.8}$$

where $\langle\,.\,\rangle$ denotes the average over time, $\theta_{ij}(t) = \theta_i(t) - \theta_j(t)$, and $\theta_i(t)$ is the instantaneous phase of the ith IMF obtained by Hilbert transform. In this case, γ_{ij} will be zero if the phases are not synchronized at all and will be one when the phase difference is constant (perfect synchronization). The key feature of γ_{ij} is that it is only sensitive to phases, irrespective of the amplitude of each signal.

The above technique has been applied to a set of real data. The real EEG data used here belong to one of the subjects that participated in a continuous visual experiment. The stimuli were presented in the centre of a computer screen positioned at a viewing distance of 80 cm. In each trial, the participants were presented with a horizontal array of three uppercase letters, the central one of which was the target letter and the remaining letters were the flankers. The participants were instructed to make a quick left-hand response if the central letter were an H and a right-hand response if the central letter were an S. The experimental sessions lasted three hours. EEG was recorded from 22 scalp sites, using Sn electrodes attached to an electrode cap (ElectroCap International). Standard 10–20 sites were F7, F3, Fz, F4, F8, T7, C3, Cz, C4, T8, P7, P3, Pz, P4, P8, O1, Oz, and O2. Additional intermediate sites were FC5, FC1, FC2, and FC6. The two separate segments of data from the first half an hour and last half an hour are extracted. These segments are the representation of alert and fatigue states. Each segment is analysed separately to measure the linear synchronization (coherency) and nonlinear synchronization (phase synchronization) between the left and right hemispheres. The alpha rhythm is extracted from F7 and F4, the beta rhythm is extracted from FC5 and FC6 and the theta rhythm is extracted from P7 and P8. Overlapped windows with a length of 4 s are considered for applying the Hilbert transform and estimating the coherency and phase synchronization. The choice of channels in each hemisphere for extracting each rhythm is based on the Hilbert transform of the resulting IMFs in several electrodes in the right and left part of the brain. The IMF is selected from the channel with IMF (in a certain frequency band) having more continuous frequency traces in time–frequency distribution. Because the resulting beta rhythm was rather noisy, the ALE was used to enhance it. The results of the estimated coherency and phase synchronization values for the two segments of the data at three more effective frequency bands are shown in Figure 10.1 and 10.2, respectively. The algorithm is repeated over more trials for alert and fatigue states. In almost all of the trials the coherency of beta rhythms between the left and right hemispheres decreases and the coherencies for alpha and theta rhythms increase from the alert to the fatigue state. The estimated phase synchronization values between the rhythms have been shown to be different across different trials, especially for the theta rhythm. However, in most trials, the phase synchronization of beta rhythms decreases while for alpha rhythms it increases from the alert to the fatigue state.

10.3   Evaluation of ERP for Mental Fatigue

Assessment of the changes in ERP parameters due to mental fatigue is another general approach to fatigue analysis. As mentioned in previous chapters, the most popular method, which also happens to be the simplest approach, to extract the ERP components from single-trial measurements is averaging over time-locked single trials. In this approach the characteristics of the ERP wave (amplitude, latency, and width) are considered constant across trials and averaging over single trials leads to the attenuation of background EEG which is considered as a random process. In many real applications, this assumption is not true. For example, when

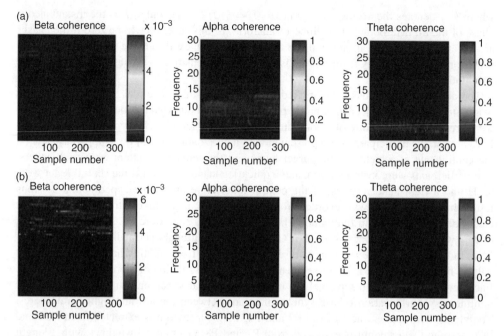

Figure 10.1 Inter-hemisphere coherency of beta, alpha and theta rhythms; (a) the fatigue state, (b) the alert state

there is a change in the degree of mental fatigue, habituation, or the level of attention, the ERP waveform changes from trial to trial. This variability of ERP components is ignored by averaging the ERP over a number of trials. Inter-trial variability of ERP components is an important parameter in order to investigate brain abnormalities. Several methods have been developed for single trial estimation of ERP components based on statistical signal processing. These methods include Wiener [19], maximum a posteriori (MAP) [17] and Kalman filtering [13, 18] approaches. The objective of all these methods is to estimate ERP components from single trial measurements and track the variability of their parameters across trials. However, these methods may fail in situations where the ERP signal to background noise ratio is very

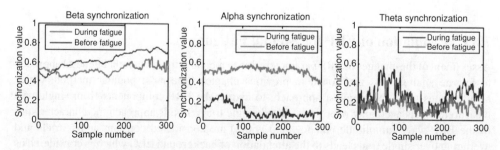

Figure 10.2 Inter-hemisphere phase synchronization of beta, alpha, and theta rhythms

low or where the measured ERPs vary from trial to trial. An effective analysis of ERPs should thus be based on a robust single trial estimation in which inter-trial variability of ERPs is also taken into account. A recent work [20] formulates wavelet coefficients of the time-locked measured ERPs in a state space and then estimates the ERP components using particle filtering. It is shown that the formulated particle filtering outperforms Kalman filtering. The method, however, fails to estimate ERP subcomponents. In this case particle filtering is a better solution to this problem. The proposed method here considers the temporal correlations between the subcomponents recorded from different sites of the brain and enables separation and identification of the ERP subcomponents.

Variation of different ERPs with time has been shown in some experiments. In these experiments the subjects were unable to prevent automatic shifting of attention [1] to irrelevant stimuli, reflected by a larger negativity in the N1 latency range (160–220 ms) for irrelevant, compared to relevant stimuli. This difference in negativity was unaffected by the time spent on a task. However, N1 and N2b (350–410 ms) amplitude did change with time: N1 amplitude decreased and the difference in N2b amplitudes between relevant and irrelevant stimuli (larger N2b amplitude evoked by relevant stimuli) decreased with time on task. The results of these experiments indicated that the effects of mental fatigue on goal-directed (top-down) and stimulus driven (bottom-up) attention are disassociated: mental fatigue results in a reduction in goal-directed attention, leaving subjects performing in a more stimulus-driven fashion.

The P300 component has been found useful in identifying the depth of cognitive information processing. It has been reported that the P300 amplitude elicited by mental task loading decreases with an increase in the perceptual or cognitive difficulty of the task, and its latency increases when the stimulus is cognitively difficult to process [10, 16, 15, 19, 21, 22]. P300 contains two subcomponents; P3a and P3b. These subcomponents usually overlap over the scalp and have temporal correlation [19]. The P3a is an unintentional response of the brain to a salient stimulus independent of the task [7, 17]. Prefrontal, frontal, and anterior temporal brain regions play an important role in generating P3a, giving it a frontocentral distribution. P3b is more distributed over the centroparietal region as it is mainly generated by posterior temporal, parietal, and posterior cingulate mechanisms [18]. Moreover, P3a is different from p3b in its shorter latency and more rapid habituation [7, 18]. P3a has a more frontal cortical distribution while P3b has a more parietal cortical distribution [13]. It has been suggested in [20,21] to use the P300 subcomponents for mental fatigue analysis since the averaged P300 does not always correspond to manifestation or appearance of mental fatigue. Evaluation of mental fatigue based on ERP is therefore recommended to be conducted from multiple perspectives: using not only P300 amplitude and latency but also feature parameters related to its subcomponents such as P3a and P3b.

This has been investigated by a number of researchers. In the work by Jarchi *et al.* [23] and the references therein the ERP components and their variations in amplitude, latency, and duration have been evaluated. A new coupled particle filtering for tracking variability of P300 subcomponents, that is, P3a and P3b, across trials has been developed in this work. The latency, amplitude, and width of each subcomponent, as the main varying parameters, have been modelled using the state-space system. In this model the observation is modelled as a linear function of amplitude and a nonlinear function of latency and width. Two Rao-Blackwellised particle filters (RBPF) are then coupled and employed for recursive estimation of the state of the system across trials. By including some simple physiologically-based constraints, the proposed technique prevents generation of invalid particles during estimation of the state of the system. The constraints exploit the relative geometrical topographies of P3a and P3b. The main advantage of the algorithm compared with other single trial based methods is its robustness in the low signal-to-noise ratio situations.

In this approach the variables describing the states of the state-space approach have been defined as the amplitudes, latencies, and widths of both P3a and P3b subcomponents. Empirically, the amplitude varies linearly and the other two variables change nonlinearly. Based on the RBPF concept a Kalman filter tracks the linear component whereas the changes in the nonlinear parameters are tracked by the particle filter.

To formulate the problem using RBPF consider that the nonlinear state is denoted by $\mathbf{x}_k^1$ and the linear (under some possible conditions) state by $\mathbf{x}_k^2$. RBPF may be used for tracking the ERP in a single channel signal. In the initialization stage, instead of generating random particles, it is feasible to use the averaged ERP of several consecutive trials and generate some particles which contribute more to the posterior density. In the subsequent trials, due to the existence of noise and artefact in the data, the correct track of ERP subcomponent parameters may be lost. Following this work [23], some physiologically meaningful constraints on the state variables have been imposed in order to remove invalid particles while they contribute to estimation of the posterior density. The constraints can be set using the prior knowledge about the ERP subcomponents' specifications. This can be expanded to more EEG channels. This allows incorporation of more statistical and physiological constraints within the estimation process.

The constrained BRPF (CBRPF) consists of two inter-connected co-operating BRPFs. This co-operation allows the discarding of irrelevant particles and refining the state transitions [23].

P300 subcomponents (P3a and P3b) are spatially disjoint. They usually overlap temporally over the scalp [18]. We recall that P3a is stronger over the frontal electrodes whereas P3b has larger amplitude over the parietal electrodes. The Fz electrode in the frontal and the Pz electrode in the parietal site are therefore good choices for this analysis. Two inter-connected RBPFs are developed. The first (RBPF1) is applied to the Fz channel and the second (RBPF2) to the Pz channel. Such assumptions and using the above channels is in line with our objectives in characterizing or tracking the P300 subcomponents.

A simple approach to mathematically describe the state variables of RBPF1 and RBPF2 is as follows:

$$\mathbf{x}_k^1 = \mathbf{R}_k^i = [\mathbf{b}_k^i \quad \mathbf{s}_k^i \quad \tilde{\mathbf{b}}_k^i \quad \tilde{\mathbf{s}}_k^i]^T \tag{10.9}$$

and

$$\mathbf{x}_k^2 = \mathbf{A}_k^i = [\mathbf{a}_k^i \quad \tilde{\mathbf{a}}_k^i]^T \tag{10.10}$$

Where $\mathbf{a}_k^i$, $\mathbf{b}_k^i$, and $\mathbf{s}_k^i$; $i = 1, 2$, are, respectively, the amplitude, latency, and width of P3a and $\tilde{\mathbf{a}}_k^i$, $\tilde{\mathbf{b}}_k^i$, and $\tilde{\mathbf{s}}_k^i$; $i = 1, 2$, are, respectively, the amplitude, latency, and width of P3b for the kth trial of the ith RBPF. For such a system the state-space equations are therefore,

$$\begin{aligned} \mathbf{R}_k^i &= \mathbf{R}_{k-1}^i + \mathbf{w}_{k-1}^{vi} \\ \mathbf{A}_k^i &= \mathbf{A}_{k-1}^i + \mathbf{w}_{k-1}^{Ai} \end{aligned} \tag{10.11}$$

Where $\mathbf{w}^{vi}_{k-1}$, and $\mathbf{w}^{Ai}_{k-1}$ are zero mean GWN noises with known covariance matrices. It is seen that tracking two temporally correlated P300 subcomponents from a single channel does not guarantee the correct estimation of the components. This is mainly because the formulated RBPF considers sum of two Gaussians plus noise and tracks the changes of Gaussian parameters across trials.

Therefore, the variations in the parameters may not be well tracked if any sudden change in any one of the parameters of the Gaussians occurs. There is also the possibility of modelling a segment of noise as a Gaussian. Therefore, we should use two RBPFs. Exploiting the logical links between these RBPFs, it is possible to prevent deviation of the estimation from the actual values. In order to make a connection between the two RBPFs, it is assumed that the latencies and widths of P3a and P3b in the brain at both the Fz and Pz channels are the same and only the amplitudes can be different. This assumption may not be exactly true in the real case, but it helps to couple two RBPFs in order to simultaneously track the changes in a system where there is an uncertainty in occurrence of the signals and their changes across trials because of the very low amplitude of the signal and inter-trial variability. Therefore, the benefit of simultaneous tracking is to reduce the uncertainty in the system by considering some meaningful physiological-based constraints.

In addition, we can assume a small delay in the latencies of P3a and P3b at the Fz and Pz channels. In the initialization stage, considering the averaged ERP we generate a rather large number of particles with relatively large variances for the latency, amplitude, and width. These initial particles are the same for RBPF1 and RBPF2. In the subsequent iterations, the particles with small weights are replaced by the particles with large weights.

Furthermore, it should be noted that in each iteration the required relation between the two RBPFs should be taken into account for drawing samples from the prior density. Since the shapes of P3a or P3b are the same for the Fz and Pz channels, the pairs of particles with the same width for P3a and P3b can be coupled. The basis of the designed CRBPF is in coupling particle pairs.

Initializations of the particles for RBPF1 and RBPF2 are the same [23]. In particle pairs, the width is the same for P3a in both RBPF1 and RBPF2, and also the same for P3b. Therefore, it may be sufficient to consider the same width for every subcomponent in both RBPFs in each iteration. However, the amplitudes of the subcomponents are different and the latencies differ in small delay. The widths and latencies of the subcomponents can be generated from the prior density. Then, Kalman filtering can be applied for estimation of amplitudes of the subcomponents. In this stage, some invalid particles may be generated. It is effective to detect these invalid particles and remove them by setting their weights to zero so that they do not have any contribution in the estimation of the state of the system. One class of invalid particles is those where the estimated amplitude of P3a at the Pz channel is larger than the amplitude of P3a at the Fz channel. Also, the particles in which the estimated amplitude of P3b at the Fz channel is larger than the amplitude of P3b at the Pz channel are considered invalid. Usually P3a and P3b overlap over the scalp but nearly always P3a has a shorter latency than P3b. Therefore, it is possible to have P3a and P3b latencies close to each other and the new estimated latencies iteratively substituted for P3a and P3b. The particles in which the latency of P3b is shorter than the latency of P3a are marked invalid. Mutual influence of the p3a and p3b is also used to adjust the weights accordingly [23].

The performance of this system has been examined using EEG recorded data from subjects who were required to be non-smokers, to have normal sleep patterns, not to work night shifts, and not to be under medication. As described in Section 10.1.2, in each trial horizontal arrays of three uppercase letters, in which the central one was the target letter and the remaining letters were the flankers, were presented to the subjects. The subjects were instructed to make a left-hand response as quickly as possible if the central letter was H and a right-hand response if the central letter was S. The letter array remained on the screen until a response was given by the subject or disappeared after 1200 ms. The subjects were also allowed to correct their incorrect responses within 500 ms following the initial response. In 50% of the trials, the target letter was presented on a red or green background while in the other half of the trials it was presented on a black background. In half of the trials, the flankers had the same type and colour as the target letter (e.g. HHH or SSS: compatible) while in the other half of the trials they had a different type and colour than the target letter (e.g. SHS or HSH: incompatible). The stimuli of different types were presented randomly with equal probability. About 1000 ms before appearance of the three letters, a pre-cue was presented for 150 ms, specifying either the colour of the target letter (the Dutch word for 'red' or 'green') or the response hand (the Dutch word for 'left' or 'right'). The hand and colour cues were presented randomly with equal probability. In 80% of the trials, the cues were valid. During each trial, a fixation mark remained visible on the screen (an asterisk of 0.5×0.5 cm).

The interval between the initial response to one trial and the beginning of pre-cue presentation in the following trial varied randomly between 900 and 1100 ms. The subjects were instructed to respond quickly and accurately and to minimize eye movements and blinking during the task performance. The task comprised a2-hours experimental block without breaks followed by a practice block of 80 trials. EEG data were recorded from 22 scalp positions, as mentioned in Section 10.2.2. Eye-blink removal and parameter settings were the same as those in Section 10.2.2. In some previous research using this dataset, the P300 was detected from the averaged ERP [20]. The CRBPF was applied to Fz and Pz channels considering 50 trials from the first and fourth half an hour. In the first half an hour the subject was not under fatigue, while in the fourth half an hour the subject was under fatigue. In both cases, before and during fatigue, the averaged ERP of 20 trials was used to initialize the particles of the CRBPF. The results of tracking P3a and P3b variability are shown in Figure 10.3. This figure illustrates the variability of P3a and P3a amplitudes, latencies and widths before and during the fatigue state. Based on the results, it is possible to draw some conclusions about the effect of mental fatigue on P300.

Generally, in some other researches on mental fatigue [10, 16, 15, 21, 22], the decrease in the overall P300 amplitude and increase in its latency have been reported. However, using the above method, we could separate P300 into its constituent subcomponents and evaluate the effect of fatigue on each subcomponent separately. Based on the results, the latencies of both P3a and P3b increase with time on task interval. During the fatigue state, the increase in the latency of P3a is slightly more than that for the P3b. Indeed, the amplitude of P3a decreases more than that of the P3b. The width of P3b remains approximately constant in the fatigue state while the width of P3a decreases. The P3a amplitude seems to become smaller with time on the task, while that of the P3b seems to increase somewhat.

During the first half an hour, a clear difference between P3a and P3b scalp positions is observed; P3a is more prominent at Fz, as expected. During the last half an hour the P3a became very small at Pz, while the P3b is prominent at this position. P3a is related to novelty. However, it is possible that the decrease in the P3a amplitude with time on the task is related to practice or habituation to the task. Therefore, mental fatigue can be related to the increase in P3a and P3b latencies, and decrease in P3a amplitude and width. In order to investigate the

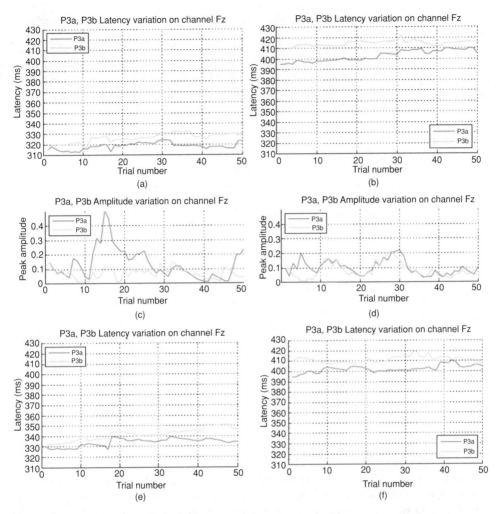

Figure 10.3 Tracking variability of P3a and P3b before and during fatigue; the left column corresponds to the parameters measured before fatigue (a), (c), (e), (g), (i), (k) while the right column to the parameters measured during the fatigue (b), (d), (f), (h), (j), (l)

effect of fatigue in more detail, it is useful to apply the method to more signals and also to more electrodes in the frontal and posterior scalp regions.

Although, monitoring ERPs influenced by mental fatigue is a valuable diagnostic tool it may not be robust enough to mark the details of gradual changes. This can be true for evaluation of coherency and synchronization too. However, for the sake of robustness, it would be extremely useful if both approaches could be applied simultaneously to the same data sets recorded from the same subjects. This requires a suitable recording protocol which elicits both ERP and the changes in the brain rhythms. One way is to stimulate the brain after a longer time interval to allow complete settling of the brain after each stimulus and to enable evaluation of synchronization and coherency within these intervals.

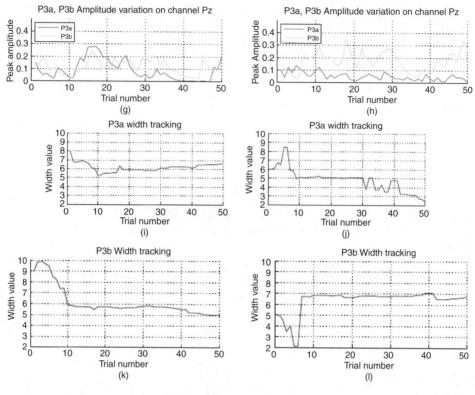

Figure 10.3 (*Continued*)

10.4   Separation of P3a and P3b

To enable evaluation of the parameters of P3a and P3b one may attempt to separate them as two different source components. The most important issue in dealing with this problem is the fact that the two components originate from the same activation or one is the consequence of the other. P3a reflects an unintentional response of the brain to a novel or salient stimulus that is independent of task relevance [24, 25]. Prefrontal, frontal, and anterior temporal brain regions play an important role in generating P3a, giving it a fronto-central distribution [24]. P3b is mainly generated by posterior temporal, parietal, and posterior cingulate mechanisms and mostly distributed over the centro–parietal region. Moreover, P3a has a shorter latency and more rapid habituation than P3b.

By visual inspection of the components it is seen that they overlap in both time and the space of electrodes. Therefore, it is not easy to employ conventional source separation systems to separate them. In an attempt by Jarchi *et al.* [26] a method for separating correlated components such as $\mathbf{X}$ has been developed. The problem may be expressed in the following simple constrained optimization form:

$$\min \left\| \mathbf{w}_1^{\mathrm{T}} \mathbf{X} - \mathbf{r}_1 \right\|_2^2 \quad \text{subject to} \quad \mathbf{w}_1^{\mathrm{T}} \mathbf{a}_2 = 0 \qquad (10.12)$$

Where $\mathbf{w}$ is the separating vector, $\mathbf{r}_1$ is a template for the first component (e.g. P3a) and $\mathbf{a}_2$ is the second component (e.g. P3b). $\mathbf{r}_1$ is modelled as a Gamma function and the above constrained problem is changed to an unconstrained one using a Lagrange multiplier. The new cost function, then, has the form

$$J = \left\| \mathbf{w}_1^\mathsf{T} \mathbf{X} - \mathbf{r}_1 \right\|_2^2 + \mathbf{w}_1^\mathsf{T} \mathbf{a}_2 q = 0 \tag{10.13}$$

where q is the Lagrange multiplier. This parameter has also been accurately estimated as [26]:

$$q = \frac{2\mathbf{r}_1 \mathbf{X}^\mathsf{T} \mathbf{C}_x^{-1} \mathbf{a}_2}{\mathbf{a}_2^\mathsf{T} \mathbf{C}_x^{-1} \mathbf{a}_2} \tag{10.14}$$

where $\mathbf{C}_x^{-1}$ is the covariance matrix of $\mathbf{X}$. Suitable choices for $\mathbf{a}_1$ and $\mathbf{a}_2$ have been considered as the projections of $\mathbf{r}_1$ and $\mathbf{r}_2$ onto $\mathbf{X}$ as $\mathbf{r}_1 \mathbf{X}$ and $\mathbf{r}_2 \mathbf{X}$, respectively. Minimizing J with respect to $\mathbf{w}_1$ and $\mathbf{w}_2$ (as for the second source, P3b, in a similar fashion) using the defined parameters for finding P3a and P3b, respectively, results in [26]

$$\mathbf{w}_1 = \mathbf{C}_x^{-1} \mathbf{X} \mathbf{r}_1^\mathsf{T} - 0.5 q \mathbf{C}_x^{-1} \mathbf{X} \mathbf{r}_2^\mathsf{T} \tag{10.15}$$

$$\mathbf{w}_2 = \mathbf{C}_x^{-1} \mathbf{X} \mathbf{r}_2^\mathsf{T} - 0.5 q \mathbf{C}_x^{-1} \mathbf{X} \mathbf{r}_1^\mathsf{T} \tag{10.16}$$

Also, the Gamma functions for both $\mathbf{r}_1$ and $\mathbf{r}_2$ have the following general form:

$$r(t) = \beta t^{k-1} \exp\left(\frac{-t}{\theta}\right) \tag{10.17}$$

where $k > 0$ is the shape parameter, $\theta > 0$ is a scale parameter, and β is a normalizing constant.

To examine the method, two components have been generated in two different hypothetical brain locations using Gamma functions with different correlation levels. Using these sources, a 20-channel dataset has been produced. Each channel included 40 trials. In all of the trials, the latency of P3a was fixed at 150 ms and the latency of P3b was fixed at 200 ms. The amplitudes of P3a and P3b change in different trials. The noise variance was fixed for all the trials. In the first attempt the method was applied to the simulated data considering the reference signal as the actual synthetic source. This is called *exact match*. On another trial the reference signals for P3a and P3b were not exactly the same as the actual synthetic sources. These waves for P3a and P3b were considered as Gamma waves with different parameters [26]. The method considering these references is called *mismatch*. Since in this work the temporal correlation is high, it is reasonable to compare the results with spatial PCA rather than temporal PCA [27, 28]. Therefore, spatial PCA (from ERP PCA Toolkit [29]) was used in which Infomax [30] was employed as the rotation algorithm for whitening the signal mixtures. It was shown that the above method performs better than spatial PCA. On the other hand the *mismatch* system performs very closely to the *exact match* system where the source model is identical to the template. The results were also compared in terms of the correlations between the original and estimated sources for 0 and −5 dB SNRs in Figure 10.4.

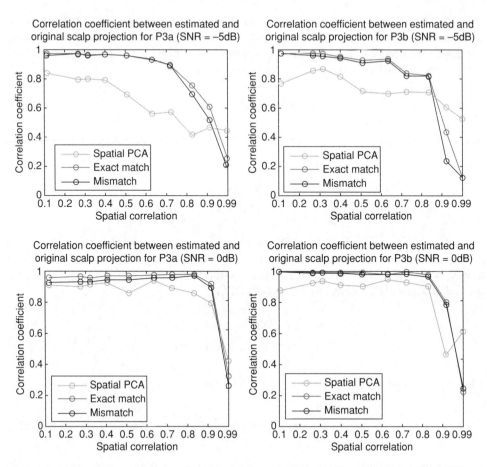

Figure 10.4 Comparison of three methods (spatial PCA, exact match and mismatch) versus the correlation coefficients between the original and estimated scalp projections of P3a and P3b in different spatial correlations and two SNR levels of 0 and −5 dB

In an application EEG data were recorded. The stimuli were presented through ear plugs inserted in the ear. Tones of 1 kHz were randomly and infrequently played for 40 times amongst tones of 2 kHz which appeared more frequently for 160 times. Their intensity was 65 dB with 10- and 50-ms duration for rare and frequent tones, respectively. The subjects were asked to press a button as soon as they heard a low tone (1 kHz). ERP components measured in this task included N100, P200, N200, P3a, and P3b. Forty trials related to the infrequent (rare) tones were selected and the above method for estimation of latency, amplitude, and scalp projections of P3a and P3b was employed. These trials and their average from channel Fz is shown in Figure 10.5a. The estimated amplitudes are also depicted in Figure 10.5b.

For selection of reference signals, the averaged P300, as shown in Figure 10.6, has been used.

In the top row, the selected reference signals have little overlap with the average P300. The normalized estimated scalp projection is depicted for both P3a and P3b. It can be seen that the

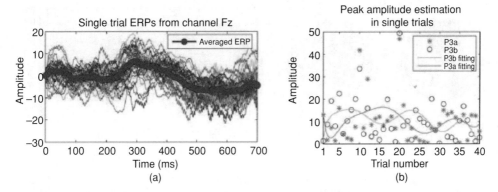

Figure 10.5 (a) Single-trial ERPs (40 trials related to the infrequent tones) and their average from channel Fz, and (b) estimated amplitudes for P3a and P3b in different trials (see Plate 7 for the coloured version)

scalp projection values for P3a are negative or very close to zero. By increasing the correlation the estimated scalp projections change. In the top row, the reference signal for P3a seems to have no correlation (or very little correlation) with actual P3a so its estimated scalp projection has negative or very close to zero values in all entries; however, it seems that there are some correlations between the reference signal and the actual signal for P3b. In the bottom row,

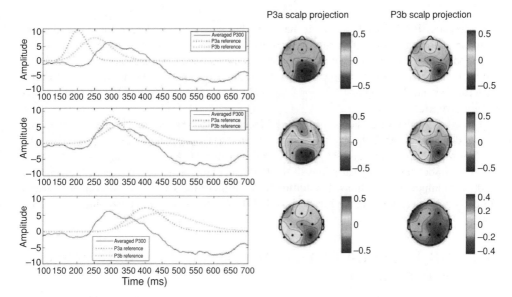

Figure 10.6 Selection of reference signals for P3a and P3b. In each row, the reference signals for P3a and P3b with their estimated scalp projections are shown. By sliding the reference signals toward and away from the averaged P300 and varying their shapes and observing the changes in their estimated scalp projections an appropriate reference can be selected. The reference signals shown in the middle row which have high correlation with the averaged P300 are used as good candidates for approximating the actual P3a and P3b

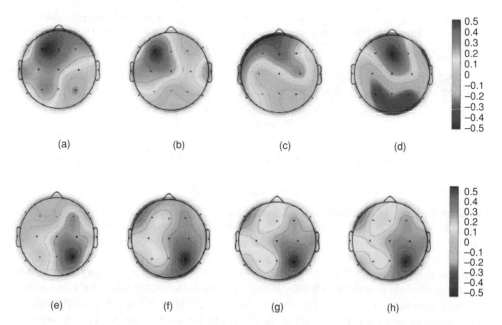

Figure 10.7   Scalp projections of (a–d) P3a and (e–h) P3b in four selected progressive trials

the reference signal for P3a seems to have overlap with the actual P3b; therefore, the first estimated subcomponent P3b scalp projections have appeared. The reference signal for P3b seems to have no correlation with actual P3b and the estimated scalp projection has negative values in all entries. So, it is logical to use the reference signals in the middle row of Figure 10.6 for P3a and P3b. Figure 10.7 shows the scalp projections of P3a and P3b in four selected progressive trials.

The average ERP often lacks the early ERPs such as N100 and P200; though these ERPs can be detected by using the average of trials related to the frequent tones.

The proposed method can be applied in order to investigate mental fatigue based on trial-to-trial amplitude and latency variations of the P300 subcomponents and the relative changes of P300 subcomponent variations. In addition, estimation of the scalp projections can be useful for detecting the changes in the locations of P300 subcomponents for schizophrenic patients.

10.5   A Hybrid EEG-ERP-Based Method for Fatigue Analysis Using an Auditory Paradigm

It seems to be favourable to introduce a unified method using both background EEG and ERP data to more efficiently check the state of mental fatigue. An auditory-based paradigm may be implemented when EEG data are recorded in two states of alert and fatigue.

The data are then examined by both approaches (i.e. EEG and ERP-based) simultaneously. Such a recording paradigm allows a sufficient time interval between the stimulations so the state of the brain and its response to the stimulus can be evaluated.

For this test the experiment was run in a quiet room that is illuminated normally. The subject was sited comfortably in an armchair and the EEG data were recorded using a 32-channel QuickAmp amplifier and Ag/AgCl electrodes positioned according to the international 10–20 system and re-referenced to linked ears. In addition, vertical (VEOG) and horizontal (HEOG) electro-oculographic signals were recorded bipolarly using electrodes above and below the left eye and from the outer canthi. The EEG data were recorded in DC mode at 1000 Hz with respect to an average reference. The EEG signal was recorded at the start of the experiment during the auditory oddball task. The subject heard 180 tones, 40 of them were infrequent tones while 140 of them were frequent tones. The subject was asked to respond to the infrequent tones by pressing a soft push button. The trial duration was set to 4 s. The subject was instructed to perform a simple arithmetic task for approximately 2 h. After that the EEG data of the subject were recorded using the same auditory oddball task.

Elicitation of P300 is expected to be better when the infrequent tones are used. The spatiotemporal filtering method was applied to the 40 trials of infrequent tones. The average of these trials (considering 600 ms) from the Cz channel was used in order to select appropriate reference signals. Forty single trial ERPs from Cz channel and their average before and during fatigue are shown in Figure 10.8 and 10.9, respectively.

The two averaged ERPs are shown separately in Figure 10.10 for the two states of before and during fatigue. These average ERPs are used in order to select appropriate reference signals for P3a and P3b.

From the averaged ERPs in the fatigue state there is a reduced amplitude and increased latency in the P300 wave. However, the reduction in amplitude is very trivial and may not be considered as the sign of fatigue. In addition, there is not sufficient consistency in the increase

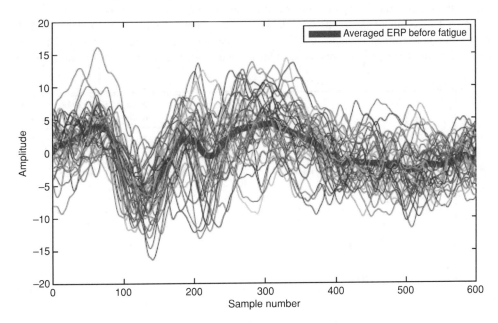

Figure 10.8 Forty single trial ERPs and their average from the Cz channel before fatigue (see Plate 8 for the coloured version)

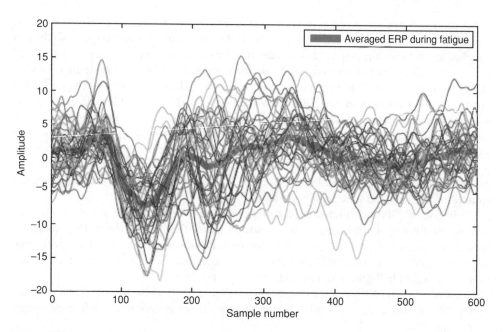

Figure 10.9   Forty single trial ERPs and their average from the Cz channel during fatigue state (see Plate 9 for the coloured version)

in latency due to fatigue across trials. Therefore, there is a need for single trial estimation of the ERPs and for that the spatiotemporal filtering method proposed in the previous section is applied to estimate the P300 subcomponent descriptors (latency, amplitude, and scalp projections). The mean latencies of P3a and P3b before fatigue were obtained as 279.6 ms and 335.5 ms, respectively. This shows that the latencies of P3a and P3b are increased by

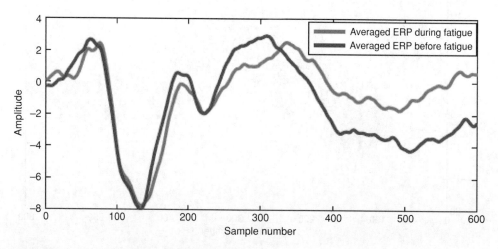

Figure 10.10   The ERP achieved by averaging 40 EEG trials before and during fatigue state from the Cz channel

(a)

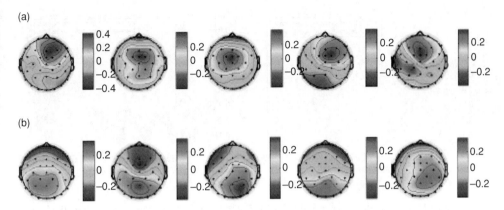

(b)

Figure 10.11   The estimated scalp projections of P3a (a) and P3b (b) before fatigue

increasing mental fatigue, as expected. The mean latencies of P3a and P3b during the fatigue state were obtained as 325.5 ms and 372.2 ms, respectively.

The scalp projections of P3a and P3b for five selected trials are shown in Figure 10.11 and 10.12 for before and during fatigue state, respectively.

Although there is not a significant difference in the estimated scalp projections for P3a and P3b before and during the fatigue state, the important issue is the separation of these subcomponents and estimation of other parameters. The results obtained by considering one subject confirm that the suggested paradigm is a good option for designing a mental fatigue detection system and the estimation of P300 subcomponent parameters can be good features for discriminating the fatigue state. Other important features can be obtained using phase synchronization measures.

From what has been discussed previously, the phase synchronization of different EEG rhythms, especially the alpha rhythm, can be used as good features for recognition of the fatigue state. To investigate this, 3 s of the data segment has been considered and the EMD

(a)

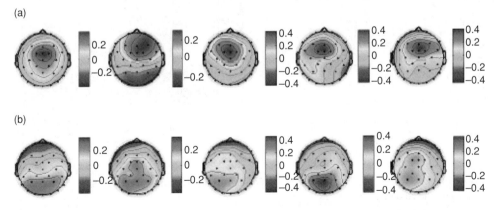

(b)

Figure 10.12   The estimated scalp projections of P3a (a) and P3b (b) during the fatigue state

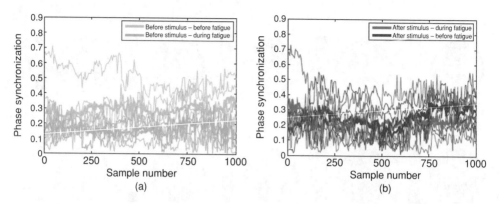

Figure 10.13   Theta phase synchronization of F3-F4; (a) before stimulus and (b) after stimulus

applied to decompose the EEG signal into its constituent oscillations. For the beta rhythm, ALE is applied to the resulting IMF. 1 s of data segment (1000 samples) before stimulus onset and 1 s after stimulus onset are considered for measuring the phase synchronization. The beta and theta rhythms are extracted from the frontal electrodes (F3 and F4 channels) and the alpha rhythm is extracted from the central electrodes (C3 and C4 channels). The phase synchronization is calculated for five trials for 1 s before and after stimulus onset. The calculated phase synchronization for theta, alpha, and beta rhythms can be seen in Figure 10.13, 10.14, and 10.15, respectively. In these figures the average phase synchronization is also depicted as a thick line. From these figures the changes in phase synchronization can be clearly seen for the alpha and theta rhythms before stimulus onset. Therefore, using the recorded EEG signal the EEG phase synchronization can be considered as a good feature for discrimination of the fatigue state.

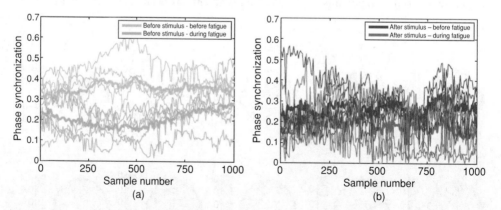

Figure 10.14   Alpha phase synchronization of C3-C4; (a) before stimulus and (b) after stimulus (see Plate 10 for the coloured version)

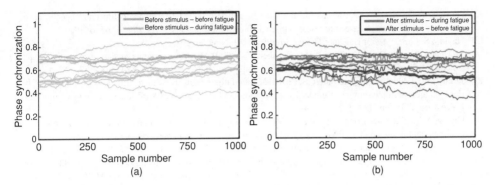

Figure 10.15 Beta phase synchronization of F3-F4; (a) before stimulus and (b) after stimulus (see Plate 11 for the coloured version)

10.6 Conclusions

Normal brain rhythms and the brain responses to various stimuli are both affected by the changes in the brain state due to mental fatigue. In addition, synchrony, coherency, and generally the connectivity of brain lobes are subject to change when the subject's brain is tired. In this chapter, these concepts have been investigated in detail. More investigations and experiments may be carried out to find more about the mutual effects of memory and fatigue for different aged subjects and those suffering from mental or physical abnormalities.

References

[1] Boksem, M.A.S., Meijman, T.F. and Lorist, M.M. (2005) Effects of mental fatigue on attention: an ERP study. *Cogn. Brain Res.*, **25**, 107–116.

[2] Yeo, M.V., Li, X. and Wilder-Smith, E.P. (2007) Characteristic EEG differences between voluntary recumbent sleep onset in bed and involuntary sleep onset in a driving simulator. *Clin. Neurophysiol.*, **118**, 1315–1323.

[3] Yeo, M.V.M., Li, X.P., Shen, K.Q. *et al.* (2004) EEG spatial characterization for intentional & unintentional sleep onset. *J. Clin. Neurosci.*, **11**(Suppl. 1), 70.

[4] Lorist, M.M., Maarten, T., Boksem, A.S. and Ridderinkhof, K.R. (2005) Impaired cognitive control and reduced cingulate activity during mental fatigue.*J. Brain Res., Cogn. Brain Res.*, **24**, 199–205.

[5] Zhang, L.Y., Zheng, C.X., Li, X.P. and Shen, K.Q. (2005) Feasibility study of mental fatigue grade based on Kolmogorov entropy. *Space Med. Med. Eng.*, **18**(5), 375–380.

[6] Breiman, L. (2001) Random forests. *Machine Learning*, **45**, 5–32.

[7] Shen, K.Q., Li, X.P., Ong, C.J. *et al.* (2008) EEG-based mental fatigue measurement using multi-class support vector machines with confidence estimate. *Clin. Neurophysiol.*, **119**, 1524–1533.

[8] Shen, K.Q., Ong, C.-J., Li, X.-P. *et al.* (2007) A feature selection method for multilevel mental fatigue EEG classification. *IEEE Trans. Biomed. Eng.*, **54**, 1231–1237.

[9] Yeo, M.V.M., Li, X.-P., Shen, K.-Q. and Wilder-Smith, E.P.V. (2009) Can SVM be used for automatic EEG detection of drowsiness during car driving? *Safety Science*, **47**(1), 115–124.

[10] Zhang, C., Zheng, C., Yu, X. and Ouyang, Y. (2008) Estimating VDT mental fatigue using multichannel linear descriptors and KPCA-HMM. *EURASIP J. Adv. Signal Process.* doi: 10.1155/2008/185638

[11] Maarten, A.S., Boksem, A.S., Meijman, T.F. and Lorist, M.M. (2005) Effects of mental fatigue on attention: An ERP study. *Cogn. Brain Res.*, **25**, 107–116.

[12] Quiroga, R.Q., Kraskov, A., Kreuz, T. and Grassberger, P. (2002) Performance of different synchronization measures in real data: A case study on electroencephalographic signals. *Phys. Rev. E*, **65**, 041903.

[13] Haykin, S. (2002) *Adaptive Filter Theory*, Prentice-Hall, New Jersey.

[14] Mormann, F., Lehnertz, K., David, P. and Elger, C. (2000) Mean phase coherence as a measure for phase synchronization and its application to the EEG of epilepsy patients. *Physica D*, **144**, 358–369.

[15] Jarchi, D., Makkiabadi, B. and Sanei, S. (2010) Mental fatigue analysis by measuring synchronization of brain rhythms incorporating empirical mode decomposition. 2nd International Workshop on Cognitive Information Processing (CIP), pp. 423–427.

[16] Huang, N.E., Shen, Z., Long, S.R. *et al.* (1998) The empirical mode decomposition and hilbert spectrum for nonlinear and non-stationary time series analysis. *Proc. Roy. Soc. London Ser.A*, **454**, 903–995.

[17] Widrow, B. (1975) Adaptive noise cancellation: Principles and applications. *Proc. IEEE*, **63**, 1692–1716.

[18] Kalman, R.E. (1960) A new approach to linear filtering and prediction problems. *J. Basic Eng.*, **82**(1), 35–45.

[19] Wiener, N. (1949) *Extrapolation, Interpolation, and Smoothing of Stationary Time Series*, Wiley, New York.

[20] Mohseni, H.R., Nazarpour, K., Wilding, E. and Sanei, S. (2009) Application of particle filters in single-trial event related potential estimation. *Physiol. Meas.*, **30**(10), 1101–1116.

[21] Murata, A., Uetake, A. and Takasawa, Y. (2005) Evaluation of mental fatiguenext term using feature parameter extracted from event-related potential. *Int. J. Ind. Ergonom.*, **35**(8), 761–770.

[22] Ullsperger, P., Metz, A.M., Yu, X. and Gille, H.G. (1988) The P300 component of the event-related brain potential and mental effort. *Ergonomics*, **31**, 1127–1137.

[23] Jarchi, D., Sanei, S., Mohseni, H.R. and Lorist, M.M. (2011) Coupled particle filtering: A new approach for P300-based analysis of mental fatigue. *J. Biomed. Signal Process. Control*, **6**(2), 175–185.

[24] Friedman, D., Cycowicz, Y.M. and Gaeta, H. (2001) The novelty p3: An event related brain potential (ERP) sign of the brain's evaluation of novelty. *Neurosci. Biobehav. Rev.*, **25**(4), 355–373.

[25] Comerchero, M.D. and Polich, J. (1999) P3a and P3b from typical auditory and visual stimuli. *Clin. Neurophysiol.*, **110**(1), 24–30.

[26] Jarchi, D., Sanei, S., Makkiabadi, B. and Principe, J. (2011) A new spatiotemporal filtering method for single-trial estimation of correlated ERP subcomponents. *IEEE Trans. Biomed. Eng.*, **58**(1), 132–143.

[27] Dien, J. (1998) Addressing misallocation of variance in principal components analysis of event-related potentials. *Brain Topogr.*, **11**(1), 43–55.

[28] Dien, J., Khoe, W. and Mangun, G.R. (2007) Evaluation of PCA and ICA of simulated ERPs: Promax vs. infomax rotations. *Hum. Brain Map.*, **28**(8), 742–763.

[29] Dien, J. (2010) The ERP PCA toolkit: An open source program for advanced statistical analysis of event-related potential data. *J. Neurosci. Methods*, **187**(1), 138–145.

[30] Delorme, A. and Makeig, S. (2004) EEGLAB: An open source toolbox for analysis of single-trial EEG dynamics including independent component analysis. *J. Neurosci. Methods*, **134**(1), 9–21.

11

Emotion Encoding, Regulation and Control

No matter how basic or complex, an emotion is a mental, brain-related physiological state associated with a wide variety of feelings, thoughts and behaviour. Emotions are subjective experiences often associated with mood temperament, personality, and disposition. The word 'emotion' originates from the French word *émouvoir* which is based on the Latin *emovere*, where *e-* (variant of *ex-*) means 'out' and *movere* means 'move'. Theories about emotions stretch back at least as far as the ancient Greek Stoics, as well as Plato and Aristotle. Numerous articles refer to physiological and biological aspects of emotions and their correlation with other states of the body and brain. Most of these theories go far beyond the scope of this chapter.

The programme of the HUMAINE summer school in 2004 held in Belfast listed 55 different emotions, namely admiration, affection, amusement, annoyance, anxiety, approval, boredom, calm, cold anger, coldness, confidence, contentment, contempt, cruelty, despair, determination, disagreeableness, disappointment, disapproval, disgust, distraction, effervescent, embarrassment, excitement, fear, friendliness, greed, guilt, happiness, hopeful, hot anger, hurt, impatience, indifference, interest, jealousy, mockery, nervousness, neutrality, panic, pleasure, pride, relaxation, relief, resentment, sadness, satisfaction, serenity, shame, shock, stress, surprise, sympathy, wariness, weariness, and worry.

Emotion is central to human daily experience, influencing cognition, perception and everyday tasks, such as learning, communication and even rational decision-making. However, the large number of emotion states and the overlaps between the corresponding brain regions make analysis of emotion very challenging for technologists and neuroscience researchers.

Symptoms such as body movement, facial expression, change of body temperature, heart rate variability, breathing, blood pressure, muscle contraction, and variation in brain rhythms may be required to accurately discriminate between the aforementioned emotion types. Hence it is extremely difficult to perform such classification using a single modality biometric.

A number of theories, at least for some particular emotions, have been proposed by neuroscientists for a better understanding of how the brain acts or react to emotions. These theories also identify the right brain regions which might be activated under each emotion.

Adaptive Processing of Brain Signals, First Edition. Saeid Sanei.
© 2013 John Wiley & Sons, Ltd. Published 2013 by John Wiley & Sons, Ltd.

11.1 Theories and Emotion Classification

Some theories are somatic and define emotions such as human expression and body movements [1]. William James and Carl Lange developed the James–Lange theory, a hypothesis on the origin and nature of emotions. It states that the human autonomic nervous system provokes physiological events, such as muscular tension, rise in heart rate, perspiration and dryness of mouth in response to the world's events. Emotions, hence, are feelings that come about as a result of physiological changes, rather than being their cause.

Other theories based on neurobiological changes describe emotion as a pleasant or unpleasant mental state organized in the limbic system of the mammalian brain. Emotions would then be mammalian elaborations of general vertebrate arousal patterns, in which neurochemicals (e.g. dopamine, noradrenaline and serotonin) step-up or step-down the brain's activity level, as visible in body movements, gestures and postures. In mammals, primates, and human beings, feelings are displayed as emotion cues.

Emotion is believed to be related to the limbic system [2]. The limbic system is a set of primitive brain structures located on top of the brainstem on both sides of the thalamus, just under the cerebrum and involved in many of our emotions and motivations, especially those related to survival. Fear, anger and emotions related to sexual behaviour are those originating from this area of the brain. It includes the hypothalamus, hippocampus, amygdala (also called amygdale) and some other brain regions. The amygdala is an almond-shaped mass of nuclei located deep within the temporal lobe of the brain. It is a limbic system involved in emotions, motivations and memory. Figure 11.1 illustrates the main regions of the limbic system.

In previous research it was claimed that an entire brain limbic system is involved in the development of emotions. However, recent research has shown that some of these limbic structures are not as directly related to emotion as others and that some non-limbic structures are of greater emotional relevance.

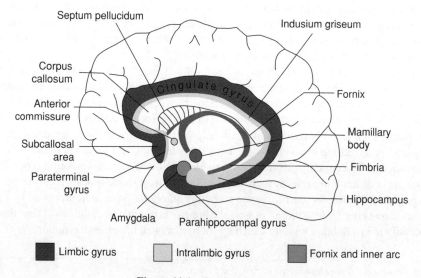

Figure 11.1 The limbic system

Often, terms such as *affect, emotion* and *mood* are used interchangeably [3]. Generally, *affect* may be used to refer to both emotions and moods. E*motion* often has an identifiable cause – a stimulus or antecedent thought, usually a spasmodic, intense experience of short duration and the person is well aware of it (i.e. emotions typically have high cognitive involvement and elaborate content). On the other hand, a *mood* tends to be more subtle, longer-lasting, less intense, more in the background, often like a frame of mind, casting a positive glow or negative shadow over experiences. Moods for healthy subjects are often nonspecific (e.g. pleasant/unpleasant; energetic/lethargic; anxious/relaxed) compared to emotions, which are usually specific, linked to clear-cut consciously available cognitive representations about their antecedents and are, therefore, typically focused on an identifiable person, object or event. In contrast, people may not be aware of their mood unless their attention is drawn to it. Mood variations may be caused by subtle factors, such as diurnal fluctuations in brain neurotransmitters, sleep/waking biorhythms, sore or tense muscles, visiting or losing a friend, sunny or rainy weather and an accumulation of pleasant or unpleasant events [3].

Based on neurobiological theories, models for emotion have been presented with contradictory observation deductions in some cases. Amongst them are two neurobiological models of emotion making opposing predictions. The Valence Model predicted that anger, a negative emotion, would activate the right prefrontal cortex, in contrast to the Direction Model which predicted that anger, an approach emotion, would activate the left prefrontal cortex. The second model was supported by further experiments [4]. However it remains questionable whether the opposite of approach in the prefrontal cortex is better described as moving away (Direction Model), unmoving but with strength and resistance (Movement Model) or unmoving with passive yielding (Action Tendency Model). Support for the Action Tendency Model (passivity related to right prefrontal activity) comes from research on shyness [5] and on behavioural inhibition research [6]. Some research examined the competing hypotheses generated by all four models and supported the Action Tendency Model [7, 8].

Another neurological approach [9] distinguishes between two classes of emotion; first, "classical emotions" including lust, anger and fear, generally evoked by *environmental stimuli*, which motivate us (to copulate/fight/flee respectively); second "homeostatic emotions" including feelings evoked by *internal body states*, which modulate our behaviour. Thirst, hunger, feeling hot or cold (core temperature), feeling sleep deprived, salt hunger and air hunger are all examples of homeostatic emotion. Homeostatic emotion onset occurs when an imbalance arises in any one of these systems hence prompting us to react and restore the balance to the system. Pain is a homeostatic emotion alerting us of an abnormal state/condition [9].

Another argument on emotions is based on the theory that cognitive activity in the form of judgments, evaluations or thoughts is necessary for an emotion to occur. Argued by Richard Lazarus, emotions are about something or have intentionality. Such cognitive activity may be conscious or unconscious and may or may not take the form of conceptual processing. It has also been suggested that emotions are often used as shortcuts to process information and influence behaviour [10].

A hybrid of the somatic and cognitive theories of emotion is the perceptual theory. Based on so-called neo-Jamesian theory, bodily responses are central to emotions. In this respect, emotions are held to be analogous to faculties such as vision or touch, providing information about the relation between the subject and the world in various ways.

Another theory called Affective Event Theory is a communication-based theory [11] which looks at the causes, structures, and consequences of emotional experience. This theory suggests that emotions are influenced and caused by events which in turn influence attitudes and behaviours. It also emphasizes on so-called emotion episodes. This theory has been utilized by numerous researchers to better understand emotion from a communicative perspective, and was reviewed further in [12].

Singer–Schachter theory is another cognitive theory. This is based on experiments purportedly showing that subjects can have different emotional reactions despite being placed into the same physiological state with an injection of adrenaline. Subjects were observed to express anger or amusement if another person in such a situation displayed that emotion.

A recent model called Component Process model considers emotions as the synchronization of many different bodily and cognitive components. Therefore, symptoms and physiological signatures may be used to evaluate emotions. Emotions are identified with an overall process where low-level cognitive appraisals, in particular the processing of relevance, trigger bodily reactions, behaviours, feelings, and actions occur. Based on this model it is clear that a thorough evaluation of emotion requires assessment of many symptoms, facial and body movement features, together with physiological signals.

More common and popular emotions, such as fear, anger, disgust, happiness, sadness, surprise, interest, shame, contempt, suffering, love, tension, and mirth, have been investigated by researchers in emotion understanding, control, and regulation mainly by analysing the brain signals and images.

Traditionally, few neuroimaging, imaging, signal processing and data analysis techniques have been used for detection and recognition of emotions. These modalities include facial pattern and gestures, respiration, blood pressure, skin impedance, body temperature, muscle activity, heart rate, and brain activity [13].

11.2   The Effects of Emotions

Respiration is a peripheral body process. There are significant differences between the states of respiration in emotions such as calmness and fear. Slow and uniform respiration occurs in the calm state as opposed to fast and abrupt respiration when in fear [14, 15]. A respiration belt is often used for measuring this biometric. Generally, breathing rhythm significantly changes with emotion. Frustration causes an increase in the breathing frequency and vice versa.

Although respiration is primarily regulated for metabolic and homeostatic changes in the brainstem, breathing can also change in response to changes in emotions [16]. They also deduced that final respiratory output is influenced by a complex interaction between the brainstem and higher centres, including the limbic system and cortical structures. The important and interesting conclusion is the coexistence of emotion and respiration, which is important in maintaining physiological homeostasis [15]. Relationships between emotions and respiration have shown more rapid breathing during an arousal state. It has been shown that the breathing frequency corresponds approximately linearly to trait anxiety [15]. They have gone even further to investigate olfactory function and the piriform–amygdala complex in relation to respiration, including oscillations of piriform–amygdala complex activity and respiratory rhythm [17]. Final respiratory output involves a complex interaction between the brainstem and higher centres, including the limbic system and cortical structures.

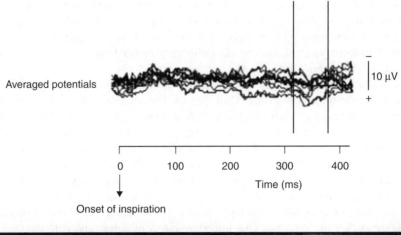

Averaged potentials

10 µV

0   100   200   300   400

Time (ms)

Onset of inspiration

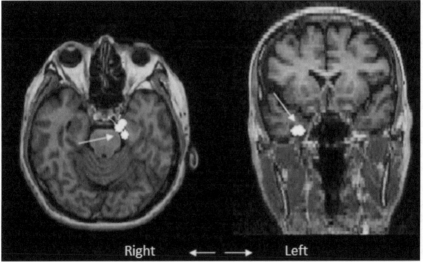

Right ← → Left

Figure 11.2 The generators of respiration-related anxiety potentials are in the right temporal pole in subjects with low anxiety and in the temporal pole and amygdala in very anxious subjects. Taken from [18]

Using both EEG and fMRI it has been observed that the generators of respiration-related anxiety potentials are in the right temporal pole in subjects with low anxiety and for the most anxious subjects in the temporal pole and amygdala, as depicted in Figure 11.2 [18].

The temporal pole located within paralimbic regions is involved in the evaluation of environmental uncertainty or danger [16]. If the respiratory rate is increased by anxiety, these regions may be activated before the onset of inspiration. It is assumed that an increase in respiratory rate is caused by unconscious evaluation in the amygdala and that these two activities occur in parallel. Therefore, stimulation of the amygdala results in a rapid increase in respiratory rate followed by a feeling of fear and anxiety [19].

Blood pressure is another human factor affected by various changes in the body metabolism, such as those caused by emotion changes. Blood pressure increases with emotions such as anger or fear and decreases with happiness or calmness. A plethysmograph is used to measure this quantity for investigation of different physiological abnormalities. Also, it is a well known phenomenon that blood pressure is directly related to stress.

Another biometric, which changes with emotion, is skin impedance. Skin impedance can be checked using a galvanic skin response (GSR) sensor. The emotional state of human beings is a physiological mechanism which affects the body and manifests itself by changing the face impression. Stress and other emotional states are controlled by the hypothalamus, a region in the brain, and the hormones released by the adrenal gland situated on the kidney. Sympathetic and para-sympathetic nervous systems are also involved. While adrenaline is responsible for creating stress, emotional feelings such as fear and anxiety stimulate the hypothalamus which in turn increases adrenaline secretion. Adrenaline causes an increase in heart beat, blood pressure, breathing rate, sweating and fainting, in extreme cases. These events prepare and alert the body to face the situation. One important effect of adrenaline is to increase blood flow to the skin to remove excess heat from the body through sweating. That is why burning occurs in shocks. When in stress, blood flow to the skin increases, vessels become porous and sweating occurs due to water secretion. This in effect removes heat from the body through evaporation of sweat. This lowers the skin resistance to facilitate removal of water more easily. A moist skin also increases electrical activity. Hence, one can conclude that the skin resistance is directly proportional to the emotional state. Consequently, body temperature is subjected to changes when the subject is consistently involved in emotion. This can be due to changes in other physiological and metabolic factors, such as heart rate, blood pressure, or muscle activities.

Electromyography (EMG) is used to measure muscle activities of the face, neck and shoulder which are very likely to change due to certain emotions. Correlation between facial expression and facial muscle activities has been investigated and verified by many researchers; see for example [20–23]. Facial pattern analysis and recognition can be performed using frontal or lateral view images taken by still or video cameras. In this direction, there has been much work done giving rise to numerous algorithms and solutions. Many databases, such as IAPS (international affective picture system) [24] have also been provided with emotional facial images.

Security, superior automatic speaker recognition (ASR) system performance, life-like agents, language etiquette, student modelling and many other applications have motivated the study of emotion-influenced language recognition. Fluctuation in voice characteristics, such as energy, speaking rate, formant frequencies, and formant bandwidths triggered by the changes in emotions have been studied by many researchers. As an example, Oudeyer [25] developed algorithms that allow a robot to express its emotions by modulating the intonation of its voice. In a different approach emotion recognition in spontaneous speech has been addressed [26].

The latest research has revealed the relations between light and emotions, colour and emotions, and more importantly music and emotions [27]. Light is used to treat mood disorders but the logic behind it is not well understood. While rod and cone eye photoreceptors process visible light, a third type of photoreceptor, particularly sensitive to blue light, mediates non-visual responses such as sleep cycles and alertness. So, light may make us feel better because it helps regulate circadian rhythms.

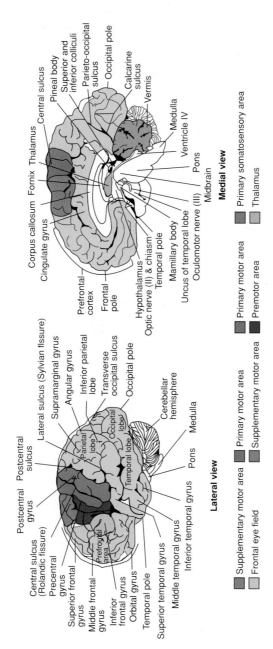

Lateral view

Central sulcus (Rolandic fissure)
Precentral gyrus
Superior frontal gyrus
Middle frontal gyrus
Inferior frontal gyrus
Orbital gyrus
Temporal pole
Superior temporal gyrus
Middle temporal gyrus
Inferior temporal gyrus

Postcentral gyrus
Postcentral sulcus
Lateral sulcus (Sylvian fissure)
Supramarginal gyrus
Angular gyrus
Inferior parietal lobe
Transverse occipital sulcus
Occipital pole

Prefrontal area
Parietal lobe
Occipital lobe
Temporal lobe
Cerebellar hemisphere
Medulla
Pons

Medial view

Corpus callosum   Fornix   Thalamus
Cingulate gyrus
Central sulcus
Pineal body
Superior and inferior colliculi
Parieto-occipital sulcus
Occipital pole
Calcarine sulcus
Vermis

Prefrontal cortex
Frontal pole
Hypothalamus
Optic nerve (II) & chiasm
Temporal pole
Mamillary body
Uncus of temporal lobe
Oculomotor nerve (III)
Midbrain
Pons
Ventricle IV
Medulla

Supplementary motor area
Frontal eye field

Primary motor area
Supplementary motor area

Primary motor area
Premotor area

Primary somatosensory area
Thalamus

(Plate 1) Figure 1.5   Diagrammatic representation of the major parts of the brain

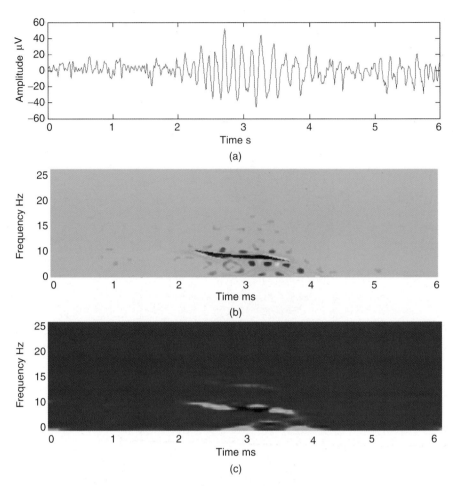

(Plate 2) Figure 4.2 TF representation of an epileptic waveform in (a) for different time resolutions using a Hanning window of (b) 1 ms, and (c) 2 ms duration

(Plate 3) Figure 6.5 The CSP patterns related to (a) right-hand movement and (b) left-hand movement. The EEG channels are indicated by numbers which correspond to three rows of channels (electrodes) within the central and centro-parietal regions

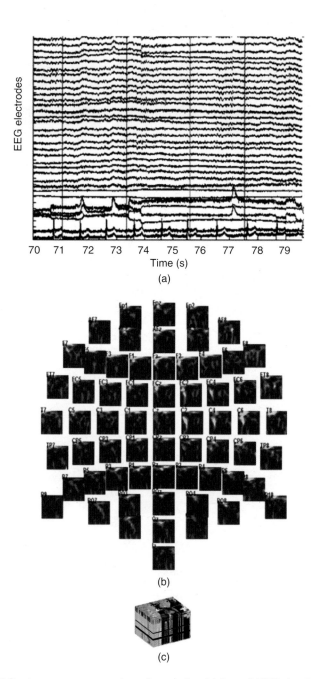

(Plate 4) Figure 7.9 A tensor representation of a set of multichannel EEG signals; (a) time domain representation, (b) time–frequency representation of each electrode signal, and (c) the constructed tensor

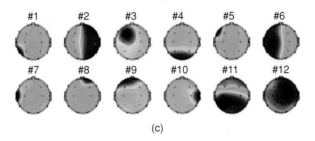

#1 #2 #3 #4 #5 #6

#7 #8 #9 #10 #11 #12

(c)

(Plate 5) Figure 7.13 The results of application of the FOBSS algorithm to a set of scalp EEGs in (a); (b) the separated sources, and (c) the corresponding topoplots

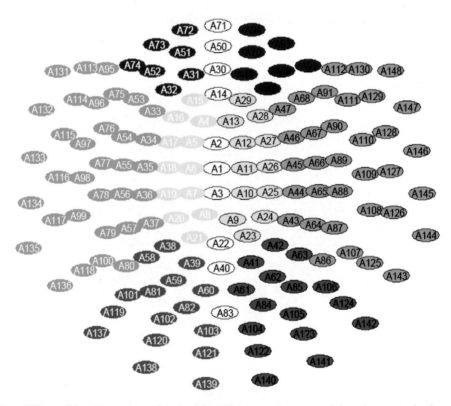

(Plate 6) Figure 8.3 Nine regions of the brain used for measuring connectivity using spectral coherency

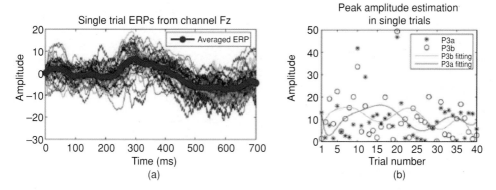

(Plate 7) Figure 10.5 (a) Single-trial ERPs (40 trials related to the infrequent tones) and their average from channel Fz, and (b) estimated amplitudes for P3a and P3b in different trials

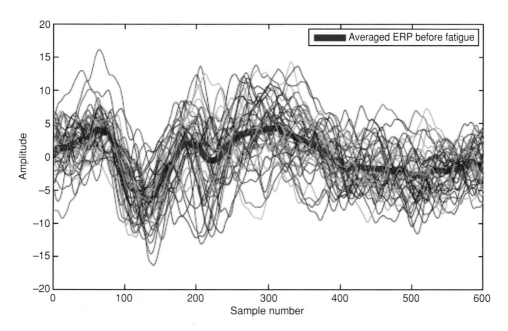

(Plate 8) Figure 10.8 Forty single trial ERPs and their average from the Cz channel before fatigue

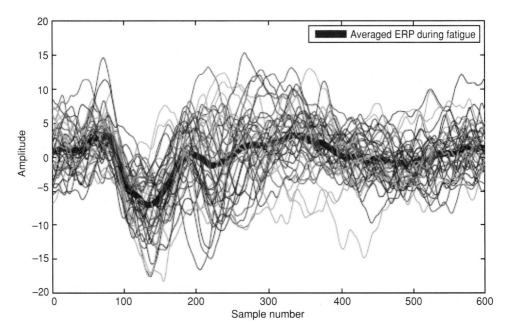

(Plate 9) Figure 10.9 Forty single trial ERPs and their average from the Cz channel during fatigue state

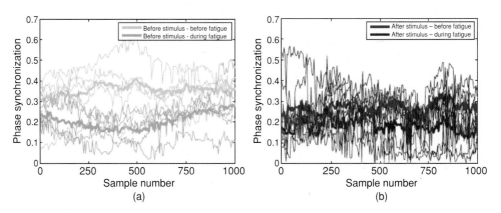

(Plate 10) Figure 10.14 Alpha phase synchronization of C3-C4; (a) before stimulus and (b) after stimulus

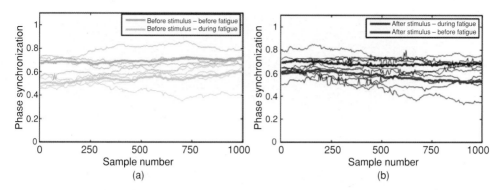

(Plate 11) Figure 10.15 Beta phase synchronization of F3-F4; (a) before stimulus and (b) after stimulus

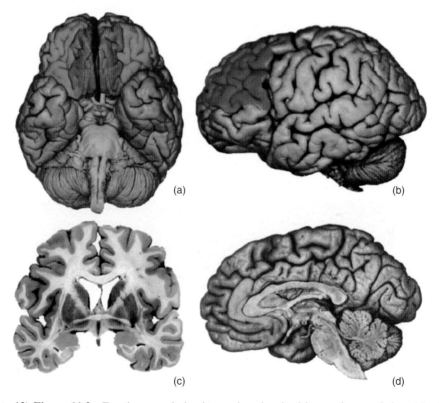

(Plate 12) Figure 11.3 Emotion neural circuitry regions involved in emotion regulation; (a) orbital prefrontal cortex (green regions) and the ventromedial prefrontal cortex (red regions), (b) dorsolateral prefrontal cortex, (c) amygdala and (d) anterior cingulate cortex; adapted from [36], © Oxford University Press

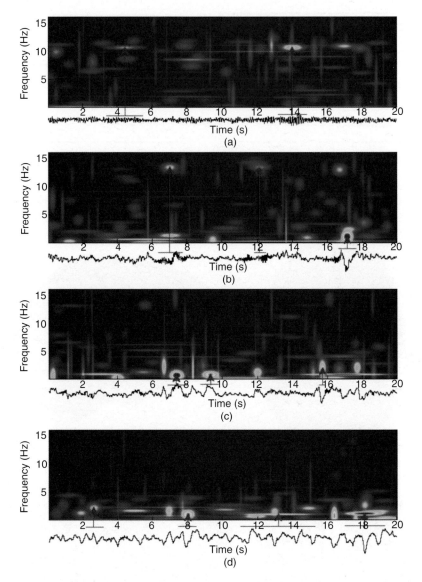

(Plate 13) Figure 12.6 Time–frequency energy map of 20 s epochs of sleep EEG in different stages. The arrows point to the corresponding blobs for (a) awake alpha, (b) spindles and K-complex related to Stage II, and (c) and (d) SWAs related, respectively, to Stages III and IV. Reprinted from [32], © IEEE

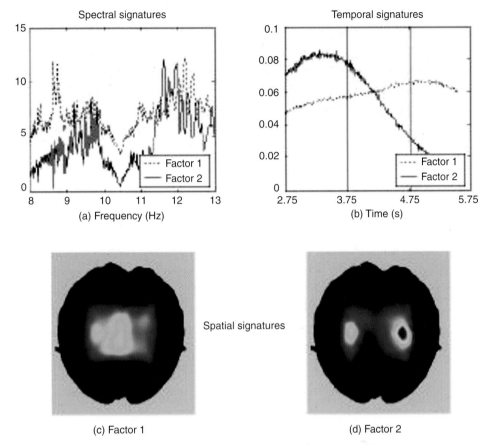

(Plate 14) Figure 13.5 Sample space–time–frequency decomposition of the 15 channel EEG signal recorded during left index movement imagination. The factor demonstrated with the solid line indicates a clear ERD in the contralateral hemisphere: (a) spectral contents of the two identified factors, (b) temporal signatures of the factors, the onset of preparation and execution cues are shown in light and dark patches, respectively, (c) and (d) represent topographic mapping of EEG for the two factors

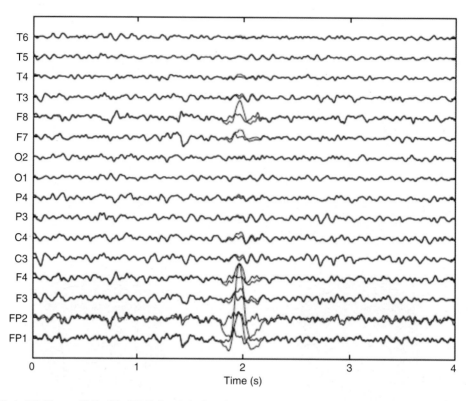

(Plate 15) Figure 13.7   The EEG signals before the removal of eye-blinking artefacts in red (pale) and after removing the artefact in blue (dark) colour

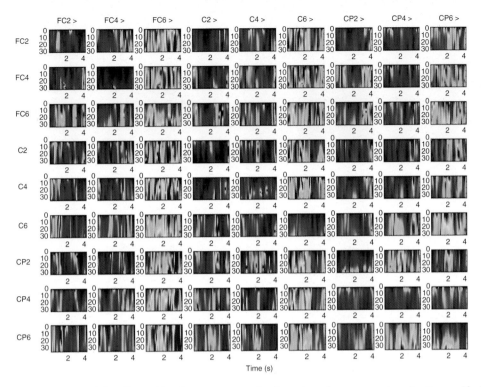

(Plate 16) Figure 13.8 Illustration of source propagation from the coherency spectrum for the specified EEG channels for the left hand

(Plate 17) Figure 13.9 Illustration of source propagation from the coherency spectrum for the specified EEG channels for the right hand

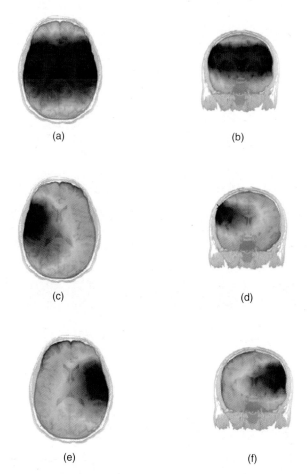

(a)

(b)

(c)

(d)

(e)

(f)

(Plate 18) Figure 14.11 Topographies or power profiles of real MEG data obtained using (a) BF method in axial view, (b) BF method in coronal view, (c) deflation BF while deflating the second source in axial view, (d) deflation BF while deflating the second source in coronal view, (e) deflation BF while deflating the first source in axial view, (f) deflation BF while deflating the first source in coronal view [59]

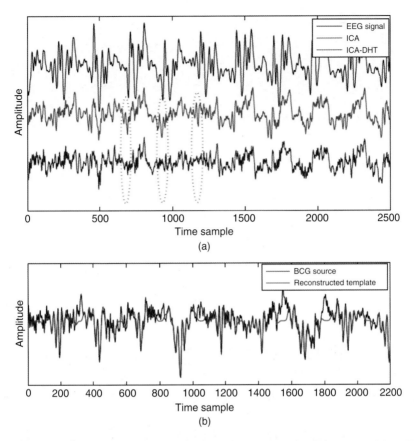

(Plate 19) Figure 16.8   (a) Comparison between ICA and ICA-DHT for BCG removal from EEG and (b) The BCG source, not deflated in ICA, together with corresponding model obtained using ICA-DHT

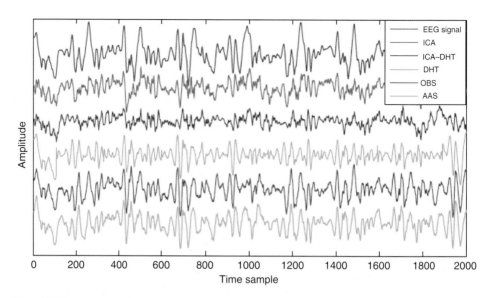

(Plate 20) Figure 16.9 Results of artefact removal from C_Z channel using ICA, combined ICA-DHT, DHT, OBS, and AAS (see key on figure)

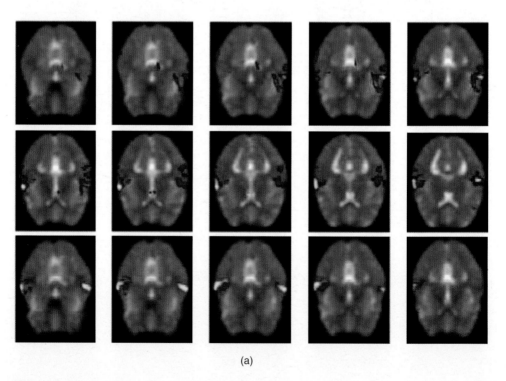

(a)

(Plate 21) Figure 16.14a The spatial maps for auditory data analysis results obtained from KL I-Divergence method

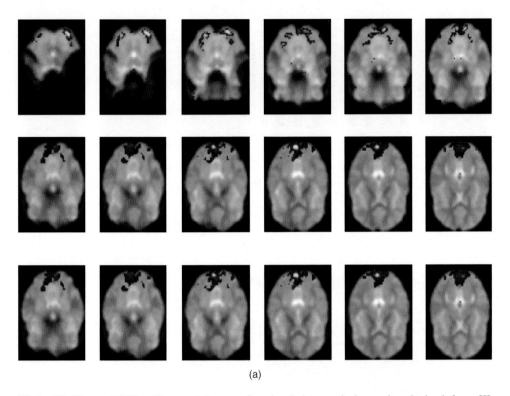

(a)

(Plate 22) Figure 16.15a The spatial maps for visual data analysis results obtained from KL I-Divergence method

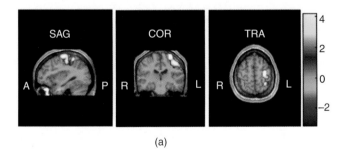

(a)

(Plate 23) Figure 16.16a Detected BOLD using PARAFAC2. SAG, COR, and TRA refer, respectively, to sagittal, coronal, and transverse planes

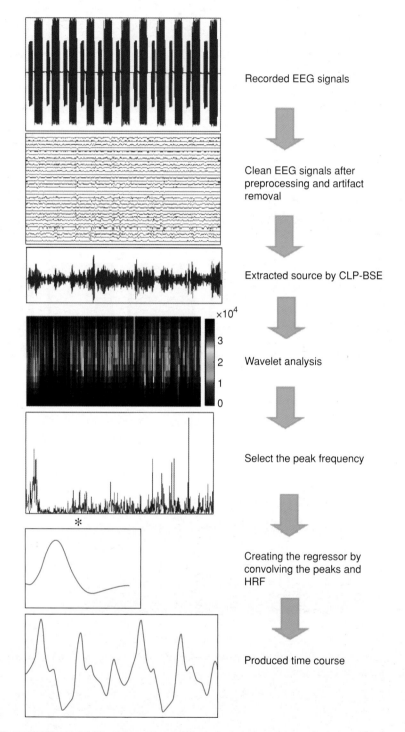

(Plate 24) Figure 16.17 Schematic of different steps of model-based EEG–fMRI analysis

To find out how this pathway directly affects our emotional state, Vandewalle and his colleagues at the University of Liège, Belgium scanned the brains of volunteers exposed to green or blue light while a neutral or angry voice recited meaningless words. As expected, the brain areas responsible for processing emotion responded more strongly to the angry voice, and this effect was amplified further by blue light [26]. Vandewalle suggests blue light is likely to amplify emotions in both directions.

The main focus of this chapter is to study the brain response for different emotions. Although some studies have used functional near-infrared spectroscopy (fINRS) to capture brain activity, mainly from the prefrontal cortex, the most important modality for this purpose would be the EEG. fINRS is a low-cost, user-friendly, and practical measurement modality. It detects the infrared light (photon count) travelling through the cortex and is used to monitor the hemodynamic changes during cognitive and/or emotional activity. A big advantage of this modality to EEG is that fINRS is not affected by the facial muscles EMG during motion expression. However, its application is limited to the close-to-cortex area and having a small number of sensors. Therefore, it is less practical when the study of various emotions is undertaken.

The study of emotion using EEG has become more attractive due to many algorithms developed by the signal processing community. Specially, the localization of brain segments involved in a particular emotion, the connectivity of those regions for expression, control and regulation of emotions, and synchronisation of brain lobes to evaluate the extent the brain is influenced and stimulated. Generally speaking, there is no doubt that the brain is involved in all physical or physiological changes in the human body.

11.3   Psychology and Psychophysiology of Emotion

Undoubtedly, emotions pervade our daily life. They can help or prevent us guide our choices, avoid dangers and also play a key role in non-verbal communication.

Cornelius [28] introduced three emotion viewpoints; Darwinian, cognitive, and Jamesian. Based on the Darwinian viewpoint, emotions are selected by nature in terms of their survival value, that is, fear exists because it helps avoid anger. The cognitive theory considers the brain as the centre of the emotions, and particularly concentrates on direct and non-reflective processes called appraisal, by which the brain judges a situation or event to be good or bad. Both of these points refer to the psychophysiology of emotion. Finally, the Jamesian theory suggests that emotions are only peripheral (bodily) perception changes, such as heart rate or dermal responses. For example, I am afraid because I shiver. This example refers to the role of physiological responses in the study of emotions.

Before any quantification or assessment of emotions three principles or goals in the analysis of emotions have to be recognised, namely emotion awareness, emotional arousal and expression, and emotion regulation. Emotion regulation skills – including those which identify and label emotions, allow and tolerate emotions, establish a working distance, increase positive emotions, reduce vulnerability to negative emotions, self-sooth, breath and distract – are also found to help when in high distress [29].

Quantitatively, emotions are often presented in a two-dimensional space of valence–arousal [30]. Valence represents the way one judges a situation, from unpleasant to pleasant; arousal expressesthe degree of excitement felt by people, from calmness to agitation. In a two-dimensional space of valence–arousal, it is possible to map a point to a categorical feeling label, although verbal description of the emotion state cannot be achieved.

Alexithymia and anhedonia – *Alexithymia* is an impaired ability to experience and express emotions. It is a prominent feature amongst neurological patients with hemispheric commissurotomy. Thus, the division in awareness affecting such patients includes an inability to communicate emotion arising from centres in the right hemisphere, through the language centres of the left hemisphere. In other words, the left hemisphere might not be aware and able to communicate emotions the right hemisphere is perfectly aware of – awareness that might be revealed if the right hemisphere possessed the same language skills as the left. In any event, the alexithymic patient's inability to discriminate between such feelings as anger and sadness suggests a rather marked deficit in explicit emotion. The question, then, is whether one can find evidence of *implicit* emotion in these patients, in terms of behavioural or physiological indices. This disorder may lead to psychophysiological and somatoform disorders if not controlled or regulated. Alexithymia can be distinguished from *anhedonia,* an inability to experience positive emotions [31]. However, the alexithymic inability to communicate emotions to others is correlated with social anhedonia, or a preference for solitary rather than social activities [32]. Physical anhedonia affects explicit (subjective) components of positive emotion, leaving implicit (behavioural and physiological) components of positive emotion intact. Alexithymia is known to be associated with hypertension.

Alexithymia and related emotion-processing variables have been examined as a function of hypertension type [33]. The hypothesis here is that if dysregulated emotional processes play a key neurobiological role in essential hypertension, they would be less present in hypertension due to specific medical causes or secondary hypertension. The results achieved are consistent with a contribution of an emotional or psychosomatic component in essential hypertension and may have practical implications for non-pharmacological management of hypertension. Such results also demonstrate the usefulness of complementary measures for emotion processing in medically ill patients.

11.4   Emotion Regulation

Emotional self-regulation, also known as emotion regulation refers to a state where one is able to properly regulate one's emotions. It is a complex process which involves the initiation, inhibition or modulation of the following functions [34]:

1. Internal feeling states (i.e. the subjective experience of emotion)
2. Emotion-related cognitions (e.g. thought reactions to a situation)
3. Emotion-related physiological processes (e.g. heart rate, hormonal, or other physiological reactions)
4. Emotion-related behaviour (e.g. reactions or facial expressions related to emotion).

As stated above, in the human brain, emotion is normally regulated by a complex physiological circuit consisting of the orbital frontal cortex, amygdala, anterior cingulated cortex and several other interconnected regions. There are both genetic and environmental contributions to the structure and function of this circuitry. Diverse regions of the prefrontal cortex (PFC), amygdala, hippocampus, hypothalamus, anterior cingulate cortex (ACC), insular

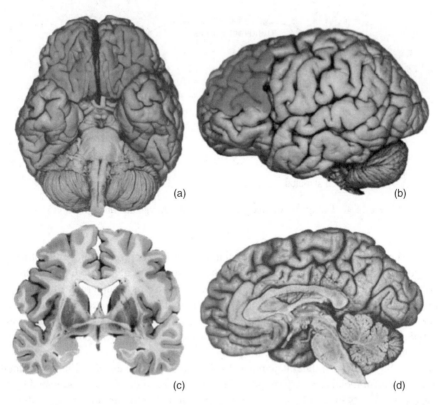

Figure 11.3 Emotion neural circuitry regions involved in emotion regulation; (a) orbital prefrontal cortex (green regions) and the ventromedial prefrontal cortex (red regions), (b) dorsolateral prefrontal cortex, (c) amygdala and (d) anterior cingulate cortex; adapted from [36], © Oxford University Press (see Plate 12 for the coloured version)

cortex, ventral striatum and other interconnected structures are involved in this complex circuitry, as depicted in Figure 11.3. Each of these interconnected structures plays a role in different aspects of emotion regulation. Abnormalities in one or more of these regions or in the interconnections amongst them are associated with failures of emotion regulation, increased propensity for impulsive aggression and violence [35].

Emotion regulation involves processes that amplify, attenuate or maintain an emotion. Here our focus is on the associated effective phenomena of anger, general negative effect, and impulsive aggression.

The amygdala is involved in the learning process. It associates stimuli with primary events that are intrinsically punishing or rewarding [37, 38]. In human neuroimaging studies, the amygdala is activated in response to cues that connote threat (such as facial signs of fear) [39, 40], as well as during induced fear (e.g. fear conditioning) [41, 42] and generalized negative affect (for example, negative effect provoked by watching unpleasant pictures) [43].

The pathways of the brain to the amygdala are depicted in Figure 11.4 [44]. An emotional stimulus, something that looks like a snake, is first processed in the brain of a hiker by the

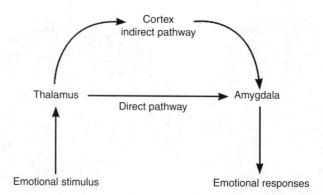

Figure 11.4   Direct and indirect pathways to the amygdala. Taken from [45]

thalamus. The thalamus passes on an immediate, but crude representation of the stimulus to the amygdala. This transmission allows the brain to start to respond to a potentially dangerous object, which could be a snake. Therefore, the hiker prepares for a dangerous situation, before he even knows what the stimulus is. Meanwhile, the thalamus also sends information to the cortex for more detailed examination. The cortex then creates a more accurate impression of the stimulus. The hiker realizes that he mistook a root for a snake. This outcome is fed to the amygdala as well and is transmitted to the body.

Patients with selective bilateral damage to the amgydala have a specific impairment in recognizing fearful facial expressions [46]. The amygdala is more strongly activated by facial expressions of fear than it is by other facial expressions, including anger. For example, an increase in intensity of fearful facial expressions is associated with activation of the amygdala. In contrast, the increase in intensity of angry facial expressions is associated with increased activation of the orbitofrontal cortex (OFC) and the ACC [47]. In some neuroimaging studies undertaken to induce anger, specifically those presented in [48, 49], normal subjects showed increased activation in the OFC and ACC. Normally, such activations may be part of an automatic regulatory response that controls the intensity of expressed anger.

It is expected that in individuals prone to aggression and violence, the increase in OFC and ACC activation, usually observed in such conditions, is attenuated [35].

In humans, when subjects view unpleasant pictures, there is an increase in the magnitude of the eye-blink reflex (measured from surface EMG recording of orbicularis oculi muscle) in response to a brief burst of noise. Moreover, the magnitude of the eye-blink reflex to the same stimulus during viewing of pleasant pictures is smaller than that during the viewing of neutral pictures [24, 50].

By triggering a startle response at different times during affective processing, information about the time course of emotion can be gleaned [51–54].

In one experiment [53], normal subjects viewed unpleasant or neutral pictures for 8 s. 4 s after the picture appeared, a digitized human voice instructed the subject to regulate the emotion they were experiencing in response to the picture. For unpleasant pictures, subjects were asked to suppress, enhance or maintain their emotional response. Subjects were instructed to continue voluntary regulation of their emotional response even after the picture disappeared. During and after the presentation of the pictures, brief noise bursts were presented to probe the time course of the emotion before and after the instruction. When subjects were requested to suppress their negative effect, they showed significantly diminished startle magnitude during and after unpleasant pictures compared with both maintain and enhance conditions. Most importantly, the subjects varied considerably in their skills at suppressing negative emotion. By inserting startle stimuli before presenting the instruction, it was confirmed that this variation in the ability to regulate emotions could not be accounted for by differences in the initial reaction to negative stimuli. In a more recent work [55], it has been found that baseline levels of regional brain activation inferred from high-density scalp-recorded EEG [55] predicted the ability of subjects to suppress emotions. Those subjects with greater relative left-sided activation in prefrontal scalp regions presented greater startle attenuation in response to suppression instruction.

Suppressing negative effect in response to a stimulus that previously aroused such emotion can be conceptualized as a form of reversal learning. Patients with lesions in the OFC and those vulnerable to impulsive aggression should particularly be deficient in such a task, although they would still show basic enhancement of startle magnitude in response to negative stimuli. In [52] it has been proposed that the mechanism underlying suppression of negative emotion is via an inhibitory connection from regions of the prefrontal cortex [56].

Based on these experiments a number of important conclusions were drawn [35] first, individual differences in the capacity to regulate emotion are objectively measurable. Second, individual differences in patterns of prefrontal activation predict the ability to perform this task and thus reflect differences in aspects of emotion regulation and third, individual differences in emotion regulation skills, particularly as they apply to suppression of negative effect, may especially be important in determining vulnerability to aggression and violence.

A greater left pre-frontal activation than right is associated with increased reactivity to positively valenced emotional stimuli [57], increased ability to recover from negative affective challenge [58], better voluntary suppression of negative affect [55], and higher scores on scales measuring psychological well-being [59]. Davidson has interpreted these findings as the left and right PFC playing differential roles in emotional processing. A recent publication [60] reveals that there are hemispheric differences in goal-directed tendencies (approach versus avoidance) beyond those captured by positive or negative effects. It is proposed that the left PFC is active in response to stimuli evoking the experience of positive affect. That is because these stimuli induce a fundamental tendency to approach the source of stimulation.

Research on the neural mechanism regarding the frontal asymmetry [61] and others on the neural mechanism of executive function, suggests that the right PFC is involved in monitoring and checking the environment, while the left PFC is primarily engaged in generating strategies for action. Effective interaction with the environment which is likely to result from initiating appropriate activity may well be associated with increased positive affect. Conversely, vigilance required for monitoring and checking the environment may be associated with anxiety, that is, negative effect.

In terms of the regions involved in different emotions there are hypotheses or experimental results which suggest the following [62]:

Agency and Intentionality — Humans often attribute agency and intentionality to others. This is a pervasive cognitive mechanism that allows one to predict the actions of others based on spatiotemporal configurations and mechanical inferences, and on their putative internal states, goals, and motives. Despite a high degree of overlap that may exist between attributing agency and intentionality, they are not the same thing.

Implicated brain regions in agency and intentionality include the parietal cortex, insula, motor cortex, medial PFC and the STS region [63, 64].

Norm Violation — Norms are abstract concepts firmly encoded in the human mind. They differ from most kinds of abstractions because the code for behavioural standards and expectations is often associated with emotional reactions when violated [65]. This effect, which marks a violation of an arbitrary norm, has been demonstrated to elicit brain responses in regions linked to conflict monitoring, behavioural flexibility, and social response reversals such as the anterior cingulate cortex, anterior insula, the lateral OFC [66]).

Guilt — Guilt emerges prototypically from (i) recognizing or envisioning a bad outcome to another person, (ii) attributing the agency of such an outcome to oneself, and (iii) being attached to the damaged person (or abstract value). Recent neuroimaging data showed the involvement of the anterior PFC, anterior temporal cortex, insula, anterior cingulate cortex, and STS region in guilt experience [67]).

Shame — While there are no available brain imaging studies on shame, brain regions similar to those demonstrated for embarrassment should be involved. The ventral region of the anterior cingulate cortex, also known as the subgenual area, has been associated with depressive symptoms, [68]). Hence, such brain region might also play a more specific role in neural representation of shame.

Embarrassment — Embarrassment has traditionally been viewed as a variant of shame [69]. The related neural structures include the medial PFC, anterior temporal cortex, STS region and lateral division of OFC.

Pride — The polar opposite of shame and embarrassment is pride. So far, there is no clear evidence for neural representation of pride, although it has been shown that patients with OFC lesions may experience this emotion inappropriately [70]. We hypothesize that brain regions involved with mapping the intentions of other persons (e.g. the medial PFC and the STS region) and regions involved in reward responses (OFC, hypothalamus, septal nuclei, ventral striatum) may play a role in this emotion.

Indignation and Anger — Indignation and anger are elicited by (i) observing a norm violation in which (ii) another person is the agent, especially if (iii) the agent acted intentionally. Indignation relies on (iv) engagement of aggressiveness, following an observation of (v) bad outcomes to self or a third party. We and others have shown that indignation evokes activation of the OFC (especially its lateral division), anterior PFC, anterior insula and anterior cingulate cortex [71].

Contempt — Contempt has been considered a blend of anger and disgust, and sometimes is considered as a more subtle form of interpersonal disgust [72]. For this reason here it is considered together with disgust. Neural representations of disgust have been shown to include the anterior insula, the anterior cingulate and temporal cortices, basal ganglia, amygdala, and OFC [71, 73, 74].

Pity and Compassion — These feelings are described as "being moved by another's suffering" [75, 76]. Preliminary functional imaging results in normal subjects point to the involvement of the anterior PFC, dorsolateral PFC, OFC, the anterior insula and anterior temporal cortex in pity or compassion [77]. Further studies with more advanced imaging techniques may test if certain limbic regions, such as hypothalamus, septal nuclei and ventral striatum could also be involved.

Awe and Elevation — Awe and elevation are self-transcendent emotions, possibly triggered by witnessing acts of human moral beauty or virtue often giving people the desire to improve themselves and work toward the greater good. Awe and elevation are poorly understood emotions that have received more attention recently [78]. While the neuroanatomy of awe is still obscure, it is likely to involve limbic regions associated with reward mechanisms, including the hypothalamus, ventral striatum, medial OFC and cortical regions linked to perspective taking and perceiving social cues, such as the anterior PFC and the STS region [77].

Gratitude — This emotion is elicited by detecting a good outcome to oneself, attributed to the agency of another person, who acted in an intentional manner to achieve the outcome. Gratitude is associated with a feeling of attachment to another agent and often promotes the reciprocation of favors. Recent studies using game-theoretic methods show that the activated brain regions encompass the ventral striatum, OFC, and anterior cingulate cortex [79, 80].

For a more natural human machine interaction emotions have to be registered, conveyed and the emotional relevance of events has to be well understood.

11.5   Emotion-Provoking Stimuli

Based on the above hypothesis, experiments, facts, assessing emotions, their impacts, and the level of regulation by each subject are essential to the understanding and control of human behaviour.

Based on the results of positron emission tomography (PET) it has been revealed and confirmed that transient sadness significantly stimulates bilateral limbic and paralimbic structures, the brain stem, thalamus and caudate/putamen, while transient happiness is associated with marked, widespread reductions in cortical cerebral blood flow (CBF) in the right prefrontal and bilateral temporal-parietal regions [81]. CBF increases during sad mood induction. The frontal pole is the only region differentiating sad states from happy states. Emotion expression in infants is also known to be associated with frontal activity asymmetries which reflect specific regulation strategies [81]. Moreover, the left frontal region corresponds to regulation involving actions that try to maintain continuity and stability.

Four models for brain lateralization of emotional processing have been proposed by Demaree *et al.* [82]; the right hemisphere, valence, approach–withdrawal, and behavioural inhibition system–behavioural activation system models.

Findings from the first model indicate that wide regions of the occipital and temporal cortices, particularly within the right hemisphere, are subject to increased activity during dynamical facial expressions viewing. Following this model, it can be deduced that facial expression represents another lateralized motoric function, like handedness and footedness, which might be controlled by the dominant cerebral hemisphere. Thus, facial expression would be right-sided for right-handers and left-sided for left-handers. The second assumption stems from emotional processing literature available in the early 1970s ([82] and references therein) and proposed that facial expression of emotion might be mediated by the right hemisphere. Thus, facial expression would be left-sided in right-handers, but not necessarily predictable in left-handers.

Based on the "valence model" the right hemisphere is specialised for negative emotion and the left hemisphere is specialized for positive emotion [83, 84]. In this model, as confirmed by other experimental results, a variant of the valence hypothesis contends that differential specialization exists for expression and experience of emotion as a function of valence, whereas *perceptual* processing of both positive and negative affective stimuli is a right cerebral function [84]. Correspondingly, this variant of valence hypothesis indicates that the left and right anterior regions are specialized for the expression and experience of positive and negative valence, respectively. On the other hand, the right posterior parietal, temporal, and occipital regions are dominant for the perception of emotion [85, 86].

Valence hypothesis was largely subsumed by the "approach–withdrawal model" of emotion processing, which establishes that emotions associated with approach and withdrawal behaviours are processed within the left- and right-anterior brain regions, respectively. The overlap between the valence and approach withdrawal models is extensive, with most negative emotions (e.g. fear, disgust) eliciting withdrawal, and most positive emotions (e.g. happiness, amusement) eliciting approach behaviours.

Two anatomical pathways underlying emotional/motivational systems have been termed the behavioural activation (BAS) and behavioural inhibition systems (BIS). BAS appears to activate behaviour in response to conditioned, rewarding stimuli and in relieving non-punishment. Thus, this system is responsible for both approach and active avoidance behaviours. Also, emotions associated with these behaviours are generally positive in nature [87].

Numerous theories about the relationship between the strength of BIS and BAS and affective disorders have been published. For example, although the BIS and BAS are thought to be rather stable over time, the existing variability around a strong BAS produces mania while variability around a weak BAS may yield depression.

The above lateralization models of emotional processing appear to have strengths. The right-hemisphere model, for example, emphasizes emotional perception at least as much as expression/experience. The valence model, as experienced, largely gave way to the approach–withdrawal model, which appears to be a better fit for the majority of data (e.g. anger). The BIS/BAS model, however, is essentially identical to the approach–withdrawal model but focuses on relatively stable emotional traits (BIS and BAS strength) instead of states (primary focus of the approach–withdrawal model).

Cognitive neuroscience of emotion has been, therefore, under development and has grown significantly [88]. Perception and evaluation of emotional stimuli (emotional processing) and the effects of emotion on memory formation (emotional memory) have possibly been more researched.

Emotional *arousal* and *valence* have been under much debate recently. *Arousal* refers to a continuum that varies from calm to excitement, whereas *valence* refers to a continuum that varies from pleasant to unpleasant, with neutral being an intermediate value.

Electrophysiological and functional neuroimaging study achievements support the role of the PFC and amygdala in evaluating the emotional content of stimuli although the roles of other structures, such as the ventral striatum, anterior, cingulate, posterior parietal, and insula regions, should not be ignored.

The amygdala is strongly associated with emotional processing by both lesion and functional neuroimaging studies. As an example, lesion studies show that patients with bilateral amygdala damage are impaired in detecting emotional facial expression and imaging studies on neurologically intact people have reported amygdala activations associated with processing of both pleasant and unpleasant stimuli.

11.6 Change in the ERP and Normal Brain Rhythms

11.6.1 ERP and Emotion

As for many other brain-related studies, ERPs have become an indicator of emotional effects. Consistent and robust modulation of specific ERP components can be achieved by viewing emotional images [89]. These modulations represent different stages of stimulus processing including perceptual encoding, stimulus representation in working memory, and elaborate stimulus evaluation. Selective processing of emotional cues by the brain is a function of stimulus novelty, emotional prime pictures, learned stimulus significance, and in the context of explicit attention tasks. Therefore, ERP measures are useful to assess the emotion–attention interface at the level of distinct processing stages.

The first ERP component reflecting differential processing of emotional compared to neutral stimuli is the early posterior negativity (EPN). A pronounced ERP difference for processing of emotionally arousing and neutral pictures developed around 150 ms and was maximally pronounced around 250–300 ms. This differential ERP appeared as negative deflection over temporo-occipital sensor sites and a corresponding polarity reversal over fronto-central regions. Despite differences in the overall topography, research presenting pictures discretely with longer presentation times (1.5 s and 1.5 s intertrial interval) demonstrated a more pronounced negative potential for emotional pictures in the same latency range over temporo-occipital sensors [89]. In addition, from the biphasic view of emotion the differential processing of pleasant and unpleasant cues varies as a function of emotional arousal. Correspondingly, EPN changes with arousal level of emotional pictures. In particular, highly arousing picture contents of erotic scenes and mutilations cause a more pronounced posterior negativity compared to less arousing categories of the same valence. Figure 11.5 illustrates the time course of an early posterior negativity and the corresponding topoplots [89].

Instead of modulation during perceptual encoding, it is observed that emotional (pleasant and unpleasant) visual stimulations elicit increases late positive potentials (LPPs) over centro-parietal regions, most apparent around 400–600 ms poststimulus, namely a P3b wave [90–92]. This LPP modulation appears sizable and can be observed in almost every individual when calculating simple difference scores (emotional–neutral). In addition, LPP is also specifically enhanced for more emotionally intense pictures (i.e. described by viewers as more arousing, indicating a heightened skin conductance response). It has been found that picture contents of high evolutionary significance, such as pictures of erotica and sexual contents and contents of threat and mutilations further enhance the LPP amplitudes [90]. The effect of stimulation repetition has not shown any significant effect on the results [89]. Figure 11.6 shows the brain's

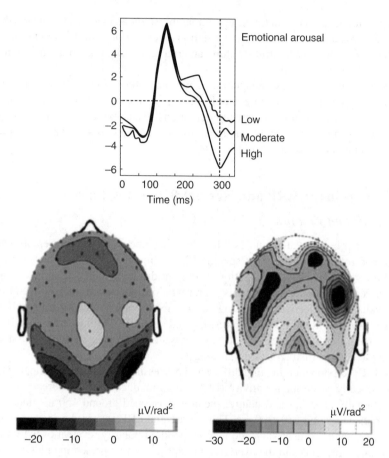

Figure 11.5 Time course of an early posterior negativity and its corresponding topoplots. Reprinted from [89], © Elsevier

time course during LPP. Topoplots of pleasant, unpleasant and neutral visual stimuli do not present noticeable difference.

In addition to LPP, in a sustained stimulation scenario, the LPP is followed by a positive slow wave. Therefore, the positive slow wave reflects sustained attention to visual emotional stimuli.

Generally, ERP modulations, induced by emotional cues, are reliable and consistently observed early in the processing stream. Additionally, emotional attentions of motivated and instructed attentions look alike.

Emotion effect has been found in different ERP components, including the P300 component (e.g. [93, 94]), N300 component [95]; and the slow wave (SW) component [94, 96]), but most consistently, emotion effect is expressed in a P300–SW complex, as discussed above. More research has shown that the P300 component may be more sensitive to emotion under intentional emotional processing (e.g. [93, 97]), whereas the N300 component appears to be more sensitive during incidental emotional processing but the evidence is not conclusive [93]. Other studies discuss variations in P3a and P3b for arousal and valence [98].

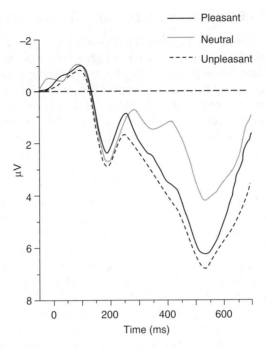

Figure 11.6 Time course of the late positive potential; grand-averaged ERP waveforms of a right parietal sensor while viewing pleasant, neutral, and unpleasant pictures. Reprinted from [89], © Elsevier

There are still two main issues concerning the emotion ERP effect which have not been completely solved. First, it is unclear whether the emotion effect is sensitive to arousal only (emotional vs neutral) or to both arousal and valence (pleasant vs unpleasant). Most studies have only found differences owing to arousal. A few recent studies have reported differences that could attribute to emotional valence [94, 96]). Many of these studies have used a small number of electrodes and, therefore, due to lack of accurate localization, no significant differences could be noted between the topographies of pleasant and unpleasant stimulations.

11.6.2 Changes in Normal Brain Waves with Emotion

In terms of changes in the normal brain rhythms, the correlation between neural activity and emotional changes has been supported by lesion [99], electrophysical [100], and functional neuroimaging [101, 102]. The same techniques support the role of the PFC and amygdala in the evaluation of emotional content of the stimuli. In the study of emotions with respect to pleasant, unpleasant, and neutral pictures the role of the PFC regions in emotional processes has been examined [102] and it was deduced that the right hemisphere is specialized for perception, expression, and experience of emotion. According to the definition of valence and based on electrophysiological and functional neuroimaging evidences, again it has been concluded that the left hemisphere is primarily associated with processing pleasant emotions, while the right hemisphere processes unpleasant emotions [101].

11.7   Perception of Odours and Emotion: Why Are They Related?

Odours may be considered as human perception after smelling. There are various odours with different strengths and humans may have different impressions about the odours. For perceptions of sensations influencing or producing emotion, the perception of odours is dependent on respiration; our sense of smell is enhanced by inhalation or inspiration. Olfactory cells have cilia (dendrites) extending from the cell body into the nasal mucosa. The axons carry impulses from activated olfactory receptors to the olfactory bulb [15]. The sensors within the olfactory bulb (Figure 11.7) send signals to the prepiriform and piriform cortex, which include the primary olfactory cortex, anterior olfactory nucleus, amygdaloid nucleus, olfactory tubercle and entorhinal cortex.

Olfactory information ascends directly to the limbic areas and is not relayed through the thalamus, so that each breath activates these areas directly. Direct stimulation of olfactory limbic areas, unconsciously, alters the respiratory pattern. Unpleasant odours increase the respiratory rate and induce shallow breathing while pleasant odours induce deep breathing. It is interesting to note that the respiratory rate increases even before the subjects discern whether a smell is unpleasant. It is likely that physiological outputs respond more rapidly than cognition [103]. Positron emission tomography (PET) and fMRI studies in humans have linked olfactory-related brain regions that are related to higher-order olfactory processing, such as discrimination [104] and emotion [105]. EEG studies have also shown three to four negative and positive waves present during inspiration and the waves are phase-locked to inspiration. These waveforms, referred to as an inspiration phase-locked α-band oscillation (I-α), are not observed during the expiratory phase or during breathing of normal air. The EEG dipole tracing method has estimated the location of I-α source generators to the entorhinal cortex, amygdala, hippocampus and orbitofrontal cortex. At 300–400 ms, the source dipole converges most in the orbitofrontal cortex. During such an event, recognition and identification of odour occurs after evaluation in the olfactory limbic areas [15]

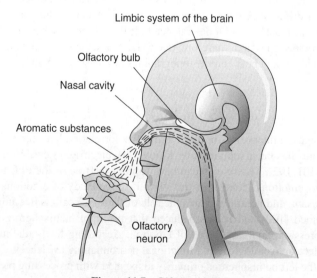

Figure 11.7   Olfactory bulb

Experiments have shown odour molecules reach olfactory receptors by inspiration, and odour information is directly projected to limbic structures but not through the thalamus. Odour stimuli thus cause respiratory changes, and simultaneously induce emotion and memory recognition via stimulation of olfactory-related limbic structures [106].

Relations between respiration and olfaction have been reviewed by Masaoka and Homma. The same authors have also reviewed the relation between brain rhythm and respiration for both normal subjects and those suffering from Parkinson's disease with olfactory impairment [106].

11.8   Emotion-Related Brain Signal Processing

Research in computer science, engineering, psychology and neuroscience has aimed at developing devices that recognize human affect, monitor and model emotions. In computer science, affective computing is a branch of the study developing artificial intelligence that deals with the design of systems and devices for recognition, interpretation and processing of human emotions. While the origins of the field may be traced back to early philosophical enquiries into emotion the more modern branch of computer science originated with Rosalind Picard's 1995 paper on affective computing [107].

It might be easy to discriminate between two contradictory types of emotions (such as sadness and happiness), but a learning system capable of classifying all or a large collection of emotions is yet far from reality. This is mainly because a wide range of emotions is tied to certain brain regions and unified by a certain structure.

Detection of emotional information, as for almost all types of physiological signals and data, incorporates the use of passive sensors to capture the user's physical state or behaviour without interpreting the input. The collected data are analogous to the cues humans use to perceive emotions in others. The more accurate the analysis of these data, the better the corresponding emotional state is recognized.

There are often debates on the differences between brain dynamic activity and emotion. Although both refer to neuron activations and neural processes within the brain, the brain engagements might be different [108].

Another facet of affective computing is the design of computational devices proposed either to exhibit innate emotional capabilities or to be capable of simulating emotions convincingly. Therefore, creating new techniques to indirectly assess frustration, stress and mood, through natural interaction and conversation, or making the computers more emotionally intelligent, especially responding to a person's frustration in a way that reduces negative feelings, can be a desired research direction in emotion recognition and regulation.

EEG has traditionally been used for the investigation and study of cortical activity and deep brain signal sources have not been considered widely. Advances in source separation, however, facilitate tracking and localization of different brain activities.

These signals may be corrupted by and are sensitive to electrical signals emanating from facial muscles while emotions are expressed. Therefore, it is logical to fuse these two modalities in the study of emotion.

In the majority of emotion studies with EEG, changes in the background activity of the brain have been observed or assessed. A simple approach for analysis of three emotion states of pleasant, unpleasant and neutral in different brain regions, based on the classification of brain responses to visual stimuli (by looking at IAPS images) within different frequency bands (4–45 Hz), has been given. The procedures, selected frequency bands, and selected electrodes are similar to those in [109]. Six features, subbands frequency powers from 12 sites (six for each side of the brain) have been measured and used for classification.

The estimation of brain lobes connectivity has been shown to be an effective means of assessing various brain abnormalities and functionalities. Although estimates for connectivity represent communications between cortical sources, additionally, they can show how sensitive the brain can be to emotional stimuli. As an example, the insula is believed to process convergent information from several sensory modalities monitoring the state of the body and related to pain and other basic emotion experiences, including disgust, anger and fear [110, 111]. Some researchers have even suggested that its role in mapping visceral states and emotions could be the basis for conscious feelings [112]. In that sense, high levels of connectivity found here could support an integrative role for this brain structure. Similarly, the amygdala, which is linked to emotions, to their role in learning modulation [113, 114] and its enhanced coherence is probably related to that in the insula functionally.

Different connectivity measures, such as mutual information (MI) [115] have been used. For a more clear assessment of brain connectivity, different conventional EEG frequency bands are often used.

Given a pathophysiological theory of a specific disease, connectivity models might allow one to define an *endophenotype* of that disease, that is, a biological marker at intermediate levels between genome and behaviour, which enables a more precise and physiologically motivated categorization of patients [116]. Such an approach has received particular attention in the field of schizophrenia research where a recent focus has been on abnormal synaptic plasticity leading to dysconnectivity in neural systems concerned with emotional and perceptual learning [117, 118]. A major challenge will be to establish neural systems models sensitive enough, and using their connectivity parameters reliably for diagnostic classification and treatment response prediction of individual patients. Ideally, such models should be used in conjunction with paradigms, minimally dependent on patient compliance and not confounded by factors like attention or performance. Given established validity and sufficient sensitivity and specificity of such a model, one could use its analogy to biochemical tests in internal medicine, that is, compare a particular model parameter (or combinations thereof) against a reference distribution derived from a healthy population [118]. Such procedures could help to decompose current psychiatric entities like schizophrenia into more well-defined subgroups characterized by common pathophysiological mechanisms which may facilitate the search for genetic underpinnings.

Brain region connectivity may also be influenced during sleep as a result of previous emotional events. It has been demonstrated that REM sleep physiology is associated with an overnight dissipation of amygdala activity in response to previous emotional experiences [61]. This alters functional connectivity and reduces next-day subjective emotionality.

11.9 Other Neuroimaging Modalities Used for Emotion Study

The advent of fMRI and other imaging technology has spawned a deluge of research examining how the brain functions at the macroscopic level. This work, which treats each voxel as a basic unit of analysis, has yielded tremendous insights as to how cortical regions networks cooperate to produce emotion [119].

fMRI has been reported as a method for determination of regions of activity within the brain during emotion processing [120]. However, spatial sampling of the signals from the amygdala is a challenging problem. It most probably depends on the emotion intensity. It has been claimed that an unjustified application of proportional signal scaling (PGSS) leads to an attenuated effect of emotional activation in structures with a positive correlation between local and global BOLD signal [120].

In a research by Eryilmaz *et al.* [121] a wavelet correlation approach has been developed to investigate the impact of transient emotions on functional connectivity during subsequent resting and how emotionally positive and negative information influences subsequent brain activity differentially at rest.

The connectivity analysis has been estimated using temporal correlations between regional brain activities in predefined wavelet sub-bands. In this application orthogonal cubic B-spline wavelet transform in the temporal domain [122, 123] has been used. The correlations between wavelet coefficients of a given sub-band for all regions of interest (ROIs) identified in GLM contrasts have been measured. Also, cross-correlation matrices between these regions for each condition and for four different wavelet sub-bands, including the typical resting state frequency range ($\sim$0.01–0.1 Hz), have been obtained [121].

To compute the normalized correlation between any pair of regions, the time courses corresponding to all blocks of a specific condition (e.g. fearful movies) are extracted for both regions. Then, by taking DWT the signal is separated into four different frequency bands: (i) 0.03–0.06 Hz, (ii) 0.06–0.11 Hz, (iii) 0.11–0.23 Hz, and (iv) 0.23–0.45 Hz. The first two frequency intervals are known to correspond to typical default mode bands, constituted our main 'bands of interest' and, therefore, the first two bands are considered here. The cross-correlations between wavelet coefficients of all possible pairs of regions are calculated and the correlation matrices for each block are constructed.

In order to statistically test the differences between two resting conditions (e.g. rest post-fearful > rest post-neutral), corresponding correlation matrices to blocks pertaining to two conditions (twelve matrices per condition for each participant) are compared by nonparametric permutation testing [65].

A threshold was defined for each condition and each sub-band and applied to wavelet correlation matrices, using a false discovery rate at 0.2. Several node measures based on small-world models were computed after applying a threshold to the matrices. In a so-called small-world network model, each region was considered as a "node" and, after applying an appropriate threshold, various measures, such as clustering coefficient, mean minimum path length and degree, can be computed for each node [124].

Clustering the wavelet coefficients led to grouping of the existing links connecting a number of neighbouring nodes. Minimum path length is the number of connections that forms the shortest path between two nodes, whereas mean minimum path length for a specific node is the average of all minimum path lengths between that node and all other nodes. The degree of a node was defined as the number of connections it makes with other network nodes. Based on these measures, network hubs were defined as nodes with the largest degree. Eventually, these parameters can be averaged over different participants to obtain a single value per condition and sub-band.

In an experiment two movies, one joyful and one fearful, were presented to the participants. As shown in Figure 11.8, different activity profiles can be observed amongst different brain regions. The posterior cingulate cortices (PCC) show no significant variation in activity

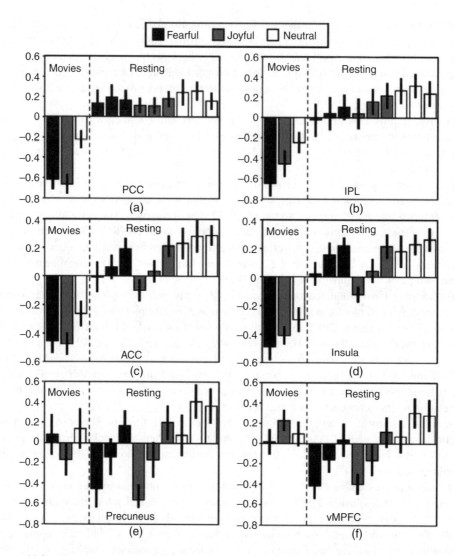

Figure 11.8   Negative and positive magnitudes of activity for the main regions differentially activated during resting periods, across three successive 30-s periods and three emotion conditions during movie-watching, with different emotion types, and during resting periods, with each successive time bin of 30 s shown for periods following fearful (three leftmost black bars), joyful (three central grey bars), and neutral movies (three rightmost white bars). These three successive time bins in each emotion condition correspond to 0–30 s, 30–60 s and 60–90 s of rest following the end of the movie. Activity in (a) the posterior cingulate cortex and (b) the inferior parietal lobule show global increases during rest but regardless of preceding emotion condition. Activity in (c) the ACC and (d) the right insula, show marked deactivations during movies, with a steep increase in activity during rest after neutral movies but more gradual restoration of rest activity after joyful and fearful movies. Unlike typical default mode regions, activity in (e) the left precuneus and (f) the vMPFC is not reduced during movies, but during the first part of resting periods following emotional movies then is progressively increased to a complete restoration level. Reprinted from [121]

between the neutral and emotional rest conditions, as can be seen in Figure 11.8a. This region shows similar increases during rest regardless of the preceding movie type. The inferior parietal lobule (IPL) bilaterally shows strong increases during the resting state, although the return to resting activity appeared slightly less complete or less rapid after emotional rather than neutral movies, as in Figure 11.8b. In contrast, a much stronger emotional effect is observed within the level and time-course of activity during rest for the insula and anterior cingulate cortices (ACC). Both the ACC and insula bilaterally show a typical default mode (DM) pattern with a strong deactivation during both movies, as shown in Figure 11.8c and d. However, activation during resting periods clearly differs, depending on whether the preceding movie is emotional or not. After neutral movies, resting state activity recovers immediately to a stable level, whereas a more gradual restoration after both joyful and fearful movies is exhibited. After emotional movies, the ventral medial prefrontal cortices (vMPFC) and precuneus exhibit a marked initial reduction during rest. Then, they recover gradually during the remaining period of rest, as illustrated in Figure 11.8e and f. These two regions show no reduction following neutral movies. All regions present a symmetric pattern in both hemispheres, but only one side is illustrated for the left IPL and right insula while other activation clusters spanned across the midline [121].

This investigation shows that the resting activity in the ACC and the insula as well as their coupling are strongly enhanced by preceding emotions, while coupling between the ventral-medial prefrontal cortex and the amygdala was selectively reduced. These effects were more pronounced for higher frequency bands after watching fearful rather than joyful movies. Initial suppression of resting activity in the ACC and insula after emotional stimuli was followed by a gradual restoration over time. Emotions did not affect IPL average activity but increased its connectivity with other regions.

11.10 Applications

Emotion assessment is a rapidly growing research field, especially in the human–computer interface community where assessing the emotional state of a user can greatly improve interaction quality by bringing it closer to human to human communication.

Changes in emotions can influence brain activities during the awake to sleep transition. They can also affect the states of mental fatigue and many other task- or target-related brain activities.

Another important impact is on memory formation. According to [94] two conclusions have been revealed. First, there is dissociation between arousal and valence. The results of emotion effect illustrate that processing of emotional content of pictures is sensitive to both arousal and valence and that their effects can be dissociated over the scalp: ERPs from parietal areas were sensitive to arousal, whereas ERPs from frontocentral areas were sensitive to both arousal and valence. Second, this is the first study in which the subsequent memory effect of emotional pictorial stimuli has been investigated. The *subsequent memory effect* results show an earlier occurrence of this effect for pleasant and unpleasant pictures than for neutral pictures and suggest that emotional stimuli have privileged access to processing resources, possibly resulting in better encoding. These results are compatible with previous evidence regarding the role of different brain areas in emotional information processing and with evidence about the beneficial effect of emotion on memory formation. This study emphasises the fact that there is a link between two seemingly unrelated ERP phenomena: the emotion effect and the subsequent memory effect.

There are plenty of external effects and applications, mainly in the areas of health and economy.

The basal ganglia role, in facial emotions recognition of Parkinson's disease (PD) has been investigated [125]. This work was to determine the extent to which visual processing of emotions and objects differs in PD and assess the impact of cognitive load on the processing of these types of information.

People with verbal communication difficulty (such as autism) sometimes send nonverbal messages that do not match with what is happening inside them. For example, a child might appear calm and receptive to learning – but have a heart rate over 120 bpm and be about to collapse. This mismatch can lead to misunderstandings such as "he became aggressive for no reason." In order to address this fundamental communication problem, advance technologies have to be developed to enable analysis of emotion-related physiological signals.

New tools and techniques in physiological arousal measurement in children with sensory challenges such as ASD and ADHD, can help them understand and control what makes them overexcited.

11.11   Conclusions

It is important to quantify and mark the level of impression and self-regulation of a subject against various emotions. Different recording modalities, such as facial impressions, body movement, eye movement, body temperature, heart rate and breathing may be used. EEG, however, allows localization, connectivity and synchronization of the brain for different emotions. It is then expected to extract and process highly valuable emotion-related information from EEGs. The role of a particular brain region such as the amygdala in emotion, evaluation of brain asymmetricity for positive and negative emotions and connectivity of the brain regions due to emotion types and levels are probably the most important topics in emotion-based brain signal processing. It is important to note that the sources within the limbic brain structure are considered as deep sources. Another problematic issue is the overlap between emotion and cognition. The current view of brain organization verifies a considerable degree of functional specialization and that many regions can be conceptualized as either 'affective' or 'cognitive'; the amygdala is involved with emotion and the lateral prefrontal cortex is the area of neuro-generators for cognition. This prevalent view is the source of many problems in emotion signal processing. Complex cognitive–emotional behaviours have their basis in dynamic coalitions of networks of brain areas, none of which should be conceptualized as specifically affective or cognitive. Central to cognitive–emotional interactions are brain areas with a high degree of connectivity, called hubs; these are critical for regulating the flow and integration of information between regions [45].

Direct detection and classification of emotion signal sources are generally complex and advanced methods in signal processing are needed. The high impact and enormous applications in this direction are well acknowledged.

References

[1] James, W. (1884) What is emotion. *Mind*, **9**, 188–205.
[2] Ten Donkelaar, H.J. (2011) *Clinical Neuroanatomy, Brain Circuitry and Its Disorders*, Springer-Verlag.

[3] Killstrom, E.J.F., Bower, G.H., Forgas, J.P. and Niedenthal, P.M. (2000) *Cognition and Emotion*, University Press.

[4] Harmon-Jones, E., Vaughn-Scott, K. and Mohr, S. (2004) The effect of manipulated sympathy and anger on left and right frontal cortical activity. *Emotion*, **4**, 95–101.

[5] Scmidt, L.A. (1999) Frontal brain electrical activity in shyness and sociability. *Psychol. Sci.*, **10**, 316–320.

[6] Garavan, H., Ross, T.J. and Stein, E.A. (1999) Right hemispheric dominance of inhibitory control: An event-related functional MRI study. *Proc. Natl. Acad. Sci. U S A*, **96**, 8301–8306.

[7] Drake, R.A. and Myers, L.R. (2006) Visual attention, emotion, and action tendency: Feeling active or passive. *Cognition.Emotion*, **20**, 608–622.

[8] Wacker, J., Chavanon, M.-L., Leue, A. and Stemmler, G. (2008) Is running away right? The behavioral activation–behavioral inhibition model of anterior asymmetry. *Emotion*, **8**, 232–240.

[9] Craig, A.D. (2003) A new view of pain as a homeostatic emotion. *Trends Neurosci.*, **26**(6), 303–307.

[10] Lazarus, R. (1991) *Emotion and Adaptation*, Oxford University Press, New York.

[11] Weiss, H.M. and Cropanzano, R. (1996) Affective events theory: A theoretical discussion of the structure, causes and consequences of affective experiences at work, in *Research in Organizational Behavior: An Annual Series of Analytical Essays and Critical Reviews* (eds B.M. Staw and L.L. Cummings), JAI Press, Greenwich, CT, pp. 1–74.

[12] Weiss, H.M. and Beal, D.J. (2005) Reflections on affective events theory, in *The Effect of Affect in Organizational Settings*, (eds N.M. Ashkanasy, W.J. Zerbe and C.E.J. Härtel) (Research on Emotion in Organizations, Volume 1), Emerald Group Publishing Ltd, pp. 1–21.

[13] Savran, A., Ciftci, K., Chanel, G. *et al.* (2006) Emotion detection in the loop from brain signals and facial images. eNTERFACE'06, Croatia.

[14] Haruki, Y., Homma, I., Umezawa, A. and Masaoka, Y. (eds.) (2001) *Respiration and Emotion*, Springer.

[15] Homma, I. and Masaoka, Y. (2008) Breathing rhythms and emotions. *Exp. Physiol.*, **93**(9), 1011–1021.

[16] Reiman, E.M., Fusselman, M.J., Fox, P.T. and Raichle, M.E. (1989) Neuroanatomical correlates of anticipatory anxiety. *Science*, **243**, 1071–1074.

[17] Rolls, E.T. (2001) The rules of formation of the olfactory representations found in the orbitofrontal cortex olfactory areas in primates. *Chem. Senses*, **26**, 595–604.

[18] Masaoka, Y. and Homma, I. (2000) The source generator of respiratory-related anxiety potential in the human brain. *Neurosci. Lett.*, **283**, 21–24.

[19] Masaoka, Y. and Homma, I. (2005) Amygdala and emotional breathing. *J. Adv.Exp. Med. Biol.*, **551**, 9–14.

[20] Larsen, J.T., Norris, C.J. and Cacioppo, J.T. (2003) Effects of positive and negative affect on electromyographic activity over zygomaticus major and corrugator supercilii. *Psychophysiology*, **40**(5), 776–785.

[21] Sato, W., Fujimura, T. and Suzuki, N. (2008) Enhanced facial EMG activity in response to dynamic facial expressions. *Int. J. Psychophysiol.*, **70**(1), 70–74.

[22] Dimberg, U. (1990) Facial electromyography and emotional reactions. *Psychophysiology*, **27**(5), 481–494.

[23] Oberman, L.M., Winkielman, P. and Ramachandran, V.S. (2009) Slow echo: facial EMG evidence for the delay of spontaneous, but not voluntary, emotional mimicry in children with autism spectrum disorders. *Dev. Sci.*, **12**(4), 510–520.

[24] Lang, P.J., Bradly, M.M. and Cuthbert, B.N. (2005) International affective picture system (IAPS): Digitized photographs, instruction manual and affective ratings. Technical Report A-6, University of Florida, Ginesville, FL.

[25] Oudeyer, P.-Y. (2003) The production and recognition of emotions in speech: features and algorithms. *Int. J. Human-Computer Studies*, **59**, 157–183.

[26] Neiberg, D., Elenius, K., Karlsson, I. and Laskowski, K. (2006) Emotion Recognition in Spontaneous Speech. Lund University, Centre for Languages & Literature, Dept. of Linguistics & Phonetics Working Papers 52, pp. 101–104.

[27] LeDoux, J.E. (2011) Music and the brain, literally. *Front. Hum. Neurosci.*, **5**, 54, doi: 10.3389/fnhum. 2011.00049.

[28] Cornelius, R.R. (2000) Theoretical approaches to emotion. Proceedings of the ISCA Workshop on Speech and Emotion, Belfast.

[29] Lewis, M., Haviland-Jones, J.M. and Feldman, L. (eds) (2008) *Handbook of Emotion*, The Guildford Press.

[30] Chanel, G., Kronegg, J., Grandjean, D. and Pun, T. (2005) Emotion assessment: arousal evaluation using EEG's and peripheral physiological signals. Technical Report, Computer Vision Group, University of Geneva, Switzerland.

[31] Chapman, L.J., Chapman, J.P., Raulin, M.L. and Eden, W.S. (1978) Schisotypy and thought disorder as a high-risk approach to Bchizophrenia, in *Cognitive Defects in the Development of Mental Illness* (ed. G. Serban), Brunner-Mazel, New York.

[32] Prince, J.D. and Berenbaum, H. (1993) Alexithymia and hedonic capacity. *J. Res. Pers.*, **27**, 15–22.

[33] Consoli, S.M., Lemogne, C., Roch, B. *et al.* (2010) Differences in emotion processing in patients with essential and secondary hypertension. *Am. J. Hypertens.*, **23**(5), 515–521.

[34] Siegler, R. (2006) *How Children Develop, Exploring Child Develop Student Media Tool Kit & Scientific American Reader to Accompany How Children Develop*, Worth Publishers, New York.

[35] Davidson, R.J., Putnam, K.M. and Larson, C.L. (2000) Dysfunction in the neural circuitry of emotion regulation - A possible prelude to violence. *Science*, **289**(5479), 591–594.

[36] DeArrnond, S.J., Fusco, M.M. and Dewey, M.M. (1989) *Structure of the Human Brain: A Photographic Atlas*, 3rd edn, Oxford Univ. Press, New York.

[37] Rolls, E.T. (1999) *The Brain and Emotion*, Oxford Univ. Press, New York.

[38] Holland, P.C. and Gallagher, M. (1999) Amygdala circuitry in attentional and representational processes. *Trends Cogn. Sci.*, **3**, 65–73.

[39] Morris, J.S., Frith, C.D., Perrett, D.I. *et al.* (1996) A differential neural response in the human amygdala to fearful and happy facial expressions. *Nature*, **383**, 812–815.

[40] Whalen, P.J., Rauch, S.L., Etcoff, N.L. *et al.* (1998) Masked presentations of emotional facial expressions modulate amygdala activity without explicit knowledge. *J. Neurosci.*, **18**(1), 411–418.

[41] Buchel, C., Morris, J.S., Dolan, R.J. and Friston, K.J. (1998) Brain systems mediating aversive conditioning: an event-related fMRI study. *Neuron*, **20**, 947–957.

[42] LaBar, K.S., Gatenby, J.C., Gore, J.C. *et al.* (1998) Human amygdala activation during conditional fear acquisition and extinction: A mixed-trial fMRI study. *Neuron*, **20**, 937–945.

[43] Irwin, W., Davidson, R.J. and Lowe, M.J. (1996) Human amagdyla activation detected with echo-planar functional magnetic resonance imaging. *Neuroreport*, **7**, 1765–1769.

[44] LeDoux, J. (1999) *The Emotional Brain*, Phoenix (an Imprint of The Orion Publishing Group), London.

[45] Pessoa, L. (2008) On the relationship between emotion and cognition. *Nat. Rev., Neurosci.* **9**, 148–158.

[46] Adolphs, R., Tranel, D., Damasio, H. and Damasio, A. (1994) Impaired recognition of emotion in facial expression following bilateral damage to the human amygdala. *Nature*, **372**, 662–672.

[47] Blair, R.J., Morris, J.S., Frith, C.D. *et al.* (1999) "Dissociable neural responses to facial expressions of sadness and anger. *Brain*, **122**, 883–893.

[48] Dougherty, D.D., Rauch, S.L. and Deckersbach, T. (2004) Ventromedial prefrontal cortex and amygdala dysfunction during an anger induction positron emission tomography study in patients with major depressive disorder with anger attacks. *Arch. Gen. Psychiat.*, **61**, 795–804.

[49] Kimbrell, T.A., George, M.S. and Parekh, P.I. (1999) "Regional brain activity during transient self-induced anxiety and anger in healthy adults. *Biol. Psychiat.*, **46**(4), 454–465.

[50] Walla, P., Brenner, G. and Koller, M. (2011) Objective measures of emotion related to brand attitude: a new way to quantify emotion-related aspects relevant to marketing. *PLoS One*, **6**(11), e26782.

[51] Globisch, J., Hamm, A.O., Esteves, F. and Ohman, A. (1999) Fear appears fast: Temporal course of startle reflex potentiation in animal fearful subjects. *Psychophysiology*, **36**, 66–75.

[52] Larson, C.L., Ruffalo, D., Nietert, J.Y. and Davidson, R.J. (2000) Temporal stability of the emotion-modulated startle response. *Psychophysiology*, **37**, 92–101.

[53] Jackson, D.C., Malmstadt, J.R., Larson, C.L. and Davidson, R.J. (2000) Suppression and enhancement of emotional responses to unpleasant pictures. *Psychophysiology*, **37**, 515–522.

[54] Davidson, R.I. (1998) Affective style and affective disorders: perspectives from affective neuroscience. *Cognition and Emotion*, **12**, 307–330.

[55] Jackson, D.C., Burghy, C.A., Hanna, A.J. *et al.* (2000) Resting frontal and anterior temporal EEG asymmetry predicts ability to regulate negative emotion. *Psychophysiology*, **37**, S50.

[56] Schaefer, S.M., Jackson, D.C., Davidson, R.J. *et al.* (2002) Modulation of amygdala activity by conscious maintenance of negative emotion. *J. Cogn. Neurosci.*, **14**, 913–921.

[57] Heller, A.S., Johnstone, T., Shackman, A.J. *et al.* (2009) "Reduced capacity to sustain positive emotion in major depression reflects diminished maintenance of fronto-striatal brain activation. *Proc. Natl. Acad. Sci. U S A*, **106**(52), 22445–22450.

[58] Jackson, D.C., Mueller, C.J., Dolski, I. *et al.* (2003) Now you feel it, now you don't: Frontal EEG asymmetry and individual differences in emotion regulation. *Psychol. Sci.* **14**, 612–617.

[59] Urry, H.L., Nitschke, J.B., Dolski, I. *et al.* (2004) Making a life worth living: Neural correlates of well-being. *Psychol. Sci.* **15**, 367–372.

[60] Urry, H.L., van Reekum, C.M., Johnstone, T. and Davidson, R.J. (2009) Individual differences in some (but not all) medial prefrontal regions reflect cognitive demand while regulating unpleasant emotion. *NeuroImage* **47**, 852–863.

[61] van der Helm, E., Yao, J., Dutt, S. *et al.* (2011) REM sleep depotentiates amygdala activity to previous emotional experiences. *Els. J. Curr. Biol.* **21**, 1–4.

[62] Shallice, T. and Cooper, R.P. (2011) *The Organization of Mind*, Oxford University Press.

[63] Saxe, R., Xiao, D.-K., Kovacs, G. *et al.* (2004) A region of right posterior superior temporal sulcus responds to observed intentional actions. *Neuropsychologia* **42**, 1435–1446.

[64] Sirigu, A., Daprati, E., Ciancia, S. *et al.* (2004) Altered awareness of voluntary action after damage to the parietal cortex. *Nat. Neurosci.* **7**, 80–84.

[65] Nichols, T.E. and Holmes, A.P. (2002) Nonparametric permutation tests for functional neuroimaging: a primer with examples. *Human Brain Mapp.* **15**, 1–25.

[66] Hornak, J., Bramham, J., Rolls, E.T. *et al.* (2003) Changes in emotion after circumscribed surgical lesions of the orbitofrontal and cingulate cortices. *Brain* **26**, 1691–1712.

[67] Takahashi, H., Yahata, N., Koeda, M. *et al.* (2004) Brain activation associated with evaluative processes of guilt and embarrassment: an fMRI study. *NeuroImage*, **23**(3), 967–974.

[68] Fu, C.H., Williams, S.C., Cleare, A.J. *et al.* (2004) Attenuation of the neural response to sad faces in major depression by antidepressant treatment: a prospective, event-related functional magnetic resonance imaging study. *Arch. Gen. Psychiat.*, **61**(9), 877–889.

[69] Lewis, M. and Steiben, J. (2004) Emotion regulation in the brain: Conceptual issues and directions for developmental research. *Child Dev.* **75**, 371–376.

[70] Beer, J.S., Heerey, E.H., Keltner, D. *et al.* (2003) The regulatory function of self-conscious emotion: Insights from patients with orbitofrontal damage. *J. Pers. Soc. Psychol.* **85**, 594–604.

[71] Moll, J., de Oliveira-Souza, R., Moll, F. *et al.* (2005) The moral affiliations of disgust: a functional MRI study. *Cogn. Behav. Neurol.*, **18**, 68–78.

[72] Haidt, J. (2007) The new synthesis in moral psychology. *Science* **316**, 998–1002.

[73] Buchanan, T.W., Tranel, D. and Adolphs, R. (2004) Anteromedial temporal lobe damage blocks startle modulation by fear and disgust. *Behav. Neurosci.* **118**, 429–437.

[74] Fitzgerald, D.A., Posse, S., Moore, G.J. *et al.* (2004) Neural correlates of internally-generated disgust via autobiographical recall: A functional magnetic resonance imaging investigation. *Neurosci. Lett.* **370**, 91–96.

[75] Haidt, J. (2003) The moral emotions, in *Handbook of Affective Sciences* (eds R.J. Davidson, K.R. Scherer and H.H. Goldsmith), Oxford University Press, Oxford, England, pp. 852–870.

[76] Adams, R.E., Boscarino, J.A. and Figley, C.R. (2006) Compassion fatigue and psychological distress among social workers: A validation study. *Am. J. Orthopsychiat.*, **76**(1), 103–108.

[77] Moll, J. (2003) Morals and the human brain: a working model. *Neuroreport*, **14**, 299–305.

[78] Haidt, J. (2003) Elevation and the positive psychology of morality, in *Flourishing: Positive Psychology and the Life Well-Lived* (eds C.L.M. Keyes and J. Haidt), American Psychological Association, Washington DC, pp. 275–289.

[79] Rilling, J.K., Gutman, D.A., Zeh, T.R. *et al.* (2002) A neural basis for social cooperation. *Neuron*, **35**, 395–405.

[80] Singer, B.H., Kiebel, S.J., Winston, J.S. *et al.* (2004) Brain responses to the acquired moral status of faces. *Neuron*, **41**(4), 653–662.

[81] Musha, T., Terasaki, Y., Haque, H.A. and Ivanitsky, G.A. (1997) Feature extraction from EEGs associated with emotions. *Artif. Life Robotics*, **1**, 15–19.

[82] Demaree, H.A., Everhart, D.E., Youngstrom, E.A. and Harrison, D.W. (2008) Brain lateralization of emotional processing: historical roots and a future incorporating "Dominance". *Behav. Cogn. Neurosci. Rev.*, **4**(1), 3–20.

[83] Ehrlichman, H. (1987) Hemispheric asymmetry and positive-negative affect, in *Duality and Unity of the Brain*, (ed. D. Ottoson), Macmillan Press, Houndmills, Hampshire, UK, pp. 194–206.

[84] Papousek, L. (2006) Individual differences in functional asymmetries of the cortical hemisphers. *Cogn. Brain Behav.* **2**, 269–298.

[85] Borod, J.C. (1993) Cerebral mechanisms underlying facial, prosodic, and lexical emotional expression: a review of neuropsychological studies and methodological issues. *Neuropsychology* **7**, 445–463.

[86] Herrmann, M.J., Huter, T., Plichta, M.M. *et al.* (2008) Enhancement of activity of the primary visual cortex during processing of emotional stimuli as measured with event-related functional near-infrared spectroscopy and event-related potentials. *Hum Brain Mapp.*, **29**(1), 28–35.

[87] Gray, J.M., Young, A.E. and Barker, W.A. (1997) *Impaired recognition of disgust in Huntington's disease carriers.* Brain, **120**, 2029–2038.

[88] Dolcos, F. and Cabeza, R. (2002) Event-related potentials of emotional memory: Encoding pleasant, unpleasant, and neutral pictures. *Cogn. Affect. Behav. Neurosci.*, **2**, 252–263.

[89] Schupp, H.T., Flaisch, T., Stockburger, J. and Junghofer, M. (2006) Emotion and attention: Event-related brain potential studies, Chapter 2, in *Progress in Brain Research*, vol. **156** (eds S. Anders, G. Ende, M. Junghofer *et al.*), Elsevier.

[90] Schupp, H.T., Cuthbert, B.N., Bradley, M.M. *et al.* (2004) Brain processes in emotional perception: motivated attention. *Cogn. Emotion*, **18**, 593–611.

[91] Schupp, H.T., Cuthbert, B.N., Bradley, M.M. *et al.* (2004) The selective processing of briefly presented affective pictures: an ERP analysis. *Psychophysiology*, **41**, 441–449.

[92] Amrhein, C., Mühlberger, A., Pauli, P. and Wiedemann, G. (2004) Modulation of event-related brain potentials during affective picture processing: a complement to startle reflex and skin conductance response? *Int. J. Psychophysiol.*, **54**, 231–240.

[93] Hagemann, D., Naumann, F., Thayer, J.F. and Bartussek, D. (2002) Does resting EEG asymmetry reflect a biological trait? An application of latent state-trait theory. *J. Pers. Soc. Psychol.*, **82**, 619–641.

[94] Dolcos, F. and Cabeza, R. (2002) Event-related potentials of emotional memory: Encoding pleasant, unpleasant, and neutral pictures. *Cogn. Affect. Behav. Neuroscience*, **2**(3), 252–263.

[95] Carretié, L., Iglesias, J., García, T. and Ballesteros, M. (1997) N300, P300 and the emotional processing of visual stimuli. *Electroencephalogr. Clin. Neurophysiol.*, **103**(2), 298–303.

[96] Cuthbert, B.N., Schupp, H.T., Bradley, M.M. *et al.* (2000) Brain potentials in affective picture processing: covariation with autonomic arousal and affective report. *Biol. Psychol.*, **52**, 95–111.

[97] Johnston, V.S., Miller, D.R. and Burleson, M.H. (1986) Multiple P3s to emotional stimuli and their theoretical significance. *Psychophysiology*, **23**, 684–694.

[98] Delplanque, S., Silvert, L., Hot, P. *et al.* (2006) Arousal and valence effects on event-related P3a and P3b during emotional categorization. *Int. J. Psychophysiol.*, **60**, 315–322.

[99] Bechara, A., Damasio, H., Damasio, A.R. and Lee, G.P. (1999) Different contributions of the human amygdala and ventromedial prefrontal cortex to decision-making. *J. Neurosci.*, **19**, 5473–5481.

[100] Wheeler, R.E., Davidson, R.J. and Tomarken, A.J. (1993) Frontal brain asymmetry and emotional reactivity: A biological substrate of affective style. *Psychophysiology*, **30**, 82–89.

[101] Davidson, R.J. and Irwin, W. (1999) The functional neuroanatomy of emotion and affective style. *Trends Cogn. Sci.*, **3**, 11–21.

[102] Dolcos, F., Rice, H.J. and Cabeza, R. (2002) Hemispheric asymmetries and aging: Right hemisphere decline or asymmetry reduction. *Neurosci. Biobehav. Rev.*, **26**(7), 819–825.

[103] LeDoux, J.E. (1998) Cognition and emotion: Listen to the brain. in *Emotion and Cognitive Neuroscience* (ed. R. Lane), Oxford U. Press, New York.

[104] Rolls, E.T., Kringelbach, M.L. and de Araujo, I.E. (2003) Different representations of pleasant and unpleasant odours in the human brain. *Eur. J. Neurosci.*, **18**, 695–703.

[105] Royet, J.P., Zald, D., Versace, R. *et al.* (2000) Emotional responses to pleasant and unpleasant olfactory, visual, and auditory stimuli: a positron emission tomography study. *J. Neuroscience*, **20**, 7752–7759.

[106] Masaoka, Y. and Homma, I. (2010) Respiratory response toward olfactory stimuli might be an index for odor-induced emotion and recognition. *Adv. Exp. Med. Biol.*, **669**, 347–352.

[107] Diamond, D. (2003) The Love Machine; Building computers that care. Wired. http://www.wired.com/wired/archive/11.12/love.html, accessed May 13, 2008.

[108] Anders, S., Ende, G., Junghofer, M. *et al.* (2006) *Understanding Emotions*, Elsevier.

[109] Aftanas, L.I., Reva, N.V., Varlamov, A.A. *et al.* (2004) Analysis of evoked EEG synchronization and desynchronization in conditions of emotional activation in humans: temporal and topographic characteristics. *Neurosci. Behav. Physiol.*, **34**, 859–867.

[110] Adolphs, R. (2002) Neural systems for recognizing emotion. *Curr. Opin. Neurobiol.*, **12**, 169–177.

[111] Baliki, M.N., Chialvo, D.R., Geha, P.Y. *et al.* (2006) "Chronic pain and the emotional brain: specific brain activity associated with spontaneous fluctuations of intensity of chronic back pain. *J. Neurosci.*, **26**, 12165–12173.

[112] Damasio, A.R. (1999) *The Feeling of What Happens; Body and Emotion in the Making of Consciousness*, Harcourt Brace, New York.

[113] Phelps, E.A. and LeDoux, J.E. (2005) Contributions of the amygdala to emotion processing: from animal models to human behaviour. *Neuron* **48**, 175–187.

[114] Sigurdsson, T., Doyere, V., Cain, C.K. and LeDoux, J.E. (2007) Long-term potentiation in the amygdala: a cellular mechanism of fear learning and memory. *Neuropharmacology*, **52**, 215–227.

[115] Salvador, R., Martínez, A., Pomarol-Clotet, E. *et al.* (2008) A simple view of the brain through a frequency-specific functional connectivity measure. *NeuroImage*, **39**, 279–289.

[116] Irving, I., Gottesman, Ph.D., Hon, F.R.C. *et al.* (2003) The Endophenotype Concept in Psychiatry: Etymology and Strategic Intentions. *Am. J. Psychiat.*, **160**, 636–645.

[117] Morris, J., Friston, K.J., Buechel, C. *et al.* (1998) A neuromodulatory role for the human amygdala in processing emotional facial expressions. *Brain*, **121**, 47–57.

[118] Stephan, B.C., Breen, N. and Caine, D. (2006) "The recognition of emotional expression in prosopagnosia: decoding whole and part faces. *J. Int. Neuropsychol. Soc.*, **12**(6), 884–895.

[119] Canli, T., Desmond, J.E., Zhao, Z. and Gabrieli, J.D. (2002) Sex differences in the neural basis of emotional memories. *Proc. Natl. Acad. Sci.*, **99**, 10789–10794.

[120] Junghofer, M., Peyk, P., Flaisch, T. and Schupp, H.T. (2006) Understanding emotions, Chapter 7, in *Neuroimaging Methods in Affective Neuroscience: Selected Methodological Issues*, Elsevier.

[121] Eryilmaz, H., Van De Ville, D., Schwartz, S. and Vuilleumier, P. (2011) Impact of transient emotions on functional connectivity during subsequent resting state: A wavelet correlation approach. *NeuroImage*, **54**, 2481–2491.

[122] Battle, G. (1987) A block spin construction of ondelettes; Part I. Lemarié functions. *Commun. Math. Phys.*, **110**, 601–615.

[123] Mallat, S. (1989) A theory for multiresolution signal decomposition: the wavelet Decomposition. *IEEE Trans. Pattern Anal. Mach. Intell.*, **11**, 674–693.

[124] Waites, A.B., Stanislavsky, A., Abbott, D.F. and Jackson, G.D. (2005) Effect of prior cognitive state on resting state networks measured with functional connectivity. *Hum. Brain Mapp.*, **24**, 59–68.

[125] Cohen, H., Gagné, M.-H., Hess, U. and Pourcher, E. (2010) "Emotion and object processing in Parkinson's disease. *J. Brain Cogn.*, **72**, 457–463.

12

Sleep and Sleep Apnoea

12.1 Introduction

Sleep has been described by early scientists as a passive condition where the brain is isolated from the rest of the body. Alcmaeon claimed that sleep is caused by the blood receding from the blood vessels in the skin to the interior parts of the body. Philosopher and scientist Aristotle suggested that while food is being digested, vapors rise from the stomach and penetrate into the head. As the brain cools, the vapors condense, flow downward and then cool the heart which causes sleep. Some others still claim that toxins that poison the brain cause sleep [1]. With the discovery of brain waves and later the discovery of the EEG system, the way sleep was studied changed. EEG enables a researcher to record the detailed electrical activity of the brain during sleep. Sleep is the state of natural rest observed in humans and animals, and even invertebrates such as fruit fly Drosophila. Lack of sleep seriously influences our brain's ability to function. With continued lack of sufficient sleep, the part of the brain that controls language, memory, planning and sense of time is severely affected and the ability of judgement deteriorates. Sleep is an interesting and not perfectly understood physiological phenomenon. The sleep state has become an important evidence for diagnosing mental disease and psychological abnormality. Sleep is characterized by a reduction in voluntary body movement, decreased reaction to external stimuli, an increased rate of anabolism (the synthesis of cell structures), and a decreased rate of catabolism (the breakdown of cell structures). Sleep is necessary for the life of most creatures. The capability for arousal from sleep is a protective mechanism and also necessary for health and survival. In terms of physiological changes in the body and particularly changes in the state of the brain, sleep is different from unconsciousness [2]. However, in manifestation, sleep is defined as a state of unconsciousness from which a person can be aroused. In this state, the brain is relatively more responsive to internal stimuli than external stimuli. Sleep should be distinguished from coma. Coma is an unconscious state from which a person cannot be aroused.

Historically, sleep was thought to be a passive state. However, sleep is now known to be a dynamic process, and our brains are active during sleep. Sleep affects our physical and mental health and the immune system.

States of brain activity during sleep and wakefulness result from different activating and inhibiting forces that are generated within the brain. Neurotransmitters (chemicals involved in

Adaptive Processing of Brain Signals, First Edition. Saeid Sanei.
© 2013 John Wiley & Sons, Ltd. Published 2013 by John Wiley & Sons, Ltd.

nerve signalling) control whether one is asleep or awake by acting on nerve cells (neurons) in different parts of the brain. Neurons located in the brainstem actively cause sleep by inhibiting other parts of the brain that keep a person awake.

In a human being, it has been demonstrated that the metabolic activity of the brain decreases significantly after 24 hours of sustained wakefulness. Sleep deprivation results in a decrease in body temperature, a decrease in immune system function as measured by white blood cell count (the soldiers of the body), and a decrease in the release of growth hormone. Sleep deprivation can also cause increased heart rate variability [3].

Sleep is necessary for the brain to remain healthy. Sleep deprivation makes a person drowsy and unable to concentrate. It also leads to impairment of memory and physical performance and reduced ability to carry out mathematical calculations and other mental tasks. If sleep deprivation continues, hallucinations and mood swings may develop. Sleep deprivation not only has a major impact on cognitive functioning but also on emotional and physical health. Disorders such as sleep apnoea which result in excessive daytime sleepiness have been linked to stress and high blood pressure. Research has also suggested that sleep loss may increase the risk of obesity because chemicals and hormones that play a key role in controlling appetite and weight gain are released during sleep.

Release of growth hormone in children and young adults takes place during deep sleep. Most cells of the body show increased production and reduced breakdown of proteins during deep sleep. Sleep helps humans maintain optimal emotional and social functioning while we are awake by giving rest during sleep to the parts of the brain that control emotions and social interactions.

12.2   Stages of Sleep

Sleep is a dynamic process. Loomis provided the earliest detailed description of the various stages of sleep in the mid-1930s, and in the early 1950s Aserinsky and Kleitman identified rapid eye movement (REM) sleep [2]. There are two distinct states that alternate in cycles and reflect differing levels of neuronal activity. Each state is characterized by a different type of EEG activity. Sleep consists of non-rapid eye movement (NREM) and REM sleep. NREM is further subdivided into four stages of I (drowsiness), II (light sleep), III (deep sleep) and IV (very deep sleep).

During the night the NREM and REM stages of sleep alternate. Stages I, II, III, and IV are followed by REM sleep. A complete sleep cycle, from the beginning of stage I to the end of REM sleep, usually takes about one and a half hours. However, generally, the ensuing sleep is relatively short and, in practice, often 10–30 minutes duration suffices.

12.2.1   NREM Sleep

The first stage, *Stage* I, is the stage of drowsiness and very light sleep, which is considered as a transition between wakefulness and sleep. During this stage, the muscles begin to relax. It occurs upon falling asleep and during brief arousal periods within sleep, and usually accounts for 5–10% of total sleep time. An individual can be easily awakened during this stage. Drowsiness shows marked age-determined changes. Hypnagogic rhythmical 4–6/s theta activity of

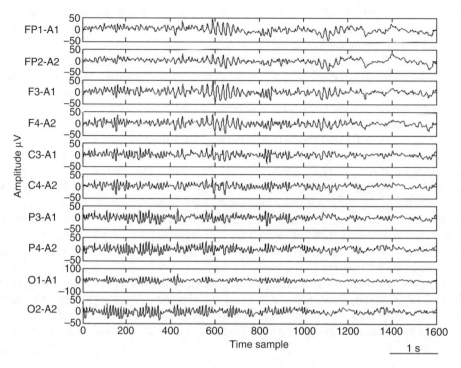

Figure 12.1   Examples of EEG signals recorded during drowsiness

late infancy and early childhood is a significant characteristic of such ages. Later in child-hood, and in several cases, in the declining years of life, the drowsiness onset involves larger amounts of slow activity mixed with the posterior alpha rhythm [4]. In adults, however, the onset of drowsiness is characterised by gradual or brisk alpha dropout [4]. The slow activity increases as the drowsiness becomes deeper. Other findings show that in light drowsiness the P300 response increases in latency and decreases in amplitude [5], and the inter- and intra-hemispheric EEG coherence alter [6]. Figure 12.1 shows a set of EEG signals recorded during the state of drowsiness. The seizure type activity within the signal is very clear.

Deep drowsiness involves the appearance of vertex waves. Before the appearance of the first spindle trains, vertex waves occur (transition from stage I to stage II). These sharp waves are also known as parietal humps [7]. The vertex wave is a compound potential; a small spike discharge of positive polarity followed by a large negative wave, which is a typical discharge wave. It may occur as an isolated event with larger amplitude than that of normal EEG. In aged individuals they may become small, inconspicuous and hardly visible. Another signal feature for deep drowsiness is the positive occipital sharp transients of sleep (POST).

Spindles (also called sigma activity), the trains of barbiturate-induced beta activity, occur independently in approximately 18–25 cycles/s predominantly in the frontal lobe of the brain. They may be identified as a "group of rhythmic waves characterized by progressively increas-ing, then gradually decreasing amplitude" [4]. However, the use of middle electrodes shows a very definite maximum of the spindles over the vertex during the early stages of sleep.

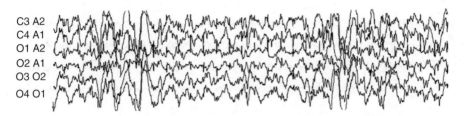

C3 A2
C4 A1
O1 A2
O2 A1
O3 O2
O4 O1

Figure 12.2 26 s of brain waves recorded during the Stage III of sleep

Stage II of sleep occurs throughout the sleep period and represents 40–50% of the total sleep time. During stage II, brain waves slow down with occasional bursts of rapid waves. Eye movement stops during this stage. Slow frequencies ranging from 0.7 to 4 cycles/s are usually predominant; their voltage is high with a very prominent occipital peak in small children and gradually falls when they become older.

K-complexes appear in stage II and constitute a significant response to arousing stimuli. As to the topographical distribution over the brain, the K-complex shows a maximum over the vertex and has presence around the frontal midline [4]. As to the wave morphology, the K-complex consists of an initial sharp component, followed by a slow component that fuses with a superimposed fast component.

In *Stage III*, delta waves begin to appear. They are interspersed with smaller, faster waves. Sleep spindles are still present at approximately 12–14 cycles/s but gradually disappear as the sleep becomes deeper. Figure 12.2 illustrates the brain waves during this stage of sleep.

In *Stage IV*, delta waves are the primary waves recorded from the brain. Delta or slow wave sleep (SWS) usually is not seen during routine EEG [8]. However, it is seen during prolonged (>24 hours) EEG monitoring.

Stages III and IV are often distinguished from each other only by the percentage of delta activity. Together, they represent up to 20% of total sleep time. During Stages III and IV all eye and muscle movement ceases. It is difficult to wake up someone during these two stages. If someone is awakened during deep sleep; he does not adjust immediately and often feels groggy and disoriented for several minutes after waking up. Generally, analysis of EEG morphology during Stage IV has been of less interest since the brain functionality cannot be examined easily.

12.2.2 REM Sleep

REM sleep including 20–25% of the total sleep follows NREM sleep and occurs 4–5 times during a normal 8- to 9-hour sleep period. The first REM period of the night may be less than 10 minutes in duration, while the last period may exceed 60 minutes.

In an extremely sleepy individual, the duration of each bout of REM sleep is very short or it may even be absent. REM sleep is usually associated with dreaming. During REM sleep, the eyeballs move rapidly, the heart rate and breathing become rapid and irregular, the blood pressure rises, and there is loss of muscle tone (paralysis), that is, the muscles of the body are virtually paralysed. The brain is highly active during REM sleep, and the overall brain metabolism may be increased by as much as 20%. The EEG activity recorded in the brain during REM sleep is similar to that recorded during wakefulness.

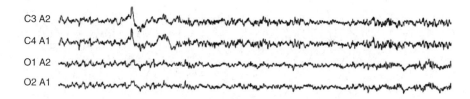

Figure 12.3   Twenty six seconds of brain waves recorded during REM state

In a patient with REM sleep behaviour disorder (RBD), the paralysis is incomplete or absent, allowing the person to act out his dreams which can be vivid, intense, and violent. These dream-acting behaviours include talking, yelling, punching, kicking, sitting, jumping from the bed, arm flailing and grabbing. Although the RBD may occur in association with different degenerative neurological conditions the main cause is still unknown.

Evaluation of REM sleep involves a long waiting period since the first phase of REM does not appear before 60–90 minutes after the start of sleep. The EEG in the REM stage shows low voltage activity with slower rate of alpha. Figure 12.3 shows the brain waves during REM sleep.

12.3   The Influence of Circadian Rhythms

Biological variations that occur in the course of 24 hours are called circadian rhythms. Circadian rhythms are controlled by the biological clock of the body. Many bodily functions follow the biological clock, but sleep and wakefulness comprise the most important circadian rhythm. Circadian sleep rhythm is one of the several body rhythms modulated by the hypothalamus.

Light directly affects the circadian sleep rhythm. Light is called *zeitgeber*, a German word meaning time-giver, because it sets the biological clock.

Body temperature cycles are also under the control of the hypothalamus. An increase in body temperature is seen during the course of the day and a decrease is observed during the night. The temperature peaks and troughs are thought to mirror the sleep rhythm. People who are alert late in the evening (i.e. evening types) have body temperature peaks late in the evening, while those who find themselves most alert early in the morning (i.e. morning types) have body temperature peaks early in the morning.

Melatonin (a chemical produced by the pineal gland in the brain and a hormone associated with sleep) has been implicated as a modulator of light entrainment. It is secreted maximally during the night. Prolactin, testosterone and growth hormone also demonstrate circadian rhythms, with maximal secretion during the night. Figure 12.4 shows a typical concentration of melatonin in a healthy adult man.

Sleep and wakefulness are influenced by different neurotransmitters in the brain. Some substances can change the balance of these neurotransmitters and affect our sleep and wakefulness. Caffeinated drinks (for example, coffee) and medicines (for example, diet pills) stimulate some parts of the brain and can cause difficulty in falling asleep. Many drugs prescribed for the treatment of depression suppress REM sleep.

Heavy smokers often sleep very lightly and have reduced duration of REM sleep. They tend to wake up after three or four hours of sleep due to nicotine withdrawal. Some people who

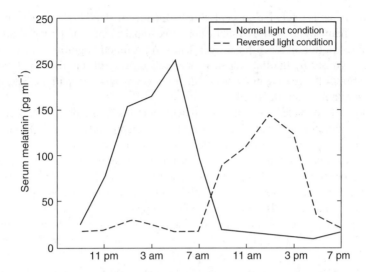

Figure 12.4 A typical concentration of melatonin in a healthy adult man (extracted from Brzezinski A. NEJM, vol. 336, pp. 186–195, 1997)

have insomnia may use alcohol. Even though alcohol may help people to fall into light sleep, it deprives them of REM sleep and the deeper and more restorative stages of sleep. Alcohol keeps them in the lighter stages of sleep from which they can be awakened easily. During REM sleep, we lose some of our ability to regulate our body temperature. Therefore, abnormally hot or cold temperatures can disrupt our REM sleep. If our REM sleep is disturbed, the normal sleep cycle progression is affected during the next sleeping time, and there is a possibility of slipping directly into REM sleep and going through long periods of REM sleep until the duration of REM sleep that is lost is caught up.

Generally, sleep disruption by any cause can be a reason for an increase in seizure frequency or severity. It can also have a negative effect on short-term memory, concentration and mood. Seizure, itself, during the night can disrupt sleep. Also, using any anticonvulsant drug may affect sleep in different ways. Approximately 19 different sleep disorders, including snoring, obstructive sleep apnoea hypopnea syndrome (OSAHS), insomnia, narcolepsy, bruxism, and restless leg syndrome have been reported by the International Classification of Sleep [9].

Both the frequency of seizure and the locality of seizure sources within the brain may change in different sleep stages and wakefulness.

12.4 Sleep Deprivation

Sleep deprivation is evaluated in terms of the tasks impaired and the average duration. In tasks requiring judgement, increasingly risky behaviours emerge as the total sleep duration is limited to five hours per night. The high cost of an action is seemingly ignored as the sleep-deprived person focuses on limited benefits. These findings can be explained by the fact that metabolism in the prefrontal and parietal associational areas of the brain decreases in individuals deprived of sleep for 24 hours. These brain areas are important for judgement, impulse control, attention and visual association.

Sleep deprivation is a relative concept. Small amounts of sleep loss (for example, one hour per night over many nights) produce subtle cognitive impairment, which may go unrecognized. More severe restriction of sleep for a week leads to profound cognitive deficits, which may also go unrecognized by the individual. If one feels drowsy during the day, falls asleep for very short periods of time (five minutes or so), or regularly falls asleep immediately after lying down, he is probably sleep-deprived.

Many studies have made it clear that sleep deprivation is dangerous. With decreased sleep, higher-order cognitive tasks are impaired early and disproportionately. On tasks used for testing coordination, sleep-deprived people perform as poorly as or worse than people who are intoxicated. Total sleep duration of seven hours per night over one week has resulted in decreased speed in tasks of both simple reaction time and more demanding computer-generated mathematical problem solving. Total sleep duration of five hours per night over one week shows both a decrease in speed and the beginning of accuracy failure.

Using sleep deprivation for detection and diagnosis of some brain abnormalities has been reported by some researchers [10–12]. It consists of sleep loss for 24–26 hours. This was used by Klingler [13] to detect the epileptic discharges that could otherwise be missed. Based on these studies it has also been concluded that sleep deprivation is a genuine activation method [14]. Its efficacy in provoking abnormal EEG discharges is not due to drowsiness. Using the information in Stage III of sleep, the focal and generalized seizure may be classified [15].

12.5 Psychological Effects

In the majority of sleep measurements and studies EEG has been used in combination with a variety of other physiological parameters. EEG studies have documented abnormalities in sleep patterns in psychiatric patients with suicidal behaviour, including longer sleep latency, increased REM time and increased phasic REM activity. Sabo *et al.* [16] compared sleep EEG characteristics of adult depressives with and without a history of suicide attempts and noted that those who attempted suicide had consistently more REM time and phasic activity in the second REM period but less delta wave counts in the fourth non-REM period. Another study [17] conducted at the same laboratory replicated the findings with psychotic patients. On the basis of two studies, the authors [17] suggest that the association between REM sleep and suicidality may cut across diagnostic boundaries and that sleep EEG changes may have a predictive value for future suicidal behaviour. REM sleep changes were later replicated by other studies in suicidal schizophrenia [18] and depression [19].

Three cross-sectional studies examined the relationship between sleep EEG and suicidality in depressed adolescents. Dahl *et al.* [20] compared sleep EEG between a depressed suicidal group, a depressed nonsuicidal group, and normal controls. Their results indicated that suicidal depressed patients had significantly prolonged sleep latency and increased REM phasic activity with a trend for reduced REM latency compared to both nonsuicidal depressed and control groups. Goetz *et al.* [21] and McCracken *et al.* [22] replicated the finding of greater REM density amongst depressive suicidal adolescents.

Study of normal ageing and transient cognitive disorders in the elderly has also shown that the most frequent abnormality in the EEG of elderly subjects is slowing of alpha frequency, whereas most healthy individuals maintain alpha activity within 9–11 Hz [23, 24].

12.6   Detection and Monitoring of Brain Abnormalities During Sleep by EEG Analysis

EEG provides important and unique information about the sleeping brain. Polysomnography (PSG) has been the well-established method of sleep analysis and the main diagnostic tool in sleep medicine, which interprets the sleep signal macrostructure based on the criteria explained by Rechtschaffen and Kales (R&K) [25]. Polysomnography or sleep study is a multiparametric test used in the study of sleep and as a diagnostic tool in sleep medicine. The test result is called a polysomnogram. PSG is a comprehensive recording of the biophysiological changes occurring during sleep. It is usually performed at night. The PSG monitors many body functions, including brain (EEG), eye movements (EOG), muscle activity or skeletal muscle activation (EMG) and heart rhythm (ECG) during sleep.

12.6.1   Analysis of Sleep Apnoea

Sleep apnoea syndrome (SAS), chronic snoring, and daytime excessive sleepiness cause problems in daily life and have short and long term physiological and psychological effects. After the identification of sleep apnoea in the 1970s, polysomnography has been used as a gold standard to combine and evaluate breathing, peripheral pulse oximetry, heartrate, and two channels of EEG for the assessment of these disorders. Later, the recorded sound of snoring has been the main focus of research. Generally, it has been shown that not all the signals recorded by the PSG are needed for the assessment of apnoea or hypopnea [26–28]. In [26] and [27] it has been shown that the snoring sound and the SpO2 value from the output of an oximeter can be utilized for the analysis of apnoea. There are also reports of using only one of these two modalities, that is, the sound signals [28] or evaluation of hypoxaemia using oximeters [29] is enough for the assessment of apnoea. However, the spindles and slow wave activities, arousals and associated activities can be detected from the EEG signals and monitored during the sleep. For analysis of apnoea a description of these activities often requires temporal segmentation of the signals into fixed segments of 20–30 s.

SAS with a high prevalence of approximately 2% in women and 4% in men between the ages of 30 to 60 years is the cause of many physiological and psychological problems which consequently change the life style of the patient [30, 17]. This syndrome is often treated by means of continuous positive airway pressure therapy or by surgery. SAS refers to sleep disorders that cause breathing pauses during sleep at night. SAS can be the result of different anatomical or physiological factors. Two common SASs, namely central sleep apnoea (CSA) and obstructive sleep apnoea (OSA) have been classified in recent research [31]. In the case of the CSA, sleep apnoea is originated by the central nervous system. If the duration of pauses in breathing is at least 10 s in the upper respiratory tract and the pulmonary system at the same time during sleep, CSA is observed. In the case of OSA, the reason for the breathing pauses is respiratory tract obstruction.

Using a standard PSG has been a popular approach for diagnosis and monitoring of this disease. The measurement is often overnight to record the sleep stage, respiratory efforts, oronasal airflow, electrocardiographic findings, and oxyhemoglobin saturation parameters in an attended laboratory setting [18]. However, PSG is not easily accessible, it is expensive and intrusive. Therefore, researchers try to extract the necessary diagnostic information from a smaller number of recording modalities as stated above. As well as snoring sound and the oxygen level, EEG is another modality used for the assessment of sleep apnoea.

In recent assessment of sleep and sleep apnoea EEG has been used for the detection of SASs. Coherence function (CF) and mutual information (MI) measures have been used as the EEG signal features in discriminating the CSA and OSA from controls. The sleep EEG series recorded from patients and healthy volunteers are classified by using a feed forward neural network (FFNN) with two hidden layers and utilizing the synchronic activities between C3 and C4 channels of the EEG recordings. Amongst the sleep stages, Stage II is considered in tests. The results show that the degree of central EEG synchronization during night sleep is closely related to sleep disorders like CSA and OSA. The MI and CF have been shown to give cooperatively meaningful information to support clinical findings [31].

12.6.2   Detection of the Rhythmic Waveforms and Spindles Employing Blind Source Separation

After the EEGs are recorded, to analyse and monitor the sleep disorders the main stage is detection of the waveforms corresponding to different stages of sleep. In order to facilitate recording of the EEGs during sleep with a small number of electrodes a method to best select the electrode positions and separate the ECG, EOG, and EMG, may be useful [19].

Recording over a long period is often necessary to investigate adequately the sleep signals and establish a night sleep profile, the electrophysiological activities manifested within the above signals have to be recognized and studied. The BSS algorithm is sought to separate the four desired signals of EEG, two EOGs and one ECG. In order to maintain the continuity of the estimated sources in the consecutive blocks of data the estimated independent components have been cross-correlated with the electrode signals and those of consecutive signal segments most correlated with each particular electrode signal are considered the segments of the same source. As a result of this work the alpha activity may not be seen consistently since the BSS system is generally underdetermined and, therefore, it cannot separate alpha activity from the other brain activities. However, the EMG complexes and REM are noticeable in the separated sources.

12.6.3   Application of Matching Pursuit

Some extensions to the R&K system using the conventional EEG signal processing methods have been proposed by Malinowska *et al.* [32]. These extensions include a finer timescale than the division into 20–30 s epochs, a measure of spindle intensity, and the differentiation of single and randomly evoked K-complexes in response to stimuli from spontaneous periodic ones. Figure 12.5 illustrates some typical spindles and K-complex waveforms.

The adaptive time–frequency (TF) approximation of signals using matching pursuit (MP) introduced initially by Mallat and Zhang [34] has been used in developing a method to investigate the above extensions [32]. MP was reviewed in Chapter 2. The required dictionary of waveforms consists of Gabor functions mainly because these functions provide optimal joint TF localization [35]. Real valued continuous time Gabor functions can be represented as:

$$g_\gamma(t) = K(\gamma) e^{-\pi\left(\frac{t-u}{s}\right)^2} \cos\left(\omega(t-u) + \varphi\right) \tag{12.1}$$

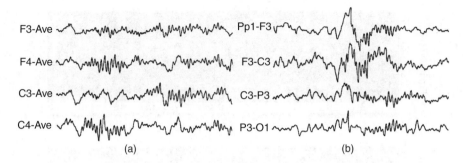

Figure 12.5 Typical waveforms for (a) spindles and (b) K-complexes, adapted from the website of Neural Networks Research Centre, Helsinki University of Technology [33]

where $K(\gamma)$ is a normalising factor of g_γ and the parameters of the Gabor function $\gamma = \{u, \omega, s\}$ provide a three-dimensional continuous space from which a finite dictionary must be chosen. In this application these parameters are drawn from uniform distributions over the signal range and correspond to the dictionary size. They are fitted to the signal by the MP algorithm and often are directly used for analysis.

Due to spectral and temporal variation of EEG during the changes in the state of sleep, Gabor functions are used to exploit the changes in the above parameters, especially the TF amplitude and phase indicators.

In this work the deep-sleep stages (III and IV) are detected from the EEG signals based on the classical R&K criteria; derivation of the continuous description of slow wave sleep fully compatible with the R&K criteria has been attempted, a measure of spindling activity has been followed, and finally a procedure for detection of arousal has been presented. MP has been shown to well separate various waveforms within the sleep EEGs. Assuming the slow wave activity (SWA) and sleep spindle to have the characteristics in Table 12.1, they can be automatically detected by MP decomposition.

As the result of decomposition and examination of the components, if between 20 and 50% of the duration of an epoch is occupied by SWA it corresponds to Stage III, and if above 50% of the duration of an epoch is occupied by SWA it corresponds to Stage IV. Therefore, Stages III and IV of sleep can be recognised by applying the R&K criteria.

The first approach to the automatic detection of arousals was based upon the MP decomposition of only the C3–A2 single EEG channel and the standard deviation of EMG by implementing the rules established by the American Sleep Disorders Association (ASDA) [36]. The MP structure shows the frequency shifts within different frequency bands. Such shifts lasting 3 s or longer are related to arousal. To score a second arousal a minimum of 10 s

Table 12.1 The time, frequency, and amplitudes of both SWA and sleep spindles [32]

	Time Duration /s	Frequency/Hz	Min. Amplitude/μV
SWA	$0.5 - \infty$	$0.5 - 4$	$0.99 \times V_{\text{EEG}} + 28.18$
Sleep spindles	$0.5 - 2.5$	$11 - 15$	15

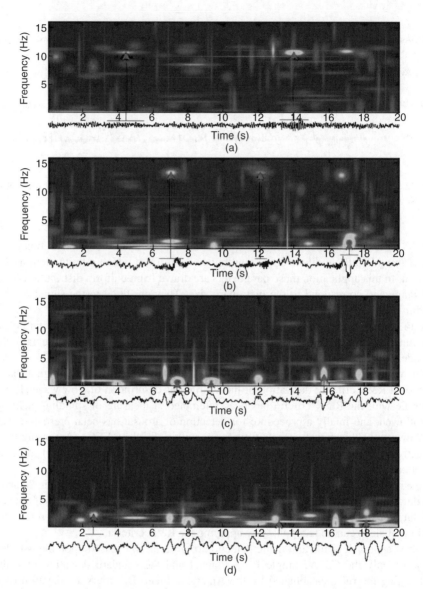

Figure 12.6 Time–frequency energy map of 20 s epochs of sleep EEG in different stages. The arrows point to the corresponding blobs for (a) awake alpha, (b) spindles and K-complex related to Stage II, and (c) and (d) SWAs related, respectively, to Stages III and IV. Reprinted from [32], © IEEE (see Plate 13 for the coloured version)

of intervening sleep is necessary [32]. In Figure 12.6 the results of applying the MP algorithm using Gabor functions for the detection of both rhythms and transients can be viewed. Each blob in the TF energy map corresponds to one Gabor function. The 8–12 Hz alpha wave, sleep spindles (and one K-complex), and SWA of Stages II and IV are presented in Figure 12.6 a–d, respectively.

The sleep spindles exhibited inversely relate to the SWA [37]. The detected arousals decrease in relation to the amount of light NREM sleep, with particular concentration before the REM episodes [38].

The MP algorithm has also been used in the differentiation of single randomly evoked K-complexes in response to stimuli from spontaneous periodic ones [32].

The tools and algorithms developed for recognition and detection of sleep stages can be applied to the diagnosis of many sleep disorders, such as apnoea and the disturbances leading to arousal.

12.6.4 Detection of Normal Rhythms and Spindles Using Higher Order Statistics

Use of long-term spectrum analysis for detection and characterisation of sleep EEG waveforms has been very popular [39, 40]. However, these methods are unable to detect transient and isolated characteristic waves, such as hump and K-complexes, accurately.

In one approach, higher order statistics (HOS) of the time domain signals together with the spectra of the EEG signals during the sleep have been utilised to characterise the dynamics of sleep spindles [41]. The spindles are considered as periodic oscillations with steady-state behaviour which can be modelled as a linear system with sinusoidal input, or nonlinear system with a limit cycle.

In this work, second and third order correlations of the time domain signals are combined to determine the stationarity of periodic spindle rhythms to detect transitions between multiple activities. The spectra (normalised spectrum and bispectrum) of the signals, on the other hand, describe frequency interactions associated with nonlinearities occurring in the EEGs.

The power spectrum of the stationary discrete signal, $x(n)$, is the power spectrum of its autocorrelation function given by:

$$P(\omega) = \sum_{n=-\infty}^{\infty} R(n)e^{-jn\omega} \cong \frac{1}{N}\sum_{i=1}^{N} X_i(\omega)X_i^*(\omega) \qquad (12.2)$$

where $X_i(\omega)$ is the Fourier transform of the ith segment of one EEG channel. Also, the bispectrum of data is defined as:

$$B(\omega_1, \omega_2) = \sum_{n_1=-\infty}^{\infty} \sum_{n_2=-\infty}^{\infty} x(n)x(n+n_1)x(n+n_2)$$

$$\cong \frac{1}{N}\sum_{i=1}^{N} X_i(\omega_1)X_i(\omega_2)X_i(\omega_1+\omega_2) \qquad (12.3)$$

where N is the number of segments of each EEG channel. Using equations (12.2) and (12.3), a normalized bispectrum (also referred to as bicoherence, second-order coherency, or bicoherency index) is defined as [42]:

$$b^2(\omega_1, \omega_2) = \frac{|B(\omega_1, \omega_2)|^2}{P(\omega_1)P(\omega_2)P(\omega_1+\omega_2)} \qquad (12.4)$$

which is an important tool for evaluating signal nonlinearities [43]. This measure (and the measure in (12.5)) have been widely used for detection of coupled periodicities. Equation (12.6) acts as the discriminant of a linear process from a nonlinear one. For example, b^2 is constant for either linear systems [43] or fully coupled frequencies [44] and $b^2 = 0$ for either Gaussian signals or random phase relations where no quadratic coupling occurs. Coupling of the frequencies occurs when the values of the normalised bispectrum vary between zero and one $(0 < b^2 < 1)$. The coherency value of one refers to quadratic interaction and an approximate zero value refers to either low or absent interactions [42].

To find the spindle periods, another method similar to the average magnitude difference function (AMDF) algorithm, often used for pitch detection from speech signals, has been applied to the short intervals of the EEG segments. The procedure has been applied to both second and third order statistical measures as [41]:

$$D_n(k) = 1 - \frac{\gamma_n(k)}{\sigma_{\gamma_n}} \qquad (12.5)$$

where

$$\gamma_n(k) = \sum_{m=-\infty}^{\infty} |x(n + m)w(m) - x(n + m - k)w(m - k)| \qquad (12.6)$$

and $\sigma_{\gamma_n} = \sqrt{\sum_i \gamma_n^2(i)}$ is the normalization factor and $w(m)$ is the window function, and

$$Q_n(k) = 1 - \frac{\varphi_n(k)}{\sigma_{\varphi_n}} \qquad (12.7)$$

where

$$\varphi_n(k) = \sum_{m=-\infty}^{\infty} |q(n + m)w(m) - q(n + m - k)w(m - k)| \qquad (12.8)$$

and $q(n)$ is the inverse two-dimensional Fourier transform of the bispectrum and $\sigma_{\varphi_n} = \sqrt{\sum_i \varphi_n^2(i)}$. The measures are used together to estimate the periodicity of the spindles. For purely periodic activities we expect these estimates to give similar results. In this case equation (12.8) manifests peaks (as in AMDF) where the first peak denotes the spindle frequency.

Based on this investigation in summary it has been shown that (i) spindle activity may not uniformly dominate all regions of the brain; (ii) during the spindle activity frontal recordings still exhibit rich mixtures in frequency contents and coupling, on the other hand, a poor coupling may be observed at the posterior regions whilst showing dominant activity of the spindles; and (iii) it is concluded that the spindle activity may be modelled using at least second-order nonlinearity.

12.6.5   Application of Neural Networks

As a nonlinear multiple-input multiple-output classifier, neural networks (NNs) can be used to classify different waveforms for recognition of various stages of sleep and also the type of mental illnesses. NNs have been widely used to analyse complicated systems without accurately modelling them in advance [45]. Often no a priori information about the data statistics, such as distribution, is necessary in developing an NN. The input to NN classifiers can be the raw data, compressed data, or the most descriptive features measured or estimated from the data in the original or transformed domains. In the case of sleep data, a number of typical waveforms from the sleep EEG can be used for training and classification. They include spindle, hump, alpha wave, hump train (although not present generally in the EEGs), and background wave. Each of these features manifests itself differently in the TF domain.

Time delay NNs (TDNN) may be used to detect the wavelets with roughly known positions on the time axis [46]. In such networks the deviation in location of the wavelet in time has to be small. For EEGs, however, a shift larger than the wavelet duration must be compensated since the occurrence times of the waveforms are not known.

In order to recognise the time-shifted pattern, another approach, named sleep EEG recognition NN (SRNN) has been proposed by Shimada *et al.* [47]. This NN has one input layer, two hidden layers, and one output layer. From the algorithmic point of view and the input–output connections, SRNN and TDNN are very similar. As the main difference, in a TDNN each row of the second hidden layer is connected to a single cell of the output layer, while in a SRNN similar cells from both layers are connected.

In order to use an SRNN the data are transformed into the TF domain. Instead of moving a sliding window over time, however, overlapped blocks of data are considered in this approach. Two-dimensional blocks with horizontal axis of time and vertical axis of frequency are considered as the inputs to the NN. Considering $y_{j,c}$ and $d_{j,c}$ to be, respectively, the jth output neuron and the desired pattern for the input pattern c then,

$$E = \frac{1}{2} \sum_{j=1}^{\substack{\text{Output} \\ \text{neurons}}} \sum_{c=1}^{\substack{\text{Input} \\ \text{patterns}}} (y_{j,c} - d_{j,c})^{1/2} \tag{12.9}$$

The learning rule, therefore, minimises the cost function by using the following gradient:

$$\Delta w_{p,q} = \mu \frac{\partial E}{\partial w_{p,q}} \tag{12.10}$$

where $w_{p,q}$ are the link weights between neurons p and q and μ is the learning rate. In the learning phase the procedure [47] performs two passes; forward and backward, through the network. In the forward pass the inputs are applied and the outputs are computed. In the backward pass, the outputs are compared with the desired patterns and an error is calculated. The error is then back projected through the network and the connection weight is changed by gradient descent of the mean squared error (MSE) as a function of weights. The optimum weights $w_{p,q}$ are obtained when the learning algorithm converges. The weights are then used to classify a new waveform, that is, to perform generalization. Further details of the performance of this scheme can be found in [47].

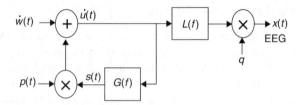

Figure 12.7   A model for neuronal slow-wave generation; $\dot{u}(t)$ is the EEG source within the brain, $G(f)$ denotes the frequency-selective feedback of the slow-wave, s(t). $p(t)$ is the feedback gain to be identified, $w(t)$ is the white noise, $L(f)$ is the lowpass filter representing the path from source to electrodes and q is the attenuation effect by the skull

12.6.6   Model-Based Analysis

Model-based approaches rely on an a priori knowledge about the mechanism of generation of the data or the data itself. Characterizing a physiological signal generation model for NREM has also been under study by several researchers [48–50]. These models describe how the depth of NREM sleep is related to the neuronal mechanism that generates slow waves. This mechanism is essentially feedback through closed loops in neuronal networks or through the interplay between ion currents in single cells. It is established that the depth of NREM sleep modulates the gain of the feedback loops [51]. According to this model, the sleep-related variations in the slow wave power (SWP) result from variations in the feedback gain. Therefore, increasing depth of sleep is related to an increasing gain in the neuronal feedback loops that generates the low frequency EEG.

In [52] a model-based estimator of the slow-wave-feedback gain has been proposed. The initial model, depicted in Figure 12.7, is analogue and a discrete time approximation of that has been built up. In the analogue model, $G(f)$ is the complex frequency transfer function of a bandpass (resonance) filter as

$$G(f) = \frac{1}{1 + j \cdot Y(f)} \tag{12.11}$$

where here $j = \sqrt{-1}$ and

$$Y(f) = \frac{f_0}{B} \cdot \left(\frac{f}{f_0} - \frac{f_0}{f} \right) \tag{12.12}$$

where the resonance frequency f_0 and the bandwidth B are approximately 1 Hz and 1.5 Hz. The closed loop equation can be written as

$$\dot{u}(t) = p(t) \cdot s(t) + \dot{w}(t) \tag{12.13}$$

where $\dot{u}(t) = du(t)/dt$ and $\dot{w}(t) = dw(t)/dt$. It is seen that the transfer function depends on $p(t)$. For $p(t)$ constant,

$$U(f) = \frac{1}{1 - p \cdot G(f)} \tag{12.14}$$

and therefore,

$$|U(f)|^2 = \frac{1 + Y^2(f)}{(1 - p)^2 + Y^2(f)} \tag{12.15}$$

In which case

$$S(f) = G(f) \cdot U(f) = \frac{G(f)}{1 - p \cdot G(f)} = \frac{1}{1 - p + j \cdot Y(f)} \tag{12.16}$$

The output $x(t)$ is the lowpass copy of $\dot{u}(t)$ attenuated by the factor of q. For $p = 0$ there is no feedback and for $p = 1$ there is an infinite peak at $f = f_0$. The lowpass filter $L(f)$ is considered known with a cut-off frequency of approximately 1.8 Hz.

The feedback gain of the model $p(t)$ represents the sleep depth. Therefore, the objective would be to estimate the feedback gain. To do that, the observation $du(t)$ is sampled by a 50 Hz sampler (a sampling interval of $\Delta = 0.02$ s). Then, define $Du(k\Delta) = u(k\Delta + \Delta) - u(k\Delta)$ over an interval $[0 \le k\Delta < N\Delta - \Delta]$, with $N\Delta = T$. Also, it is considered that $p(t) = p$, constant over the interval T. Equation (12.13) then becomes

$$Du(k\Delta) = p.s(k\Delta).\Delta + Dw(k\Delta) \tag{12.17}$$

where $Dw(k\Delta) = w(k\Delta + \Delta) - w(k\Delta)$ is the increment of the standard continuous-time Wiener process $w(t)$. Assuming the initial state for the feedback filter $G(f)$ to be G_0 and $w(k\Delta)$ to have Gaussian distribution, the likelihood of $Du(k\Delta)$ can be factorised according to Bayes' rule as:

$$P[Du(k\Delta) : 0 \le k < N - 1 | G_0, p]$$

$$= \prod_{k=0}^{N-1} [Du(k\Delta)|[Du(m\Delta) : 0 \le m < k], G_0, p]$$

$$= \prod_{k=0}^{N-1} [Du(k\Delta)|s(k\Delta), p] \tag{12.18}$$

$$= \prod_{k=0}^{N-1} \left\{ \frac{1}{\sqrt{2\pi\Delta}} \exp\left(-[Du(k\Delta) - p.s(k\Delta).\Delta]^2 / 2\Delta \right) \right\}$$

$$= \frac{1}{\sqrt{2\pi\Delta}} \exp\left\{ \sum_{k=0}^{N-1} \left(-[Du(k\Delta) - p.s(k\Delta).\Delta]^2 / 2\Delta \right) \right\}$$

To maximize this likelihood it is easy to conclude that the last term in the square brackets has to be maximised. This gives [52]

$$\hat{p} = \frac{\sum_{k=0}^{N-1} [s(k\Delta).Du(k\Delta)]}{\sum_{k=0}^{N-1} s^2(k\Delta).\Delta} \tag{12.19}$$

Hence, $\hat{p}$ approximates the amount of slow wave and is often represented as its percentage. This completes the model and, therefore, the sleep EEG may now be constructed.

12.6.7 Hybrid Methods

Diagnosis of sleep disorders and other related abnormalities may not be complete unless other physiological symptoms are studied. These symptoms manifest themselves within other physiological and pneumological signals, such as respiratory airflow, position of the patients, electromyogram (EMG) signal, hypnogram, level of SaO_2, abdominal effort, and thoracic effort, which may also be considered in the classification system.

A simple system to use the features extracted from the sleep signals in classification of the apnoea stages and detection of the SAS has been suggested and used [10]. This system processes the signals in three phases. In the first phase the relevant characteristics of the signals are extracted and a segmentation based on significant time intervals of variable length is performed. The intermediate phase consists of assigning suitable labels to these intervals and combining these symbolic information sources with contextual information in order to build the necessary structures that will identify clinically significant events. Finally, all the relevant information is collected and a rule-based system is established to exploit the above data, provide the induction, and produce a set of conclusions.

In a rule-based system for detection of the SAS, two kinds of cerebral activities are detected and characterized from the EEG signals: rhythmic (alpha, beta, theta, and delta rhythms) and transitory (K-complexes and spindles) [53]. The magnitude and the change in the magnitude (evolution) are measured and the corresponding numerical values are classified together with the other features based on clinical and heuristic [11] criteria.

To complete this classifier the slow eye movement, very frequent during sleep Stage I, and REM are measured using electro-oculogram (EOG) signals. The distinction between the above two eye movements is based on the synchrony, amplitude, and slope of the EOG signals [10].

In another approach for sleep staging of patients with obstructive sleep apnoea the EEG features are used, classified, and compared with the results from cardiorespiratory features [12].

12.7 EEG and Fibromyalgia Syndrome

Fibromyalgia syndrome (FMS) is defined by the existence of chronic, often full-body pain accompanied by a variety of additional symptoms such as widespread musculoskeletal pain, fatigue, and pain in the tendons and ligaments, and sleep disturbances. With the diversity and complexity of these symptoms, fibromyalgia is often misdiagnosed, leaving sufferers frustrated and confused. However, sleep apnoea commonly coexists with fibromyalgia. The sleep disorder, known as alpha-EEG anomaly, is prevalent in FMS patients. This condition is defined by interruption of deep sleep by awake-like brain activity.

Alpha waves indicate you are awake but relaxed. In the corresponding EEG records, a sleeping brain typically displays delta waves during Stage 3 of the sleep, which is considered "deep" sleep. In the alpha-EEG anomaly, alpha waves intrude into deep sleep, indicating that the brain is not resting like it should. Alpha wave intrusion occurs when alpha waves appear with non-REM sleep when delta activity is expected. Some researchers theorize that this anomaly may explain the unrefreshing sleep that is characteristic of fibromyalgia [54]. Therefore, an evaluation of an EEG can reveal some valuable information for diagnosis of this disorder.

12.8   Sleep Disorders of Neonates

Recently, the changes in EEG associated with apneic episodes in neonates have been studied, for example, in [55–57]. In [55] it was found that the prolonged apneic episodes accompanied by hypoxia and bradycardia can be associated with altered cerebral function in the neonate. This study can help the clinician to know the lowest limit of oxygen saturation required to suppress EEG activity. It has also been shown that only apneic events with oxygen desaturations below 20% are always associated with complete EEG suppression.

Despite the frequency of occurrence of apnoea in the neonate and the concern about adverse long-term effects [57], few studies have used EEG to examine the effects of apneic episodes [55]. From the sleep EEG it can be discovered if the apnoea is related or correlated with neonate seizure, hypoxia, or any other neurological or even physiological disorders.

12.9   Dreams and Nightmares

It is difficult to deny that we all dream. People awakened from REM periods in sleep experiments report they have been dreaming 80%–100% of the time. REM dreams are considered to be more perceptual and emotional as opposed to NREM dreams. Content of NREM dreams is often a recreation of some psychologically important event. According to Freud [58] REM dreams are like primary-process thinking which is often unrealistic and emotional, and NREM dreams are like secondary-process thinking which is more realistic. There are mainly three theories on the meaning of dreams:

Based on Freud theory, we dream to satisfy unconscious desires or wishes, especially those involving sex and aggression. If we were to fulfil these wishes during day time it would create too much anxiety. Freud stated that the wishes are represented with symbols since they would otherwise be anxiety producing. Based on this theory a therapist must interpret these symbols to help clients discover unconscious desires.

The theory states that the hindbrain transmits chaotic patterns of signals to the cerebral cortex, and then higher-level cognitive processes in the cerebral cortex try to integrate these signals into a dream plot [59].

Dreams can also be viewed as extensions of waking life, which include thoughts and concerns, especially emotional ones. Then, in a sense dreams provide clues to the person's problems, concerns, and emotions [60]. Different from nightmare, dreaming may be a sign of good sleeping. Although there are many theories and hypotheses about dreams, there are still many unanswered questions about the reasons for dreams, their exact functions, and how physiological processes are involved. Sleep EEG records are likely to present alpha attenuation as an indicator of visual activity during dreaming [61]. On the other hand, unlike dreams in general, nightmares are known to exclusively occur during REM sleep and during the longer and later REM phases in the latter part of the sleep cycle.

12.10   Conclusions

Study of sleep EEG has opened a new direction to investigate the psychology of a human being and various sleep disorders. There is, however, much more to discover from these signals. Various physiological and mental brain disorders manifest themselves differently in the sleep EEG signals. Different stages of sleep may be identified using simple established tools in signal processing. Apnoea and fibromyalgia syndrome can be studied and monitored through analysis of EEG. Detection and classification of mental diseases from the sleep EEG signals, however, require more deep analysis of the data by developing and utilizing advanced digital signal processing techniques. The analysis becomes more challenging when other parameters such as age are involved. For example, in neonates many different types of complex waveforms may be observed for which the origin and causes are still unknown. On the other hand, there are some similarities between the normal rhythms within the sleep EEG signals and the EEGs of abnormal rhythms such as epileptic seizure and hyperventilation. An efficient algorithm (based solely on EEG or combined with other physiological signals) should be able to differentiate between these signals.

References

[1]   Lavie, P. (1996) *Enchanted World of Sleep* (transl. A. Berris)), Yale University, London.

[2]   Steriade, M. (1992) Basic mechanisms of sleep generation. *Neurology*, **42** (Suppl. 6), 9–17.

[3]   Kubicki, S., Scheuler, W. and Wittenbecher, H. (1991) Short-term sleep EEG recordings after partial sleep deprivation as a routine procedure in order to uncover epileptic phenomena: an evaluation of 719 EEG recordings. *Epilepsy Res. Suppl.*, **2**, 217–230.

[4]   Niedermeyer, E. (1999) Sleep and EEG, Chapter 10, *Electroencephalography Basic Principles, Clinical Applications, and Related Fields* (eds E. Niedermeyer and F.L. Da Silva), Lippincott Williams & Wilkins, pp. 174–188.

[5]   Koshino, Y., Nishio, M., Murata, T. *et al.* (1993) The influence of light drowsiness on the latency and amplitude of P300. *Clin. Electroencephalogr.*, **24**, 110–113.

[6]   Wada, Y., Nanbu, Y., Koshino, Y. *et al.* (1996) Inter- and intrahemispheric EEG coherenceduring light drowsiness. *Clin. Electroencephalogr*, **27**, 24–88.

[7]   Niedermeyer, E. (1999) Maturation of EEG: development of walking and sleep patterns, Chapter 11, in *Electroencephalography* (eds E. Niedermeyer and F. Lopez de silva), Lippincott Williams & Wilkins.

[8]   Bonanni, E., Di Coscio, E., Maestri, M. *et al.* (2012), Differences in EEG delta frequency characteristics and patterns in slow-wave sleep between dementia patients and controls: a pilot study. *J. Clin. Neurophysiol.*, **29**(1), 50–54.

[9]   Shiomi, F.K., Pisa, I.T. and de Campos, C.J.R. (2011) Computerized analysis of snoring in sleep apnea syndrome. *Braz J. Otorhinolaryngology*, **77**(4), 488–498.

[10] Cabrero-Canosa, M., Hernandez-Pereira, E. and Moret-Bonillo, V. (2004) Intelligent dignosis of sleep apnea syndrome. *IEEE Eng. Med. Biol. Mag.*, **23**(2),72–81.

[11] Karskadon, M.A. and Rechtschaffen, A. (1989) *Priniciples and Practice of Sleep Medicine*, Saunders, Philadelphia, pp. 665–683.

[12] Redmond, S.J. and Heneghan, C. (2006) Cardiorespiratory-based sleep staging in subjects with obstructive sleep apnea. *IEEE Trans. Biomed. Eng.*, **51**(3), 485–496.

[13] Klingler, D., Tragner, H. and Deisenhammer, E. (1991) The nature of the influence of sleep deprivation on the EEG. *Epilepsy Res. Suppl.*, (**2**), 231–234.

[14] Jovanovic, U.J. (1991) General considerations of sleep and sleep deprivation. *Epilepsy Res. Suppl.*, **2**, 205–215.

[15] Naitoh, P., Kelly, T.L. and Englund, C., (1990) Health effects of sleep deprivation. *Occup. Med.*, **5**(2), 209–237.

[16] Sabo, E., Reynolds, C.F., Kupfer, D.J. and Berman, S.R. (1991) Sleep, depression, and suicide. *Psychiat. Res.*, **36**(3), 265–77.

[17] Weitzenblum, E. and Racineux, J.-L. (2004) *Syndrome d'Apnées Obstructives du Sommeil*, 2nd edn. Masson, Paris, France.

[18] Man, G.C. and Kang, B.V. (1995) Validation of portable sleep apnea monitoring device. *Chest*, **108**(2), 388–393.

[19] Porée, F., Kachenoura, A., Gavrit, H. *et al.* (2006) Blind source separation for ambulatory sleep recording. *IEEE Trans. Inf. Technol. Biomed.*, **10**(2), 293–301.

[20] Dahl, R.E. and Puig-Antich, J. (1990) EEG sleep in adolescents with major depression: the role of suicidality and inpatient status. *J. Affective Disord.*, **19**(1), 63–75.

[21] Goetz, R.R., Puig-Antich, J., Dahl, R.E. *et al.* (1991) EEG sleep of young adults with major depression: a controlled study. *J. Affect Disord.*, **22**(1–2), 91–100.

[22] McCracken, J.T., Poland, R.E., Lutchmansingh, P. and Edwards, C. (1997) Sleep electroencephalographic abnormalities in adolescent depressives: Effects of scopolamine. *Biol. Psychiat.*, **42**, 577–584.

[23] Van Swededn, B., Wauquier, A. and Niedermeyer, E. (1999) Normal aging and transient cognitive disorders in elderly, Chapter 18, in *Electroencephalography* (eds E. Niedermeyer and F. Lopez de Silva), Lipincott Williams & Wilkins.

[24] Klass, D.W. and Brenner, R.P. (1995) Electroencephalography in the elderly. *J. Clin. Neurophysiol.*, **12**, 116–131.

[25] (1986) A Manual of Standardized Terminology and Scoring System for Sleep Stages in Human Subjects, ed. A. Rechtschaffen and A. Kales, Ser. National Institutes of Health Publications. Washington DC: U.S. Government Printing Office, no. 204.

[26] Nobuyuki, A., Yasuhiro, N., Taiki, T. *et al.* (2009) Trial of measurement of sleep apnea syndrome with sound monitoring and SpO2 at home. *Healthcom*, 66–79.

[27] Ydollahi, A. and Giannouli, E. (2010) Sleep apnea monitoring and diagnosis based on pulse oximetry and tracheal signals. *Med. Biol. Eng. Comput.*, **48**, 1987–1097.

[28] Sola-Soler, J., Fiz, J.A., Morea, J. and Jane, Raiman (2012) Multiclass classification of subjects with sleep apnoea-hypopnoea syndrome through snoring analysis. *Els. J.Med. Eng. Phys.*, **JJBE-2043.** doi: 10.1016/j.medenegphy.2011.12.008

[29] Hang, L.W., Wang, J.-F., Yen, C.-W. and Lin, C.-L. (2009) EEG arousal prediction via hypoxemia indicator in patients with obstructive sleep apnea syndrome. *Internet J. Med. Update*, **4**(2), 24–28.

[30] Young, T., Palta, M., Dempsey, J. *et al.* (1993) The occurrence of sleep-disordered breathing among middle-aged adults. *N. Engl. J. Med.*, **328**, 1230–1235.

[31] Aksahin, M., Aydın, S., Fırat, H. and Erogul, O. (2012) Artificial apnea classification with quantitative sleep EEG synchronization. *J. Med. Syst.*, **36**(1), 139–144.

[32] Malinowska, U., Durka, P.J., Blinowska, K.J. *et al.* (2006) Micro- and macrostructure of sleep EEG; a universal, adaptive time-frequency parametrization. *IEEE Eng. Med. Biol. Mag.*, **25**(4), 26–31.

[33] Laboratory of Computer and Information Science (CIS), Department of Information and Computer Science (ICS) at the Helsinki University of Technology, http://www.cis.hut.fi/, accessed 8th February 2013.

[34] Mallat, S. and Zhang, Z. (1993) Matching pursuit with time-frequency dictionaries. *IEEE Trans. Signal Process.*, **41**, 3397–3415.

[35] Mallat, S. (1999) *A Wavelet Tour of Signal Processing*, 2nd edn, Academic, New York.

[36] American Sleep Disorder Association (1992) EEG arousals: Scoring rules and examples. A preliminary report from the Sleep Disorder Task Force of the American Sleep Disorder Association. *Sleep*, **15**(2), 174–184.

[37] Aeschbach, D. and Borb'ely, A.A. (1993) All-night dynamics of the human sleep EEG. *J. Sleep Res.*, **2**(2), 70–81.

[38] Terzano, M.G., Parrino, L., Rosa, A. *et al.* (2002) CAP and arousals in the structural development of sleep: An integrative perspective. *Sleep Med.*, **3**(3), 221–229.

[39] Principe, J.C. and Smith, J.R. (1986) SAMICOS-A sleep analysing microcomputer system. *IEEE Trans. Biomed. Eng.*, **33**, 935–941.

[40] Principe, J.C., Gala, S.K. and Chang, T.G. (1989) Sleep staging automation based on the theory of evidence. *IEEE Trans. Biomed. Eng.*, **36**, 503–509.

[41] Akgül, T., Sun, M., Sclabassi, R.J. and Çetin, A.E. (2000) Characterization of sleep spindles using higher order statistics and spectra. *IEEE Trans. Biomed. Eng.*, **47**(8), 997–1009.

[42] Nikias, C.L. and Petropulu, A. (1993) *Higher Order Spectra Analysis, A Nonlinear Signal Processing Framework*, Prentice Hall, Englewood Cliffs, NJ, USA.

[43] Rao, T.S. (1993) Bispectral analysis of nonlinear stationary time series, in *Handbook of Statistics 3* (eds D.R. Brillinger and P.R. Krishnaiah),Elsevier, Amsterdam

[44] Michel, O. and Flandrin, P. (1995) Higher order statistics for chaotic signal analysis, in *DSP Techniques and Applications* (ed. E.T. Leondes), Academic, New York.

[45] Lippman, R.P. (1987) An introduction to computing with neural nets. *IEEE Acoust. Speech Signal Process. Mag.*, **4**(2), 4–22.

[46] Weibel, A., Hanazawa, T., Hinton, G. and Lang, K. (1989) Phoneme recognition using time-delay neural networks. *IEEE Trans. Acoust. Speech Signal Process.*, **37**, 328–339.

[47] Shimada, T., Shiina, T. and Saito, Y. (2000) Detection of characteristic waves of sleep EEG by neural network analysis. *IEEE Trans. Biomed. Eng.*, **47**(3), 369–379.

[48] Kemp, B., Zwinderman, A.H., Tuk, B. *et al.* (2000) Analysis of a sleep-dependent neuronal feedback loop: the slow-wave microcontinuity of the EEG. *IEEE Trans. Biomed. Eng.*, **47**(9), 1185–1194.

[49] Kemp, B. (1996) NREM sleep depth = neuronal feedback = slow-wave shape. *J Sleep Res.*, **5**, S106.

[50] Merica, H. and Fortune, R.D. (2003) A unique pattern of sleep structure is found to be identical at all cortical sites: a neurobiological interpretation. *Cerebral Cortex*, **13**(10), 044–1050,

[51] Steriade, M., McCormick, D.A. and Sejnowski, T.J. (1993) Thalamocortical oscillations in the sleeping and aroused brain. *Science*, **262**, 679–685.

[52] Kemp, B., Zwinderman, A.H., Tuk, B. *et al.* (2000) Analysis of a sleep-dependent neuronal feedback loop: the slow-wave microcontinuity of the EEG. *IEEE Trans. Biomed. Eng.*, **47**(9), 1185–1194.

[53] Steriade, M., Gloor, P., Llinas, R.R. *et al.* (1990) Basic mechanisms of cerebral rhythmic activities. *Electroenceph. Clin. Neurophysiol.*, **76**, 481–508.

[54] Mueller, H. H., Donaldson, C. C. S., Nelson, D. V. and Layman, M. (2001) Treatment of fibromyalgia incorporating EEG-driven stimulation: A clinical outcomes study. *J. Clin. Psychol.*, **57**(7), 933–952; Hammond, D. C. (2001). Treatment of chronic fatigue with neurofeedback and self-hypnosis. *J. Neuro Rehab.*, **16**, 295–300.

[55] Low, E., Dempsey, E.M., Ryan, C.A. *et al.* (2012) EEG Suppression associated with apneic episodes in a neonate. *Case Report. Neurol. Med.*, **2012**. Article ID 250801, 7. doi: 10.1155/2012/250801

[56] Murray, D.M., Boylan, G.B., Ryan, C.A. and Connolly, S. (2009) Early EEG findings in hypoxic-ischemic encephalopathy predict outcomes at 2 years. *Pediatrics*, **124**(3), e459–e467.

[57] Janvier, A., Khairy, M., Kokkotis, A. *et al.* (2004) Apnea is associated with neurodevelopmental impairment in very low birth weight infants. *J. Perinatol.*, **24**(12), 763–768.

[58] Franken, R.E. (1988) *Human Motivation*, Brooks/Cole, California.

[59] Hobson, J.A. and Stickgold, R. (1995) The conscious state paradigm: A neurocognitive approach to waking, sleeping, and dreaming, in *The Cognitive Neurosciences* (ed. M.S. Gazzaniga), MIT Press, Cambridge, MA.

[60] Plotnik, R. (1995) *Introduction to Psychology*, 3rd edn, Brooks/Cole, California.

[61] Bértolo, H., Paiva, T., Pessoa, L. *et al.* (2003) Visual dream content, graphical representation and EEG alpha activity in congenitally blind subjects. *Brain Res. Cogn. Brain Res.*, **15**(3), 277–284.

13

Brain–Computer Interfacing

13.1 Introduction

Often, the effortless way in which our intentions for movement are converted into actions is taken for granted. The loss of motor function can be one of the most devastating consequences of disease or injury to the nervous system. Development of the brain–computer interfaces (BCI) during the last two decades has enabled communications or control over external devices, such as computers and artificial prostheses with the electrical activity (e.g. EEG) of the human nervous system. However, the performance of these artificial interfaces rarely matches the speed and accuracy of natural limb movements, making their clinical applications rather limited. Their practical utility, therefore, depends not only on the extent to which patients can learn to operate these devices but also on advancement of the real-time, but not necessarily complicated, algorithms that can decode motor intentions efficiently. In a conventional BCI set-up, the user is instructed to imagine movement of different body parts (e.g. right hand or leg movements) and the computer learns to recognize different patterns of the simultaneously recorded EEG activity. In the majority of cases, the signal processing effort for the progress of EEG-based BCI systems has been focused on extracting more informative features from the EEGs and refinement/reduction of the feature space.

Primarily, electromyography (EMG) measurements have been widely used in both rehabilitation and development of various robotic and haptic equipments. EMG, therefore, is still the most popular and effective method for connecting human to machine. BCI (also called brain–machine interfacing (BMI)) is a challenging problem that forms part of a wider area of research, namely human–computer interfacing (HCI), which interlinks thought to action. BCI can potentially provide a link between the brain and the physical world without any physical contact. In BCI systems the user messages or commands do not depend on the normal output channels of the brain [1]. As such the main objectives of BCI are to manipulate the electrical signals generated by the neurons of the brain and generate the necessary signals to control some external systems. The most important application is to energize the paralysed organs or bypass the disabled parts of the human body. BCI systems may appear as the unique communication mode for people with severe neuromuscular disorders, such as spinal cord injury, amyotrophic lateral sclerosis, stroke and cerebral palsy.

Adaptive Processing of Brain Signals, First Edition. Saeid Sanei.
© 2013 John Wiley & Sons, Ltd. Published 2013 by John Wiley & Sons, Ltd.

Approximately 100 years after discovery of the electrical activity of the brain the first BCI research was reported by Jacques Vidal [2, 3] during the period 1973–1977. In his research it was shown how brain signals could be used to build up a mental prosthesis. BCI has moved at a stunning pace since the first experimental demonstration in 1999 that ensembles of cortical neurons could directly control a robotic manipulator [4]. Since then there has been tremendous research in this area [5] though practically, EMG, has been more applicable and commercialized.

BCI addresses analysing, conceptualization, monitoring, measuring, and evaluating the complex neurophysiological behaviours detected and extracted by a set of electrodes over the scalp or from the electrodes implanted over the cortex or inside the brain. It is important that a BCI system be easy, effective, efficient, enjoyable to use, and user friendly. BCI is a multidisciplinary field of research since it deals with cognition, electronic sensors and circuits, machine learning, neurophysiology, psychology, sensor positioning, signal detection, signal processing, source localization, pattern recognition, clustering, and classification.

BCI has now become popular in many institutions and in different countries. However, as the pioneers in BCI, the Berlin BCI (BBCI) group, established in 2000, has followed the objective of transferring the effort of training from the human to the machine. The major focus in their work is reducing the inter-subject variability of BCI by minimizing the level of subject training. Some of their works have been reported in [6–8]; the Wadsworth BCI research group have used mainly the event-related desynchronisation (ERD) of the mu rhythm for EEG classification of real or imaginary movements, achieved after training the subject [9, 10]; the Graz BCI activity, led by Pfurtscheller, utilizes mu or beta rhythms for training and control. The expert users of their system are able to control a device based on the modulations of the precentral mu or beta rhythms of sensorimotor cortices in a similar way to the Wadsworth BCI. However, while the Wadsworth BCI directly presents the power modulations to the user, the Graz system for the first time also uses machine adaptation to control the BCI. They were also able to allow grasping by the non-functional arm of a disabled patient by functional electrical stimulation (FES) of the arm controlled by EEG signals [11–15]; the Martigny BCI started with adaptive BCI in parallel with the Berlin BCI. The researchers have proposed a neural network-based classifier based on linear discriminant analysis for classification of the static features [16]. In their approach three subjects are able to achieve 75% correct classification by imagination of left- or right-hand movement or by relaxation with closed eyes in an asynchronous environment after a few days of training [17, 18]. Finally, the thought translation device (TTD) which has been developed mainly for locked-in patients, enables the subjects to learn self-regulation of the slow cortical potentials at central scalp positions using EEG or electrocorticogram (ECoG). The subjects are able to generate binary decisions and hopefully provide a suitable communication channel to the outside world [19].

In recent years the techniques and signal processing algorithms have been further developed to increase the accuracy and robustness of BCI and to incorporate other brain data recording modalities, such as fMRI, separately or together with EEG. In this chapter the fundamental concepts and the requirement for the BCI design using EEG signals are reviewed.

13.2 State of the Art in BCI

The correspondence between EEG patterns and computer actions constitutes a machine-learning problem since the computer should learn how to recognize a given EEG pattern.

As for other learning problems, in order to solve this problem, a training phase is necessary, in which the subject is asked to perform prescribed mental activities (MAs) and a computer algorithm is in charge of extracting the associated EEG patterns. After the training phase is finished the subject should be able to control the computer actions with his thoughts. This is the major goal for a BCI system.

In terms of signal acquisition, the BCI systems are classified into invasive (intracranial) and non-invasive. Non-invasive systems primarily exploit EEGs to control a computer cursor or a robotic arm. The techniques in designing such systems have been under development recently due to their hazardless nature and flexibility [1, 19–27]. However, despite the advantage of not exposing the patient to the risks of brain surgery, EEG-based techniques provide limited information, mainly because of the existence of system and physiological noise and the interfering undesired signals and artefacts. However, despite these shortcomings, EEG-based methods can detect modulations of brain activity that correlate with visual stimuli, gaze angle, voluntary intentions and cognitive states [5]. These advantages led to development of several classes of EEG-based systems, which differ according to the cortical areas recorded, the extracted features of the EEGs, and the sensory modality providing feedback to subjects.

Although the feedback during a BCI process can be visual, audio, and haptic (tactile or somatosensory), the most effective feedbacks are visual and haptic. In a short paper by Kauhanen *et al.* [28] it has been concluded that both visual and haptic feedbacks have similar effects to the brain. Haptic feedback may be necessary for blind subjects. Less accurate results have often been reported for audio feedback experiments. In an editorial review however, it has been emphasised that haptic feedback provides more complete and immersive sensation than a purely visual environment [29] can do. Haptic feedback involves tactile and proprioceptive sensory modalities, and can thus help increase the subject's attention and motivation during repeated performance.

In non-invasive BCI approaches, although there may be different EEG electrode settings for certain BCI applications, an efficient BCI system exploits all the information content within the EEG signals, particularly if the brain connectivity is to be assessed. In all cases detection and separation of the control signals from the raw EEG signals is probably the first objective. The event-related source signals can be effectively clustered or separated if the corresponding control signals are well characterized. Given that these control sources are likely to be moving inside the brain, an exciting research area is also to localize and track these sources in real-time. The first step in developing an effective BCI paradigm is, therefore, determining suitable control signals from the EEG. A suitable control signal has the following attributes: it can be precisely characterised for an individual, it can be readily modulated or translated to express the intention, and it can be detected and tracked consistently and reliably.

There are two main approaches towards BCI; one is based on event-related potentials (ERPs) which can be captured using a small number of electrodes around Pz and Cz electrode sites, and another is based on the multiple sensor EEG activities recorded in the course of ordinary brain activity. In many cases, however, only a few electrodes are used to detect the changes in the cortical sources within the motor area. The latter approach is more comprehensive and does not require any particular stimulus.

Preparation and execution of movements (or imagination of movement) lead to short-lasting and circumscribed attenuation of the contralateral (to the side of movement) rolandic mu (μ) (8–13 Hz) and the beta (β) (14–28 Hz) rhythms of the EEGs. This phenomenon is known as event-related desynchronization (ERD). ERD is followed by a rebound amplification phase called event-related synchronization (ERS) in which μ and β rhythms re-emerge. Distinct spatial, temporal, and spectral characteristics of ERD/ERS can only be observed if several trials of EEG are averaged to cancel out the effect of (presumably) zero-mean, Gaussian and spatiotemporally independent recording background brain activity, measurement noise, or to minimize the effect of volume conduction. Therefore, much effort has been invested in development and data-driven optimization of spatiotemporal filters that can extract the ERD/ERS features from noisy single-trial EEG measurements. Several other works as in [30, 31] have examined the suitability of P300 and steady state movement related potentials [32] for BCI.

However, as mentioned before, in many cases, such as those where there is a direct connection from the electrodes to the mechanical systems, the number of recording channels, that is, electrodes, is generally limited. In such cases EEG channel selection across various subjects has become another popular research area within the BCI community [33]. The general idea behind these approaches is based on a recursive channel elimination (RCE) criterion; channels that are well known to be important (from a physiological point of view) are consistently selected, whereas task-irrelevant channels are disregarded. The corresponding spatial patterns may be recognized using different techniques. Non-negative matrix factorisation [NMF] has been used to analyse and identify local spatiotemporal patterns of neural activity in the form of sparse basis vectors [34].

An ERP appears in response to some specific stimulus. The most widely used ERP evoked potential (EP) is the P300 signal, which can be auditory, visual, or somatosensory. It has a latency of approximately 300 ms and is elicited by rare or significant stimuli, and its amplitude is strongly related to the unpredictability of the stimulus; the more unforeseeable the stimulus, the higher the amplitude [35]. Another type of visual EP (VEP) is those for which the ERPs have short latency, representing the exogenous response of the brain to a rapid visual stimulus. They are characterized by a negative peak around 100 ms (N1) followed by a positive peak around 200 ms (P2). The ERPs can provide control when the BCI produces the appropriate stimuli. Therefore the BCI approach based on ERP detection from the scalp EEG seems to be easy since the cortical activities can be easily measured non-invasively and in real-time. Also, an ERP-based BCI needs little training for a new subject to gain control of the system. However, the information achieved through ERP extraction and measurement is not accurate enough for extraction of movement-related features and they have vast variability in different subjects with various brain abnormalities and disabilities. More importantly, the subject has to wait for the relevant stimulus presentation [36, 37].

There are also two other approaches towards BCI; one is based on steady-state visual-evoked responses (SSVERs), which are natural responses for visual stimulations at specific frequencies. These responses are elicited by a visual stimulus that is modulated at a fixed frequency. The SSVERs are characterized by an increase in EEG activity around the stimulus frequency. With feedback training, subjects learn to voluntarily control their SSVER amplitude. Changes in the SSVER result in control actions occurring over fixed intervals of time [37, 38]. The second approach is slow cortical potential shifts (SCPSs) that are shifts of cortical voltage, lasting from a few hundred milliseconds up to several seconds. Subjects can learn to generate slow cortical amplitude shifts in an electrically positive or negative direction for binary control. This can be achieved if the subjects are provided with feedback on the evolution of their SCP and if they are positively reinforced for correct responses [39]. However, in both of the above methods the subject has to wait for the brain stimulus.

Generally, some BCI approaches rely on the ability of the subjects to develop control of their own brain activity using biofeedback [40–42], whereas others utilise classification algorithms that recognise EEG patterns related to particular voluntary intentions [43]. Initial attempts to enable the subjects to use the feedback from their own brain activity began in the 1960s. The system enables the subjects to gain voluntary control over brain rhythms. It has been claimed that after training with an EEG biofeedback, human subjects are able to detect their own alpha [44,45] and mu rhythms [46]. This has also been tested on cats using their mu rhythms [47] and dogs [48] to control their hippocampal theta rhythm. Classification-based approaches have also been under research recently [43]. In a demonstration, subjects navigated through a virtual environment by imagining themselves walking [49]. These works paved the path for more research in this area.

Beverina *et al.* [50] have used P300 and steady-state visual-evoked potentials (SSVEPs). They have classified the ERP feature patterns using SVM. In their SSVEP approach they have used signals from the occipital electrodes (O_z, O_2, PO_8). The stimulations have been considered random in time instants, and a visual feedback [40] has been suggested for training purposes. In normal cases it is possible to amplify the brain waves through the feedback. In another work, SSVEPs have been used in a visually elaborate immersive 3D game [51]. The SSVEP generated in response to phase-reversing checkboard patterns is used for the proposed BCI.

There are many attractions in using the normal EEGs (or spontaneous signals) for BCI. A BCI system of this kind generates a control signal at given intervals of time based on the classification of EEG patterns resulting from particular mental activity (MA). In all types of BCI systems human factors such as boredom, fatigue, stress, or various illnesses are of great influence and, therefore, motivating the subject often becomes very important.

In terms of detection and classification of brain motor responses (real or imaginary), the common spatial patterns (CSP) method [52] has been shown to be a robust technique for movement-related EEG pattern classification. Some variants of this approach have been developed recently which will be discussed later in this chapter.

Although most of the recent BCI research has been focused upon scalp EEGs, some work on invasive EEGs has also been reported. Invasive BCI approaches are based on recordings from ensembles of single brain cells or on the activity of multiple neurons. They rely on the physiological properties of individual cortical and subcortical neurons or combination of neurons that modulate their movement-related activities. This work started in the 1960s and 1970s through some experiments by Fetz and his co-researchers [53–58]. In these experiments, monkeys learnt how to control the activity of their cortical neurons voluntarily with the help of biofeedback, which indicated the firing rate of single neurons. A few years later, Schmidt [59] showed that the voluntary motor commands could be extracted from raw cortical neural activity and used them to control a prosthetic device designed to restore motor functions in severely paralysed patients.

Most of the research on invasive BCI was carried out on monkeys. These studies relied on single cortical site recordings either of local field potentials [60–63], or from small samples of neurons or multiple brain zones [64–66]. They are mostly recorded in the primary motor cortex [64,65], although some work has been undertaken on the signals recorded from the posterior parietal cortex [67].

Normally, subdural electrodes for deep brain recording are not used for BCI. They are either used for seizure or as deep brain stimulators, which have been recently used for monitoring Parkinson's patients [68].

Philip Kennedy and his colleagues [69] have presented an impressive result from implanted cortical electrodes. In another study [4] a monkey managed to remotely control a robotic-arm using implanted cortical electrodes. These studies, however, require solutions to possible risk problems, advances in robust and reliable measurement technologies, and clinical competence. On the other hand, the disadvantage of using scalp recordings lies in the very low quality of the signals, due to attenuation of the electrical activity signals on their way to the electrodes and the effect of various noises.

In the following subsections a number of features used in BCI are explained. Initially, the changes in EEG before, during, and after the externally or internally paced events are observed. These events can be divided into two categories; first, the ERPs including evoked potentials and second, event-related desynchronization (ERD) and synchronization (ERS). The main difference between the two types is that the ERP is a stimulus-locked, or more generally, a phase-locked reaction, while the ERD/ERS is a non-phase-locked response. In a finger movement process, for example, often discussed in BCI, pre-movement negativity prominent prior to movement onset (readiness potential), and post-movement beta oscillations occurring immediately after movement-offset are, respectively, phase-locked (evoked) and non-phase-locked processes [70].

In recent research CSP have been used as a robust feature detection technique for classification of brain activity related to various body movements. Application of CSP is explained in the later sections of this chapter.

13.3 BCI-Related EEG Features

13.3.1 Readiness Potential and Its Detection

An early indication of movement can be realised from the so-called readiness potential, Bereitschaftspotential (BP), which is the German word for readiness potential, or pre-motor potential. This is a transient signal hump which appears just before the movement around the brain motor region. This was discovered by Helmut Kornhuber and Lüder Deecke at the University of Freiburg in Germany in 1964. The BP is much (10 to 100 times) smaller than the EEG alpha rhythm and it can be seen only by averaging, relating the electrical potentials to the onset of the movement. Figure 13.1 illustrates a typical BP together with the onset of voluntary hand (or finger) movement.

A time–frequency–space approach, such as that in [32], may be followed to detect and characterize the BP. This method uses PARAFAC-based tensor factorization [71] to detect the BP and differentiate between left- and right-hand (or finger) movements. Detection and tracking of BP over trials is important in monitoring the rehabilitation process in humans.

13.3.2 ERD and ERS

ERD is due to blocking of alpha activity just before and during the real or imagery (imaginary) movement. A simple measure of ERD is given as:

$$\text{ERD Level} = \frac{P(f, n) - P_{\text{ref}}(f)}{P_{\text{ref}}(f)} \tag{13.1}$$

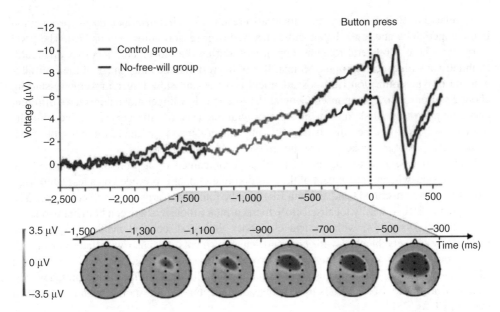

Figure 13.1 Readiness potential elicited around the finger movement time instant 0 (taken from http://neuroskeptic.blogspot.co.uk/)

where $P(f,n)$ is the value of a signal power at a given time–frequency point of an average power map, and $P_{ref}(f)$ is an average power during some reference time calculated for frequency f. This represents the level of rhythmic activity within the alpha band just before or during the movement. Any attention dramatically attenuates the alpha rhythms, while an increase in task complexity or attention results in an increased magnitude of ERD.

Increased cellular excitability in thalamo-cortical systems results in a low amplitude desynchronised EEG. So ERD may be due to the electrophysiological correlate of various activated cortical regions involved in processing sensory or cognitive information or production of motor reaction. Involvement of more neurons increases the ERD magnitude. In the BCI context, explicit learning of a movement sequence, for example, key pressing with different fingers, is accompanied by an enhancement of the ERD over the contralateral central regions. As the learning progresses and becomes more automatic the ERD decreases.

The cortical mu rhythm is of particular interest in BCI, mainly because it can be modulated/translated through imagery and can be monitored via a non-invasive technique.

The overall alpha band may be divided into lower and higher alphas. Lower alpha (6–10 Hz) is a response to any type of task and topographically is spread over almost all the electrodes. Higher alpha ERD, restricted to parieto-occipital areas, is found during visually presented stimulations.

The level of ERD is closely linked to semantic memory processes; those with good memory show a larger ERD in the lower alpha band [72].

In an auditory memory task, the absence of an ERD can be explained by the anatomical localization of the auditory cortex below the surface. Detection of the auditory ERD from the EEGs is often difficult.

As related to BCI, voluntary movement also results in a circumscribed desynchronization in the upper alpha and lower beta bands, localized over sensorimotor regions [73]. The ERD starts over the contralateral rolandic region and, during the movement, becomes bilaterally symmetrical with execution of movement. It is of interest that the time course of the contralateral mu desynchronization is almost identical to brisk and slow finger movement, starting about 2 s prior to movement onset. Generally, brisk and slow finger movements have different encoding processes. Brisk movement is pre-programmed and the afferents are delivered to the muscles as bursts. On the other hand, slow movement depends on the reafferent input from kinaesthetic receptors evoked by the movement itself.

Finger movement of the dominant hand is accompanied by a pronounced ERD in the ipsilateral side, whereas movement of the non-dominant finger is preceded by a less lateralized ERD [73]. Circumscribed hand area mu ERD can be found in nearly every subject, whereas, a foot area mu ERD is hardly localised close to the primary foot area between both hemispheres.

In another study [74] with cortical electrodes, it was discovered that mu rhythms are not only selectively blocked with arm and leg movements, but also with face movement. The ECoG captures more detailed signals from smaller cortical areas than the conventional EEG-based systems. These signals also contain low-amplitude high-frequency gamma waves. Consequently, ECoG-based BCIs have better accuracy and require shorter training time than those of EEGs [75].

In ERS, however, the amplitude enhancement is based on the cooperative or synchronised behaviour of a large number of neurons. In this case, the field potentials can be easily measured even using scalp electrodes. It is also interesting to know that approximately 85% of cortical neurons are excitatory, with the other 15% being inhibitory.

13.3.3   Transient Beta Activity after the Movement

This activity, also called post movement beta synchronization (PMBS) is another interesting robust event starting during the movement and continuing for about 600 ms [73]. It is found after finger or foot movement over both hemispheres without any significant bilateral coherence. The frequency band may vary from subject to subject; for finger movement the range is around 16–21 Hz [76] whereas for foot movement it is around 19–26 Hz [77]. The PMBS has similar amplitude for brisk and slow finger movements. This is interesting since brisk and slow movements involve different neural pathways. Moreover, this activity is significantly larger with hand as compared to finger movement [73]. Also, larger beta oscillations with wrist as compared to finger movement can be interpreted as the change of a larger population of motor cortex neurons from an increased neural discharge during the motor act to a state of cortical disfacilitation or cortical idling [73]. This means movement of more fingers results in a larger beta wave. Beta activity is also important in the generation of a grasp signal, since it has less overlap with other frequency components [78].

13.3.4   Gamma Band Oscillations

Oscillation of neural activity (ERS) within the gamma band (35–45 Hz) has also been of interest recently. Such activity is very obvious after visual stimuli or just before the movement task. This may act as the carrier for the alpha and lower beta oscillations, and relate to binding of sensory information and sensorimotor integration. Gamma, together with other activities in the above bands, can be observed around the same time after performing a movement task. Gamma ERS manifests itself just before the movement, whereas beta ERS occurs immediately after the event.

13.3.5   Long Delta Activity

Rather than other known ERS and ERD activities within alpha, beta, and gamma bands a long delta oscillation starts immediately after the finger movement and lasts for a few seconds. Although this has not been reported often in the literature, it can be a prominent feature in distinguishing between movement and non-movement states.

The main task in BCI is how to exploit the behaviour of the EEGs in the above frequency bands before, during, and after the imaginary movement, or after certain brain stimulation, in generation of the control signals. The following sections address this problem.

13.4   Major Problems in BCI

A simple and very popular BCI system set-up is illustrated in Figure 13.2. Feature extraction and classification of the features for each particular body movement is the main objective in most of the BCI systems.

As mentioned previously, the main problem in BCI is separating the control signals from the background EEG. Meanwhile, cortical connectivity, as an interesting identification of various task-related brain activities, has to be studied and exploited. Detection and evaluation of various features in different domains will then provide the control signals. To begin, however, the EEG

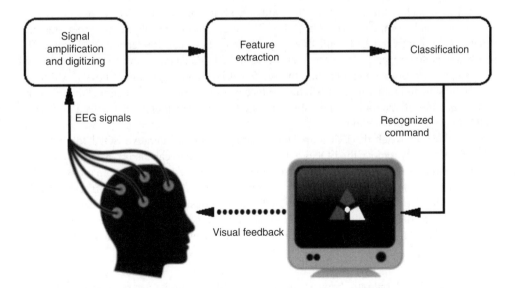

Figure 13.2   A typical BCI system using scalp EEGs when visual feedback is used

signals have to be pre-processed since the signals are naturally contaminated by various internal and external interferences. Pre-processing and conditioning the data (including noise and artefact removal) can further enhance the signals. In the case of BSS, for example, pre-whitening may also be necessary before implementation of the source separation algorithms.

13.4.1 Pre-Processing of the EEGs

In order to have an artifact-free EEG to extract the control signals, the EEGs have to be restored from the artefacts such as eye-blinking, electrocardiograms (ECGs), and any other internal or external disturbing effects.

Eye-blinking artefacts are very clear in both frontal and occipital EEG recordings. ECGs on the other hand can be seen more over the occipital electrodes. Many attempts have been made by different researchers to remove these artefacts.

Most of the noise and artefacts are filtered out by the hardware provided in new EEG machines. As probably the most dominant remaining artefact, interfering eye blinks (ocular artefact; OA) generate a signal within EEGs that is of the order of ten times larger in amplitude than cortical signals, and can last between 200 and 400 ms.

There have been some studies by researchers to remove OAs. Certain researchers have tried to estimate the propagation factors, as discussed in [79] based on regression techniques in both the time and frequency domains. In this attempt there is a need for a reference electrooculogram (EOG) channel during the EEG recordings.

PCA and SVMs have also been utilized for this purpose [80]. In these methods the EEGs and OAs are assumed statistically uncorrelated. Adaptive filtering has also been utilized [81]. This approach has considered the EEG signals individually and, therefore, ignored the mutual information amongst the EEG channels. ICA has also been used in some approaches. In these the EEG signals are separated into their constituent independent components (ICs) and the ICs are projected back to the EEGs using the estimated separating matrix after the artefact-related ICs are manually eliminated [82]. In [83] a BSS algorithm based on second-order statistics separates the combined EEG and EOG signals into statistically independent sources. The separation is then repeated for a second time with the EOG channels inverted. The estimated ICs in both rounds are compared, and those ICs with different signs are removed. Although, due to the sign ambiguity of the BSS the results cannot be justified, it is claimed that by using this method the artefacts are considerably mitigated. As noticed, there is also a need to separate EOG channels in this method.

In a robust approach the EEGs are first separated using an iterative SOBI following by SVM to effectively remove the EOG artefacts [84]. The method can also be easily extended to removal of the ECG artefacts. The proposed algorithm consists of BSS, automatic removal of the artefact ICs, and finally reprojection of the ICs to the scalp, providing artifact-free EEGs. This is depicted in Figure 13.3. Iterative SOBI as previously discussed has been effectively used to separate the ICs in the first stage. In the second stage only four features were carefully selected and used for classification of the normal brain rhythms from the EOGs. These features are as follows:

> *Feature* I: A large ratio between the peak amplitude and the variance of a signal suggests that there is an unusual value in the data. This is a typical identifier for the

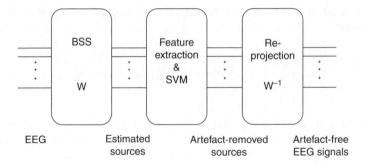

Figure 13.3   A hybrid BSS–SVM artefact removal system [82]

eye blink because it causes a large deflection on the EEG trace. This is described
mathematically as:

$$f_1 = \frac{\max\left(|\mathbf{u}_n|\right)}{\sigma_n^2} \quad \text{for} \quad n = 1, \ldots, N \tag{13.2}$$

where $\mathbf{u}_n$ is one of the N ICs, max(.) is a scalar valued function that returns
the maximum element in a vector, σ_n is the standard deviation of $\mathbf{u}_n$, and $|\,.\,|$
denotes absolute value. Normal EEG activity is tightly distributed about its mean.
Therefore, a low ratio is expected while the eye-blink signals manifest a large
value.

Feature II: This is a measure of skewness which is a third-order statistics of the
data, defined as:

$$f_2 = \left| \frac{E\left\{\mathbf{u}_n^3\right\}}{\sigma_n^3} \right| \quad \text{for} \quad n = 1, \ldots, N \tag{13.3}$$

for zero mean data. The EEG containing eye blink typically has a positive or
negative skewness since the eye-blinking signal has a considerably larger value
for this feature.

Feature III: The correlation between the ICs and the EEG signals from certain
electrodes is significantly higher than those of other ICs. The electrodes with
most contributed EOG are frontal electrodes FP_1, FP_2, F_3, F_4 and occipital lobe
electrodes O_1 and O_2 (in total, six electrode signals) The reference dataset, that is,
the EEG from the aforementioned electrodes, is distinct from the training and test
datasets. This will make the classification more robust by introducing a measure
of the spatial location of the eye-blinking artefact. Therefore, the third feature can

be an average of the correlations between the ICs and the signals from these six electrodes:

$$f_3 = \frac{1}{6} \sum_{i=1}^{6} \left(\left| E \left\{ x_i^0(n)u(n+\tau) \right\} \right| \right) \quad \text{for} \quad n = 1, \ldots, N \qquad (13.4)$$

where $x_i^0(n)$ are eye blinking reference signals, and i indexes each of the afore-mentioned electrode locations. The value of this feature will be larger for ICs containing an eye blinking artefact, since they will have a larger correlation for a particular value of τ in contrast to ICs containing normal EEG activity.

Feature IV: The statistical distance between distributions of the ICs and the electrode signals which are more likely to contain EOG is used. This can be measured using the Kullback–Leibler (KL) distance defined as:

$$f_4 = \int_{-\infty}^{\infty} p(u(n)) \ln \frac{p(u(n))}{p(x_{\text{ref}})} du(n) \quad \text{for} \quad n = 1, \ldots, N \qquad (13.5)$$

where $p(.)$ denotes the pdf. When the IC contains OA the KL distance between its pdf and the pdf of the reference IC will be approximately zero, whereas the KL distance to the pdf of a normal EEG signal will be larger.

An SVM with an RBF nonlinear kernel is then used to classify the ICs based on the above features. Up to 99% accuracy in detection of the EOG ICs has been reported [84].

After the artefact signals are marked, they will be set to zero. Then, all the estimated sources are re-projected to the scalp electrodes to reconstruct the artefact-free EEG signals.

The same idea has been used directly for extraction of the movement-related features [85] from the EEGs. In this work it is claimed that without any long-term training the decision as to whether there is any movement for a certain finger or not can be achieved by BSS followed by a classifier. A combination of a modified genetic algorithm (GA) and an SVM classifier has been used to condition and classify the selected features.

13.5   Multidimensional EEG Decomposition

All movement-related potentials are limited in duration and in frequency. In addition, each channel contains the spatial information of the EEG data. PCA and ICA have been widely used in decomposition of the EEG multiple sensor recordings. However, an efficient decomposition of the data requires incorporation of the space, time, and frequency dimensions.

Time–frequency (TF) analysis exploits variations in both time and frequency. Most of the brain signals are decomposable in the TF domain. This has been better described as sparsity of the EEG sources in the TF domain. In addition, TF domain features are much more descriptive of the neural activities. In [86], for example, the features from the subject-specific frequency bands have been determined and then classified using linear discriminant analysis (LDA).

In a more general approach the spatial information is also taken into account. This is because the majority of the events are localized in distinct brain regions. As a favourable approach, joint space–time–frequency classification of the EEGs has been studied for BCI applications [37, 87]. In this approach the EEG signals are measured with reference to *digitally linked ears* (DLE). DLE voltage can be easily found in terms of the left and right earlobes as

$$V_{\mathrm{e}}^{\mathrm{DLE}} = V_{\mathrm{e}} - \frac{1}{2}\left(V_{\mathrm{A}_1} + V_{\mathrm{A}_2}\right) \tag{13.6}$$

where V_{A_1} and V_{A_2} are, respectively, the left and right earlobe reference voltages. Therefore, the multivariate EEG signals are composed of the DLE signals of each electrode. The signals are multivariate since they are composed of the signals from multiple sources. A decomposition of the multivariate signals into univariate classifications has been carried out after the segments contaminated by eye blink artefacts are rejected [37].

There are many ways to write the general class of TF distributions for classification purposes [88]. In the above work the characteristic function (CF) $M(\theta, \tau)$ as in

$$C(t, \omega) = \frac{1}{4\pi^2} \int\limits_{\tau=-\infty}^{\infty} \int\limits_{0}^{2\pi} M(\theta, \tau) e^{-j\theta t - j\tau\omega} \, d\theta \, d\tau \tag{13.7}$$

for a single channel EEG signal, $x(t)$, assumed continuous time, (a discretized version can be used in practice) is defined as

$$M(\theta, \tau) = \phi(\theta, \tau) A(\theta, \tau) \tag{13.8}$$

where

$$
\begin{aligned}
A(\theta, \tau) &= \int_{-\infty}^{\infty} x^* \left(u - \frac{1}{2}\tau\right) x \left(u + \frac{1}{2}\tau\right) e^{j\theta u} \, du \\
&= \int_{0}^{2\pi} \hat{X}^* \left(\omega + \frac{1}{2}\theta\right) \hat{X} \left(\omega - \frac{1}{2}\theta\right) e^{j\tau\omega} \, d\omega
\end{aligned}
\tag{13.9}
$$

and $\hat{X}(\omega)$ is the Fourier transform of $x(t)$, which has been used for classification. This is a representative of the joint TF auto-correlation of $x(t)$. $\phi(\theta, \tau)$ is a kernel function which acts as a mask to enhance the regions in the TF domain so the signals to be classified are better discriminated. In [37] a binary function has been suggested as the mask.

In the context of EEGs, as multi-channel data, a multivariate system can be developed. Accordingly, the multivariate ambiguity function (MAF) of such a system is defined as

$$\mathbf{MA}(\theta, \tau) = \int_{-\infty}^{\infty} \mathbf{x}\left(t + \frac{\tau}{2}\right) \mathbf{x}^{\mathrm{H}}\left(t - \frac{\tau}{2}\right) e^{j\theta t} \, dt \tag{13.10}$$

where $(.)^H$ denotes conjugate transpose. This ambiguity function can also be written in a matrix form as:

$$\mathbf{MA}(\theta, \tau) = \begin{bmatrix} a_{11} & \cdots & a_{1N} \\ & \cdot & \\ & \cdot & \\ a_{N1} & \cdots & a_{NN} \end{bmatrix} \tag{13.11}$$

where

$$a_{ij} = \int_{-\infty}^{\infty} x_j^* \left(t - \frac{\tau}{2} \right) x_i \left(t + \frac{\tau}{2} \right) e^{j\theta t} dt \tag{13.12}$$

The diagonal terms are called auto-ambiguity functions and the off-diagonal terms are called cross-ambiguity functions. MAF can, therefore, be an indicator of the multivariate time–frequency–space autocorrelation of the corresponding multivariate system. The space dimension is taken into account by the cross-ambiguity functions.

13.5.1   Space–Time–Frequency Method

It is desirable to exploit the changes in EEG signals in time, frequency, and space (electrodes) at the same time. The disjointedness property of the brain sources may not be achievable in time, frequency or space separately due to the strong overlaps of the sources in each domain. However, in a multidimensional space the sources are more likely to be disjoint. Such a property may be exploited to separate and localize them. An early work in [89] represents this idea. The block diagram in Figure 13.4 illustrates the steps of the approach.

The above concept may be studied in the framework of tensor factorization where the signal variations in all possible dimensions can be considered at the same time. The PARAFAC and Tucker methods explained in previous chapters are the two main approaches. Here, application of PARAFAC to BCI particularly for artefact removal is discussed.

Another direction in BCI research is to evaluate the cortical connectivity and phase synchronization for characterization of continuous movement. The work on this area is, however, limited. Multivariate autoregressive (MVAR) modelling followed by directed transfer functions (DTFs) and evaluation of the diagonal and off-diagonal terms has been the main approach.

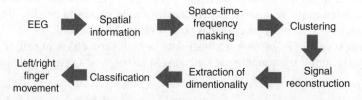

Figure 13.4   Classification of left/right finger movements using space–time–frequency decomposition

However, as comprehensively discussed in Chapter 8, there are other approaches for this evaluation. One application of the MVAR is discussed in a later section of this chapter.

The contrast of MAF is then enhanced using a multidimensional kernel function and the powers of cross-signals (cross-correlation spectra) are used for classification [37]. Using this method (as well as MVAR) the location of the event-related sources can be tracked and effectively used in BCI.

13.5.2 Parallel Factor Analysis

In this approach the events are considered sparse in the space–time–frequency domain and no assumption is made on either independence or uncorrelatedness of the sources. Therefore, the main advantage of PARAFAC over PCA or ICA is that uniqueness is ensured under mild conditions, making it unnecessary to impose orthogonality or statistical independence constraints. Harshman [90] was the first researcher to suggest that PARAFAC be used for EEG decomposition. Harshman, Carol, and Chang [91] independently proposed PARAFAC in 1970.

Möcks reinvented the model, naming it topographic component analysis, to analyse the ERP of channel × time × subjects [92]. The model was further developed by Field and Graupe [93]. Miwakeichi *et al.* eventually used PARAFAC to decompose the EEG data to its constituent space–time–frequency components [94]. In [87] PARAFAC has been used to decompose the wavelet transformed event-related EEG given by the inter-trial phase coherence. Figure 13.5 and 13.6 show, respectively, the space–time–frequency decomposition of the 15 channel EEG signals recorded during left and right index finger movement imaginations. Spectral contents, temporal profiles of the two identified factors, and the topographic mapping of EEG for the two factors are shown in these figures.

Accordingly, space–time–frequency features can be evaluated and used by a suitable classifier to distinguish between the left and right finger movements (or finger movement imagination) [95]. In an experiment the $I \times l \times K$ size $\mathbf{X}$ was formed by applying finite difference implementation of a spatial Laplacian filter [96] to 15 channels of the EEG and then transformed to the TF domain using complex Morelet's wavelets $w(n, f_0)$ as:

$$EEG_{\text{filtered}} = EEG_i(n) - \frac{1}{4}\sum_{l \in N_i} EEG_l \tag{13.13}$$

and

$$\mathbf{X}(n) = |w(n, f_0)EEG_{\text{filtered}}|^2 \tag{13.14}$$

The surface Laplacian filter may be considered as a spatial highpass filter. Combining PARAFAC with another signal processing modality, such as BSS or beamforming, can be used effectively for the removal of EEG artefacts. Figure 13.7 represents the results using PARAFAC combined with beamforming [97]. In this approach the solution to the beamforming problem is used as a spatial constraint into the PARAFAC.

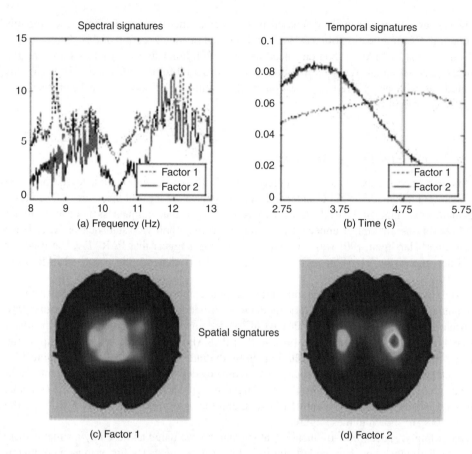

Figure 13.5 Sample space–time–frequency decomposition of the 15 channel EEG signal recorded during left index movement imagination. The factor demonstrated with the solid line indicates a clear ERD in the contralateral hemisphere: (a) spectral contents of the two identified factors, (b) temporal signatures of the factors, the onset of preparation and execution cues are shown in light and dark patches, respectively, (c) and (d) represent topographic mapping of EEG for the two factors (see Plate 14 for the coloured version)

13.6   Detection and Separation of ERP Signals

Utilization of the ERP signals provides another approach in BCI design. The ERP-based BCI systems often consider a small number of electrodes to study the movement-related potentials of certain body organs. However, in recent works multichannel EEGs have been used, followed by an efficient means of source separation algorithm in order to exploit the maximum amount of information within the recorded signals. Since the major problems in this context are related to detection, separation and classification of the ERP signals a separate chapter (Chapter 9) has been dedicated to that. As stated previously, these systems are initiated by introducing certain stimulations of the brain. As soon as the movement-related ERP components are classified the system can be used in the same way as in the previous sections. In single trial applications the ERP components can be tracked in order to evaluate the state of the brain during BCI.

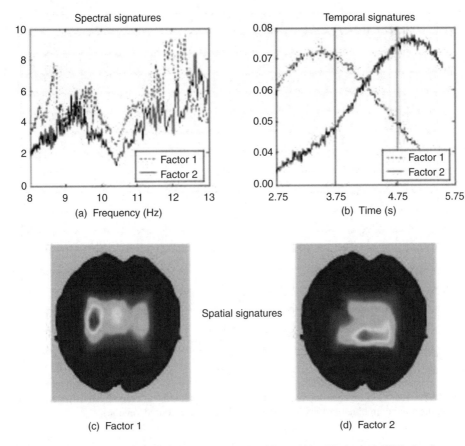

Figure 13.6 Sample space–time–frequency decomposition of the 15 channel EEG signal recorded during right index movement imagination. The factor demonstrated with the solid line indicates a clear ERD in the contralateral hemisphere: (a) spectral contents of the two identified factors, (b) temporal signatures of the factors, the onset of preparation and execution cues are shown in light and dark patches, respectively, (c) and (d) show topographic mapping of EEG for the two factors

13.7   Estimation of Cortical Connectivity

The planning and the execution of voluntary movements are related to the pre-movement attenuation and post-movement increase in amplitude of alpha and beta rhythms in certain areas of the motor and sensory cortex [98, 99]. Also it has been found that during movement planning, two rhythmical components in the alpha frequency range, namely mu1 and mu2, play different functional roles. Differentiation of these two components may be achieved by using the matching pursuit (MP) algorithm [100] based on the signal energy in the two bands [101, 102].

The MP algorithm refers to the decomposition of signals into basic waveforms from a very large and redundant dictionary of functions. MP has been utilised for many applications, such as epileptic seizure detection, evaluation, and classification by many researchers.

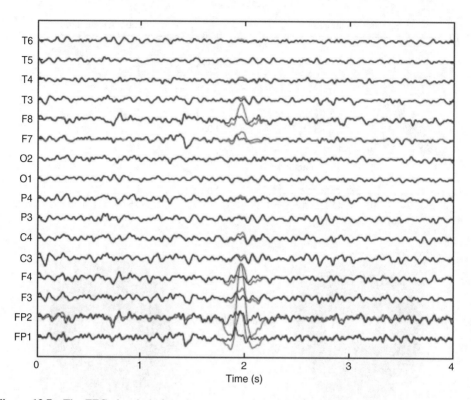

Figure 13.7 The EEG signals before the removal of eye-blinking artefacts in red (pale) and after removing the artefact in blue (dark) colour (see Plate 15 for the coloured version)

Determination of propagation of brain electrical activity, its direction, and the frequency content, is of great importance. Directed transfer functions (DTFs) using a multivariate autoregressive (MVAR) model have been employed for this purpose [103]. In this approach the signals from all EEG channels are treated as realizations of a multivariate stochastic process. A short-time DTF (SDTF) was also developed [102] for estimation and evaluation of AR coefficients for short-time epochs of the EEGs.

MP and SDTF have been performed for analysis of the EEG activity during planning of self-paced finger movements, and we will discuss the results with respect to representation of the features of cortical activities during voluntary action.

The MP has been applied to decomposition of the EEGs to obtain reference-free data, and the averaged maps of power were constructed. ERP/ERS were calculated as described in Section 13.3.2.

Model order is normally selected at the point where the model error does not decrease considerably. A well known criterion called Akaike AIC [104] has been widely used for this purpose. According to the AIC, the correct model order will be the value of p which minimizes the following criterion:

$$\text{AIC}(p) = 2 \log\left(\det(\mathbf{V})\right) + \frac{2kp}{N} \qquad (13.15)$$

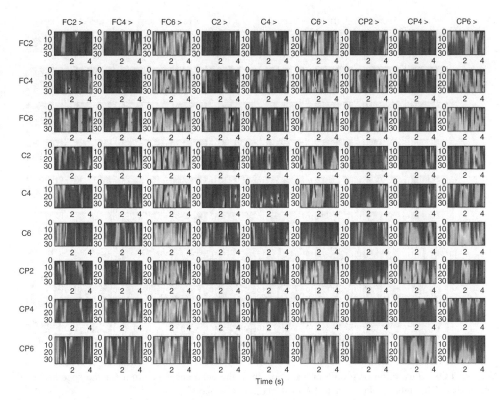

Figure 13.8 Illustration of source propagation from the coherency spectrum for the specified EEG channels for the left hand (see Plate 16 for the coloured version)

where **V** is the variance matrix of the model noise, N is the data length for each channel, and k is the number of channels.

The transitions in the DTF patterns can be illustrated for different EEG channels for left and right hands finger movements as depicted in Figure 13.8 and 13.9, respectively.

The direction of the signal source movement is realised from the cross-correlations between signals, which are computed for different time shifts in the procedure of correlation $\mathbf{R}(n)$ matrix estimation. These time shifts are translated into phase shifts by transformation to the frequency domain. The phase dependences between channels are reflected in the transfer matrix. The DTF values express the direction of a signal component in a certain frequency (not the amount of delay) [105]. Analysis of the DTF values, however, will be difficult when the number of channels increases, resulting in an increase in the number of MVAR coefficients.

The coherency of brain activities, as described in the previous section, may be presented from a different perspective, namely, brain connectivity. This concept plays a central role in neuroscience. Temporal coherence between the activities of different brain areas is often defined as functional connectivity, whereas the effective connectivity is defined as the simplest brain circuit that would produce the same temporal relationship as observed experimentally between cortical regions [106]. A number of approaches have been proposed to estimate how

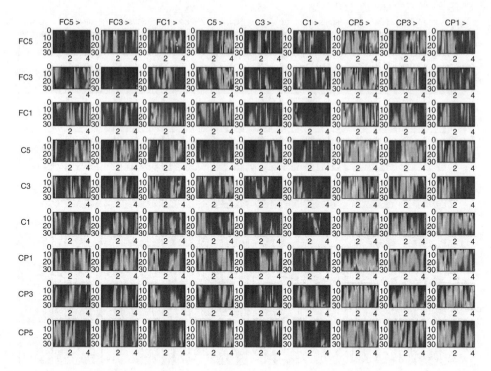

Figure 13.9 Illustration of source propagation from the coherency spectrum for the specified EEG channels for the right hand (see Plate 17 for the coloured version)

different brain areas are working together during motor and cognitive tasks from the EEG and fMRI data [107–109].

Structural equation modelling (SEM) [110], explained comprehensively in Chapter 8, has also been used to model such activities from high resolution (both spatial and temporal) EEG data. Anatomical and physiological constraints have been exploited to change an underdetermined set of equations to a determined one.

13.8 Application of Common Spatial Patterns

As stated in Chapter 6, CSPs are probably the most effective feature selection algorithms used in BCI for feature selection and classification. In early 2000 in a 2-class BCI set-up, Ramoser *et al.* [52] proposed application of common spatial patterns that learned to maximize the variance of bandpass filtered EEG signals from one class while minimizing their variance from the other class. Formally, the CSP (**w**) minimizes the Rayleigh quotient of the spatial covariance matrices to achieve the variance imbalance between the two classes of data $\mathbf{X}_1$ and $\mathbf{X}_2$, and is defined using

$$J(\mathbf{w}) = \frac{\mathbf{w}^T \mathbf{X}_1^T \mathbf{X}_1 \mathbf{w}}{\mathbf{w}^T \mathbf{X}_2^T \mathbf{X}_2 \mathbf{w}} = \frac{\mathbf{w}^T \mathbf{C}_1 \mathbf{w}}{\mathbf{w}^T \mathbf{C}_2 \mathbf{w}} \tag{13.16}$$

where T denotes transpose, $\mathbf{X}_i$ is the data matrix for the ith class (with the training samples as rows and the channels as columns) and $\mathbf{C}_i$ is the spatial covariance matrix of ith class signals, assuming a zero mean for EEG signals. As was stated in Chapter 6, the CSP problem is often solved by the generalized eigenvalue equation

$$\mathbf{C}_1\mathbf{w} = \lambda\mathbf{C}_2\mathbf{w} \tag{13.17}$$

or $C_2^{-1}\mathbf{C}_1\mathbf{w} = \lambda\mathbf{w}$. In detection and recognition of 2-class patterns in BCI the CSP is widely used as an effective approach. Most of the existing CSP-based methods exploit covariance matrices on a subject-by-subject basis so that inter-subject information is neglected. CSP and its variants have received much attention and have been one of the most efficient feature extraction methods for BCI. However, despite its straightforward mathematics, CSP overfits the data and is highly sensitive to noise. To address these shortcomings, recently it has been proposed to improve the CSP learning process with prior information in terms of regularization terms. Lotte *et al.* [111] reviewed, categorized and compared 11 different regularized CSP approaches: from regularization in the estimation of the EEG covariance matrix [112] to several different regularization methods such as composite CSP [113], regularized CSP with generic learning [112], regularized CSP with diagonal loading [113] and invariant CSP [114]. They applied these methods to the EEGs recorded from 17 patients and verified the superiority of CSP with Tikhonov regularization in which the optimization of $J(\mathbf{w})$ is penalized by minimizing $\|\mathbf{w}\|^2$, hence, minimizing the influence of artefacts and outliers. Regularization of $\mathbf{w}$ can also be used to reduce the number of EEG channels without compromising the classification score. Farquhar *et al.* [115] converted CSP into a quadratically constrained quadratic optimization problem with -norm penalty and Arvaneh *et al.* [116] used l_1/l_2 -norm constraint. Recently, in [112] a computationally expensive quasi -norm-based principle has been applied to achieve a sparse solution for $\mathbf{w}$.

The topography images in Figure 13.10 demonstrate the effectiveness of applying CSPs to distinguish between left and right hand movements from EEG recordings.

In [117] the CSP has been modified for subject-to-subject transfer, where a linear combination of covariance matrices of subjects under consideration has been exploited. In this

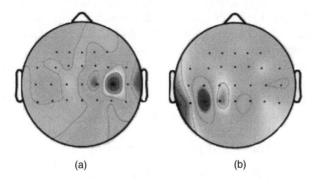

(a)   (b)

Figure 13.10   The results of applying CSP to classify the cortical activity of the brain for both (a) left and (b) right hand movements

approach a composite covariance matrix has been used that is a weighted sum of covariance matrices involving subjects, leading to composite CSP.

To alleviate the effect of noise the objective function in (13.17) may be regularized either during the estimation of covariance matrix for each class or during the minimization process of the CSP cost function by imposing priors to the spatial filters $\mathbf{w}$ [111].

The proposed method in [112] for estimating the regularized covariance matrix for class i $\hat{\mathbf{C}}_i$ is given as:

$$\hat{\mathbf{C}}_i = (1 - \gamma)\mathbf{P} + \gamma\mathbf{I} \tag{13.18}$$

where

$$\mathbf{P} = (1 - \beta)\mathbf{C}_i + \beta\mathbf{G}_i \tag{13.19}$$

In these equations $\mathbf{C}_i$ is the initial spatial covariance matrix, $\mathbf{G}_i$ is the so-called generic covariance matrix, and γ and $\beta \in [0,1]$ are the regularizing parameters. $\mathbf{G}_i$ is estimated empirically by averaging the covariance matrices of a number of trials for class i.

In regularizing the CSP objective function on the other hand, a regularization term is added to the CSP objective function in order to penalize the resulting spatial filters that do not satisfy a given prior. This results in a slightly different objective function as [111]:

$$J(\mathbf{w}) = \frac{\mathbf{w}^T\mathbf{C}_{X_1}\mathbf{w}}{\mathbf{w}^T\mathbf{C}_{X_2}\mathbf{w} + \alpha\mathbf{Q}(\mathbf{w})} \tag{13.20}$$

where the penalty function $\mathbf{Q}(\mathbf{w})$ weighted by the penalty parameter $\alpha \geq 0$ indicates how much the spatial filter $\mathbf{w}$ satisfies a given prior. This term needs to be minimized in order to maximize $J(\mathbf{w})$. Following the discussion in Chapter 6, different quadratic and non-quadratic penalty functions have been defined in the literature, such as those in [111, 115], and [118]. The results achieved in implementing 11 different regularization methods in [111] shows that in places where the data is noisy, the regularization improves the results by approximately 3% on average. However, the best algorithm outperforms CSP by about 3 to 4% in mean classification accuracy and by almost 10% in median classification accuracy. The regularized methods are also more robust, that is, they show lower variance across both classes and subjects. Amongst various approaches the weighted Tikhonov regularized CSP proposed in [111] has been reported to have the best performance.

Often not all the EEG channels are used in CSPs. In practical BCI applications the data recorded from the centro-parietal scalp sites, as highlighted in Figure 13.11, are initially bandpass filtered between 8 and 30 Hz before implementation of the CSP filter.

13.9   Multiclass Brain–Computer Interfacing

Often a good model for performing a full BCI task or detection of various body movements becomes useful and important. This cannot be achieved if a robust multiclass classifier is not in place. Most cases in the literature refer to two-class problems which can be solved using

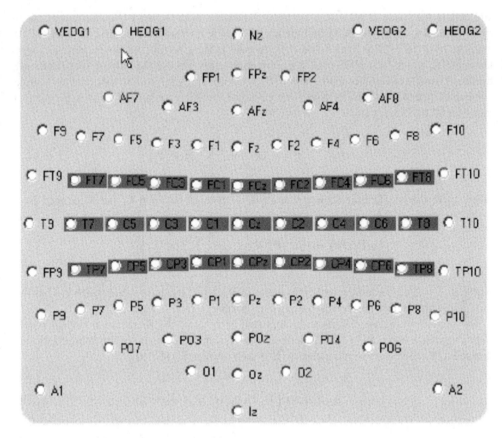

Figure 13.11 The electrodes highlighted by dark grey are those practically used in application of CSPs (as in http://www.gtec.at/)

efficient classifiers, such as SVM. For general applications multiclass classifiers such as neural networks (NNs) have always been an option. The efficiency or computational complexity of such algorithms has usually been under question.

Two new classification approaches for BCI have been introduced in [119]. One of these classifiers is based on the distance to the Riemannian mean and the other one works in the Riemannian tangent space. Obviously, the popular classification algorithms, such as LDA, SVM, and NNs, cannot be implemented directly in the Riemannian manifold since they are based on projections into hyperplanes.

In this BCI approach consider short-time segments each including T_s sample of EEG signal or trials in the form of a matrix $\mathbf{X}_i = [\mathbf{x}_{t+T_i} \ldots \mathbf{x}_{t+T_i+T_s-1}] \in \Re^{n \times T_s}$ which corresponds to the ith trial of real or imagined movement started at time T_i. Define $\mathbf{P}_i$ as the $n \times n$ sample covariance matrix as

$$\mathbf{P}_i = \frac{1}{T_s - 1} \mathbf{X}_i \mathbf{X}_i^{\mathrm{T}} \tag{13.21}$$

and denote the space of all $n \times n$ symmetric matrices in the space of square real matrices by $S(n) = \{\mathbf{S} \in M(n), \mathbf{S}^T = \mathbf{S}\}$ and the set of all $n \times n$ symmetric positive-definite (SPD) matrices by $P(n) = \{\mathbf{P} \in S(n), \mathbf{u}^T\mathbf{Pu} > 0, \forall \mathbf{u} \in \Re^n\}$. Then, to find the closest class to a new test input $\mathbf{X}$, which is an EEG trial of the unknown class, we use the training set to compute the above sample covariance matrix to find $\mathbf{P}_i$. Then, we apply this to X to calculate the sample covariance matrix $\mathbf{P}$. In the next step we calculate the Riemannian (also called geometric) mean for all the classes $k = 1, \ldots, K$ as [119]:

$$\vartheta(\mathbf{P}_1, \ldots, \mathbf{P}_I) = \arg \min_{\mathbf{P} \in P(n)} \sum_{i=1}^{I} \delta_R^2(\mathbf{P}, \mathbf{P}_i) \tag{13.22}$$

where $\delta_R(\mathrm{P}_1, \mathrm{P}_2)$ is called the Riemannian geodesic distance between $\mathbf{X}_1$ and $\mathbf{X}_2$ defined as:

$$\delta_R(\mathbf{P}_1, \mathbf{P}_2) = \left\| \log \left(\mathbf{P}_1^{-1}\mathbf{P}_2 \right) \right\|_F = \left[\sum_{i=1}^{n} \log^2 \lambda_i \right]^{1/2} \tag{13.23}$$

where $\lambda_i, i = 1, \ldots, n$ are the real eigenvalues of $\mathbf{P}_1^{-1}\mathbf{P}_2$. Class $\hat{k}$ for which $\delta_R(\mathbf{P}, \mathbf{P}_{\hat{k}})$ has the minimum value will be the corresponding class for $\mathbf{X}$.

The second multiclass classification approach is by using the Riemannian tangent space. In this approach equation (13.22) is applied first. Assuming the result of this minimization is denoted as $\mathbf{P}_\vartheta$ the corresponding class to $\mathbf{X}$ is then estimated as [119]:

$$\mathbf{s}_i = \mathrm{upper} \left(\mathbf{P}_\vartheta^{-\frac{1}{2}} \log_{\mathbf{P}_\vartheta}(\mathbf{P}_i)\mathbf{P}_\vartheta^{-\frac{1}{2}} \right) \tag{13.24}$$

where upper(.) operator keeps the upper triangular part of a symmetric matrix, vectorizing it by applying unity weight for diagonal elements and $\sqrt{2}$ weight for off-diagonal elements.

The above classification methods have been applied to a set of BCI competition IV data [120] from nine subjects who perform four kinds of motor imagery (right hand, left hand, foot, and tongue) movements. The results were compared with those of multiclass CSP (hierarchical CSP) and LDA. The overall performance of the above proposed system was proved to be superior to CSP and LDA.

13.10   Cell-Cultured BCI

Most of the information about this type of BCI comes from the public media and university websites. Researchers have built devices to interface with neural cells and entire neural networks in cultures outside animals. Neurochips powered by neuroelectronics have been developed to enable stimulating, sampling and recording from neurons directly [121].

A *neurochip* is a chip (integrated circuit/microprocessor) that is designed for the interaction with neuronal cells. It is made of silicon that is doped in such a way, that it contains EOSFETs (electrolyte-oxide-semiconductor field effect transistors) that can sense the electrical activity of the neurons (action potentials) in the above-standing physiological electrolyte solution. It also contains capacitors for the electrical stimulation of these cells.

The world's first neurochip with sixteen neurons connected was developed by Pine and Maher in 1997 [122]. Based on research by Theodore Berger in 2003, a neurochip was proposed to function as an artificial or prosthetic hippocampus. The chip was designed to function in rat brains and was intended as a prototype (an ancestor) for the eventual development of higher-brain prosthesis. The hippocampus was chosen because it is the origin of various activations and is the most ordered and structured part of the brain and has been well studied. Its main function is to encode experiences for archiving information as long-term memories in the brain [123].

Thomas DeMarse from the University of Florida used a culture of 25 000 neurons taken from a rat's brain to fly an F-22 fighter jet simulator [124]. After collection, the cortical neurons were cultured. The neurons rapidly began to reconnect themselves to form a living neural network. The cells were arranged over a grid of 60 electrodes and used to control the pitch and yaw functions of the simulator. This study was mainly to understand and demonstrate how the human brain performs and learns computational tasks at a cellular level.

13.11   Conclusions

Although EMG is still used widely for both BCI and robotics aiming at rehabilitation, BCI using EEG has been increasing during the last three decades. BCI using cortical electrodes is more effective but requires invasive operation. A review of the ongoing research has been provided in this chapter. Static features measured in different EEG conventional frequency bands have been widely used in the classification of finger, arm, and leg movements. Dynamic features, such as those characterising the motion of movement-related sources have also been considered recently. Finally, estimation of the cortical connectivity patterns provides a new tool in evaluation of the directivity of brain signals and localization of the movement-related sources.

The advances in signal processing, especially in detection, separation, and classification of brain signals, have led to very exciting results in BCI, however, as yet, not all physiological, anatomical, and functional constraints have been taken into account. Also, no gold standard has yet been defined for BCI-based application of BCI, particularly to paralysed subjects. The effectiveness of the solution depends on the type and the level of subject disabilities. Moreover, there has not been any attempt to provide BCIs for subjects suffering from mental disorders. However, in practice, common spatial patterns and their regularized variants show sufficient robustness and accuracy for the majority of the brain two-class classifications. New multiclass approaches, such as in [121], are useful when more than two motor activities are to be separated.

Often, EEG patterns change with time. Visual [40, 41], auditory [42], and other types of feedback BCI systems [125] seem to provide more robust solutions in the presence of these changes. Development of on-line adaptive BCI systems, such as that in [15] enhances the application of BCI in various areas. A complete feedback system, however, requires more thought and research to be undertaken.

Generally, to achieve a clinically useful BCI (invasive or non-invasive) system stable, low noise, and long recordings from multiple brain regions/electrodes are necessary. In addition, more computationally efficient algorithms have to be developed in order to cope with the real-time applications.

Combining BCI with virtual reality helps the subjects learn how to use brain plasticity better to incorporate prosthetic devices into the body representation. This will make the prosthetic feel like a natural part of the body of the subject.

Neurofeedback directly through visual, audio, and somatosensory stimulation and indirectly through a processor to analyse the brain function after each response to a stimulus can be part of future development in BCI.

References

[1] Wolpaw, J.R., Birbaumer, N., Heetderks, W.J. *et al.* (2000) Brain-computer interface technology: a review of the first international meeting. *IEEE Trans. Rehabil Eng.*, **8**(2), 164–73.

[2] Vidal, J.J. (1973) Direct brain-computer communication. *Ann. Rev. Biophys. Bioeng.*, **2**, 157–158.

[3] Vidal, J.J. (1977) Real-time detection of brain events in EEG. *IEEE Proc.*, **65**, 633–664.

[4] Chapin, J.K., Moxon, K.A., Markowitz, R.S. and Nicolelis, M.A. (1999) Real-time control of a robot arm using simultaneous recorded neurons in the motor cortex. *Nat. Neurosci.*, **2**, 664–670.

[5] Lebedev, M.A. and Nicolelis, M.A.L. (2006) Brain-machine interfaces: past, present and future. *Trends Neurosci.*, **29**(9), 536–546.

[6] Blankertz, B., Dorhege, D., Krauledat, M. *et al.* (2006) The Berlin brain computer interface: EEG-based communication without subject training. *IEEE Trans. Neur. Sys. Rehab. Eng.*, **14**, 147–152.

[7] Blankertz, B., Dornheg, G., Schäfer, C. *et al.* (2003) Boosting bit rates and error detection for the classification of fast-paced motor commands based on single-trial EEG analysis. *IEEE Trans. Neur. Sys. Rehab. Eng.*, **11**, 127–131.

[8] Blankertz, B., Müller, K.-R., Curio, G. *et al.* (2004) The BCI competition 2003: progress and perspectives in detection and discrimination of EEG single trials. *IEEE Trans. Biomed. Eng.*, **51**, 1044–1051.

[9] Wolpaw, J.R., McFarland, D.J. and Vaughan, T.M. (2000) Brain-computer interface research at the Wadsworth Centre. *IEEE Trans. Neur. Sys. Rehab. Eng.*, **8**, 222–226.

[10] Wolpaw, J.R. and McFarland, D.J. (2003) Control of two-dimensional movement signal by a non-invasive brain-computer interface in human. *Natl. Acad. Sci. USA*, **101**, 17849–17854.

[11] Peters, B.O., Pfurtscheller, G. and Flyvbjerg, H. (2001) Automatic differentiation of multichannel EEG signals. *IEEE Trans. Biomed. Eng.*, **48**, 111–116.

[12] Müller-Putz, G.R., Neuper, C., Rupp, R. *et al.* (2003) Event-related beta EEG changes during wrist movements induced by functional electrical stimulation of forearm muscles in man. *Neurosci. Lett.*, **340**, 143–147.

[13] Müller-Putz, G.R., Scherer, R., Pfurtscheller, G. and Rupp, R. (2005) EEG-based neuroprosthesis control: a step towards clinical practice. *Neurosci. Lett.*, **382**, 169–174.

[14] Pfurtscheller, G., Muller, G.R., Pfurtscheller, J. *et al.* (2003) Thought' - control of functional electrical stimulation to restore hand grasp in a patient with tetraplegia. *Neurosci. Lett.*, **351**, 33–36.

[15] Vidaurre, C., Schogl, A., Cabeza, R. *et al.* (2006) A fully on-line adaptive BCI. *IEEE Trans. Biomed. Eng.*, **53**(6), 1214–1219.

[16] Millan, J.D.R., Mourino, J., Franze, M. *et al.* (2002) A local neural classifier for the recognition of EEG patterns associated to mental tasks. *IEEE Neur. Net.*, **13**, 678–686.

[17] Millan, J.D.R. and Mourino, J. (2003) Asynchronous BCI and local neural classifiers: an overview of the adaptive brain interface project. *IEEE Trans. Neur. Sys. Rehab. Eng.*, **11**, 1214–1219.

[18] Millan, J.D., Renkens, F., Mourino, J. and Gerstner, W. (2004) Noninvasive brain-actuated control of a mobile robot by human EEG. *IEEE Trans. Biomed. Eng.*, **53**, 1214–1219.

[19] Birbaumer, N., Ghanayim, N., Hinterberger, T. *et al.* (1999) A spelling device for the paralysed. *Nature*, **398**, 297–298.

[20] Wolpaw, J.R., McFarland, D. and Pfurtscheller, G. (2002) Brain computer interfaces for communication and control. *Clin. Neurophysiol.*, **113**(6), 767–791.

[21] Hinterberger, T., Veit, R., Wilhelm, B. *et al.* (2005) Neuronal mechanisms underlying control of a brain-computer interface. *Eur. J. Neurosci.*, **21**, 3169–3181.

[22] Kubler, A., Kotchoubey, B., Kaiser, J. *et al.* (2001) Brain-computer communication: unlocking the loked-in. *Psychol. Bull.*, **127**, 358–375.

[23] Kubler, A., Kotchoubey, B., Kaiser, J. *et al.* (2001) Brain-computer communication: self regulation of slow cortical potentials for verbal communication. *Arch. Phys. Med. Rehabil.*, **82**, 1533–1539.

[24] Obermaier, B., Muller, G.R. and Pfurtscheller, G. (2003) Virtual keyboard controlled by spontaneous EEG activity. *IEEE Trans. Neural Syst. Rehabil. Eng.*, **11**, 422–426.

[25] Obermaier, B., Muller, G.R. and Pfurtscheller, G. (2001) Information transfer rate in a five-classes brain-computer interface. *IEEE Trans. Neural Syst. Rehabil. Eng.*, **9**, 283–288.

[26] Wolpow, J.R. (2004) Brain computer interfaces (BCIs) for communication and control: a mini review. *Suppl. Clin. Neurophysiol.*, **57**, 607–613.

[27] Birbaumer, N., Weber, C., Neuper, C. *et al.* (2006) Brain-computer interface research: coming of age. *Clin. Neurophysiol.*, **117**, 479–483.

[28] Kauhanen, L., Palomaki, T., Jylanki, P. *et al.* (2006) Haptic feedback compared with visual feedback for BCI. Proceedings of the 3rd International Brain-Computer Interface Workshop & Training Course 2006, Graz, Austria.

[29] Burdet, E., Sanguineti, V., Heuer, H. and Popovic, D.B. (2012) Motor skill learning and neuro-rehabilitation. Guest Editorial; *IEEE Trans. Neural Syst. Rehabil. Eng.*, **20**(3), 237–238.

[30] Salvaris, M., Cinel, C., Citi, L. and Poli, R. (2012) Novel protocols for P300-based brain–computer interfaces. *IEEE Trans. Neural Sys. Rehab. Eng.*, **20**(1), 8–17.

[31] Li, Y., Long, J., Yu, T. *et al.* (2010) An EEG-based BCI system for 2-D cursor control by combining Mu/Beta rhythm and P300 potential. *IEEE Trans. Biomed. Eng.*, **57**(10), 2495–2505.

[32] Nazarpour, K., Praamstra, P., Miall, R.C. and Sanei, S. (2009) Steady-state movement related potentials for brain–computer interfacing. *IEEE Trans. Biomed. Eng.*, **56**(8), 2104–2113.

[33] Schröder, M.I., Lal, T.N., Hinterberger, T. *et al.* (2005) Robust EEG Channel Selection across Subjects for Brain-Computer Interfaces. *EURASIP J. Appl. Signal Process.*, **2005**(19), 3103–3112.

[34] Kim, S-P., Rao, Y.N., Erdogmus, D. *et al.* (2005) Determining patterns in neural activity for reaching movements using nonnegative matrix factorization. *EURASIP J. Appl. Signal Process.*, **2005**(19), 3113–3121.

[35] Donchin, E., Spencer, K.M. and Wijesinghe, R. (2000) The mental prosthesis: assessing the speed of a P300-based brain-computer interface. *IEEE Trans. Rehab. Eng.*, **8**, 174–179.

[36] Bayliss, J.D. (2001) A Flexible Brain-Computer Interface. PhD thesis, University of Rochester, NY, USA.

[37] Molina, G.A., Ebrahimi, T. and Vesin, J-M. (2003) Joint time-frequency-space classification of EEG in a brain computer interface application. *EURASIP J. Appl. Signal Process.*, **7**, 713–729.

[38] Middendorf, M., Mc Millan, G., Calhoun, G. and Jones, K.S. (2000) Brain-computer interfaces based on the steady-state visual-evoked response. *IEEE Trans. Rehab. Eng.*, **8**(2), 211–214.

[39] Kübler, A., Kotchubey, B. and Salzmann, H.P. (1998) Self regulation of slow cortical potentials in completely paralysed human patients. *Neurosci. Lett.*, **252**(3), 171–174.

[40] McFarland, D.J., McCane, L.M. and Wolpaw, J.R. (1998) EEG based communication and control: short-term role of feedback. *IEEE Trans. Rehab. Eng.*, **7**(1), 7–11.

[41] Neuper, C., Schlogl, A. and Phertscheller, G. (1999) Enhancement of left-right sensorimotor imagery. *J. Clin. Neurophysiol.*, **4**, 373–382.

[42] Hinterberger, T., Neumann, N., Pham, M. *et al.* (2004) A multimodal brain-based feedback and communication system. *Exp. Brain Res.*, **154**(4), 521–526.

[43] Bayliss, J.D. and Ballard, D.H. (2000) A virtual reality testbed for brain brain-computer interface research. *IEEE Trans. Rehabil. Eng.*, **8**, 188–190.

[44] Nowlis, D.P. and Kamiya, J. (1970) The control of electroencephalographic alpha rhythms through auditory feedback and the associated mental activity. *Psychophysiology*, **6**, 476–484.

[45] Plotkin, W.B. (1976) On the self-regulation of the occipital alpha rhythm: control strategies, states of consciousness, and the role of physiological feedback. *J. Exp. Psychol. Gen.*, **105**, 66–99.

[46] Sterman, M.B., Macdonald, L.R. and Stone, R.K. (1974) Biofeedback training of the sensorimotor electroencephalogram rhythm in man: effects on epilepsy. *Epilepsia*, **15**, 395–416.

[47] Whyricka, W. and Sterman, M. (1968) Instrumental conditioning of sensorimotor cortex spindles in the walking cat. *Psychol. Behav.*, **3**, 703–707.

[48] Black, A.H. (1971) The direct control of neural processes by reward and punishment. *Am. Sci.*, **59**, 236–245.

[49] Pfurtscheller, G., Leeb, R., Keinrath, C. *et al.* (2006) Walking from thought. *Brain Res.*, **1071**, 145–152.

[50] Beverina, F., Palmas, G., Silvoni, S. *et al.* (2003) User adaptive BCIs: SSVEP and P300 based interfaces. *PsychNology J.*, **1**(4), 331–354.

[51] Lalor, E.C., Kelly, S.P., Finucane, C. *et al.* (2005) Steady-state VEP-based brain-computer interface control in an immersive 3D gaming environment. *EURASIP J. Appl. Signal Process.*, **2005**(19) 3156–3164.

[52] Ramoser, H., Muller-Gerking, J. and Pfurtscheller, G. (2000) Optimal spatial filtering of single trial EEG during imagined hand movement. *IEEE Trans Rehab. Eng.*, **8**(4), 441–446.

[53] Fetz, E.E. (1969) Operant conditioning of cortical unit activity. *Science*, **163**, 955–958.

[54] Fetz, E.E. (1972) Are movement parameters recognizably coded in activity of single neurons?. *Behav. Brain Sci.*, **15**, 679–690.

[55] Fetz, E.E. and Baker, M.A. (1973) Operantly conditioned patterns on precentral unit activity and correlated responses in adjacent cells and contralateral muscles. *J. Neurophysiol.*, **36**, 179–204.

[56] Fetz, E.E. and Finocchio, D.V. (1971) Operant conditioning of specific patterns of neural and muscular activity. *Science*, **174**, 431–435.

[57] Fetz, E.E. and Finocchio, D.V. (1972) Operant conditioning of isolated activity in specific muscles and precentral cells. *Brain Res.*, **40**, 19–23.

[58] Fetz, E.E. (1975) Correlations between activity of motor cortex cells and arm muscles during operantly conditioned response patterns. *Exp. Brain Res.*, **23**, 217–240.

[59] Schmidt, E.M. (1980) Single neuron recording from motor cortex as a possible source of signals for control of external devices. *Ann. Biomed. Eng.*, **8**, 339–349.

[60] Mehring, C., Rickert, J., Vaadia, E. *et al.* (2003) Interface of hand movements from local field potentials in monkey motor cortex. *Nat. Neurosci.*, **6**, 1253–1254.

[61] Rickert, J., Oliveira, S.C., Vaadia, E. *et al.* (2005) Encoding of movement direction in different frequency ranges of motor cortical local field potentials. *J. Neurosci.*, **25**, 8815–8824.

[62] Pezaran, B., Pezaris, J.S., Sahani, M. *et al.* (2002) Temporal structure in neuronal activity during working memory in macaque parietal cortex. *Nat. Neurosci.*, **5**, 805–811.

[63] Scherberger, H., Jarvis, M.R. and Andersen, R.A. (2005) Cortical local field potential encodes movement intentions in the posterior parietal cortex. *Neuron.*, **46**, 347–354.

[64] Serruya, M.D., Hatsopoulos, N.G., Paninski, L. *et al.* (2002) Instant neural control of a movement signal. *Nature*, **416**, 141–142.

[65] Taylor, D.M., Tillery, S.I. and Schwartz, A.B. (2002) Direct cortical control of 3D neuroprosthetic devices. *Science*, **296**, 1829–1832.

[66] Tillery, S.I. and Taylor, D.M. (2004) Signal acquisition and analysis for cortical control of neuroprosthetic. *Curr. Opin. Neurobiol.*, **14**, 758–762.

[67] Musallam, S., Corneil, B.D., Greger, B. *et al.* (2004) Cognitive control signals for neural prosthetics. *Science*, **305**, 258–262.

[68] Patil, P.G., Carmena, J.M., Nicolelis, M.A. *et al.* (2004) Ensemble recordings of human subcortical neurons as a source of motor control signals for a brain-machine interface. *Neurosurgery*, **55**, 27–35.

[69] Kennedy, P.R., Bakay, R.A.E., Moore, M.M. *et al.* (2000) Direct control of a computer from the human central nervous system. *IEEE Trans. Rehanilitation Eng.*, **8**(2), 198–202.

[70] Pfurstcheller Jr, G., Stancak, A. and Neuper, C. (1996) Post-movement beta synchronization. A correlate of an idling motor area? *Electroenceph. Clin. Neurophysiol.*, **98**, 281–293.

[71] Bro, R. (1997) Multi-way analysis in the food industry: Models, algorithms and applications. PhD thesis, University of Amsterdam, and Royal Veterinary and Agricultural University.

[72] Esch, W., Schimke, H., Doppelmayr, M. *et al.* (1996) Event related desynchronization (ERD) and the Dm effect: does alpha desynchronization during encoding predict later recall performance?. *Int. J. Psychophysiol.*, **24**, 47–60.

[73] Pfurtscheller, G. (1999) EEG even-related desynchronization (ERD) and event-related synchronization (ERS), in *Electroencephalography* (eds E. Niedermeyer and F. Lopes Da Silva), Lippincott Williams&Wilkins, pp. 958–966. Chapter 53.

[74] Arroyo, S., Lesser, R.P., Gordon, B. *et al.* (1993) Functional significance of the mu rhythm of human cortex: an electrophysiological study with subdural electrodes. *Electroenceph. Clin. Neurophysiol.*, **87**, 76–87.

[75] Leuthardt, E.C., Schalk, G., Wolpaw, J.R. *et al.* (2004) A brain-computer interface using electrocorticographic signals in humans. *J. Neural Eng.*, **1**, 63–71.

[76] Pfurtscheller, G., Stancak, A. Jr and Edlinger, G. (1997) On the existence of different types of central beta rhythms below 30 Hz. *Electroenceph. Clin. Neurophysiol.*, **102**, 316–325.

[77] Neuper, C. and Pfurtscheller, G. (1996) Post movement synchronization of beta rhythms in the EEG over the cortical foot area in man. *Neurosci. Lett.*, **216**, 17–20.

[78] Pfurtscheller, G., Müller-Putz, G.R., Pfurtscheller, J. and Rupp, R. (2005) EEG-based asynchronous BCI controls functional electrical stimulation in a tetraplegic patient. *EURASIP J. Appl. Signal Process.*, **2005**(19), 3152–3155.

[79] Gratton, G. (1969) Dealing with artefacts: the EOG contamination of the event-related brain potentials over the scalp. *Electroencephalogr. Clin. Neurophysiol.*, **27**, 546.

[80] Lins, O.G., Picton, T.W., Berg, P. and Scherg, M. (1993) Ocular artefacts in EEG and event-related potentials, i: scalp topography. *Brain Topogr.*, **6**, 51–63.

[81] Celka, P., Boshash, B. and Colditz, P. (2001) Preprocessing and time-frequency analysis of new born EEG seizures. *IEEE Eng. Med. Biol. Mag.*, **20**, 30–39.

[82] Jung, T.P., Humphies, C. and Lee, T.W. (1998) Extended ICA removes artefacts from electroencephalographic recordings. *Adv. Neur. Inf. Process. Syst.*, **10**, 894–900.

[83] Joyce, C.A., Gorodnitsky, I. and Kautas, M. (2004) Automatic removal of eye movement and blink artefacts from EEG data using blind component separation. *Psychophysiology*, **41**, 313–325.

[84] Shoker, L., Sanei, S. and Chambers, J. (2005) Artifact removal from electro-encephalograms using a hybrid BSS-SVM algorithm. *IEEE Signal Process. Lett*, **12** (10).

[85] Peterson, D.A., Knight, J.N., Kirby, M.J. *et al.* (2005) Feature Selection and Blind Source Separation in an EEG-Based Brain-Computer Interface. *EURASIP J. Appl. Signal Process.*, **2005**(19), 3128–3140.

[86] Coyle, D., Prasad, G. and McGinnity, T.M. (2005) A time-frequency approach to feature extraction for a brain-computer interface with a comparative analysis of performance measures. *EURASIP J. Appl. Signal Process.*, **2005**(19), 3141–3151.

[87] Mørup, M., Hansen, L.K., Herrmann, C.S. *et al.* (2006) Parallel factor analysis as an exploratory tool for wavelet transformed event-related EEG. *NeuroImage*, **29**(3), 938–947.

[88] Cohen, L. (1995) *Time Frequency Analysis*, Prentice Hall Signal Processing Series, Prentice Hall, Upper Saddle River, NJ, USA.

[89] Shoker, L., Nazarpour, K., Sanei, S. and Sumich, A. (2006) A novel space-time-frequency masking approach for quantification of EEG source propagation, with application to brain computer interfacing. Proceedings of the EUSIPCO, Florence, Italy.

[90] Harshman, R.A. (1970) Foundation of the PARAFAC: models and conditions for an 'explanatory' multi-modal factor analysis. UCLA Work, Pap. Phon., no. 16, pp. 1–84.

[91] Carol, J.D. and Chang, J. (2009) Analysis of individual differences in multidimensional scaling via an N-way generalization of 'Eckart-Young' decomposition. *Psychometrika*, **35**, 283–319.

[92] Möcks, J. (1988) Decomposing event-related potentials: a new topographic components model. *Biol. Psychol.*, **26**, 199–215.

[93] Field, A.S. and Graupe, D. (1991) Topographic component (parallel factor) analysis of multichannel evoked potentials: practical issues in trilinear spatiotemporal decomposition. *Brain Topogr.*, **3**, 407–423.

[94] Miwakeichi, F., Martinez-Montes, E., Valdes-Sosa, P.A. *et al.* (2004) Decomposing EEG data into space-time-frequency components using parallel factor analysis. *NeuroImage*, **22**, 1035–1045.

[95] Nazarpour, K., Shoker, L., Sanei, S. and Chambers, J. (2006) Parallel space-time-frequency decomposition of EEG signals for brain computer interfacing. Proceedings of the European Signal Proc. Conf. Italy.

[96] Hjorth, B. (1975) An on-line transformation of EEG scalp potentials into orthogonal source derivations. *Electroenceph. Clin. Neurophysiol*, **39**(5), 526–530.

[97] Nazarpour, K., Wangsawat, Y., Sanei, S. *et al.* (Sept 2008) Removal of the eye-blink artifacts from EEGs via STF-TS modeling and robust minimum variance beamforming. *IEEE Trans. Biomed. Eng.*, **55**(9), 2221–2231.

[98] Pfurtscheller, G. and Neuper, C. () Synchronization of mu rhythm in the EEG over the cortical hand area in men. *Neurosci. Lett.*, **174**, 93–96.

[99] Ginter, Jr, J., Blinowska, K.J., Kaminski, M. and Durka, P.J. (2001) Phase and amplitude analysis in time-frequency space—application to voluntary finger movement. *J. Neurosci. Methods*, **110**, 113–124.

[100] Mallat, S. and Zhang, Z. (1993) Matching pursuits with time-frequency dictionaries. *IEEE Trans. Signal Process.*, **4112**, 3397–3415.

[101] Durka, P.J., Ircha, D. and Blinowska, K.J. (2001) Stochastic time-frequency dictionaries for matching pursuit. *IEEE Trans. Signal Process.*, **49**(3), 507–510.

[102] Durka, P.J., Ircha, D., Neuper, C. and Pfurtscheller, G. (2001) Time-frequency microstructure of event-related EEG desynchronization and synchronization. *Med. Biol. Eng. Comput.*, **39**(3), 315–321.

[103] Kaminski, M.J. and Blinowska, K.J. (1991) A new method of the description of the information flow in the structures. *Biol. Cybern.*, **65**, 203–210.

[104] Bozdogan, H. (1987) Model-selection and Akaike's information criterion (AIC): The general theory and its analytical extensions. *Psychometrika*, **52**, 345–370.

[105] Ginter, J. Jr, Bilinowska, K.J., Kaminski, M. and Durka, P.J. (2001) Phase and amplitude analysis in time-frequency-space—application to voluntary finger movement. *J. Neurosci. Methods*, **110**, 113–124.

[106] Astolfi, L., Cincotti, F., Babiloni, C. *et al.* (2005) Estimation of the cortical connectivity by high resolution EEG and structural equation modelling: Simulations and application to finger tapping data. *IEEE Trans. Biomed. Eng.*, **52**(5), 757–767.

[107] Gerloff, C., Richard, J., Hardley, J. *et al.* (1998) Functional coupling and regional activation of human cortical motor areas during simple, internally paced and externally paced finger movement. *Brain*, **121**, 1513–1531.

[108] Urbano, A., Babiloni, C., Onorati, P. and Babiloni, F. (1998) Dynamic functional coupling of high resolution EEG potentials related to unilateral internally triggered one-digit movement. *Electroencephalogr. Clin. Neurophysiol.*, **106**(6), 477–487.

[109] Jancke, L., Loose, R., Lutz, K. *et al.* (2000) Cortical activations during paced finger-tapping applying visual and auditory pacing stimuli. *Cogn. Brain Res.*, **10**, 51–56.

[110] Bollen, K.A. (1989) *Structural Equations with Latent Variable*, John Wiley & Sons Inc., New York.

[111] Lotte, F. and Guan, C.T. (2011) Regularizing common spatial patterns to improve BCI designs: unified theory and new algorithms. *IEEE Trans. Biomed. Eng.*, **58**(2), 355–362.

[112] Lu, H., Plataniotis, K. and Venetsanopoulos, A. (2009) Regularized common spatial patterns with generic learning for EEG signal classification. Proceedings of the EMBC, pp. 6599–6602.

[113] Kang, H., Nam, Y. and Choi, S. (2009) Composite common spatial pattern for subject-to-subject transfer. *IEEE Signal Process. Lett.*, **16**(8), 683–686.

[114] Blankertz, B., Kawanabe, M., Tomioka, R. *et al.* (2008) Invariant common spatial patterns: Alleviating nonstationarities in brain-computer interfacing. Proceedings of the NIPS 20, 2008.

[115] Farquhar, J., Hill, N., Lal, T. and Schölkopf, B. (2006) Regularised CSP for sensor selection in BCI. Proceedings of the 3rd International BCI Workshop, 2006.

[116] Arvaneh, M., Guan, C.T., Kai, A.K. and Chai, Q. (2011) Optimizing the Channel Selection and Classification Accuracy in EEG-Based BCI. *IEEE Trans. Biomed.Eng.*, **58**(6), 1865–1873.

[117] Goksu, F., Ince, N.F. and Tewfik, A.H. (2011) Sparse common spatial patterns in brain computer interface applications. Proceedings of the IEEE International Conference on Acoustics, Speech, and Signal Processing, ICASSP, pp. 533–536.

[118] Yong, X., Ward, R. and Birch, G. (2008) Sparse spatial filter optimization for EEG channel reduction in brain-computer interface. Proceedings of IEEE International Conference on Acoustic, Speech, and Signal Processing, ICASSP, Taiwan, pp. 417–420.

[119] Barachant, A., Bonnet, S., Congedo, M. and Jutten, C. (2012) Multiclass brain-computer interface classification by Riemannian Geometry. *IEEE Trans. Biomed. Eng.*, **59**(4), 920–928.

[120] Leeb, R., Brunner, C., Müller-Putz, G.R. *et al.* (2008) BCI Competition 2008-Graz dataset B. Graz University of Technology, Austria.

[121] Mazzatenta, A., Giugliano, M., Campidelli, S. *et al.* (2007) Interfacing neurons with carbon nanotubes: Electrical signal transfer and synaptic stimulation in cultured brain circuits. *J. Neurosci.*, **27**(26), 6931–6936.

[122] Press release, Caltech, Oct. 27, 1997.

[123] Coming to a brain near you, Wired News, Oct. 22, 2004.

[124] 'Brain' in a dish flies flight simulator, CNN, Nov. 4, 2004.

[125] Kauhanen, L., Palomaki, T., Jylamki, P. *et al.* (2006) Haptic feedback compared with visual feedback for BCI. Proceedings of the 3rd International BCI Workshop & training Course, pp. 66–67.

14

EEG and MEG Source Localization

14.1 Introduction

The human brain consists of a large number of regions each of which when active generates a local magnetic field or synaptic electric current. The brain activities can be considered to constitute signal sources which are either spontaneous, corresponding to the normal rhythms of the brain, a result of brain stimulation, or are related to physical movements. Localization of brain signal sources from EEGs has been an active area of research during the last two decades.

Source localization is necessary to study brain physiological, mental, pathological, and functional abnormalities, and even problems related to various body disabilities, and ultimately to specify the cortical and deep sources of abnormalities.

Radiological imaging modalities have also been widely used for this purpose. However, these techniques are often unable to locate the abnormalities when they stem from malfunctions. Moreover, they are costly and some may not be accessible to all patients at the time they need them.

fMRI is able to show the location of anatomical disorders, such as tumour and functional changes in the brain, as a result of the changes in blood oxygen level. This can be due to event-related and movement-related stimulations or some abnormalities, such as epileptic seizure. From the fMRI, one may detect the effect of blood-oxygen-level-dependence (BOLD) during metabolic changes, such as those caused by interictal seizure, in the form of white patches. The major drawbacks of fMRI, however, are its poor temporal resolution and its limitations in detecting the details of functional and mental activities. As a result, despite the cost limitation for this imaging modality, it has been reported that in 40–60% of the cases with interictal activity in EEG, fMRI cannot locate any foci, even when the recording is simultaneous with EEG.

Functional brain imaging and source localization based on scalp potentials require a solution to an ill-posed inverse problem with many possible solutions. Selection of a particular solution

---

Adaptive Processing of Brain Signals, First Edition. Saeid Sanei.
© 2013 John Wiley & Sons, Ltd. Published 2013 by John Wiley & Sons, Ltd.

often requires a priori knowledge acquired from the overall physiology of the brain and the status of the subject.

Although, in general, localization of the brain sources is a difficult task, there are some simple situations where the localization problem can be simplified and accomplished:

(a) In places where the objective is to find the proximity of the actual source locations over the scalp. A simple method is just to attempt to somehow separate the sources using PCA or ICA, and back-project the sources of interest onto the electrode space and look at the scalp topography.
(b) Situations, where the aim is to localize a certain source of interest within the brain, for example, a source of evoked potential (EP) or a source of movement-related potential (as widely used in the context of brain–computer interfacing). It is easy to fit a single dipole at various locations over a coarse spatial sampling of the source space, and then choose the location producing the best match to the electrode signals (mixtures of the sources) as the focus of a spatially constrained, but more finely sampled search. The major problem in this case is that the medium is not linear and this causes error, especially for sources deep inside the brain or where there is no prior knowledge about the shape of the source signal.

14.2   General Approaches to Source Localization

In order to localize multiple sources within the brain using EEG or MEG two general approaches have been proposed by researchers:

1. Equivalent current dipole (ECD) and
2. Linear distributed (LD) approaches.

In the ECD approach the signals are assumed to be generated by a relatively small number of focal sources [1–4]. In the LD approaches all possible source locations are considered simultaneously [5–12].

In the inverse methods using the dipole source model the sources are considered as a number of discrete magnetic dipoles located in certain places in a three-dimensional space within the brain. The dipoles have fixed orientations and variable amplitudes.

On the other hand, in the current distributed-source reconstruction (CDR) methods, there is no need for any knowledge about the number of sources. Generally, this problem is considered as an underdetermined inverse problem. An L_p norm solution is the most popular regulation operator to solve this problem. This regularized method is based upon minimizing the cost function

$$\psi = \|\mathbf{L}\mathbf{x} - m\|_p + \lambda \|\mathbf{W}\mathbf{x}\|_p \qquad (14.1)$$

where $\mathbf{x}$ is the vector of the source currents, $\mathbf{L}$ is the lead field matrix; m is the EEG measurement, $\mathbf{W}$ is a diagonal location weighting matrix, λ is the regulation parameter, and $1 \leq p \leq 2$, the norm, is the measure in complete normed vector (Banach) space [13]. A minimum L_p norm method refers to the above criterion when $\mathbf{W}$ is equal to the identity matrix.

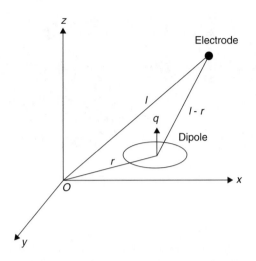

Figure 14.1 The magnetic field **B** at each electrode is calculated with respect to the moment of the dipole and the distance between the centre of the dipole and the electrode

14.2.1 Dipole Assumption

For a dipole at location **L**, the magnetic field observed at electrode i at location $R(i)$ is achieved as (Figure 14.1)

$$\mathbf{B}(i) = \frac{\mu}{4\pi} \frac{\mathbf{Q} \times (\mathbf{R}(i) - \mathbf{L})}{|\mathbf{R}(i) - \mathbf{L}|} \quad \text{for } i = 1, \ldots, n_e \tag{14.2}$$

where **Q** is the dipole moment, | . | denotes absolute value, and × represents an outer vector product. This is frequently used as the model for magneto-encephalographic (MEG) data observed by magnetometers. This can be extended to the effect of a volume containing m dipoles at each one of the n_e electrodes (Figure 14.2).

In the m-dipole case the magnetic field at point j is obtained as:

$$\mathbf{B}(i) = \frac{\mu_0}{4\pi} \sum_{j=1}^{m} \frac{\mathbf{Q}_j \times (\mathbf{R}(i) - \mathbf{L}_j)}{|\mathbf{R}(i) - \mathbf{L}_j|} \quad \text{for } i = 1, \ldots, n_e \tag{14.3}$$

where n_e is the number of electrodes and $\mathbf{L}_j$ represents the location of the jth dipole. The matrix **B** can be considered as $\mathbf{B} = [B(1), B(2) \ldots B(n_e)]$. On the other hand, the dipole moments can be factorised into the product of their unit orientation moments and strengths, that is, $\mathbf{B} = \mathbf{GQ}$ (normalised with respect to $\mu_0/4\pi$) where $\mathbf{G} = [\mathbf{g}(1), \mathbf{g}(2), \ldots \mathbf{g}(m)]$ is the propagating medium (mixing matrix) and $\mathbf{Q} = [Q_1, Q_2, \ldots Q_m]^{\mathrm{T}}$. The vector $\mathbf{g}(i)$ has dimension 1×3 (thus **G** is $m \times 3$). Therefore, $\mathbf{B} = \mathbf{GMS}$. **GM** can be written as a function of location and orientation, such as $\mathbf{H}(\mathbf{L},\mathbf{M})$, and therefore $\mathbf{B} = \mathbf{H}(\mathbf{L},\mathbf{M})\mathbf{S}$. The initial solution to this problem

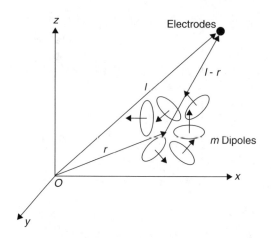

Figure 14.2 The magnetic field **B** at each electrode is calculated with respect to the accumulated moments of the *m* dipoles and the distance between the centre of the dipoles' volume and the electrode

was by using a least-squares search that minimises the difference between the estimated and the measured data;

$$J_{ls} = \|X - H(L, M)S\|_F^2 \tag{14.4}$$

where X is the magnetic (potential) field over the electrodes. The parameters to be estimated are location, dipole orientation, and magnitude for each dipole. This is subject to knowing the number of the sources (dipoles). If too few dipoles are selected then the resulting parameters are influenced by the missing dipoles. On the other hand, if too many dipoles are selected the accuracy will decrease since some of them are not valid brain sources. Also, the computation cost is high due to optimising a number of parameters simultaneously. One way to overcome this is by converting this problem into a projection minimisation problem as:

$$J_{ls} = \|X - H(L, M)S\|_F^2 = \left\|P_H^{\perp} X\right\|_F^2 \tag{14.5}$$

The matrix $P_H^{\perp}$ projects the data onto the orthogonal complement of the column space of $H(L,M)$. X can be reformed by singular value decomposition (SVD), that is, $X = U\Sigma V^T$. Therefore

$$J_{ls} = \left\|P_H^{\perp} U\Sigma V^T\right\|_F^2 \tag{14.6}$$

In this case orthogonal matrices preserve the Frobenius norm. The matrix $Z = U\Sigma$ is $m \times m$ unlike X which is $n_e \times T$; T is the number of samples. Generally, the rank of X satisfies rank$(X) \leq m$, and $T \gg m$, and Σ can have only m non-zero singular values. This means that the overall computation cost has been reduced by a large amount. The SVD can also be used to reduce the computation cost for $P_H^{\perp} = (I - HH^{\dagger})$. The pseudoinverse $H^{\dagger}$ can be decomposed as $V_H \Sigma_H^{\dagger} U_H^T$, where $H = U_H \Sigma_H^{\dagger} V_H^T$.

The dipole model, however, requires a priori knowledge about the number of sources, which is usually unknown.

Many experimental studies and clinical experiments have examined the developed source localization algorithms. Yao and Dewald [14] have evaluated different cortical source localization methods, such as the moving dipole (MDP) method [15], minimum L_p norm [16], and low-resolution tomography (LRT) as for LORETA [17] using simulated and experimental EEG data. In their study, only the scalp potentials have been taken into consideration.

In this study some other source localization methods, such as the cortical potential imaging method [18] and the three-dimensional resultant vector method [19], have not been included. These methods, however, follow similar steps as minimum norm methods and dipole methods, respectively.

14.3 Most Popular Brain Source Localization Approaches

Although most of the methods described here have been used for both EEG and MEG signals, those specifically used for dipole source localization perform much more accurately for MEG signals since the brain is more linear and homogenous against magnetic field compared with electric field.

14.3.1 ICA Method

In a simple though not very accurate approach by Spyrou *et al.* [20] ICA is used to separate the EEG sources. The correlation values of the estimated sources and the mixtures are then used to build up the model of the mixing medium. The LS approach is then used to find the sources using the inverse of these correlations that is, $d_{ij} \approx 1/(C_{ij})^{0.5}$, where d_{ij} shows the distance between source i and electrode j, and C_{ij} shows the correlation of their signals. The method has been applied to separate and localize the sources of P3a and P3b subcomponents for five healthy subjects and five schizophrenic patients. The study of ERP components, such as P300 and its constituent subcomponents, is very important in analysis, diagnosing, and monitoring of mental and psychiatric disorders. Source location, amplitude, and latency of these components have to be quantified and used in the classification process. Figure 14.3 illustrates the results overlaid on two MRI templates for the above two cases. It can be seen that for the healthy subjects the subcomponents are well apart whereas for the schizophrenic patients they are geometrically mixed.

14.3.2 MUSIC Algorithm

Multiple signal classification (MUSIC) [21] has been used for localization of the magnetic dipoles within the brain [22–25] using EEG signals. In an early development of this algorithm a single-dipole model within a three-dimensional head volume is scanned and projections onto an estimated signal subspace are computed [26]. To locate the sources, the user must search the head volume for multiple local peaks in the projection metric. In an attempt to overcome the exhaustive search by the MUSIC algorithm a recursive MUSIC algorithm was developed [22]. Following this approach, the locations of the fixed, rotating, or synchronous

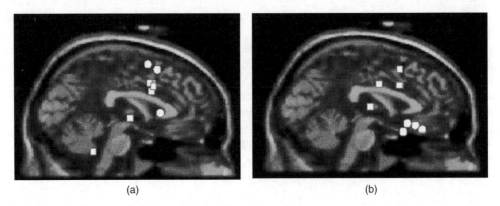

Figure 14.3 Localization results for (a) the schizophrenic patients and (b) the normal subjects. The circles represent P3a and the squares represent P3b

dipole sources are automatically extracted through a recursive use of subspace estimation. This approach tries to solve the problem of how to choose the locations, which give the best projection on to the signal (EEG) space. In the absence of noise and using a perfect head and sensors model, the forward model for the source at the correct location projects entirely into the signal subspace. In practice, however, there are estimation errors due to noise. In finding the solution to the above problem there are two assumptions which may not be always true; first, the data are corrupted by additive spatially white noise and second, the data are produced by a set of asynchronous dipolar sources. These assumptions are waved in the proposed recursive MUSIC algorithm [22].

The original MUSIC algorithm for the estimation of brain sources may be described as follows. Consider the head model for transferring the dipole field to the electrodes to be $\mathbf{A}(\rho, \theta)$, where ρ and θ are respectively the dipole location and direction parameters. The relationship between the observations (EEG), $\mathbf{X}$, the model $\mathbf{A}$, and the sources $\mathbf{S}$ is given as:

$$\mathbf{X} = \mathbf{A}\mathbf{S}_T + \mathbf{E} \tag{14.7}$$

where $\mathbf{E}$ is the noise matrix. The goal is to estimate the parameters $\{\rho, \theta, \mathbf{S}\}$, given the data set $\mathbf{X}$. The correlation matrix of $\mathbf{X}$, that is, $\mathbf{R}_x$ can be decomposed as:

$$\mathbf{X}_F = [\boldsymbol{\Phi}_s, \boldsymbol{\Phi}_e] \begin{bmatrix} \boldsymbol{\Lambda} + n_e \sigma_e^2 \mathbf{I} & 0 \\ 0 & n_e \sigma_e^2 \mathbf{I} \end{bmatrix} [\boldsymbol{\Phi}_s, \boldsymbol{\Phi}_e]^T \tag{14.8}$$

or

$$\mathbf{R}_x = \boldsymbol{\Phi}_s \boldsymbol{\Lambda}_s \boldsymbol{\Phi}_s^T + \boldsymbol{\Phi}_e \boldsymbol{\Lambda}_e \boldsymbol{\Phi}_e^T \tag{14.9}$$

where $\boldsymbol{\Lambda}_s = \boldsymbol{\Lambda} + n_e \sigma_e^2 \mathbf{I}$ is the $m \times m$ diagonal matrix combining both the model and noise eigenvalues, and $\boldsymbol{\Lambda}_e = n_e \sigma_e^2 \mathbf{I}$ is the $(n_e - m) \times (n_e - m)$ diagonal matrix of noise-only eigenvalues. Therefore, the signal subspace span ($\boldsymbol{\Phi}_s$) and noise-only subspace span ($\boldsymbol{\Phi}_e$)

are orthogonal. In practice, T samples of the data are used to estimate the above parameters that is,

$$\hat{\mathbf{R}}_x = \mathbf{X}\mathbf{X}^{\mathrm{T}} = \hat{\boldsymbol{\Phi}}_s\hat{\boldsymbol{\Lambda}}_s\hat{\boldsymbol{\Phi}}_s^{\mathrm{T}} + \hat{\boldsymbol{\Phi}}_e\hat{\boldsymbol{\Lambda}}_e\hat{\boldsymbol{\Phi}}_e^{\mathrm{T}} \tag{14.10}$$

where the first m left singular vectors of the decomposition are designated as $\hat{\boldsymbol{\Phi}}_s$ and the remaining eigenvectors as $\hat{\boldsymbol{\Phi}}_e$. Accordingly, the diagonal matrix $\hat{\boldsymbol{\Lambda}}_s$ contains the first m eigenvalues and $\hat{\boldsymbol{\Lambda}}_e$ contains the remainder. To estimate the above parameters the general rule using least-squares fitting is:

$$\{\hat{\rho}, \hat{\boldsymbol{\theta}}, \hat{\mathbf{S}}\} = \arg\min_{\rho,\theta,\mathrm{S}} \left\| \mathbf{X} - \mathbf{A}(\rho, \theta)\mathbf{S}^{\mathrm{T}} \right\|_{\mathrm{F}}^2 \tag{14.11}$$

where $\| \, . \, \|_{\mathrm{F}}$ denotes the Frobenius norm. By optimal substitution [27] we achieve:

$$\{\hat{\rho}, \hat{\boldsymbol{\theta}}\} = \arg\min_{\rho,\theta} \left\| \mathbf{X} - \mathbf{A}\mathbf{A}^\dagger\mathbf{X} \right\|_{\mathrm{F}}^2 \tag{14.12}$$

where $\mathbf{A}^\dagger$ is the Moore–Penrose pseudoinverse of $\mathbf{A}$ [28]. Given that the rank of $\mathbf{A}(\rho, \theta)$ is m and the rank of $\hat{\boldsymbol{\Phi}}_s$ is at least m, the smallest subspace correlation value, $C_m = \mathrm{subcorr}\{\mathbf{A}(\rho, \theta), \hat{\boldsymbol{\Phi}}_s\}_m$, represents the minimum subspace correlation (maximum principal angle) between principal vectors in the column space of $\mathbf{A}(\rho, \theta)$ and the signal subspace $\hat{\boldsymbol{\Phi}}_s$. In MUSIC, the subspace correlation of any individual column of $\mathbf{A}(\rho, \theta)$ that is, $\mathbf{a}(\rho_i, \theta_i)$ with the signal subspace must therefore equal or exceed this minimum subspace correlation;

$$C_i = \mathrm{subcorr}\{\mathbf{a}(\rho_i, \theta_i), \hat{\boldsymbol{\Phi}}_s\} \geq C_m \tag{14.13}$$

$\hat{\boldsymbol{\Phi}}_s$ approaches $\boldsymbol{\Phi}_s$ when the number of samples increases or SNR becomes higher. Then the minimum correlation approaches unity when the correct parameter set is identified, such that the m distinct sets of parameters (ρ_i, θ_i) have subspace correlations approaching unity. Therefore, a search strategy is followed to find the m peaks of the metric [22]

$$\mathrm{subcorr}^2\{\mathbf{a}(\rho, \theta), \hat{\boldsymbol{\Phi}}_s\} = \frac{\mathbf{a}^{\mathrm{T}}(\rho, \theta)\hat{\boldsymbol{\Phi}}_s\hat{\boldsymbol{\Phi}}_s^{\mathrm{T}}\mathbf{a}(\rho, \theta)}{\left\| \mathbf{a}^{\mathrm{T}}(\rho, \theta) \right\|_2^2} \tag{14.14}$$

where $\|\cdot\|_2^2$ denotes a squared Euclidean norm. For a perfect estimation of the signal subspace m global maxima equal to unity are found. This requires searching for both sets of parameters ρ and θ. In a quasilinear approach [22] it is assumed that $\mathbf{a}(\rho_i, \theta_i) = \mathbf{G}(\rho_i)\theta_i$ where $\mathbf{G}(.)$ is called the gain matrix. In the EEG source localization application, therefore, we first find the dipole parameters ρ_i which maximize the $\mathrm{subcorr}\{\mathbf{G}(\rho_i), \hat{\boldsymbol{\Phi}}_s\}$ and then extract the corresponding quasilinear θ_i which maximize this subspace correlation. This avoids explicitly searching for these quasilinear parameters, reducing the overall complexity of the nonlinear search. Therefore, the overall localization of the EEG sources using classic MUSIC can be summarized to the following steps: (i) Decompose $\mathbf{X}$ or $\mathbf{X}\mathbf{X}^{\mathrm{T}}$ and select the rank of the signal subspace to obtain $\hat{\boldsymbol{\Phi}}_s$. Slightly overspecifying the rank has little effect on the performance, whereas underspecifying it can dramatically reduce the performance; (ii) form the gain matrix **G** at each point (node) of a dense grid of dipolar source locations and calculate the subspace correlations subcorr$\{\mathbf{G}(\rho_i), \hat{\mathbf{\Phi}}_s\}$; (iii) find the peaks of the plot $\sqrt{1 - C_1^2}$, where C_1 is the maximum subspace correlation. Locate m or fewer peaks in the grid. At each peak, refine the search grid to improve the location accuracy and check the second subspace correlation. A large second subspace correlation is an indication of a rotating dipole [22]. Unfortunately, there are often errors in estimating the signal subspace and the correlations are computed at only a finite set of grid points. A recursive MUSIC algorithm overcomes this problem by recursively building up the independent topography model. In this model the number of dipoles is initially considered as one and the search is carried out to locate a single dipole. A second dipole is then added and the dipole point that maximizes the second subspace correlation, C_2, is found. At this stage there is no need to recalculate C_1. The number of dipoles is increased and the new subspace correlations are computed. If m topographies comprise m_1 single-dipolar topographies and m_2 2-dipolar topographies, then, the recursive MUSIC will first extract the m_1 single dipolar models. At the $(m_1 + 1)$th iteration, no single dipole location that correlates well with the subspace will be found. By increasing the number of dipole elements to two, the searches for both have to be carried out simultaneously such that the subspace correlation is maximised for C_{m_1+1}. The procedure continues to build the remaining m_2 2-dipolar topographies. After finding each pair of dipoles that maximises the appropriate subspace correlation, the corresponding correlations are also calculated.

Another extension of MUSIC-based source localization for diversely polarized sources, namely recursively applied and projected (RAP) MUSIC [24], uses each successively located source to form an intermediate array gain matrix (similar to the recursive MUSIC algorithm) and projects both the array manifold and the signal subspace estimate into its orthogonal complement. In this case the subspace is reduced. Then, the MUSIC projection to find the next source is performed. This method was initially applied to localization of MEG sources [24, 25].

In another study by Xu *et al.* [29], an approach to EEG three-dimensional (3D) dipole source localization by using a non-recursive subspace algorithm, called FINES, has been proposed. The approach employs projections onto a subspace spanned by a small set of particular vectors in the estimated noise-only subspace, instead of the entire estimated noise-only subspace in the case of classic MUSIC. The subspace spanned by this vector set is, in the sense of principal angle, closest to the subspace spanned by the array manifold associated with a particular brain region. By incorporating knowledge of the array manifold in identifying the FINES vector sets in the estimated noise-only subspace for different brain regions, this approach is claimed to be able to estimate sources with enhanced accuracy and spatial resolution, thus enhancing the capability of resolving closely spaced sources and reducing estimation errors. The simulation results show that, compared to classic MUSIC, FINES has a better resolvability of two closely spaced dipolar sources and also a better estimation accuracy of source locations. In comparison with RAP-MUSIC, The performance of the FINES is also better for the cases where the noise level is high and/or correlations amongst dipole sources exist [29].

A method for using a generic head model, in the form of an anatomical atlas, has also been proposed to produce EEG source localization [30]. The atlas is fitted to the subject by a nonrigid warp using a set of surface landmarks. The warped atlas is used to compute a finite element model (FEM) of the forward mapping or lead-fields between neural current generators and the EEG electrodes. These lead-fields are used to localize current sources from the EEG data of the subject and the sources are then mapped back to the anatomical atlas. This approach provides a mechanism for comparing source localizations across subjects in an atlas-based coordinate system.

14.3.3   LORETA Algorithm

The low-resolution electromagnetic tomography algorithm (LORETA) for localization of brain sources has already been commercialised. In this method, the electrode potentials and matrix **X**, are considered to be related as

$$\mathbf{X} = \mathbf{L}\mathbf{S} \qquad (14.15)$$

where **S** is the actual (current) source amplitudes (densities) and **L** is an $n_e \times 3m$ matrix representing the forward transmission coefficients from each source to the array of sensors. **L** has also been referred to as the system response kernel or the lead-field matrix [31]. Each column of **L** contains the potentials observed at the electrodes when the source vector has unit amplitude at one location and orientation, and is zero at all others. This requires the potentials to be measured linearly with respect to the source amplitudes based on the superposition principle. Generally, this is not true and, therefore, such assumption inherently creates some error. The fitted source amplitudes, $\mathbf{S}_n$, can be approximately estimated using an exhaustive search through the inverse least-squares (LS) solution, that is,

$$\mathbf{S}_n = (\mathbf{L}^T\mathbf{L})^{-1}\mathbf{L}^T\mathbf{X} \qquad (14.16)$$

L may be approximated by 3D simulations of current flow in the head, which requires a solution to the well-known Poisson equation [32]:

$$\nabla \cdot \sigma \nabla \mathbf{X} = -\rho \qquad (14.17)$$

where σ is the conductivity of the head volume $(\Omega\ \text{m})^{-1}$ and ρ is the source volume current density (A m^{-3}). An FEM or boundary-element method (BEM) may be used to solve this equation. In such models the geometrical information about the brain layers and their conductivities [33] has to be known. Unless some a priori knowledge can be used in the formulation, the analytic model is ill-posed and a unique solution is hard to achieve.

On the other hand the number of sources, m, is typically much larger than the number of sensors, n_e, and the system in (14.15) is underdetermined. Also, in the applications where more concentrated focal sources are to be estimated such methods fail. As we see later, using a minimum norm approach, some researchers choose the solution which satisfies some constraints such as the smoothness of the inverse solution.

One approach is the minimum norm solution, which minimises the norm of S under the constraint of the forward problem:

$$\min \|\mathbf{S}\|_2^2, \text{ subject to } \mathbf{X} = \mathbf{L}\mathbf{S} \qquad (14.18)$$

with a solution as

$$\mathbf{S} = \mathbf{L}^{\mathrm{T}}(\mathbf{L}\mathbf{L}^{\mathrm{T}})^{\dagger}\mathbf{X} \tag{14.19}$$

The motivation of the minimum norm solution is to create a sparse solution with zero contribution from most of the sources. This method has the serious drawback of poor localization performance in 3D space. An extension to this method is the weighted minimum norm (WMN) solution, which compensates for deep sources and hence performs better in 3D space. In this case the norms of the columns of $\mathbf{L}$ are normalized. Hence, the constrained WMN is formulated as:

$$\min \|\mathbf{W}\mathbf{S}\|_2^2, \text{ subject to } \mathbf{X} = \mathbf{L}\mathbf{S} \tag{14.20}$$

with a solution as

$$\mathbf{S} = \mathbf{W}^{-1}\mathbf{L}^{\mathrm{T}}(\mathbf{L}\mathbf{W}^{-1}\mathbf{L}^{\mathrm{T}})^{\dagger}\mathbf{X} \tag{14.21}$$

where $\mathbf{W}$ is a diagonal $3m \times 3m$ weighting matrix, which compensates for deep sources in the following way:

$$\mathbf{W} = \mathrm{diag}\left[\frac{1}{\|\mathbf{L}_1\|_2}, \frac{1}{\|\mathbf{L}_2\|_2}, \ldots, \frac{1}{\|\mathbf{L}_{3m}\|_2}\right] \tag{14.22}$$

where $\|\mathbf{L}_i\|_2$ denotes the Euclidean norm of the ith column of $\mathbf{L}$, that is, $\mathbf{W}$ corresponds to the inverse of the distances between sources and electrodes.

In another similar approach a smoothing Laplacian operator is employed. This operator produces a spatially smooth solution agreeing with the physiological assumption mentioned earlier. The function of interest is then:

$$\min \|\mathbf{B}\mathbf{W}\mathbf{S}\|_2^2, \text{ subject to } \mathbf{X} = \mathbf{L}\mathbf{S} \tag{14.23}$$

where $\mathbf{B}$ is the Laplacian operator. This minimum norm approach produces a smooth topography in which the peaks representing the source locations are accurately located.

An FEM has been used to achieve a more anatomically realistic volume conductor model of the head, in an approach called adaptive standardized LORETA/FOCUSS (ALF) [34]. It is claimed that using this application-specific method a number of different resolution solutions using different mesh intensities can be combined to achieve the localization of sources with less computational complexities. Initially, FEM is used to approximate solutions to (14.23) with a realistic representation of the conductor volume based on magnetic resonance (MR) images of the human head. The dipolar sources are presented using the approximate Laplace method [35].

14.3.4 FOCUSS Algorithm

The FOCUSS [34] algorithm is based on a high-resolution iterative WMN method that uses the information from the previous iterations as:

$$\min \|\mathbf{CS}\|_2^2, \text{ subject to } \mathbf{X} = \mathbf{LS} \tag{14.24}$$

where $\mathbf{C} = (\mathbf{Q}^{-1})^{\mathsf{T}}\mathbf{Q}^{-1}$ and $\mathbf{Q}_i = \mathbf{WQ}_{i-1}\left[\text{diag}(\mathbf{S}_{i-1}(1)\cdots\mathbf{S}_{i-1}(3m)\right]$ and the solution at iteration i becomes:

$$\mathbf{S}_i = \mathbf{Q}_i\mathbf{Q}_i^{\mathsf{T}}\mathbf{L}^{\mathsf{T}}(\mathbf{LQ}_i\mathbf{Q}_i^{\mathsf{T}}\mathbf{L}^{\mathsf{T}})^{+}\mathbf{X} \tag{14.25}$$

The iterations stop when there is no significant change in the estimation. The result of FOCUSS is highly dependent on the initialisation of the algorithm. In practice, the algorithm converges close to the initialisation point and may easily be stuck in some local minima. A clever initialisation of FOCUSS has been suggested to be the solution to LORETA [34].

14.3.5 Standardised LORETA

As another option referred to as standardised LORETA (sLORETA), a unique solution to the inverse problem is found. It uses a different cost function, which is:

$$\min_{\mathbf{S},\mathbf{L}} \left(\|\mathbf{X} - \mathbf{LS}\|_2^2 + \lambda\|\mathbf{S}\|_2^2\right) \tag{14.26}$$

Hence, sLORETA uses a zero-order Tikhonov–Phillips regularization [36,37], which provides a possible solution to the ill-posed inverse problems:

$$\mathbf{s}_i = \mathbf{L}_i^{\mathsf{T}}\left[\mathbf{L}_i\mathbf{L}_i^{\mathsf{T}} + \lambda_i\mathbf{I}\right]^{-1}\mathbf{X} = \mathbf{R}_i\mathbf{S} \tag{14.27}$$

where $\mathbf{s}_i$ indicates the candidate sources and $\mathbf{S}$ are the actual sources. $\mathbf{R}_i$ is the resolution matrix defined as

$$\mathbf{R}_i = \mathbf{L}_i^{\mathsf{T}}\left[\mathbf{L}_i\mathbf{L}_i^{\mathsf{T}} + \lambda_i\mathbf{I}\right]^{-1}\mathbf{L}_i \tag{14.28}$$

The reconstruction of multiple sources performed by the final iteration of sLORETA is used as an initialisation for the combined ALF and weighted minimum norm (WMN or FOCUSS) algorithms [35]. The number of sources is reduced each time and (14.19) is modified to

$$\mathbf{s}_i = \mathbf{W}_i\mathbf{W}_i^{\mathsf{T}}\mathbf{L}_f^{\mathsf{T}}\left[\mathbf{L}_f\mathbf{W}_i\mathbf{W}_i^{\mathsf{T}}\mathbf{L}_f^{\mathsf{T}} + \lambda I\right]^{-1}\mathbf{X} \tag{14.29}$$

$\mathbf{L}_f$ indicates the final $n \times m$ lead-field returned by the sLORETA. $\mathbf{W}_i$ is a diagonal $3m_f \times 3m_f$ matrix, which is recursively refined based on the current density estimated by the previous step:

$$\mathbf{W}_i = \text{diag}\left(s_{i-1}(1), s_{i-1}(2), \ldots, s_{i-1}(3n_f)\right) \tag{14.30}$$

and the resolution matrix in (14.28) after each iteration changes to

$$\mathbf{R}_i = \mathbf{W}_i \mathbf{W}_i^{\mathrm{T}} \mathbf{L}_{\mathrm{f}}^{\mathrm{T}} \left[\mathbf{L}_{\mathrm{f}} \mathbf{W}_i \mathbf{W}_i^{\mathrm{T}} \mathbf{L}_{\mathrm{f}}^{\mathrm{T}} + \lambda I \right]^{-1} \mathbf{L}_{\mathrm{f}} \qquad (14.31)$$

Iterations are continued until the solution does not significantly change. In another approach by Liu *et al.* [38] called shrinking standard LORETA-FOCUSS (SSLOFO), sLORETA is used for initialisation. Then, it uses the re-WMN of FOCUSS. During the process the localization results are further improved by involving the above standardisation technique. However, FOCUSS normally creates an increasingly sparse solution during iteration. Therefore, it is better to eliminate the nodes with no source activities or recover those active nodes which might be discarded by mistake. The algorithm proposed in [39] shrinks the source space after each iteration of FOCUSS, hence reducing the computational load [38]. For the algorithm not to be trapped in a local minimum a smoothing operation is performed. The overall SSLOFO is therefore summarized into the following steps [38, 39].

1. Estimate the current density $\hat{\mathbf{S}}_0$ using sLORETA,
2. Initialise the weighting matrix as $\mathbf{Q}_0 = \mathrm{diag}[\hat{\mathbf{S}}_0(1), \hat{\mathbf{S}}_0(2) \dots \hat{\mathbf{S}}_0(3m)]$,
3. Estimate the source power using standardised FOCUSS,
4. Retain the prominent nodes and their neighbouring nodes. Adjust the values on these nodes through smoothing,
5. Redefine the solution space to contain only the retained nodes, that is, only the corresponding elements in $\mathbf{S}$ and the corresponding column in $\mathbf{L}$,
6. Update the weighting matrix
7. Repeat the steps 3–6 until a stopping condition is satisfied,
8. The final solution is the result of the last step before smoothing.

The stopping condition may be by defining a threshold, or when there is no negligible change in the weights in further iterations.

14.3.6   Other Weighted Minimum Norm Solutions

In an LD approach by Phillips *et al.* [5] a WMN solution (or Tikhonov regularization) method [40] has been proposed. The solution has been regularized by imposing some anatomical and physiological information upon the overall cost function in the form of constraints. The squared error costs are weighted based on spatial and temporal properties. The information, such as hemodynamic measures of brain activity from other imaging modalities such as fMRI, is used as constraints (or priors) together with the proposed cost function. In this approach it is assumed that the sources are sufficiently densely distributed and the sources are orientated orthogonal to the cortical sheet.

The instantaneous EEG source localization problem using a multivariate linear model and the observations, $\mathbf{X}$, as electrode potentials, is generally formulated on the basis of the observation model

$$\mathbf{X} = \Im(\mathbf{r}, \mathbf{J}) + \mathbf{V} \qquad (14.32)$$

where, $\mathbf{X} = [\mathbf{x}_1, \mathbf{x}_2, \ldots, \mathbf{x}_{n_e}]^T$ has dimension $n_e \times T$, where T represents the length of the data in samples and n_e is the number of electrodes, $\mathbf{r}$ and $\mathbf{J} = [\mathbf{j}_1, \mathbf{j}_2, \ldots, \mathbf{j}_m]$ are, respectively, source locations and moments of the sources, and $\mathbf{V}$ is the additive noise matrix. $\mathfrak{I}$ is the function linking the sources to the electrode potentials. In calculation of $\mathfrak{I}$ a suitable three-layer head model is normally considered [41,42]. A structural MR image of the head may be segmented into three isotropic regions namely, brain, skull, and scalp, of the same conductivity [43] and used as the model. Most of these models consider the head as a sphere for simplicity.

However, in [5] the EEG sources are modelled by a fixed and uniform three-dimensional grid of current dipoles spread within the entire brain volume. Also, the problem is an under-determined linear problem as

$$\mathbf{X} = \mathbf{LJ} + \mathbf{V} \tag{14.33}$$

where, $\mathbf{L}$ is the head field matrix, which inter-relates the dipoles to the electrode potentials. To achieve a unique solution for the above underdetermined equation some constraints have to be imposed. The proposed regularization method constrains the reconstructed source distribution by jointly minimizing a linear mixture of some weighted norm $\|\mathbf{Hj}\|_2$ of the current source $\mathbf{j}$ and the main cost function of the inverse solution. Assuming the noise is Gaussian with a covariance matrix $\mathbf{C}_v$ then

$$\hat{\mathbf{J}} = \arg \min_{\mathbf{j}} \left\{ \left\| \mathbf{C}_v^{-1/2}(\mathbf{Lj} - \mathbf{x}) \right\|_2^2 + \lambda^2 \|\mathbf{Hj}\|_2^2 \right\} \tag{14.34}$$

where the Lagrange multiplier λ has to be adjusted to make a balance between the main cost function and the constraint $\|\mathbf{Hj}\|_2$. The covariance matrix is scaled in such a way to have $\mathrm{trace}(\mathbf{C}_v) = \mathrm{rank}(\mathbf{C}_v)$ (recall that trace of a matrix is sum of its diagonal elements, and rank of a matrix refers to its number of independent columns (rows)). This can be stated as an overdetermined least-squares problem [18]. The solution to the minimisation of (14.34) for a given λ is in the form of

$$\hat{\mathbf{J}} = \mathbf{BX} \tag{14.35}$$

where

$$\begin{aligned} \mathbf{B} &= \left[\mathbf{L}^T \mathbf{C}_v^{-1} \mathbf{L} + \lambda^2 (\mathbf{H}^T \mathbf{H}) \right]^{-1} \mathbf{L}^T \mathbf{C}_v^{-1} \\ &= (\mathbf{H}^T \mathbf{H})]^{-1} \mathbf{L}^T \left[\mathbf{L}(\mathbf{H}^T \mathbf{H})^{-1} \mathbf{L}^T + \lambda^2 \mathbf{C}_v \right]^{-1} \end{aligned} \tag{14.36}$$

These equations describe the WMN solution to the localization problem. However, this is not complete unless a suitable spatial or temporal constraint is imposed. Theoretically, any number of constraints can be added to the main cost function in the same way and the hyperparameters, such as Lagrange multipliers, λ, can be calculated by expectation maximisation [44]. However, more assumptions, such as those about the covariance of the sources, have to be implied in order to find $\mathbf{L}$ effectively, which includes information about the locations and the moments. One assumption, in the form of constraint, can be based on the fact that $(\mathrm{diag}(\mathbf{L}^T\mathbf{L}))^{-1}$, which is proportional to the covariance components, should be normalized. Another constraint is based on the spatial fMRI information which appears as BOLD when the sources are active. Evoked responses can also be used as temporal constraints.

14.3.7   Evaluation Indices

The accuracy of inverse solution using the simulated EEG data has been evaluated by three indices: (i) the error distance (ED) that is, the distance between the actual and estimated locations of the sources, (ii) the undetected source number percentage (USP), and (iii) the percentage of falsely-detected source number (FSP). Obviously, these quantifications are based on the simulated models and data. For real EEG data it is hard to quantify and evaluate the results obtained by different inverse methods.

ED between the estimated source locations, $\tilde{\mathbf{s}}$, and the actual source locations, $\mathbf{s}$, is defined as:

$$ED = \frac{1}{N_d} \sum_{i=1}^{N_d} \min_j \left\{ \left\| \tilde{\mathbf{s}}_i - \mathbf{s}_j \right\| \right\} + \frac{1}{N_{ud}} \sum_{j=1}^{N_{ud}} \min_i \left\{ \left\| \tilde{\mathbf{s}}_i - \mathbf{s}_j \right\| \right\} \tag{14.37}$$

where i and j are the indices of locations of the estimated and actual sources, and N_d and N_{ud} are, respectively, the total numbers of estimated and undetected sources.

USP and FSP are respectively defined as $USP = N_{un}/N_{real} \times 100\%$ and $FSP = N_{false}/N_{estimated} \times 100\%$ where $N_{un}, N_{real}, N_{false}$, and $N_{estimated}$ are, respectively, the numbers of undetected, real, falsely detected, and estimated sources.

In practice, three types of head volume conductor models can be used: a homogeneous sphere head volume conductor model, a BEM or a FEM model. Since the FEM is computationally very intensive, the subject-specific BEM model, albeit for an oversimplifying sphere head model, is currently used [45].

In terms of ED, USP, and FSP, LRT1 (i.e. $p = 1$) has been verified to give the best localization results.

Use of the temporal properties of brain signals to improve the localization performance has also been attempted. An additional temporal constraint can be added, assuming that for each location the change in the source amplitude with time is minimal. The constraint to be added is $\min \|\mathbf{s}(n) - \mathbf{s}(n-1)\|_F^2$, where n denotes the time index.

14.3.8   Joint ICA-LORETA Approach

In another study [46] Infomax ICA-based BSS has been implemented as a pre-processing scheme before the application of LORETA to localize the sources underlying the mismatch negativity (MMN). MMN is an involuntary auditory ERP, which peaks at 100–200 ms when there is a violation of a regular pattern. This ERP appears to correspond to a primitive intelligence. MMN signals are mainly generated in the supramental cortex [47–51].

The LORETA analysis was performed with the scalp maps associated with selected ICA components to find the generators of these maps. Only values greater than 2.5 times the standard deviation of the standardised data (in the LORETA spatial resolution) were accepted as activations.

The inverse problem has also been tackled within a Bayesian framework [52]. In such methods some information about the prior probabilities are normally essential. Again we may consider the EEG generation model as:

$$\mathbf{x}(n) = \mathbf{H}\mathbf{s}(n) + \mathbf{v}(n) \tag{14.38}$$

where $\mathbf{x}(n)$ is an $n_e \times 1$ vector containing the EEG sample values at time n, $\mathbf{H}$ is an $n_e \times m$ matrix representing the head medium model, $\mathbf{s}(n)$ are the $m \times 1$ vector sample values of the sources at time n and $\mathbf{v}(n)$ is the $n_e \times 1$ vector of noise samples at time n. The a priori information about the sources imposes some constraints on their locations and their temporal properties. The estimation may be performed using a maximum a posteriori (MAP) criterion in which the estimator tries to find $\mathbf{s}(n)$ that maximises the probability distribution of $\mathbf{s}(n)$ given the measurements $\mathbf{x}(n)$. The estimator is denoted as:

$$\hat{\mathbf{s}}(n) = \max\left[p(\mathbf{s}(n)|\mathbf{x}(n))\right] \tag{14.39}$$

and following Bayes' rule where the posterior probability is:

$$p(\mathbf{s}(n)|\mathbf{x}(n)) = p(\mathbf{x}(n)|\mathbf{s}(n))\, p(\mathbf{s}(n))/p(\mathbf{x}(n)) \tag{14.40}$$

where $p(\mathbf{x}(n)|\mathbf{s}(n))$ is the likelihood, $p(\mathbf{x}(n))$ is the marginal distribution of the measurements, and $p(\mathbf{s}(n))$ is the prior probability. The posterior can be written in terms of energy functions that is,

$$p(\mathbf{s}(n)|z(n)) = \frac{1}{z(n)}\exp\left[-U(\mathbf{s}(n))\right] \tag{14.41}$$

and $U(\mathbf{s}(n)) = (1 - \lambda)U_1(\mathbf{s}(n)) + \lambda U_2(\mathbf{s}(n))$ where U_1 and U_2 correspond to the likelihood and the prior, respectively, and $0 \leq \lambda \leq 1$. The prior may be separated into two functions, spatial priors, U_s, and temporal priors, U_t. The spatial prior function can take into account the smoothness of the spatial variation of the sources. A cost function that determines the spatial smoothness is:

$$\Phi(u) = \frac{u^2}{1 + (u/K)^2} \tag{14.42}$$

where K is the scaling factor which determines the required smoothness. Therefore, the prior function for the spatial constraints can be written as:

$$U_s(\mathbf{s}(n)) = \sum_{k=1}^{n_e}\left[\Phi_k^x\left(\nabla_x\mathbf{s}(n)|k\right) + \Phi_k^y\left(\nabla_y\mathbf{s}(n)|k\right)\right] \tag{14.43}$$

where the indices x and y correspond to horizontal and vertical gradients, respectively.

The temporal constraints are imposed by assuming that the projection of $\mathbf{s}(n)$ to the space perpendicular to $\mathbf{s}(n-1)$ is small. Thus, the temporal prior function, as the second constraint, can be written as:

$$U_t(\mathbf{s}(n)) = \left\| \mathbf{P}_{n-1}^{\perp} \mathbf{s}(n) \right\|^2 \tag{14.44}$$

where $\mathbf{P}_{n-1}^{\perp}$ is the projection onto the perpendicular space to $\mathbf{s}(n-1)$. Therefore, the overall minimisation criterion for estimation of $\mathbf{s}(n)$ looks like:

$$\hat{\mathbf{s}}(n) = \arg\min_{\mathbf{S}} \left\{ \|\mathbf{x}(n) - \mathbf{H}\mathbf{s}(n)\|^2 + \alpha \sum_{k=1}^{n_e} \left[\Phi_k^x \left(\nabla_x \mathbf{s}(n) | k \right) + \Phi_k^y \left(\nabla_y \mathbf{s}(n) | k \right) \right] \right.$$

$$\left. + \beta \left\| \mathbf{P}_{n-1}^{\perp} \mathbf{s}(n) \right\|^2 \right\} \tag{14.45}$$

where α and β are the penalty terms (regularization parameters).

According to the results of this study the independent components can be generated by one or more spatially separated sources. This confirms that each dipole is somehow associated with one dipole generator [53]. In addition, it is claimed that a specific brain structure can participate in different components, working simultaneously in different observations. The combination of ICA and LORETA exploits spatiotemporal dynamics of the brain as well as localization of the sources.

In [38] four different inverse methods namely, WMN, sLORETA, FOCUSS, and SSLOFO have been compared (based on a spherical head assumption and in the absence of noise). It has been shown that in their original forms SSLFO gives more accurate localization when WWN and sLORETA fail in achieving unique and accurate solutions.

14.3.9   Partially Constrained BSS Method

In a recent work [54] the locations of the known sources, such as some normal brain rhythms, have been used as a prior in order to find the locations of the abnormal or the other brain source signals using constrained BSS. The cost function of the BSS algorithm is constrained by this information and the known sources are iteratively calculated. Consider $\tilde{\mathbf{A}} = [\mathbf{A}_k \vdots \mathbf{A}_{uk}]$ is the mixing matrix including the geometrical information about the known, $\mathbf{A}_k$, and unknown, $\mathbf{A}_{uk}$, sources. $\mathbf{A}_k$ is an $n_e \times k$ matrix and $\mathbf{A}_{uk}$ is an $n_e \times (m - k)$ matrix. Given $\mathbf{A}_k$, $\mathbf{A}_{uk}$ may be estimated as follows:

$$\mathbf{A}_{uk_{n+1}} = \mathbf{A}_{uk_n} - \zeta \nabla_{A_{uk}}(J_c) \tag{14.46}$$

where

$$\nabla_{A_{uk}}(J_c) = 2([\mathbf{A}_k \vdots \mathbf{A}_{uk_n}] - \mathbf{R}_{n+1}\mathbf{W}_{n+1}^{-1}) \tag{14.47}$$

$$\mathbf{R}_{n+1} = \mathbf{R}_n - \gamma \nabla_R(J_c) \tag{14.48}$$

and

$$J_c = \left\| \tilde{\mathbf{A}}_n - \mathbf{R}_{n+1} \mathbf{W}_{n+1}^{-1} \right\|_{\mathrm{F}}^2 \tag{14.49}$$

$$\nabla_R(J_c) = 2 \left(\mathbf{W}_{n+1}^{-1} \mathbf{A}_k + R_n \mathbf{W}_{n+1}^{-1} \left(\mathbf{W}_{n+1}^{-1} \right)^{\mathrm{T}} - \mathbf{W}_{n+1}^{-1} \mathbf{A}_{uk_n} \right) \tag{14.50}$$

with

$$\mathbf{W}_{n+1} = \mathbf{W}_n - \mu \nabla_W J \tag{14.51}$$

where $J(\mathbf{W}) = J_{\mathrm{m}}(\mathbf{W}) + \lambda J_c(\mathbf{W})$; J_{m} is the main BSS cost function. The parameters μ, γ and ζ are either set empirically or changed iteratively; they decrease when the convergence error decreases, and recall that $\mathbf{A}_k$ is known and remains constant.

14.3.10 Constrained Least-Squares Method for Localization of P3a and P3b

Often some information about the shape of the source helps in its localization. ERPs look like positive and negative humps in the EEG signals. Such humps may be approximated by Gaussian or Laplacian functions. A method based on least squares (LS) can be followed [55]. Using this method, the scalp maps (the column vectors of the forward matrix $\mathbf{H}$) are estimated. Consider

$$\mathbf{R} = \mathbf{X}\mathbf{Y}^{\mathrm{T}} = \mathbf{H}\mathbf{S}\mathbf{Y}^{\mathrm{T}} \tag{14.52}$$

where $\mathbf{Y}$ is a matrix with rows equal to y_c and $\mathbf{Y} = \mathbf{D}\mathbf{S}$, where $\mathbf{D}$ is a diagonal scaling matrix;

$$\mathbf{D} = \begin{bmatrix} d_1 & 0 \ldots 0 \\ 0 & d_2 \ldots 0 \\ \vdots & \ddots \vdots \\ 0 & 0 \quad d_{\mathrm{m}} \end{bmatrix} \tag{14.53}$$

Post-multiplying $\mathbf{R}$ by $\mathbf{R}_y^{-1} = \mathbf{Y}\mathbf{Y}^{-1}$;

$$\mathbf{R}\mathbf{R}_y^{-1} = \mathbf{H}\mathbf{S}\mathbf{Y}^{\mathrm{T}}(\mathbf{Y}\mathbf{Y}^{\mathrm{T}})^{-1} = \mathbf{H}\mathbf{D}^{-1}\mathbf{Y}\mathbf{Y}^{\mathrm{T}}(\mathbf{Y}\mathbf{Y}^{\mathrm{T}})^{-1} = \mathbf{H}\mathbf{D}^{-1} \tag{14.54}$$

The order of the sources is arbitrary. Therefore, the permutation does not affect the overall process. Hence, the ith scaled scalp map corresponds to the scaled ith source. An LS method may exploit the information about the scalp maps to localize the ERP sources within the brain.

Assuming an isotropic propagation model of the head, the sources are attenuated with the third power of the distance [56] that is, $d_j = 1/h_j^{1/3}$, where h_j is the jth element of a specific column of the $\mathbf{H}$ matrix. The source locations $\mathbf{q}$ are found as the solution to the following LS problem:

$$\min_{\mathbf{q},M} E(\mathbf{q}, M) = \min_{\mathbf{q},M} \sum_{j=1}^{n_e} \left[M \left\| \mathbf{q} - \mathbf{a}_j \right\|_2 - \mathbf{d}_j \right]^2 \tag{14.55}$$

where $\mathbf{a}_j$ are the positions of the electrodes, $\mathbf{d}_j$s are the scaled distances, and M (scalar) is the scale to be estimated together with $\mathbf{q}$. M and $\mathbf{q}$ are iteratively estimated according to

$$\mathbf{q}_{\rho+1} = \mathbf{q}_\rho - \mu_1 \nabla_{\mathbf{q}} E|_{\mathbf{q}=\mathbf{q}_\rho} \tag{14.56}$$

and

$$M_{\rho+1} = M_\rho - \mu_2 \nabla_M E|_{M=M_\rho} \tag{14.57}$$

where μ_1 and μ_2 are the learning rates and $\nabla_{\mathbf{q}}$ and ∇_M are respectively the gradients with respect to $\mathbf{q}$ and M which are computed as [56]

$$\nabla_{\mathbf{q}} E(\mathbf{q}, M) = 2 \sum_{j=1}^{n_e} (\mathbf{q} - \mathbf{a}_j) \left(M^2 - M \frac{\mathbf{d}_j}{\left\| \mathbf{q} - \mathbf{a}_j \right\|_2} \right) \tag{14.58}$$

$$\nabla_M E(\mathbf{q}, M) = 2 \sum_{j=1}^{n_e} M \left\| \mathbf{q} - \mathbf{a}_j \right\|_2^2 - \left\| \mathbf{q} - \mathbf{a}_j \right\|_2 \mathbf{d}_j \tag{14.59}$$

The solutions to $\mathbf{q}$ and M are unique given $n_e \geq 3$ and $n_e > m$. Using the above localization algorithm ten sets of EEGs from five patients and five controls, each divided into 20 segments, are examined. Figure 14.4 illustrates the locations of P3a and P3b sources for the patients. From this figure it is clear that the clusters representing the P3a and P3b locations are distinct. Figure 14.5 on the other hand, presents the sources for the control subjects. Unlike in Figure 14.4, the P3a and P3b sources are randomly located within the brain.

14.3.11   Spatial Notch Filtering Approach

A spatial notch filter (SNF), also called a null beamformer, has been designed to maximise the similarity between the extracted ERP source and a generic template which approximates an ERP shape, and at the same time enforce a null at the location of the source [57]. This method is based on a head and source model, which describes the propagation of the brain sources. The sources are modelled as magnetic dipoles, potentially located at some vertices within the brain, and their propagation to the sensors is mathematically described by an appropriate forward model. The considered 3D grid can be fine or coarse depending on the required accuracy of the results. In this model the EEG signal is considered as an $n_e \times T$ matrix, where

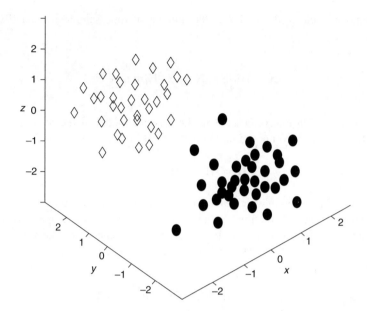

Figure 14.4 The locations of the P3a and P3b sources for five patients in a number of trials. The diamonds ◊ are the locations of the P3a sources and the circles • show the locations of P3b sources. The x-axis denotes right to left, the y-axis shows front to back, and z-axis denotes up and down. It is clear that the classes are distinct

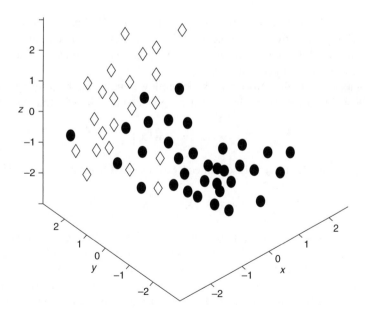

Figure 14.5 The locations of the P3a and P3b sources for five healthy individuals in a number of trials. The diamonds ◊ are the locations of the P3a sources and the circles • show the locations of P3b sources. The x-axis denotes right to left, the y axis shows front to back, and z-axis denotes up and down

n_e is the number of electrodes and T is the number of time samples for an EEG channel signal block:

$$X = HMS + N = \sum_{i=1}^{m} H_i m_i s_i + N \tag{14.60}$$

where H is an $n_e \times 3m$ matrix describing the forward model of the m sources to the n_e electrodes. H is further decomposed into m matrices H_i, $i = 1, \ldots, m$ as:

$$H = [H_1 \cdots H_i \cdots H_m] \tag{14.61}$$

where H_i are $n_e \times 3$ matrices in which each column describes the potential at the electrodes due to the ith dipole for each of the three orthogonal orientations. For example, the first column of H_i describes the forward model of the x component of the ith dipole when the y and z components are zero. Similarly, M is a $3m \times m$ matrix describing the orientation of the m dipoles and is decomposed as:

$$M = \begin{bmatrix} m_1 & 0 & 0 & 0 & 0 \\ 0 & \cdots & 0 & 0 & 0 \\ 0 & 0 & m_i & 0 & 0 \\ 0 & 0 & 0 & \cdots & 0 \\ 0 & 0 & 0 & 0 & m_m \end{bmatrix} \tag{14.62}$$

where m_i is a 3×1 vector describing the orientation of the ith dipole; s_i, a $1 \times T$ vector, is the time signal originated from ith dipole, and N is the measurement noise and the modelling error. In addition, the ERP signal r_i for the ith source has been modelled as a Gaussian-shaped template with variable width and latency.

A constrained optimisation procedure is then followed in which the primary cost function is the Euclidean distance between the reference signal and the filtered EEG [57], that is,

$$f_d(w) = \left\| r_i - w^T X \right\|_2^2 \tag{14.63}$$

The minimum point for this can be obtained by the classic LS minimisation and is given by

$$w_{opt} = \left(XX^T \right)^{-1} Xr^T \tag{14.64}$$

Following this method an $n_e \times 1$ filter w_{opt} is designed which gives an estimate of the reference signal when applied to EEGs. However, this procedure does not include any spatial information unless the w_is for all the sources are taken into account. In this way, we can construct a matrix W, similar to the separating matrix in an ICA framework, which could be converted to the forward matrix H. To proceed with this idea a constraint function is defined as:

$$f_c(w) = w^T H(p) = 0 \tag{14.65}$$

where $\mathbf{H}(p)$ is the forward matrix of a dipole at location p, and then we perform a grid search over a number of locations. The constrained optimisation problem is then stated as [57]:

$$\min f_d(\mathbf{w}) \text{ subject to } f_c(\mathbf{w}) = 0 \qquad (14.66)$$

Without going through the details the SNF $\mathbf{w}$ for extraction of a source at location j, as the desired source (ERP component), when there is neither noise or correlation amongst the components, is described as

$$\mathbf{w}^T = \mathbf{r} \sum\nolimits_{i \neq j} \mathbf{s}_i^T \mathbf{m}_i^T \mathbf{H}_i^T \mathbf{C}_x^{-1} \mathbf{H}_j \left(\mathbf{H}_j^T \mathbf{C}_x^{-1} \mathbf{H}_j \right)^{-1} \mathbf{H}_j^T \mathbf{C}_x^{-1} \qquad (14.67)$$

where $\mathbf{C}_x = \mathbf{X}\mathbf{X}^T$. In the case of correlation amongst the sources, the beamformer, $\mathbf{w}_c$, for extraction of a source at location j, includes another term as [57]:

$$\mathbf{w}_c^T = \mathbf{w}^T + \mathbf{r}\tilde{\mathbf{X}}^T \mathbf{C}_x^{-1} \mathbf{H}(p) \left(\mathbf{H}^T(p) \mathbf{C}_x^{-1} \mathbf{H}(p) \right)^{-1} \mathbf{H}^T(p) \mathbf{C}_x^{-1} \qquad (14.68)$$

where $\tilde{\mathbf{X}}$ includes all the sources correlated with the desired source and $\mathbf{w}$ is given in equation (14.67). A similar expression can also be given for when noise is involved [57].

The SNF algorithm finds the location of the desired brain ERP components (by manifesting a sharp null at the position of the desired source) with a high accuracy even in the presence of interference and noise and where the correlation between the components is considerable.

The ability of the algorithm for correct localization of the sources in various scenarios has been investigated. A forward model has also been built up using the BrainStorm software [58]. Brainstorm is a collaborative, open-source application for MEG and EEG data analysis (visualization, processing and advanced source modelling).

A three-layer spherical head model with conductivities of 0.33, 0.0042, 0.33 μS cm^{-1}, for scalp, skull, and brain, respectively has been used. 32 Gaussian pulses in 32 different locations with random orientations were created, peaking at different latencies and using 30 electrodes. This is to have an underdetermined system. Several different cases have been examined in order to evaluate the effect of noise and correlation between the sources. 2700 voxels have been considered. The source of interest is originally placed at location numbered 1000. For the simple case of no noise and uncorrelated 1sources, an accurate localization of the source has been achieved, as depicted in Figure 14.6. In this figure, the numbers on the horizontal axis show the location of the vertices within the grid inside the brain. Also, to apply the proposed spatial notch filtering method to real data a Gaussian-shaped (which could be a Laplacian-shaped) template has been selected and used as $\mathbf{r}$.

The method has been compared to the standard LCMV beamformer for various levels of noise and correlation between the ERP and he background EEG. The results of this comparison can be viewed in Figure 14.7 and 14.8.

To examine the algorithm for real EEG data the reference electrodes were linked to the earlobes, the sampling frequency set to $F_s = 2$ kHz, and the data were subsequently bandpass filtered (0.1–70 Hz). The EEG data were recorded for control and schizophrenic patients. The stimuli were presented through ear plugs inserted in the ear. Forty rare tones (1 kHz) were randomly distributed amongst 160 frequent tones (2 kHz). Their intensity was 65 dB with 10 and 50 ms duration for rare and frequent tones, respectively. The subject was asked to

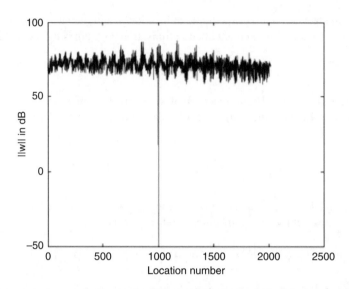

Figure 14.6 Localization plot for one source uncorrelated with other sources in a noise free environment. The location number refers to a geometrical location in a 3D grid within the brain [57]

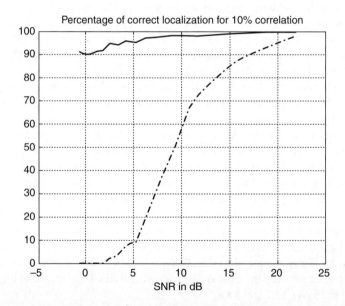

Figure 14.7 Percentage of successful localization for various SNRs for the SNF algorithm (bold) and the LCMV (dashed). The purpose is to evaluate the performance of the algorithm for different orientations of the sources. The same noise sequence has been used for 1000 different orientations and various SNR values. Here, the correlation is 10% [57]

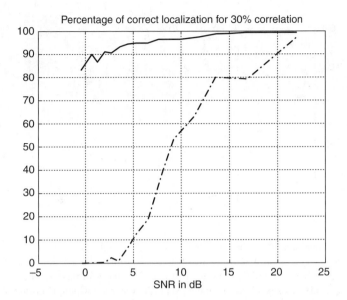

Figure 14.8 Percentage of successful localization for various SNRs for the SNF algorithm (bold) and the LCMV (dashed). The purpose is to evaluate the performance of the algorithm for different orientations of the sources. The same noise sequence has been used for 1000 different orientations and various SNR values. Here, the correlation is 30% [57]

press a button as soon as he heard a low tone (1 kHz). The ability to distinguish between low and high tones was confirmed before the start of the experiment. The task is designed to assess basic memory processes for both healthy subjects and schizophrenic patients. ERP components measured in this task included N100, P200, N200 and P3a and P3b. However, the results for the P3a and P3b have been demonstrated [57]. Figure 14.9 and 14.10 show the results of P3a and P3b source localization for the control and schizophrenic patients. In these figures the numbers next to the circles and squares show the patient number.

It can be seen that the locations of the P3a and P3b for the schizophrenic patients are less distinct than the locations for the healthy individuals [57].

14.3.12 Deflation Beamforming Approach for EEG/MEG Multiple Source Localization

Poor spatial resolution of EEG/MEG motivates research into methods that can more accurately localize the sources from the recordings using these modalities. A popular strategy in brain source localization is to use the dipole source assumption [59].

Physiologically, this assumption may not always be true. Under this assumption, electric current dipoles are specified by their 3D locations and their 3D moments. Linearly constrained minimum variance (LCMV) beamformer (BF) is a well-established method in MEG dipole source localization [60]. BF provides an adaptive method which places nulls, using some linear constraints, at positions corresponding to the noise sources. The transient and often correlated

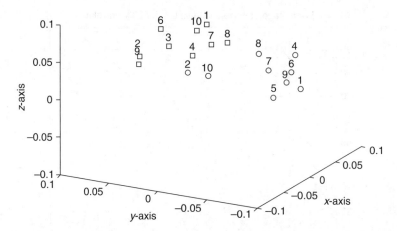

Figure 14.9 Localization plot for the P3a, circles, ○, and P3b, squares, □, for ten normal people. The numbers correspond to the control subject's number (e.g. ○1 shows the location of the P3a for control subject no. 1). The three axes refer to the geometrical coordinates in meters. The y-axis determines front–back of the head, x-axis is left–right, and z-axis is the vertical position. Units are in metres [57]

nature of the neural activation in different parts of the brain, however, often limits the BF performance. As an example, since MEG has variable sensitivity to source locations, the noise gain of the filter varies with the changes in location. One strategy to account for this effect is to normalize the output power of BF with respect to the estimated power in the presence of noise only [61]. Beamspace transformations for dimension reduction which preserve source

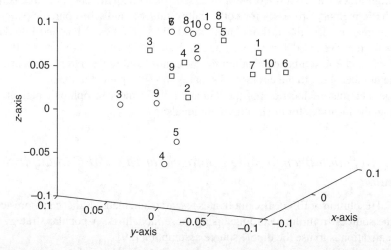

Figure 14.10 Localization plot for the P3a, circles, ○, and P3b, squares, □, for ten schizophrenic patients. The numbers correspond to the control subject's number (e.g. ○1 shows the location of the P3a for control subject no. 1). The three axes refer to the geometrical coordinates in meters. The y-axis determines front–back of the head, x-axis is left–right, and z-axis is the vertical position. Units are in metres [57]

activity located within a given region of interest have also been presented in [62]. For more details of different methods for MEG source localization, see [63, 64].

In [59] the BF approach has been generalized to include more constraints. The first constraint additionally stops the noise power at the output of the BF. This method, so called *deflation beamformer*, assumes the covariance matrix of the noise is known a priori. The second constraint places nulls at known locations where other dipoles have been detected previously to improve the detection of the unidentified dipoles. In this approach, a dipole is located by finding the grid node which has the maximum power, while deflating the power of any dipoles that have already been identified. Using a regularized optimization approach by means of penalty functions, such as Lagrange multipliers, the multiple-constraint problem is changed to an unconstrained problem and solved. During this process the power at each location is estimated and normalized with respect to the power in the presence of noise only. The deflation BF method helps to overcome two main problems of multiple dipole source localization using the BF, that is, its shortcoming of dipole localization to the dominant sources, and its inaccurate performance in the presence of highly correlated sources. In addition, to improve the performance of the method further, an iterative approach for deflation and localization of dipoles has been developed [59]. To formulate the problem let $\mathbf{y}$ be one of the measurements from N scalp electrodes. Each dipole is specified by its 3D location ρ and its 3D moment $\mathbf{m}$. The medium between the sources and the electrodes is assumed to be homogeneous, and the MEG field $\mathbf{y}$ to be a superposition of the fields from q dipoles. Therefore, we may write

$$\mathbf{y} = \sum_{i=1}^{q} \mathbf{H}(\rho_i)\mathbf{m}(\rho_i) + \mathbf{n} \tag{14.69}$$

where $\mathbf{n}$ is the noise mutually uncorrelated with the signals. For a two-source problem, assuming source ρ_1 is already localized, the deflation beamforming problem for locating source ρ_2 is stated as the following constrained optimization problem for $\mathbf{W}$ [59]:

$$\min_{\mathbf{W}^{\mathrm{T}}} \left\| \mathbf{W}^{\mathrm{T}}\mathbf{y} \right\| \text{ subject to } \min_{\mathbf{W}^{\mathrm{T}}} \left\| \mathbf{W}^{\mathrm{T}}\mathbf{n} \right\|, \mathbf{W}^{\mathrm{T}}\mathbf{H}(\rho_1) = 0, \text{ and } \mathbf{W}^{\mathrm{T}}\mathbf{H}(\rho_2) = \mathbf{I} \tag{14.70}$$

The term $\mathbf{W}^{\mathrm{T}}\mathbf{H}(\rho_1)$ deflates the first source. For a higher number of sources more sources should be deflated before the new source can be detected. The optimum solution for $\mathbf{W}$ can then be found as [59]:

$$\mathbf{W}^{\mathrm{T}} = \tilde{\mathbf{I}}_{\mathrm{d}}^{\mathrm{T}} \left(\tilde{\mathbf{H}}^{\mathrm{T}}(\mathbf{C}_y + \lambda\mathbf{C}_{\mathrm{n}})^{-1}\tilde{\mathbf{H}} \right)^{-1} \tilde{\mathbf{H}}^{\mathrm{T}}(\mathbf{C}_y + \lambda\mathbf{C}_{\mathrm{n}})^{-1} \tag{14.71}$$

where $\tilde{\mathbf{I}}_{\mathrm{d}}^{\mathrm{T}} = \begin{bmatrix} 0 & \mathbf{I} \end{bmatrix}$ and $\tilde{\mathbf{H}}^{\mathrm{T}} = \begin{bmatrix} \mathbf{H}(\rho_1) & \mathbf{H}(\rho_2) \end{bmatrix}$. $\mathbf{C}_{\mathrm{n}}$ and $\mathbf{C}_y$ are the covariance of noise and background EEG including the first source.

In many practical cases, we have no prior knowledge about the true source locations. Under these circumstances, if we locate the first source incorrectly, localization of other sources may also fail. To circumvent this problem, the sources are found via a number of iterations such that in each round the other sources are deflated while searching for the current source location; the sources then swap and source 1 is estimated while previously estimated source 2 is deflated.

To examine the method on real MEG, event-related fields (ERF) were recorded in an auditory paradigm. 1500 auditory stimuli were delivered every 0.5 s bilaterally to the subject.

The stimulus was a broadband noise lasting 0.1 s. This data set was selected because the stimulus train generates activity in the left and right primary auditory cortices. The data was acquired from a 28 year old male subject. Whole head MEG recordings were made using a 275-channel radial gradiometer system. The sampling rate was 1000 Hz and recordings were filtered off-line with a bandpass of 0.03 to 40 Hz. After visual rejection of trials containing eyeblink and movement artefacts, the remaining trials were averaged. The estimated noise covariance matrix $\mathbf{C}_n$ used in both the BF and deflation BF approaches, was calculated from the 0.1 s pre-stimulus segments. In this experiment, two dipoles were considered; one in the right and one in the left hemisphere of the brain).

Figure 14.11a and b show the power profiles of the BF and estimated locations in axial and coronal views. The estimated locations are shown with cross markers. The estimated power

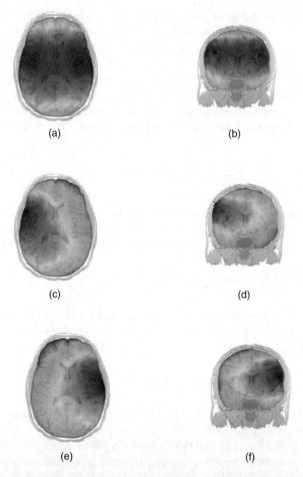

(a)   (b)

(c)   (d)

(e)   (f)

Figure 14.11 Topographies or power profiles of real MEG data obtained using (a) BF method in axial view, (b) BF method in coronal view, (c) deflation BF while deflating the second source in axial view, (d) deflation BF while deflating the second source in coronal view, (e) deflation BF while deflating the first source in axial view, (f) deflation BF while deflating the first source in coronal view [59] (see Plate 18 for the coloured version)

profile and estimated locations using the deflation BF method are shown in Figure 14.11c–f. Figure 14.11c and d show the location of the first dipole while deflating the second dipole, and Figure 14.11e and f show the location of the second dipole while deflating the first dipole. The deflation BF method converged after three iterations. It is seen from Figure 14.11a and b that for the BF the sources are bleeding towards the centre of the head due to the partial correlation between the sources. On the other hand, the deflation BF places the sources at biologically plausible locations in the primary auditory cortices in the left and right hemisphere. The power obtained using the deflation BF is also more focal than the power obtained using the BF approach. Furthermore, unlike for the BF, no spurious activations near the centre of the sphere model were observed when deflation BF was implemented [59].

14.3.13   Hybrid Beamforming – Particle Filtering

In EEG or MEG source localization the number of unknown parameters is often higher than the number of known measurements and, therefore, the system is underdetermined. *Dipole source localization* assumes that one or multiple current dipoles represent the electric sources [65]. To express and formulate the problem, suppose the measured multichannel signals $\mathbf{y}_k$ from M sensors at time k are produced by q dipoles, so we can model $\mathbf{y}_k$ as:

$$\mathbf{y}_k = \sum_{i=1}^{q} \mathbf{F}(\rho_k(i))\,\mathbf{m}_k(i) + \mathbf{v}_k \qquad (14.72)$$

where $\rho_k(i)$ is a 3D location vector, $\mathbf{m}_k(i)$ is a 3D moment vector of the ith dipole and $\mathbf{v}_k$ is the observation noise. It is assumed here that the number of dipoles is known *a priori* and their locations are to be found. Hence, the states are defined as [65]:

$$\mathbf{L}_k = [\rho_k(1), \ldots, \rho_k(q)] \qquad (14.73)$$

By introducing an $M \times 3q$ matrix of location $\mathbf{F}(\mathbf{L}_k)$ and a $3q \times 1$ vector of moments $\underline{\mathbf{m}}_k$ as [65]:

$$\mathbf{F}(\mathbf{L}_k) = [\mathbf{F}(\rho_1), \ldots, \mathbf{F}(\rho_q)] \qquad (14.74)$$

$$\underline{\mathbf{m}}_k = [\mathbf{m}_k^{\mathrm{T}}(1), \ldots, \mathbf{m}_k^{\mathrm{T}}(q)]^{\mathrm{T}} \qquad (14.75)$$

In that case a matrix form of equation (14.72) is given as:

$$\mathbf{y}_k = \mathbf{F}(\mathbf{L}_k)\,\underline{\mathbf{m}}_k + \mathbf{v}_k$$

where $\mathbf{F}$ is a nonlinear function of q dipoles. The state-space equations of particle filter (PF) take the following form:

$$\begin{aligned} \mathbf{L}_k &= \mathbf{L}_{k-1} + \mathbf{w}_{k-1} \\ \mathbf{y}_k &= \mathbf{F}(\mathbf{L}_k)\,\underline{\mathbf{m}}_k + \mathbf{v}_k \end{aligned} \qquad (14.76)$$

In which $\mathbf{w}_k$ is the state noise. In these equations to estimate $\mathbf{L}_k$, however, $\underline{\mathbf{m}}_k$ has to be known. The vector of moments corresponds linearly to the EEG (or MEG) measured data and under no noise assumption, $\underline{\mathbf{m}}_k$, can be recursively calculated as [65]:

$$\underline{\mathbf{m}}_k = \mathbf{F}^\dagger (\mathbf{L}_{k-1}) \mathbf{y}_k \tag{14.77}$$

where $\mathbf{F}^\dagger = (\mathbf{F}^T \mathbf{F})^{-1} \mathbf{F}^T$ which is the pseudo-inverse of $\mathbf{F}$. Note that in equation (14.77), $\underline{\mathbf{m}}_k$ is estimated from the location matrix $\mathbf{L}_{k-1}$ at the previous step and measurements $\mathbf{y}_k$ at the current step. This method is a grid-based method meaning that the sources are considered to be in one of the predefined vertices of a 3D grid inside the head. Therefore, after the prediction stage, the locations might need to be adjusted to the nearest cells' locations.

To use a beamformer, assume that there are G grid points out of which q of them are the source locations and the EEG/MEG signals can be decomposed as [65]:

$$\mathbf{y}_k = \sum_{i=1}^{G} \mathbf{F}(g_i) \mathbf{m}_k(i) \tag{14.78}$$

To select the source of interest coming from location $\breve{\rho}$, a linear spatial filter $\mathbf{W}_{\breve{\rho}}$ is used to have the following ideal response:

$$\mathbf{W}_{\breve{\rho}}^T \mathbf{F}(g_i) = \begin{cases} \mathbf{F}(g_i) & g_i \in \breve{\rho} \\ \mathbf{O} & \text{else} \end{cases} \tag{14.79}$$

where $\mathbf{O}$ is a $3 \times M$ null matrix. Using two new matrices $\underline{\mathbf{F}} = [\mathbf{F}(g_1) \dots \mathbf{F}(g_G)]$ and $\mathbf{F_O} = [\mathbf{O} \dots \mathbf{F}(\rho(1)) \dots \mathbf{F}(\rho(q)) \dots \mathbf{O}]$ which has zero entries except at locations of the dipoles we can have

$$\mathbf{W}_{\breve{\rho}}^T \underline{\mathbf{F}} = \mathbf{F_O} \tag{14.80}$$

From this, an optimum solution can be found as [65]:

$$\mathbf{W}_{\breve{\rho}}^T = \underline{\mathbf{F}}^{\dagger T} \mathbf{F_O}^T \tag{14.81}$$

To construct the beamforming PF the spatially filtered data is used instead of the original measurements to compute the moment vector $\underline{\mathbf{m}}_k$. The filter in (14.81) now can be applied to the measurements to have

$$\underline{\mathbf{m}}_k = \mathbf{F}^\dagger (\mathbf{L}_{k-1}) \mathbf{W}_{\breve{\rho}}^T \mathbf{y}_k \tag{14.82}$$

Matrix $\mathbf{F_O}$ in equation (14.81) is constructed for each particle using locations indicated by particles from the previous step. Moreover, since matrix $\mathbf{F}$ is the matrix of all gains and is independent of the desired locations, for computational efficiency, its pseudo-inverse can be calculated only one time. This algorithm has been applied to synthetic and real sources and its good performance demonstrated [65].

14.4   Determination of the Number of Sources from the EEG/MEG Signals

For the majority of applications in source localization a priori information about the number of sources is required. The problem of detection of the number of independent (or uncorrelated) sources can be defined as analysing the structure of the covariance matrix of the observation matrix. This matrix can be expressed as $\mathbf{C} = \mathbf{C}_{\text{sig}} + \mathbf{C}_{\text{noise}}$, where $\mathbf{C}_{\text{sig}}$ and $\mathbf{C}_{\text{noise}}$ are, respectively, the covariance of source signals and the covariance of noise. PCA and SVD may perform well if the noise level is low. In this case the number of dominant eigenvalues represents the number of sources. In the case of white noise, the covariance matrix $\mathbf{C}_{\text{noise}}$ can be expressed as $\mathbf{C}_{\text{noise}} = \sigma_n^2 \mathbf{I}$, where σ_n^2 is the noise variance and $\mathbf{I}$ is the identity matrix. In the case of coloured noise some similar methods can be implemented if the noise covariance is known apart from σ_n^2. The noise covariance matrix is a symmetric positive definite matrix $\mathbf{C}_{\text{noise}} = \sigma_n^2 \boldsymbol{\Psi}$. Then, a non-singular square matrix $\boldsymbol{\psi}$ $(m \times m)$ exists such that $\boldsymbol{\Psi} = \boldsymbol{\psi}\boldsymbol{\psi}^{\mathrm{T}}$ [66].

For both white and coloured noises the eigenvalues can be calculated from the observation covariance matrices, and then analysed by implementing the information theoretic criterion to estimate the number of independent sources [67].

However, it has been shown that this approach is suboptimal when the sources are temporally correlated [68]. Selection of an appropriate model for EEG analysis and source localization has been investigated by many researchers, and several criteria have been established to solve this problem. In most of these methods the amplitudes of the sources are tested to establish whether they are significantly larger than zero, in which case the sources are included in the model. Alternatively, the locations of the sources can be tested to determine whether they differ from each other significantly, in which case these sources should also be included in the model.

PCA and ICA may separate the signals into their uncorrelated and independent components, respectively. By back-projecting the individual components to the scalp electrodes both the above criteria may be tested. Practically, the number of distinct active regions within the back-projected information may denote the number of sources. The accuracy of the estimation also increases when the regions are clustered based on their frequency contents. However, due to the existence of noise with unknown distribution the accuracy of the results is still under question.

In [69] the methods based on residual variance (RV), Akaike information criterion (AIC), Bayesian information criterion (BIC), and Wald tests on amplitudes (WA) and locations (WL) have been discussed. These methods have been later examined on MEG data [70] for both pure white error and coloured error cases. The same methods can be implemented for the EEG data too. In this test the MEG data from m sensors and T samples are collected for each independent trial $j = 1, \ldots, n_e$ in the $m \times T$ matrix $\mathbf{Y}_j = (\mathbf{y}_{1j}, \ldots, \mathbf{y}_{Tj})$, with $\mathbf{y}_{ij} = (y_{1ij}, \ldots, y_{mij})^{\mathrm{T}}$. Considering the average over a number of trials n as $\bar{\mathbf{Y}} = \frac{1}{n} \sum_{j=1}^{n} \mathbf{Y}_j$, the model for the averaged data can be given as:

$$\bar{\mathbf{Y}} = \mathbf{GA} + \mathbf{E} \qquad (14.83)$$

where $\mathbf{G}$ includes the sensor gains of the sources of unit amplitudes. Matrix $\mathbf{G}$ depends on the location and orientation parameters of the dipolar sources. Based on this model the tests for model selection are as follows:

The RV test defined as [69]:

$$RV = 100 \frac{\text{tr}[(\bar{\mathbf{Y}} - \mathbf{GA})(\bar{\mathbf{Y}} - \mathbf{GA})^{\text{T}}]}{\text{tr}[\bar{\mathbf{Y}}\bar{\mathbf{Y}}]} \tag{14.84}$$

compares the squared residuals to the squared data for all sensors and samples simultaneously. The RV decreases as a function of the number of parameters, and therefore over-fits easily. The model is said to fit if the RV is below a certain threshold [71].

The AIC method penalizes the log-likelihood function for additional parameters required to describe the data. These parameters may somehow describe the sources [71]. The number of sources has been kept limited for this test since at some point any additional source hardly decreases the log-likelihood function, but increases the penalty. The AIC is defined as:

$$\text{AIC} = nmT \ln\left(\frac{\pi s^2}{n}\right) + \frac{1}{ns^2}\text{tr}[(\bar{\mathbf{Y}} - \mathbf{GA})(\bar{\mathbf{Y}} - \mathbf{GA})'] + 2p \tag{14.85}$$

In this equation s^2 is the average of diagonal elements of the spatial covariance matrix [69].

The BIC test resembles the AIC method with more emphasis on the additional parameters. Therefore less over-fitting is expected when using BIC. This criterion is defined as [72]:

$$\text{BIC} = n_e mT \ln\left(\frac{\pi s^2}{n_e}\right) + \frac{1}{n_e s^2}\text{tr}[(\bar{\mathbf{Y}} - \mathbf{GA})(\bar{\mathbf{Y}} - \mathbf{GA})^{\text{T}}] + p \ln(mT) \tag{14.86}$$

Similarly, the model with the minimum BIC is selected.

The Wald test is another important criterion, which gives the opportunity to test a hypothesis on a specific subset of the parameters [73]. Both amplitudes and locations of the sources can be tested using this criterion. If $\mathbf{r}$ is a q vector function of the source parameters (i.e. the amplitude and location), $\mathbf{r}_h$ the q vector of fixed hypothesized value of $\mathbf{r}$, $\mathbf{R}$ the $q \times k$ Jacobian matrix of $\mathbf{r}$ with respect to k parameters, and $\mathbf{C}$ the $k \times k$ covariance matrix of source parameters, then the Wald test is defined as [74]:

$$\mathbf{W} = \frac{1}{q}(\mathbf{r} - \mathbf{r}_h)^{\text{T}}(\mathbf{RC}^{-1}\mathbf{R}')^{-1}(\mathbf{r} - \mathbf{r}_h) \tag{14.87}$$

An advantage of using the WA technique in spatio-temporal analysis is the possibility of checking the univariate significance levels to determine at which samples the sources are active.

The tests carried out for two synthetic sources and different noise components [70] showed that the WL test has superior overall performance, and the AIC and WA perform well when the sources are close together.

These tests have been based on simulations of the sources and noise. It is also assumed that the locations and the orientations of the source dipoles are fixed and only the amplitudes change. For real EEG (or MEG) data, however, such information may be subject to change and generally unknown.

Probably the most robust approach for detection of the number of sources is that developed by Bai and He [66]. In this approach an information criterion method in which the penalty functions are selected based on the spatio-temporal source model, has been developed to estimate the number of independent dipole sources from EEG or MEG In this approach the following steps are followed:

1. Calculating the covariance matrix $\mathbf{C_X}$ from the measured data matrix $\mathbf{X}$.
2. Applying SVD to decompose $\mathbf{C_X}$ to obtain the eigenvalues λ_i, where $\lambda_1 > \cdots > \lambda_{\mathrm{m.}}$
3. Calculating the information criterion value using the eigenvalues. The information criterion (IC_k) can be calculated using either

$$IC = w(n_e - k)\log\frac{1}{n_e - k}\sum_{i=k+1}^{n_e}\lambda_1 - w\sum_{i=k+1}^{n_e}\log\lambda_1 + 2d(k, n_e)\beta(w) \quad (14.88)$$

when the noise information is accurate. In equation (14.88) $d(k, m) = k(2m - k + 1)/2$. In places where the noise statistics is unknown or

$$IC = -\left(w - 1 - k - \frac{2(n_e - k)^2 + n_e - k + 2}{6(n_e - k)} + \sum_{i=1}^{k}\frac{\bar{\lambda}_{n_e-k}^2}{(\lambda_i - \bar{\lambda}_{n_e-k})^2}\right) \cdot$$

$$\qquad\qquad\qquad\qquad\qquad\qquad\qquad\qquad\qquad\qquad\qquad (14.89)$$

$$\log\left(\bar{\lambda}_{n_e-k}^{-(n_e-k)}\prod_{i=k+1}^{n_e}\lambda_i\right) + 2d(k, n_e)\beta(w - 1)$$

Where in this equation $d(k,n_e) = k(n_e - k + 2)(n_e - k - 1)/2$, $\bar{\lambda}_{n_e-k}$ is the average of the n_e -k smallest eigenvalues. In both equations n_e is the number of electrode signals, k is the number of assumed dipole sources to be estimated, w is the number of time points to form a spatio-temporal data matrix, and $\beta(w)$ is the penalty function which can have constant or logarithmic values [66].

4. According to the rule of information criterion [66] the number of sources with minimum IC is selected as the estimated number of sources.

In various experiments it has been shown that in a moderate noise environment, which is the case for EEG signals, the accuracy of the method is above 80%.

14.5 Conclusions

Source localization, from only EEG and MEG signals, is an ill-posed optimisation problem. This is mainly because the number of sources is unknown. This number may change from time to time, especially when the objective is to investigate the EP or movement-related sources. Most of the proposed algorithms fall under one of the two methods of equivalent current dipole and linear distributed approaches. Some of the above methods, such as sLORETA, have been commercialised and reported to have reasonable outcome for many applications. A hybrid system of different approaches seems to give better results. Localization may also be more accurate if the proposed cost functions can be constrained by some additional information stemming from clinical findings or from certain geometrical boundaries. Non-homogeneity of the head medium is another major problem for EEG source localization. This is less troublesome for MEG; comprehensive medical and physical experimental studies have to be carried out to find an accurate model of the head. By combining another neuroimaging modality, such as fMRI, with EEG, the accuracy of the localization process increases significantly. There are many potential applications for brain source localization, such as for localization of the ERP signals [20] and movement-related potentials useful for brain–computer interfacing [75], and seizure source localization [76]. Recent advances in EEG and MEG source localization such as those described in this chapter allow more accurate localization of multiple sources.

References

[1] Miltner, W., Braun, C., Johnson, R.E. and Rutchkin, A.D.S. (1994) A test of brain electrical source analysis (BESA): A simulation study. *Electroencephal. Clin. Neurophysiol.* **91**, 295–310.

[2] Scherg, M. and Ebersole, J.S. (1994) Brain source imaging of focal and multifocal epileptiform EEG activity. *Clin. Neurophysiol.* **24**, 51–60.

[3] Scherg, M., Best, T. and Berg, P. (1999) Multiple source analysis of interictal spikes: goals, requirements, and clinical values. *J. Clin. Neurophysiol.* **16**, 214–224.

[4] Aine, C., Huang, M., Stephen, J. and Christopher, R. (2000) Multistart algorithms for MEG empirical data analysis reliably characterize locations and time courses of multiple sources. *NeuroImage* **12**, 159–179.

[5] Phillips, C., Rugg, M.D. and Friston, K.J. (2002) Systematic regularization of linear inverse solutions of the EEG source localization problem. *Neuroimage* **17**, 287–301.

[6] Backus, G.E. and Gilbert, J.F. (1970) Uniqueness in the inversion of inaccurate gross earth data. *Philos. Trans. R. Soc.*, **266**, 123–192.

[7] Sarvas, J. (1987) Basic mathematical and electromagnetic concepts of the biomagnetic inverse problem. *Phys. Med. Biolog.*, **32**, 11–22.

[8] Hamalainen, M.S. and Llmoniemi, R.J. (1994) Interpreting magnetic fields of the brain: minimum norm estimates. *Med. Biol. EMG Comput.* **32**, 35–42.

[9] Menendez, R.G. and Andino, S.G. (1999) Backus and Gilbert method for vector fields. *Hum. Brain Mapp.* **7**, 161–165.

[10] Pascual-Marqui, R.D. (1999) Review of methods for solving the EEG inverse problem. *Int. J. Bioelectromagn.* **1**, 75–86.

[11] Uutela, K., Hamalainen, M.S. and Somersalo, E. (1999) Visualization of magnetoencephalographic data using minimum current estimates. *NeuroImage* **10**, 173–180.

[12] Phillips, C., Rugg, M.D. and Friston, K.J. (2002) "Anatomically informed basis functions for EEG source localisation: combining functional and anatomical constraints. *NeuroImage* **16**, 678–695.

[13] Banach, S. (1932) Théorie des opérations linéaires, vol. 1. Virtual Library of Science Math. – Phys. Collection, Warsaw.

[14] Yao, J. and Dewald, J.P.A. (2005) Evaluation of different cortical source localization methods using simulated and experimental EEG data. *NeuroImage* **25**, 369–382.

[15] Yetik, I.S., Nehorai, A., Lewine, J.D. and Muravchik, C.H. (2005) Distinguishing between moving and stationary sources using EEG/MEG measurements with an application to epilepsy. *IEEE Trans. Biomed. Eng.*, **52**(3), 471–479.

[16] Xu, P., Tian, Y., Chen, H. and Yao, D. (2007) Lp norm iterative sparse solution for EEG source Localization. *IEEE Trans. Biomed. Eng.*, **54**(3), 400–409.

[17] Pascual-Marqui, R.D., Michel, C.M. and Lehmann, D. (1994) Low resolution electromagnetic tomography; a new method for localizing electrical activity in the brain. *Int. J. Psychophysiol.* **18**, 49–65.

[18] He, B., Zhang, X., Lian, J. *et al.* (2002) Boundary element method-based cortical potential imaging of somatosensory evoked potentials using subjects' magnetic resonance images. *NeuroImage*, **16**, 564–576.

[19] Ricamato, A., Dhaher, Y. and Dewald, J. (2003) Estimation of active cortical current source regions using a vector representation scanning approach. *J. Clin. Neurophysiol.* **20**, 326–344.

[20] Spyrou, L., Jing, M., Sanei, S. and Sumich, A. (2007) Separation and localization of P300 sources and their subcomponents using constrained blind source separation. *EURASIP J. Appl. Signal Process.*, Article ID 82912, 10 pages.

[21] Schmit, R.O. (1986) Multiple emitter location and signal parameter estimation. *IEEE Trans. Antennas Propag.*, **34**, 276–280; reprint of the original paper presented at RADC Spectrum Estimation Workshop, 1979.

[22] Mosher, J.C. and Leahy, R.M. (1998) Recursive Music: A framework for EEG and MEG Source Localization. *IEEE Trans. Biomed. Eng.*, **45**(11), 1342–1354.

[23] Mosher, J.C., Leahy, R.M. and Lewis, P.S. (1999) EEG and MEG: Forward solutions for inverse methods. *IEEE Trans. Biomed. Eng.*, **46**(3), 245–259.

[24] Mosher, J.C. and Leahy, R.M. (1999) Source localization using recursively applied and projected (RAP) MUSIC. *IEEE Trans. Biomed. Eng.*, **47**(2), 332–340.

[25] Ermer, J.J., Mosher, J.C., Baillet, S. and Leahy, R.M. (2001) Rapidly recomputable EEG forward models for realisable head shapes. *J. Phys. Med. Biol.*, **46**, 1265–1281.

[26] Mosher, J.C., Lewis, P.S. and Leahy, R.M. (1992) Multiple dipole modelling and localization from spatio-temporal MEG data. *IEEE Trans. Biomed. Eng.*, **39**, 541–557.

[27] Golub, G.H. and Pereyra, V. (1973) The differentiation of pseudo-inverses and nonlinear least squares problems whose variables separate. *SIAM J. Numer. Anal.*, **10**, 413–432.

[28] Golub, G.H. and Van Loan, C.F. (1984) *Matrix Computations*, 2nd edn, John Hopkins Univ. Press, Baltimore, MD.

[29] Xu, X.-L., Xu, B. and He, B. (2004) An alternative subspace approach to EEG dipole source localization. *Phys. Med. Biol.*, **49**, 327–343.

[30] Darvas, F., Ermer, J.J., Mosher, J.C. and Leahy, R.M. (2006) Generic head models for atlas-based EEG source analysis. *Hum. Brain Mapp.*, **27**(2), 129–143.

[31] Buchner, H., Knoll, G., Fuchs, M. *et al.* (1997) Inverse localization of electric dipole current sources in finite element models of the human head. *Electroenceph. Clin. Neurophys.*, **102**(4), 267–278.

[32] Steele, C.W. (1996) *Numerical Computation of Electric and Magnetic Fields*, Kluwer Academic Publishers.

[33] Geddes, A. and Baker, L.E. (1967) The specific resistance of biological material—A compendium of data for the biomedical engineer and physiologist. *Med. Biol. Eng.*, **5**, 271–293.

[34] Schimpf, P.H., Liu, H., Ramon, C. and Haueisen, J. (2005) Efficient electromagnetic source imaging with adaptive standardized LORETA/FOCUSS. *IEEE Trans. BioMed. Eng.*, **52**(5), 901–908.

[35] Gorodnitsky, I.F., George, J.S. and Rao, B.D. (1999) Neuromagnetic source imaging with FOCUSS: A recursive weighted minimum norm algorithm. *J. Clinc. Neurophysiol.*, **16**(3), 265–295.

[36] Pascual-Marqui, R.D. (2002) Standardized low resolution brain electromagnetic tomography (sLORETA): Technical details. *Method Find. Exp. Clin. Pharmacol.*, **24D**, 5–12.

[37] Hanson, P.C. (1998) *Rank-Efficient and Discrete Ill-Posed Problems*, SIAM, Philadelphia, PA.

[38] Liu, H., Schimpf, P.H., Dong, G. *et al.* (2005) Standardized shrinking LORETA-FOCUSS (SSLOFO): A new algorithm for spatio-temporal EEG source reconstruction. *IEEE Trans. Biomed. Eng.*, **52**(10), 1681–1691.

[39] Liu, H., Gao, X., Schimpf, P.H. *et al.* (2004) A recursive algorithm for the three-dimensional imaging of brain electric activity: shrinking LORETA-FOCUSS. *IEEE Trans. Biomed. Eng.*, **51**(10), 1794–1802.

[40] Tikhonov, A.N. and Arsenin, V.Y. (1997) *Solution of Ill Posed Problems*, New York, Wiley.

[41] Ferguson, A.S. and Stronik, G. (1997) Factors affecting the accuracy of the boundary element method in the forward problem: I. Calculating surface potentials. *IEEE Trans. Biomed. Eng.*, **44**, 440–448.

[42] Buchner, H., Knoll, G., Fuchs, M. *et al.* (1997) Inverse localisation of electric dipole current sources in finite element models of the human head. *Electroenceph. Clin. Neurophysiol.*, **102**(4), 267–278.

[43] Ashburner, J. and Friston, K.J. (1997) multimodal image coregistration and partitioning – A unified framework. *NeuroImage* **6**, 209–217.

[44] Vapnic, V. (1998) *Statistical Learning Theory*, John Wiley & Sons Inc.

[45] Fuchs, M., Drenckhahn, R., Wichmann, H.A. and Wager, M. (1998) An improved boundary element method for realistic volume-conductor modelling. *IEEE Trans. BioMed. Eng.*, **45**, 980–977.

[46] Macro-Pallares, J., Grau, C. and Ruffini, G. (2005) Combined ICA-LORETA analysis of mismatch negativity. *NeuroImage* **25**, 471–477.

[47] Alain, C., Woods, D.L. and Night, R.T. (1998) A distributed cortical network for auditory sensory memory in humans. *Brain Res.* **812**, 23–27.

[48] Rosburg, T., Haueisen, J. and Kreitschmann-Andermahr, I. (2004) The dipole location shift within the auditory evoked neuromagnetic field components N100m and mismatch negativity (MMNm). *Clin. Neurosci. Lett.* **308**, 107–110.

[49] Jaaskelainen, I.P., Ahveninen, J., Bonmassar, G. and Dale, A.M. (2004) Human posterior auditory cortex gates novel sounds to consciousness. *Proc. Natl. Acad. Sci.* **101**, 6809–6814.

[50] Kircher, T.T.J., Rapp, A., Grodd, W. *et al.* (2004) Mismach negativity responses in schizophrenia: a combined fMRI and whole-head MEG study. *Am. J. Psychiat.* **161**, 294–304.

[51] Muller, B.W., Juptner, M., Jentzen, W. and Muller, S.P. (2002) Cortical activation to auditory mismatch elicited by frequency deviant and complex novel sounds: a PET study. *NeuroImage* **17**, 231–239.

[52] Serinagaoglu, Y., Brooks, D.H. and Macleod, R.S. (2005) Bayesian solutions and performance analysis in bioelectric inverse problems. *IEEE Trans. Biomed. Eng.*, **52**(6), 1009–1020.

[53] Makeig, S., Debener, S., Onton, J. and Delorme, A. (2004) Mining event related brain dynamic trends. *Cogn. Sci.* **134**, 9–21.

[54] Latif, M.A., Sanei, S., Chambers, J.A. and Shoker, L. (2006) Localization of abnormal EEG sources using blind source separation partially constrained by the locations of known sources. *IEEE Signal Process. Lett.*, **13**(3), 117–120.

[55] Spyrou, L., Sanei, S. and Cheong Took, C. (2007) Estimation and location tracking of the P300 subcomponents from single-trial EEG. Proceedings of IEEE, International Conference on Acoustics, Speech, and Signal Processing, ICASSP, USA, 2007.

[56] Sarvas, J. (1987) Basic mathematical and electromagnetic concepts of the biomagnetic inverse problem. *Phys. Med. Biol.*, **32**(1), 11–22.

[57] Spyrou, L. and Sanei, S. (2008) Source localisation of event related potentials incorporating spatial notch filters. *IEEE Trans. Biomed. Eng.*, **55**(9), 2232–2239.

[58] Tadel, F., Baillet, S., Mosher, J.C. *et al.* (2011) Brainstorm: A user-friendly application for MEG/EEG Analysis. *Comput. Intell. Neurosci.*, **2011**, 13. doi: 10.1155/2011/879716

[59] Mohseni, H.R. and Sanei, S. (2010) A new beamforming-based MEG dipole source localization method. Proceedings of IEEE, International Conference on Acoustics, Speech and Signal Processing, ICASSP, USA, 2010.

[60] Van Veen, B.D., Van Dronglen, W., Yuchtman, M. and Suzuki, A. (1997) Localization of brain electrical activity via linearly constrained minimum variance spatial filtering. *IEEE Trans. Biomed. Eng.*, **44**, 867–880.

[61] Robinson, S.E. and Vrba, J. (1999) Functional neuroimaging by synthetic aperture magnetometry (SAM), in *Recent Advances in Biomagnetism* (eds T. Yoshimoto, M. Kotani, S. Kuriki *et al.*), Tohoku University Press, Japan, pp. 302–305.

[62] Gutierrez, D., Nehorai, A. and Dogandzic, A. (2006) Performance analysis of reduced-rank beamformers for estimating dipole source signals using EEG/MEG. *IEEE Trans. Biomed. Eng.*, **53**(5), pp. 840–844.

[63] Baillet, S., Mosher, J.C. and Leahy, R.M. (2001) Electromagnetic brain mapping. *IEEE Signal Process. Magn.*, **18**, 14–30.

[64] Michela, C.M., Murraya, M.M., Lantza, G. *et al.* (2004) EEG source imaging. *Clin. Neurophysiol.*, **115**, 2195–2222.

[65] Mohseni, H.R., Ghaderi, F., Wilding, E. and Sanei, S. (2009) A beamforming particle filter for EEG dipole source localization. Proceedings of IEEE, International Conference on Acoustics, Speech and Signal Processing, ICASSP, Taiwan, 2009.

[66] Bai, X. and He, B. (2006) Estimation of number of independent brain electric sources from the scalp EEGs. *IEEE Trans. Biomed.*, **53**(10), 1883–1892.

[67] Knösche, T., Brends, E., Jagers, H. and Peters, M. (1998) Determining the number of independent sources of the EEG: A simulation study on information criteria. *Brain Topography*, **11**, 111–124.

[68] Stoica, P. and Nehorai, A. (1989) MUSIC, maximum likelihood, and Cramer-Rao bound. *IEEE Trans. Acoust. Speech, Signal Process.*, **37**(5), 720–741.

[69] Waldorp, A, Huizenga, H.M., Nehorai, A. and Grasman, R.P. (2002) Model selection in electromagnetic source analysis with an application to VEFs. *IEEE Trans. Biomed. Eng.*, **49**(10), 1121–1129.

[70] Waldorp, L.J., Huizenga, H.M., Nehorai, A. and Grasman, R.P. (2005) Model selection in spatio-temporal electromagnetic source analysis. *IEEE Trans. Biomed. Eng.*, **52**(3), 414–420.

[71] Akaike, H. (1992). Information theory and an extension of the maximum likelihood principle in *Breakthroughs in Statistics, Vol. 1*, (eds S. Kotz, and N.L. Johnson), Springer-Verlag, London, pp. 610–624.

[72] Chow, G.C. (1981) A comparison of information and posterior probability criteria for model selection. *J. Econometrics*, **16**, 21–33.

[73] Huzenga, H.M., Heslenfeld, D.J., and Molennar, P.C.M. (2002) Optimal measurement conditions for spatiotemporal EEG/MEG source analysis. *Psychometrika*, **67**, 299–313.

[74] Seber, G.A.F. and Wild, C.J. (1989) *Nonlinear Regression*, John Wiley & Sons, Toronto, Canada.

[75] Wentrup, M.G., Gramann, K., Wascher, E. and Buss, M. (2005) EEG source localization for brain-computer-interfaces. Proceedings of IEEE EMBS Conference, 2005, pp. 128–131.

[76] Ding, L., Worrell, G.A., Lagerlund, T.D. and He, B. (2006) 3D source localization of interictal spikes in epilepsy patients with MRI lesions. *Phys. Med. Biol.*, **51**, 4047–4062.

15

Seizure and Epilepsy

15.1 Introduction

Epilepsy is a known, mostly chronic, neurological disorder affecting more than one in a thousand humans worldwide. The word "epilepsy" is derived from the Greek word "epilambanein", which means "to seize or attack". Originally, it comes from the ancient Indian medicine during the Vedic period of 4500–1500 BC. In the Ayurvedic literature of Charaka Samhita (around 400 BC and the oldest existing description of the complete Ayurvedic medical system), epilepsy is described as "*apasmara*" which means "*loss of consciousness*". We now know, however, that seizures are the result of sudden, usually brief, excessive electrical discharges in a group of brain cells (neurones). We are also aware of the fact that different regions of the brain can be the sites of such discharges. The clinical manifestations of seizures, therefore, vary and depend on where in the brain the disturbance first starts and how far it spreads. Transient symptoms can occur, such as loss of awareness or consciousness and disturbances of movement, sensation (including vision, hearing and taste), mood or mental function.

The literature of Charaka Samhita contains abundant references to all aspects of epilepsy, including symptomatology, aetiology, diagnosis and treatment. Another ancient and detailed account of epilepsy is on a Babylonian tablet in the British Museum in London. This is a chapter from a Babylonian textbook of medicine comprising 40 tablets dating as far back as 2000 BC. The tablet accurately records many of the different recognised seizure types. In contrast to the Ayurvedic medicine, however, it emphasizes the supernatural nature of epilepsy, with each seizure type associated with the name of a spirit or god – usually evil. Treatment was, therefore, largely a spiritual matter. The Babylonian view was the forerunner of the Greek concept of "*the sacred disease*", as described in the famous treatise by Hippocrates (dated to the fifth Century BC). The term "*seleniazetai*" was also often used to describe people with epilepsy because they were thought to be affected by the phases of the moon or by the moon god (Selene), and hence the notion of "*moonstruck*" or "*lunatic*" (the Latinized version) arose. Hippocrates, however, believed that epilepsy was not sacred, but a disorder of the brain. He recommended physical treatments and stated that if the disease became chronic, it was incurable. However, the perception that epilepsy was a brain disorder did not begin to take root until the eighteenth and nineteenth Centuries AD. The intervening 2000 years were dominated by more supernatural views. In Europe, for example, St Valentine has been the patron saint of

Adaptive Processing of Brain Signals, First Edition. Saeid Sanei.
© 2013 John Wiley & Sons, Ltd. Published 2013 by John Wiley & Sons, Ltd.

people with epilepsy since medieval times. During this time people with epilepsy were viewed with fear, suspicion and misunderstanding and were subjected to enormous social stigma.

During the nineteenth Century, as neurology emerged as a new discipline distinct from psychiatry, the concept of epilepsy as a brain disorder became more widely accepted. This helped to reduce the stigma associated with the disorder. Bromide, introduced in 1857 as the world's first effective anti-epileptic drug, became widely used in Europe and the USA during the second half of the last century. The first hospital centre for the *"paralysed and epileptic"* was established in London, England, in 1857. At the same time a more humanitarian approach to the social problems of epilepsy resulted in the establishment of epilepsy 'colonies' for care and employment.

As a major campaign for the treatment of epilepsy, the International League Against Epilepsy (ILAE) and the International Bureau for Epilepsy (IBE) joined forces with the World Health Organization in 1997 to establish the Global Campaign Against Epilepsy to address these issues. Also, currently, many organizations such as Epilepsy Action in the UK try to increase the public awareness about epilepsy and raise money for research in this area and treatment of the disease.

The new understanding of epilepsy (pathophysiology) was also established in the nineteenth Century with the work of neurologist Hughlings Jackson in 1873 who proposed that seizures were the result of sudden brief electrochemical discharges in the brain. Soon afterwards the electrical excitability of the brain in animals and human was discovered by David Ferrier in London, Gustav Theodor Fritsch and Eduard Hitzig in Germany. Hans Berger, who discovered EEG, worked on epilepsy in the 1930s onwards. The EEG revealed the presence of electrical discharges in the brain. It also showed different patterns of brainwave discharges associated with different seizure types. The EEG also helped to locate the site of seizure discharges and expanded the possibilities of neurosurgical treatments, which became much more widely available from the 1950s onwards in London, Montreal and Paris.

Neuroimaging techniques, such as fMRI and PET, boost the success in the diagnosis of epilepsy. Such technology has revealed many of the more subtle brain lesions responsible for epilepsy. Further diagnosis of epilepsy is carried out by implanting subdural electrodes around the seizure generating sites of the brain. Several brain lesions such as trauma, congenital, developmental, infection, vascular, and tumour might lead to epilepsy in some people.

Epilepsy is a sudden and recurrent brain malfunction and is a disease which reflects an excessive and hypersynchronous activity of the neurons within the brain and is probably the most prevalent brain disorder amongst adults and children, second only to stroke. Over 50 million people worldwide are diagnosed with epilepsy whose hallmark is recurrent seizures [1]. The prevalence of epileptic seizures changes from one geographic area to another [2]. The seizures occur at random to impair the normal function of the brain. Epilepsy can be treated in many cases and the most important treatment today is pharmacological. The patient takes anticonvulsant drugs on a daily basis trying to achieve a steady-state concentration in the blood, chosen to provide the most effective seizure control. Surgical intervention is an alternative for carefully selected cases that are refractory to medical therapy. However, in almost 25% of the total number of patients diagnosed with epilepsy, seizures cannot be controlled by any available therapy. Furthermore, side effects from both pharmacological and surgical treatments have been reported.

An epileptic seizure can be characterized by paroxysmal occurrence of synchronous oscillations. Such seizures can be classified into two main categories depending on the extent of involvement of various brain regions; focal (or partial) and generalized. Generalized seizures involve most areas of the brain whereas focal seizures originate from a circumscribed region of the brain, often called epileptic foci [3]. Figure 15.1 shows two segments of EEG signals involving generalized and focal seizures, respectively.

Successful surgical treatment of focal epilepsies requires exact localization of the epileptic focus and its delineation from functionally relevant areas [4]. The physiological aspects of seizure generation and the treatment and monitoring of a seizure, including pre-surgical examinations, have been well established [3] and the medical literature provided.

EEG, MEG and, recently, functional MRI (fMRI) are the major neuroimaging modalities used for seizure detection. The blood-oxygenation-level-dependent (BOLD) regions in fMRI of the head may show the epileptic foci if the exact onset time points for generation of the haemodynamic response function (HRF) and consequently modelling the fMRI time course can be found. However, the number of fMRI machines is limited in each area, they are costly, and a multiple/sequence of full head scan is time consuming. Often it is not convenient enough for seizure patients to remain steady in an fMRI tube during the seizure onset either. Therefore, using fMRI for all patients at all times is not feasible. MEG on the other hand, is noisy and since the patient under care has to be steady during the recording, it is hard to achieve clear data for moderate and severe cases using current MEG machines.

On the other hand, surface electromyography (sEMG) has also been used for seizure detection [5]. In this method the EMG signals are highpass filtered first and then the zero-crossings are counted. It has been shown that the onset of tonic-clonic seizures can be detected with very high sensitivity and with short latencies after the onset.

Therefore, EEG remains the most useful and cost effective modality for the study of epilepsy. Although for generalized seizure the duration of seizure can be easily detected using a naked eye, for most focal epilepsies, however, such intervals are difficult to recognise. Automatic and blind segmentation of the EEG signals [6] may isolate the segments of the data related to the ictal periods.

15.2   Types of Epilepsy

Epilepsy may be described in two ways depending on the type of epilepsy and type of seizure. The type of epilepsy refers to the cause of epilepsy, such as symptomatic epilepsy or idiopathic epilepsy. The former means there is a known cause, for example, brain injury, and the latter means the disease is generic or inherited. The type of seizure depends on the symptoms. Classification of epilepsy, therefore, refers to the latter case.

From the pathological point of view, there are clear classification schemes for seizures. "Partial" is used to describe isolated phenomena that reflect focal cortical activity, either clinically evident or by EEG. The term "simple" indicates that consciousness is not impaired. For example, a seizure visible as a momentarily twitching upper extremity, which subsides, would be termed a simple partial seizure with motor activity. Partial seizures may have motor, somatosensory, psychic, or autonomic symptoms [7].

The term "complex" defines an alteration of consciousness associated with the seizure. "Generalization" is a term used to denote spread from a focal area of the cortex, which could be evident clinically by EEG, and involves all areas of the cortex with resulting generalized motor convulsion. It is known that in adults the most common seizure type is that of initial

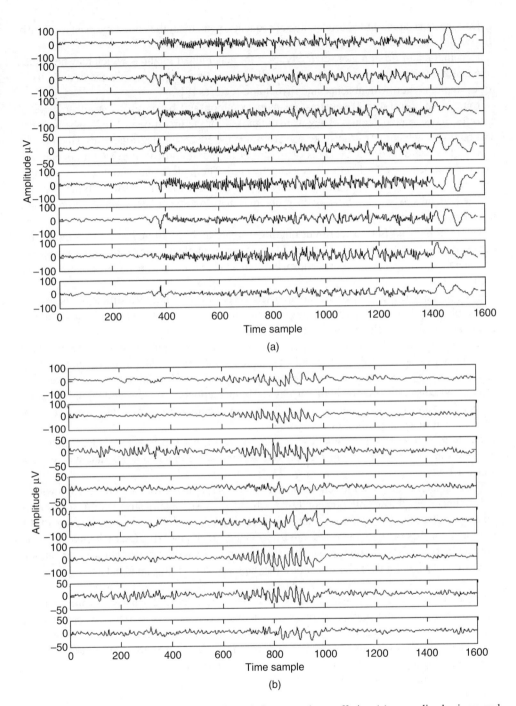

Figure 15.1 Two segments of EEG signals each from a patient suffering (a) generalized seizure and (b) focal seizure onset

activation of one area of the cortex with subsequent spread to all the cortex regions; frequently this occurs too quickly to be appreciated by bedside observation.

The other major grouping of seizure types is for generalized seizures, which may be termed convulsive or nonconvulsive. For this type, all areas of the cortex are activated at once. This, for example, is seen with absence and myoclonic seizures [8].

Tonic-clonic (grand mal) seizures are more common and because of that other types of seizures may escape detection. They were diagnosed for the first time in 1827 by Bravais [9]. The recurrent behaviour of the spike waveforms in the EEG signals triggers investigation for a possible seizure disorder. Seizures of frontal or temporal cortical origin with nonclassical motor movements are fairly commonly encountered. The patient may show some seemingly organized motor activity without the usually in-phase jerking movements more typical of generalized seizures. Also complicating the problem is that clouding or alteration of consciousness may occur without complete loss of consciousness.

The term *aura* is used to represent ongoing seizure activity limited to a focal area of the cortex; in this case the abnormal electrical activity associated with the seizure does not spread or generalize to the entire cerebral cortex but remains localized and persists in one abnormal focus.

One seizure type may evolve into another seizure type. For example, a simple motor seizure may evolve into a complex partial seizure with altered consciousness; the terminology for this would be "partial complex status epilepticus" [8].

Absence seizures (also known as petit mal) are a primarily generalized seizure type involving all cortical areas at once; this is typically a seizure disorder of childhood with a characteristic EEG pattern [8]. At times, absence seizures may persist with minimal motor movements and altered consciousness for hours or days. Absence epileptic seizure and complex partial epileptic seizure are often grouped under the term "nonconvulsive status epilepticus" and are referred to at times as twilight or fugue states.

The term "subtle status epilepticus" is more correctly used to indicate patients that have evolved from generalized convulsive epileptic seizure or are in a comatose state with epileptiform activity.

The character of an epileptic seizure is determined based on the region of the brain involved and the underlying basic epileptic condition, which are mostly age-determined. A rare type of epilepsy, called *Reading epilepsy*, often originates from the occipito-temporal junction of the brain and consists of myoclonic jerks mostly related to the jaw [10]. Reading epilepsy is an uncommon type of reflex epilepsy and has two variants; jaw jerk variant and posterior variant.

As a conclusion, clinical classification of epileptic seizures is summarised as

- Partial (Focal) seizures; these seizures arise from an electric discharge of one or more localised areas of the brain regardless of whether the seizure is secondarily generalized. Depending on their type, they may or may not impair consciousness. Whether seizures are partial or focal, they begin in a localized area of the brain, but then may spread to the whole brain causing a generalized seizure. They are divided to:
  - simple partial including Reading epilepsy
  - complex partial
    * simple partial onset followed by alteration of consciousness
    * partial evolving to secondarily generalized.

- Generalized seizures (convulsive and nonconvulsive); the electrical discharge which leads to these seizures involves the whole brain and may cause loss of consciousness and/or muscle contractions or stiffness. They include what used to be known as "grand mal" convulsion and also the brief "petit mal" absence of consciousness. These seizures are further divided into:
  - o absence (typical and atypical)
  - o clonic, tonic, or tonic-clonic (grand mal)
  - o myoclonic
  - o atonic (astatic).
- Epileptic seizures with unknown waveform patterns
- Seizures precipitated by external triggering events [8].

Moreover, epileptic seizures may be chronic or acute. However, so far no automatic classification of these seizures based on EEG waveforms has been reported.

Most traditional epilepsy analysis methods, based on the EEG, are focused on the detection and classification of epileptic seizures, amongst which the best method of analysis is still the visual inspection of EEG by a highly skilled electroencephalographer or neurophysiologist. However, with the advent of new signal processing methodologies several computerised techniques have been proposed to detect and localize epileptic seizures. In addition, based on the mathematical theory of nonlinear dynamics [11], there has been an increased interest in the analysis of the EEG for prediction of epileptic seizures.

Seizure detection and classification using signal processing methods has been an important research field for the past two decades [12–15]. Researchers have tried to highlight different signal characteristics within various domains, and classify the signal segments based on the measured features. Adult seizure can be better characterized than neonate (newborn) seizure. Therefore, its detection, labelling, and classification are not as difficult as those of neonates. On the other hand, neonate seizure is more chaotic and although some methods have been suggested for the detection of such events the problem still remains open. Therefore, various automated spike detection approaches have been developed.

Recently, predictability of seizure from long EEG recordings has attracted many researchers. It has been shown that epileptic sources gradually tend to be less chaotic from a few minutes before the seizure onset. This finding is clinically very important since the patients do not need to be under anticonvulsant administration permanently, but from just a few minutes before seizure. In the following sections the major research topics in the areas of epileptic seizure detection and prediction are discussed.

15.3   Seizure Detection

15.3.1   Adult Seizure Detection

In clinics, for patients with medically intractable partial epilepsies, time-consuming video-EEG monitoring of spontaneous seizures is often necessary [14]. Visual analysis of interictal EEG is, however, time intensive. On the other hand, application of invasive methods for monitoring the seizure signals and identification of an epileptic zone is hazardous, requires considerable clinical care and effort, and involves risk for the patient.

Before designing any automated seizure detection system the characteristics of the EEG signals before, during, and after the seizure have to be identified. The features may represent the static behaviour of the signals within a short time interval, such as signal energy, or the dynamic behaviour of the signals, such as chaoticity and the change in frequency during the seizure onset.

Automatic seizure detection, quantification, and recognition has been an area of interest and research by clinicians, physiologists, neurologists, and engineers since the 1970s [16–29]. In some early works in spike detection [15, 23–25] a number of parameters, such as relative amplitude, sharpness, and duration of EEG waves, were measured from the EEG signals and evaluated. These methods are sensitive to various artefacts. In these attempts different states, such as active wakefulness or desynchronised EEG, were defined, in which typical nonepileptic transients were supposed to occur [24, 25]. A multistage system to detect the epileptiform activities from the EEGs was developed by Dingle *et al.* [20]. They combined a mimetic approach with a rule-based expert system and, thereby, considered and exploited both the spatial and temporal systems. In another approach [21] multichannel EEGs were used and a correlation-based algorithm was attempted to reduce the muscle artefacts. Following this method approximately 67% of the spikes can be detected. By incorporating both multichannel temporal and spatial information, and including the ECG, EMG, and EOG information into a rule-based system [22] a higher detection rate was achieved. A two-stage automatic system was developed by Davey *et al.* [19]. In the first stage a feature extractor and in the second stage a classifier were introduced. A 70% sensitivity was claimed for this system.

Artificial neural networks (ANNs) have been used for seizure detection by many researchers [28, 30]. The Kohonen self-organizing feature map ANN [31, 32] was used for spike detection by Kurth *et al.* [30]. In this work for each patient three different-sized NN have been examined. The training vector included a number of signals with typical spikes, a number of eye-blink artefact signals, some signals of muscle artefacts, and also background EEG signals. The major problem with these methods is that the epileptic seizure signals do not follow similar patterns. Presenting all types of seizure patterns to the ANN, on the other hand, reduces the sensitivity of the overall detection system. Therefore, a clever feature detection followed by a robust classifier often provides an improved result.

Amongst recent works, time–frequency (TF) approaches effectively use the fact that the seizure sources are localized in the TF domain. Most of these methods are mainly for detection of neural spikes [33] of different types. Different TF methods following different classification strategies have been proposed by many researchers [34, 35] in this area. The methods are especially useful since the EEG signals are statistically nonstationary. The DWT obtains a better TF representation than the TF based on the short-term Fourier transform due to its multiscale (multilevel) characteristics; that is, it can model the signal according to its coarseness. The DWT analyses the signal over different frequency bands, with different resolutions, by decomposing the signal into a coarse approximation and detailed information. In a recent approach by Subasi [35], a DWT-based TF method followed by an ANN has been suggested. The ANN classifies the energy of various resolution (detail) levels. Using this technique, it is possible to detect more than 80% of adult epileptic seizures. Other TF distributions such as pseudo Wigner–Ville can also be used for the same purpose [36].

In [37] and [38] an algorithm for seizure detection has been proposed and implemented. The system is capable of accurate real-time detection, quantitative analysis, and very short-term prediction of clinical onset of seizure. This system computes a measure, namely "foreground", of the median signal energy in the frequencies between 8 and 42 Hz in a short window of specific length (e.g. 2 s). The foreground is calculated through the following steps: (i) decomposing the signals into epileptiform (containing epileptic seizures) and nonepileptiform (without any seizure) components using a 22-coefficient wavelet filter (DAUB4, level 3) which separates the frequency subbands from 8 to 40 Hz; (ii) the epileptiform component is squared; and (iii) the squared components are median filtered. On the other hand, a "background" reference signal is obtained as an estimate of the median energy of a longer time (approximately 30 min) of the signal. A large ratio between the foreground and background then shows the event of seizure [39]. A system using an analogue wavelet was later developed to improve the technical drawbacks of the above system, such as speed and robustness to noise [39].

A number of static features derived from single or multichannel EEG signals have often been used for seizure analysis. Often seizures increase the average energy of the signals during the onset. For a windowed segment of the signal this can be measured as:

$$E(n) = \frac{1}{L} \sum_{p=n-L/2}^{n-1+L/2} x^2(p) \tag{15.1}$$

where L is the window length and the time index n is the window centre. The seizure signals have a major cyclic component and, therefore, generally exhibit a dominant peak in the frequency domain. The frequency of this peak, however, decays with time during the onset of seizure. Therefore the slope of decay is a significant factor in the detection of seizure onset. Considering $X(f,n)$ the estimated spectrum of the windowed signal $x(n)$ centred at n, the peak at time n will be

$$f_d(n) = \arg\max_f \left(|X(f, n)| \right) \tag{15.2}$$

The spectrum is commonly estimated using autoregressive modelling [40]. From the spectrum the peak frequency is measured. The slope of the decay in the peak frequency can then be measured and used as a feature. The cyclic nature of the EEG signals can also be measured and used as an indication of seizure. This can be best identified by incorporating certain higher order statistics of the data. One such indicator is related to the second- and fourth-order statistics of the measurements as follows [41]:

$$I = \left| C_2^0(0) \right|^{-4} \sum_{\alpha \neq 0} |P^\alpha|^2 \tag{15.3}$$

where $P^\alpha = C_4^\alpha(0, 0, 0)$ represents the Fourier coefficients of the fourth-order cyclic cumulant at zero lag and can be estimated as follows:

$$\hat{C}_4^\alpha(0, 0, 0) = \frac{1}{N} \sum_{n=0}^{N-1} x_c^4(n) e^{-\frac{j2\pi n\alpha}{N}} - 3 \sum_{\beta=0}^{\alpha} C_2^{\alpha-\beta}(0) C_2^\beta(0) \tag{15.4}$$

where $x_c(n)$ is the zeroed mean version of $x(n)$, and an estimation of $C_2^\alpha(0)$ is calculated as:

$$\hat{C}_2^\alpha(0) = \frac{1}{N} \sum_{n=0}^{N-1} x_c(n)e^{-\frac{j2\pi n\alpha}{N}} \tag{15.5}$$

This indictor is also measured with respect to the time index n since it is calculated for each signal window centred at n. For a frame centred at n, the value I measures the spread of the energy. Over the range of frequencies before seizure onset, the EEG is chaotic and no frequency appears to control its trace. During seizure, the EEG becomes more ordered (rhythmic) and therefore the spectrum has a large peak.

The above features have been measured and classified using an SVM classifier [40]. It has been illustrated that for both tonic-clonic and complex partial seizures the classification rate can be as high as 100%.

In a robust detection of seizure, however, all statistical measures from the EEGs together with some other biometrics, such as blood morphology, body movement, pain, changes in metabolism, heart rate variability, respiration, before, during, and after seizure, have to be quantified and effectively taken into account. This is a challenge for future signal processing/data fusion-based approaches.

In another attempt at seizure detection a cascade of classifiers based on ANNs has been used [42]. The features feed two perceptrons to classify peaks into definite epileptiform, definite nonepileptiform, and unknown waveforms. The features are selected based on the expected shape of an epileptic spike. Since three outputs are needed after the first stage (also called preclassifier) two single layer perceptrons are used in parallel. One perceptron is trained to give +1 for *definite nonepileptiform* and –1 otherwise. The second network produces +1 for *definite epileptiform* and –1 for otherwise. A segment, which produces –1 at the output of both networks is assigned to the *unknown* group.

The updated equation to find $\mathbf{w}$, the vector of the weights of the perceptron ANNs, at the kth iteration can be simply presented as:

$$\mathbf{w}_k = \mathbf{w}_{k-1} + \mu(\mathbf{d} - \mathbf{y})\mathbf{x} \tag{15.6}$$

where $\mathbf{x}$ is the input feature vector, $\mathbf{d}$ is the expected feature vector, $\mathbf{y}$ is the output calculated as $\mathbf{y} = \text{sign}(\mathbf{w}^T\mathbf{x} - T_r)$, where T_r is an empirical threshold, and μ is the learning rate set empirically or adaptively.

In the second stage the *unknown* waveforms (spikes) are classified using a radial-basis function (RBF) neural network. An RBF has been empirically shown to have a better performance than other ANNs for this stage of classification. The inputs are the segments of actual waveforms. The output is selected by hard-limiting the output of the last layer (after normalization).

Finally, in the third stage a multidimensional SVM with a nonlinear (RBF) kernel is used to process the data from its multichannel input. An accuracy of up to 100% has been reported for detection of generalized and most focal seizures [42].

Epilepsy is often characterised by the sudden occurrence of synchronous activity within relatively large brain neuronal networks that disturb the normal working of the brain. Therefore, some measures of dynamical change have also been used for seizure detection. These measures significantly change in the transition between preictal and ictal states or even in the transition between interictal and ictal states. In the latter case the transition can occur either as a continuous change in phase, such as in some cases of mesial temporal lobe epilepsy (MTLE), or as a sudden leap, for example, in most cases of absence seizures. In the approaches based on chaos measurement by estimation of the attractor dimension (as discussed later in this chapter), for the first case, the attractor of the system gradually deforms from an interictal to an ictal attractor. In the second case, where a sharp critical transition takes place, we can assume that the system has at least two simultaneous interictal and ictal attractors all the time. In a study by Lopes da Silva *et al.* [43] three states (routes) have been characterised as illustrative models of epileptic seizures: (i) an abrupt transition of the bifurcation type, caused by a random perturbation. An interesting example of which is the epileptic syndrome characterized by proxysmal spike-and-wave discharges in the EEG and nonconvolsive absence type of seizures; (ii) a route where a deformation of the attractor is caused by an external perturbation (photosensitive epilepsy); and (iii) a deformation of the attractor leading to a gradual evolution into the ictal state, for example, temporal lobe epilepsy (TLE). The authors concluded that under these routes it is possible to understand those circumstances where the transition from the ongoing (interictal) activity mode to the ictal (seizure) mode may, or may not, be anticipated. Also, any of the three routes is possible, depending on the type of the underlying epilepsy. Seizures may be generally unpredictable, as most often in absence-type seizures of idiopathic (primary) generalised epilepsy, or predictable, preceded by a gradual change in dynamics, detectable some time before the manifestation of seizure, as in TLE.

Chaotic behaviour of the EEGs is discussed in Section 15.4. Although this is mainly used for prediction of seizure, in some publications such as [44] chaos has also been used as a robust measure for seizure and epilepsy detection. In this work chaos has been measured for different conventional EEG frequency bands of delta, theta, alpha, beta, and gamma subbands. In this work two indexes have been defined for the measurement of chaos; one is Cao's embedding function defined as [45]:

$$E(d) = \frac{1}{N - m_0 d} \sum_{i=1}^{N - m_0 d} a_i(d) \tag{15.7}$$

Where m_0 is the optimum lag value and d, the embedding dimension, is gradually increased till the number of false neighbours decreases to zero or the above $E(d)$ remains constant. Two points are true neighbours if for a true embedding dimension they are close to each other in the d-dimensional phase space and remain close in the $d + 1$-dimensional phase space. In (15.7) [46]

$$a_i(d) = \frac{\left\| \mathbf{Y}_i(d + 1) - \mathbf{Y}_{n(i,d)}(d + 1) \right\|}{\left\| \mathbf{Y}_i(d) - \mathbf{Y}_{n(i,d)}(d) \right\|} \tag{15.8}$$

where $\mathbf{Y}_i(d) = \{x_i, x_{i+m_0}, \ldots, x_{i+m_0}(d - 1)\}$ for the EEG signal $\mathbf{X} = \{x_1, x_2, \ldots, x_N\}$. Often $E_n(d) = E(d)/E(d + 1)$ is used instead of $E(d)$, which converges to 1 in the case of a finite dimensional attractor [45]. The coherence dimension, on the other hand, is estimated using Takens estimator. This estimator uses a finite length signal of length N_C samples, where

$N_C = N - m_0 d_M$ for minimum embedded dimension d_M. The coherence dimension, CD, is therefore estimated as [44]:

$$CD = -\left[\frac{2}{N_C(N_C - 1)} \sum_{i=1}^{N_C} \sum_{j=1}^{N_C} \log\left(\frac{|\mathbf{Y}_i(d_M) - \mathbf{Y}_i(d_M)|}{\varepsilon} \right) \right]^{-1} \tag{15.9}$$

where ε is the radius of the boundary of the measuring unit. Based on the CD measure and also the largest Lyaponuv exponents (LLEs), explained in Chapter 5 it has been found that the CD differentiates between the three groups of healthy, epileptic but measured during seizure-free interval, and epileptic measured during seizure, when applied to the beta and gamma band signals and the LLE when applied to the entire band-limited signal.

In another attempt a coherency measure based on the phase-slope index (PSI), explained in Chapter 5, and largest Lyaponuv exponent (LLE), pointed out in Chapter 5, have been implemented for seizure detection [46]. The way these metrics used for seizure detection purpose is as follows. For segment k the so called outward PSI is defined as:

$$\Lambda_i^k = \sum_{\substack{j=1 \\ j\neq i}}^{M} P_{ij}^k \mathbf{1}_{\{P_{ij}^k \geq 2\}} \tag{15.10}$$

where P_{ij}^k is the PSI for the kth segment of the signal and

$$\mathbf{1}_{\{P_{ij}^k \geq\}} = \begin{cases} 1, & \text{if } P_{ij}^k \geq 2 \\ 0, & \text{otherwise} \end{cases} \tag{15.11}$$

This shows the influence of other channels on channel i. The global level of interaction between the channels in the kth segment is also given as [46]:

$$\Lambda^k = \sum_{\substack{i,j=1 \\ j\neq i}}^{M} P_{ij}^k \mathbf{1}_{\{P_{ij}^k \geq 2\}} \tag{15.12}$$

Then, seizure is declared in segment k if Λ^k becomes larger than a threshold Γ^k. Γ^k is selected as

$$\Gamma^k = \bar{\Lambda}^k + \beta \cdot \sigma_{\Lambda^k} \tag{15.13}$$

Where the moving $\bar{\Lambda}^k$ is the mean and σ_{Λ^k} is the variance over segment k. The PSI and LLE measures, however, have been taken from ECoG signals of five child patients. It has been shown that this method is able to indicate the onset of seizure better than the method which decides based on the signal power only.

15.3.2 Detection of Neonate Seizure

Seizure occurs in approximately 0.5% of newborn (the first four weeks of life) infants. It represents a distinctive indicator of abnormality in the central nervous system (CNS). There are many causes for this abnormality with the majority due to biochemical imbalances within the CNS, intracranial haemorrhage and infection, developmental (structural) defects, lack of oxygen, meningitis, and passive drug addiction and withdrawal [46]. Analysis of neonatal EEG is a very difficult issue in the biomedical signal processing context. Unlike in adults, the presence of spikes may not be the indication of seizure. The clinical signs in the newborn are not always as obvious as those for an adult, where seizure is often accompanied by uncontrollable, repetitive, or jerky movement of the body, or the tonic flexion of muscles. The less obvious symptoms in the newborn, that is, subtle seizures, may include sustained eye opening with ocular fixation, repetitive blinking or fluttering of eyelids, drooling, sucking, or other slight facial expressions or body movements. Therefore, detection of epileptic seizures for the newborn is far more complex than for adults and so far only a few approaches have been attempted. These approaches are briefly discussed in this section.

Although there is significant spectral range for the newborn seizure signal [48] in most of the cases the seizure frequency band lies within the delta and theta bands (1–7 Hz). However, TF approaches are more popular due to the statistical nonstationarity of the data, but they can also be inadequate since the spikes may have less amplitude than the average amplitude of a normal EEG signal.

In most cases the signals are pre-processed before application of any seizure detection algorithm. Eye blinking and body movements are the major sources of artefacts [49]. Conventional adaptive filtering methods (with or without a reference signal) may be implemented to remove interference [50]. This may be followed by calculation of a TF representation of seizure [51], or used in a variable-time model of the epileptiform signal.

A model-based approach was proposed in [52] to model a seizure segment and the model parameters were estimated. The steps are as follows:

1. The periodogram of the observed vector of an EEG signal, $\mathbf{x} = [x(1), x(2), \ldots, x(N)]^T$, is given as:

$$I_{xx}(k) = \frac{1}{2\pi N} \left| \sum_{n=1}^{N} e^{-j\lambda_k n} x(n) \right|^2 \tag{15.14}$$

where $\lambda_k = \frac{2\pi k}{N}$.

2. A discrete approximation to the log-likelihood function for estimation of a parameter vector of the model, $\boldsymbol{\theta}$, is computed as:

$$L_N(\mathbf{x}, \boldsymbol{\theta}) = - \sum_{k=0}^{\lfloor (N-1)/2 \rfloor} \left[\log(2\pi)^2 S_{xx}(\lambda_k, \boldsymbol{\theta}) + \frac{I_{xx}(\lambda_k)}{S_{xx}(\lambda_k, \boldsymbol{\theta})} \right] \tag{15.15}$$

where, $S_{xx}(\lambda_k, \boldsymbol{\theta})$ is considered as the spectral density of a Gaussian vector process $\mathbf{x}$ with parameters $\boldsymbol{\theta}$. L_N needs to be maximized with respect to $\boldsymbol{\theta}$. As $N \to \infty$ this approximation approaches that of Whittle [53].

The parameter estimate of Whittle is given by

$$\hat{\theta} = \arg\max_{\theta} (L_N(\mathbf{x}, \theta); \ \theta \in \Theta) \tag{15.16}$$

The parameter space Θ may include any property of either the background EEG or the seizure, such as nonnegative values for post-synaptic pulse shaping parameters. This is the major drawback of the method since these parameters are highly dependent on the model of both seizure spikes and the background EEG. A self-generating model [54] followed by a quadratic discriminant function was used to distinguish between the spectrum density of background EEG segments ($S_k^{\text{Background}}$) and seizure segments (S_k^{Seizure}); k represents the segment number. This model has already been explained in Chapter 2.

3. Based on the model above, the power of the background EEG and the seizure segment is calculated [52] and their ratio is tested against an empirical threshold level, say γ, that is,

$$\gamma = \frac{P_{\text{Seizure}}}{P_{\text{Background}}} \tag{15.17}$$

where, $P_{\text{Background}} = \sum_{k=0}^{N-1} S_k^{\text{Background}}$ and $P_{\text{Seizure}} = \sum_{k=0}^{N-1} S_k^{\text{Seizure}}$.

It is claimed that for some well-adjusted model parameters a false alarm percentage as low as 20% can be achieved [52].

A TF approach for neonate seizure detection has been suggested [55] in which a template in the TF domain is defined as

$$Z_{\text{ref}}(\tau, v) = \sum_{i=1}^{L} \exp\left(\frac{-(v - \alpha_i \tau)^2}{2\sigma^2}\right) \tag{15.18}$$

where the time scales α_is and variance σ^2 can change, respectively, with the position and width of the template. Z_{ref}, with variable position and variance in the TF domain, resembles a seizure waveform and is used as a template. This template is convolved with the TF domain EEG, $X(t,k)$, in both the time and frequency domains, that is,

$$\eta_{\alpha,\sigma^2}(n, k) = X(n, k) ** Z_{\text{ref}}(n, k) \tag{15.19}$$

where, $**$ represents 2D convolution with respect to discrete time and frequency, k is discrete frequency and $X(n,k)$ for a real discrete signal x is defined as:

$$X(n, k) = \sum_{q=n/2}^{N-n/2} x(q + n/2)x(q - n/2)e^{-j2\pi kq} \tag{15.20}$$

where N is the signal length in terms of samples. Assuming the variance is fixed, the α_is may be estimated in order to maximise a test criterion to minimise the difference between Z_{ref} and the desired waveform. This criterion is tested against a threshold level to decide whether there is a seizure. The method has been applied to real neonate EEGs and a false detection rate (FDR) as low as 15% in some cases has been reported [55].

The neonate EEG signals do not, unfortunately, manifest any distinct TF pattern at the onset of seizure. Moreover, they include many seizure type spikes due to immature mental activities of newborns and their rapid changes in brain metabolism. Therefore, although the above methods work well for some synthetic data, due to the nature of neonate seizures, as described before, they often fail to detect effectively all types of neonate seizures.

In a recent work by Karayiannis *et al.* a cascaded rule-based NN algorithm for detection of epileptic seizure segments in neonatal EEG has been developed [56]. In this method it is assumed that the neonate seizures are manifested as subtle but somehow stereotype repetitive waveforms that evolve in amplitude and frequency before eventually decaying. Three different morphologies of seizure patterns for, respectively, pseudosinusoidal, complex morphology, and rhythmic runs of spike-like waves have been considered. These patterns are illustrated in Figure 15.2 [56].

The automated detection of neonate seizure is then carried out in three stages. Each EEG channel is treated separately, and spatial and inter-channel information are not exploited. In the first stage the spectrum amplitude is used to separate bursts of rhythmic activities. In the second stage the artefacts, such as the results of patting, sucking, respiratory function, and EKG (ECG) are mitigated, and in the last stage a clustering operation is performed to distinguish between epileptic seizures, nonepileptic seizures and the normal EEG affected by different artefacts. As a result of this stage, isolated and inconsistent candidate seizure segments are eliminated, and the final seizure segments are recognised. The performances of conventional feed-forward neural networks (FFNN) [57] as well as quantum neural networks (QNN) [57] have been compared for classification of some frequency domain features in all the above stages. These features are denoted as first dominant frequency, second dominant frequency, width of dominant frequency, percentage of power contributed to the first dominant frequency, percentage of power contributed to the second dominant frequency, peak ratio, and stability ratio (a time domain parameter that measures the amplitude stability of the EEG segment). It has also been shown that there is no significant difference in using an FFNN or QNN and the results are approximately the same [56]. The overall algorithm is very straightforward to implement and both its sensitivity and specificity have been shown to be above 80%.

A few other methods have been developed recently. In [59] a multistage system was proposed for neonate seizure detection from the EEG signals. In [60] a singular-spectrum approach has been followed. Following this method, individual channels can be analysed in the subspace domain. A human observer has been mimicked in developing another automated neonate seizure detection system [61]. The applications of these methods, however, have been limited in clinical use [62]. In another recent method a number of channels (of the conventional 10–20 EEG system) have been used in differential mode to record the neonate EEGs for seizure detection. Figure 15.3 illustrates the conventional 10–20 electrode system which has been modified for neonates. The montage is bipolar and the connections are F4-C4, C4-O2, F3-C3, C3-O1, T4-C4, C4-Cz, Cz-C3, and C3-T3 [62].

In this work a measure of bipolar channel importance has been proposed. The recorded signals are weighted and the weights are computed based on the integrated synchrony of classifier probabilistic outputs for the channels which share a common electrode. Figure 15.4 shows a block diagram of the neonate seizure detection in [62].

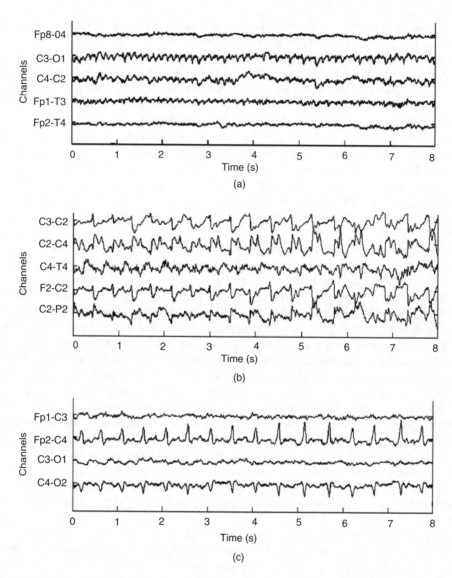

Figure 15.2 The main three different neonate seizure patterns: (a) low amplitude depressed brain type discharge around the channels C3-O1 and FP1-T3, (b) repetitive complex slow waves with superimposed higher frequencies in all the channels, and (c) repetitive or periodic runs of sharp transients in channels FP2-C4 and C4-O2

In the weighting stage of his algorithm channel l is weighted by a weighting factor $w_l(t)$ at time t using some probabilistic models as follows [62]:

$$w_l(t) = \frac{\exp(k P_t(e_l|\mathbf{x}) P_t(e_l|S))}{\sum_{j=1}^{N} \exp(k P_t(e_j|\mathbf{x}) P_t(e_j|S))} \tag{15.21}$$

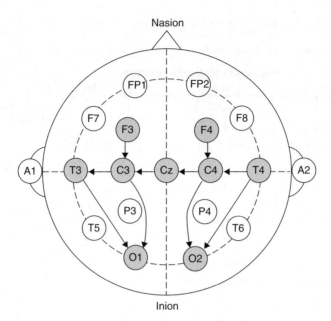

Figure 15.3 The bipolar montage for recording the EEG of neonate for seizure detection. The connections are F4-C4, C4-O2, F3-C3, C3-O1, T4-C4, C4-Cz, Cz-C3, and C3-T3 [62]

where $P_t(e_l|S)$ is the probability that, given a seizure occurring, it is visible in electrode e_l at time t. $P_t(e_l|\mathbf{x})$ represents the *importance* of the e_lth electrode. To calculate $P_t(e_l|\mathbf{x})$, consider $\mathbf{y}_l(r)$ is the output vector of channel l at time r of selected channels which model the importance of the lth channel. Then, the $P_t(e_l|\mathbf{x})$ is expressed as[62]:

$$P_t(e_l|\mathbf{x}) = \frac{\sum_{r=1}^{t} \mathbf{y}_l^{\mathrm{T}}(r)\mathbf{Q}_l\mathbf{y}_l(r)/|\mathbf{y}_l(r)|}{\sum_{j=1}^{N} (\sum_{r=1}^{t} \mathbf{y}_j^{\mathrm{T}}(r)\mathbf{Q}_j\mathbf{y}_j(r)/|\mathbf{y}_j(r)|)} \tag{15.22}$$

where $N = 9$ is the number of electrodes used in a recording, $|\mathbf{y}_k(r)|$ denotes the cardinality (here the number of channels associated with the kth electrode,) and $\mathbf{Q}_k$ is the $|\mathbf{y}_k(r)| \times |\mathbf{y}_k(r)|$ square matrix of the form [62]

$$\mathbf{Q}_k = \begin{bmatrix} 0 & 1 & 1 & . & . & . & 1 \\ 0 & 0 & 1 & . & . & . & 1 \\ \vdots & \vdots & \vdots & . & . & . & \vdots \\ 0 & 0 & 0 & . & . & . & 0 \end{bmatrix} \tag{15.23}$$

It has been claimed that by weighting the channels this way, the performance of the seizure detection system is highly increased [62].

Raw multichannel EEG signals

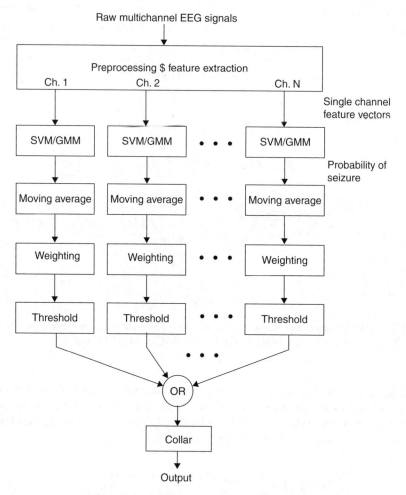

Figure 15.4   Block diagram of the neonatal seizure detector proposed in [62]

15.4   Chaotic Behaviour of EEG Sources

Nonlinear analysis techniques provide insights into many processes, which cannot be directly formulated or exactly modelled using state machines. This requires time series analysis of long sequences. A state-space reconstruction of the chaotic data may be performed based on embedding methods [63, 64]. Using the original time series and its time-delayed copies that is, $\mathbf{x}(n) = [x(n), x(n+T), \ldots, x(n+(d_E-1)T)]$, an appropriate state space can be reconstructed. Here, $\mathbf{x}(n)$ is the original one-dimensional data, T is the time delay, and d_E is the embedding dimension. The time delay T is calculated from the first minimum of the average mutual information (AMI) function [65]. The minimum point of the AMI function provides adjacent delay coordinates with a minimum of redundancy. Embedding dimension (d_E) can be computed from a global false nearest neighbours (GFNN) analysis [66], which compares the distances between neighbouring trajectories at successively higher dimensions. The false neighbors occur when trajectories that overlap in dimension d_i are distinguished in dimension d_{i+1}. As i increases, the total percentage of false neighbours declines and d_E is chosen where this percentage approaches zero.

The stationarity of the EEG patterns may be investigated and established by evaluating recurrence plots generated by calculation of the Euclidean distances between all pairs of points $\mathbf{x}(i)$ and $\mathbf{x}(j)$ in the embedded state-space and then such points are plotted in the (i,j) plane where $\delta_{i,j}$ is less than a specific radius ρ,

$$\delta_{ij} = \|\mathbf{x}(i) - \mathbf{x}(j)\|_2 < \rho \qquad (15.24)$$

where $\| \, . \, \|_2$ denotes the Euclidean distance. Since i and j are time instants, the recurrence plots convey natural and subtle information about temporal correlations in the original time series [67]. Nonstationarities in the time series are manifested as gross homogenities in the recurrent plot. The value of ρ for each time series is normally taken as a small percentage of the total data set size.

The scale-invariant self-similarity, as one of the hallmarks of low dimensional deterministic chaos [68], results in a linear variation of the logarithm of the correlation sum, $\log[C(r,N)]$, with respect to $\log(r)$ as $r \to 0$. In places where such similar dynamics exist, the *correlation dimension*, D, is defined as

$$D = \lim_{N \to \infty} \lim_{r \to 0} \frac{\log[C(r, N)]}{\log(r)} \qquad (15.25)$$

where $C(r,N)$ is calculated as

$$C(r, N) = \frac{1}{(N - n)(N - n - 1)} \times \sum_{i=1}^{N} \sum_{j=1}^{N} H\left(r - \|\mathbf{x}(i) - \mathbf{x}(j)\|_2\right) \qquad (15.26)$$

where r represents the volume of points being considered and $H(.)$ is a heaviside step function. Computation of $C(r,N)$ is susceptible to noise and nonstationarities. It is also dominated by the finite length of the data set. The linearity of the relationship between $\log[C(r,N)]$ and $\log(r)$ can be examined from the local slopes of $\log[C(r,N)]$ versus $\log(r)$.

Generally, traditional methods, such as the Kolmogorov entropy, the correlation dimension, or Lyapunov exponents [7] can be used to quantify the dynamical changes of the brain.

Finite-time Lyapunov exponents quantify the average exponential rate of divergence of neighbouring trajectories in state-space, and thus provide a direct measure of the sensitivity of the system to infinitesimal perturbations.

A method based on the calculation of the largest Lyapunov exponent has often been used for evaluation of chaos in the intracranial EEG signals. In a p-dimensional system there are p different Lyapunov exponents, λ_i. They measure the exponential rate of convergence or divergence of the different directions in the phase-space. If one of the exponents is positive, the system is chaotic. Thus, two close initial conditions will diverge exponentially in the direction defined by that positive exponent. Since these exponents are ordered, $\lambda_1 \geq \lambda_2 \ldots \geq$

λ_d, to study the chaotic behaviour of a system it is sufficient to study the changes in the largest Lyapunov exponent, λ_1. Therefore, we focus on the changes in the value of λ_1 as the epileptic brain moves from one state to another.

The maximum Lyapunov exponent (MLE) (λ_1) for a dynamical system can be defined from [69]

$$d(n) = d_0 e^{\lambda_1 n} \tag{15.27}$$

where $d(n)$ is the mean divergence between neighbouring trajectories in state space at time n and d_0 is the initial separation between neighbouring points. *Finite-time* exponents (λ^*) are distinguished from true Lyapunov exponents λ_1, which are strictly defined only in the dual limit as $n \to \infty$ and $d_0 \to 0$ in Equation (15.24). For the finite length observation λ^* is the average of the Lyapunov exponents.

A practical procedure for the estimation of λ_1 from a time series was proposed by Wolf *et al.* [70]. This procedure gives a global estimate of λ_1 for stationary data. Since the EEG data are nonstationary [71], the algorithm to estimate λ_1 from the EEG should be capable of automatically identifying and appropriately weighting the transients of the EEG signals. Therefore, a modification of Wolf's algorithm, proposed in [72], which modifies mainly the searching procedure to account for the nonstationarity of the EEG data, may be used. This estimate is called the short-term largest Lyapunov exponent, STL_{max}. The changes in the brain dynamics can be studied by the time evolution of the STL_{max} values at different electrode sites. Estimation of the STL_{max} for time sequences using Wolf's algorithm has already been explained in Chapter 5.

15.5 Predictability of Seizure from the EEGs

Although most seizures are not life threatening, they are sources of annoyance and embarrassment. They occur when a massive group of neurons in the cerebral cortex begins to discharge in a very organized way, leading to a temporary synchronized electrical activity that disrupts the normal activity of the brain. Sometimes, such disruption manifests itself in a brief impairment of consciousness, but it can also produce a more or less complex series of abnormal sensory and motor manifestations.

The brain is assumed to be a dynamical system, since epileptic neuronal networks are essentially complex nonlinear structures and their interactions are thus expected to exhibit nonlinear behaviour. These methods have substantiated the hypothesis that quantification of the brain's dynamical changes from the EEG might enable prediction of epileptic seizures, while traditional methods of analysis have failed to recognize specific changes prior to seizure.

Iasemidis *et al.* [73] were the first to apply nonlinear dynamics to clinical epilepsy. The main concept in their studies is that a seizure represents a transition of the epileptic brain from chaotic to a more ordered state and, therefore, the spatiotemporal dynamical properties of the epileptic brain are different for different clinical states. Further studies of the same group, based on the temporal evolution of the short-term largest Lyapunov exponent (LLE) (a modification of the LLE to account for the nonstationarity of the EEG) for patients with temporal lobe epilepsy (TLE) [72], suggested that the EEG activity becomes progressively less chaotic as the seizure approaches. Therefore, the idea that seizures were abrupt transitions in and out of an abnormal state was substituted by the idea that the brain follows a dynamical transition to seizure for at least some kinds of epilepsy. Since these pioneering studies, nonlinear methods derived from the theory of dynamical systems have been employed to quantify the changes in the brain dynamics before the onset of seizures, providing evidence for the hypothesis of a *route* to seizure. Lehnertz *et al.* [74] focused their studies on the decrease of complexity in neuronal networks prior to seizure. They used the information provided by changes in the *neuronal complexity loss* that summarizes the complex information content of the correlation dimension profiles in just a single number. Lerner [75] observed that changes in the correlation integral could be used to track accurately the onset of seizure for a patient with TLE. However, Osorio *et al.* [76] demonstrated that these changes in the correlation integral could be perfectly explained by changes in the amplitude and frequency of the EEG signals. Van Quyen *et al.* [77] found a decrease in the dynamical similarity during the period prior to seizure and that this behaviour became more and more pronounced as the onset of seizure approached. Moser *et al.* [78] employed four different nonlinear quantities within the framework of the chaos using Lyaponuv exponents and found strongly significant preictal changes. Litt *et al.* [79] demonstrated that the energy of EEG signals increases as seizure approaches. In their later works, they provided evidence of seizure predictability based on the selection of different linear and nonlinear features of the EEG [80]. Iasemidis *et al.* [81,82], by using the spatiotemporal evolution of the short-term LLE, demonstrated that minutes or even hours before seizure, multiple regions of the cerebral cortex progressively approach a similar degree of chaoticity of their dynamical states. They called it *dynamical entrainment* and hypothesized that several critical sites have to be locked with the epileptogenic focus over a common period of time in order for a seizure to take place. Based on this hypothesis they presented an adaptive seizure prediction algorithm that analyses continuous EEG recordings for prediction of temporal lobe epilepsy when only the occurrence of the first seizure is known [83].

Most of these studies for prediction of epilepsy are based on intracranial EEG recordings. Two main challenges face the previous methods in their application to scalp EEG data: (i) the scalp signals are more subject to environmental noise and artefacts than the intracranial EEG, and (ii) the meaningful signals are attenuated and mixed in their propagation through soft tissue and bone. Traditional nonlinear methods (TNMs), such as the Kolmogorov entropy or the Lyapunov exponents, may be affected by the above two difficulties and, therefore, they may not distinguish between slightly different chaotic regimes of the scalp EEG [84]. One approach to circumvent these difficulties is based on the definition of different nonlinear measures that yield better performance over the TNM for the scalp EEG. This is the approach followed by Hively *et al.* [85]. They proposed a method based on the *phase-space dissimilarity measures* (PSDM) for forewarning of epileptic events from scalp EEG. The approach of Iasemidis *et al.* of dynamical entrainment has also been shown to work well on scalp unfiltered EEG data for seizure predictability [86–88].

In principle, a nonlinear system can lie in a high-dimensional or infinite-dimensional phase-space. Nonetheless, when the system comes into a steady state, portions of the phase-space are revisited over time and the system lies in a subset of the phase-space with a finite and generally small dimension, called an *attractor*. When this attractor has sensitive dependence to initial conditions (it is chaotic), it is termed as a *strange* attractor and its geometrical complexity is reflected by its dimension, D_a. In practice, the system's equations are not available and we only have discrete measurements of a single observable, $u(n)$, representing the system. If the system comes into such a steady state, a p-dimensional phase space can be reconstructed by generating p different scalar signals, $x_i(n)$, from the original observable, $u(n)$, and embedding them into a p-dimensional vector,

$$\mathbf{x}(n) = [x_1(n), x_2(n), \ldots, x_p(n)]^{\mathrm{T}} \qquad (15.28)$$

According to Takens [89], if p is chosen large enough, we shall generally obtain a good phase portrait of the attractor and, therefore, good estimates of the nonlinear quantities. In particular, Takens' theorem states that the embedding dimension, p, should be at least equal to $2 \times D_a + 1$. The easiest and probably the best way to obtain the embedding vector $x(n)$ from $u(n)$ is by *the method of delays*. According to this method, p different time delays, $n_0 = 0$, $n_1 = \tau, n_2 = 2\tau, \ldots, n_{p-1} = (p-1)\tau$, are selected and the p different scalar signals are obtained as $x_i(n) = x(n + n_i)$ for $i = 0, \ldots, p-1$. If τ is chosen carefully, we obtain a good phase portrait of the attractor and, therefore, good estimates of the parameters of nonlinear behaviour.

Since the brain is a nonstationary system, it is never in a steady state in the strictly dynamical sense. However, it can be considered as a dynamical system that constantly moves from one stable steady state to another. Therefore, local estimates of nonlinear measures should be possible and the changes in these quantities should be representative of the dynamical changes in the brain.

Previous studies have demonstrated a more ordered state of the epileptic brain during seizure than before or after it. The correlation dimension has been used to estimate the dimension, d, of the ictal state [73]. The values obtained ranged between 2 and 3, demonstrating the existence of a low-dimensional attractor. Therefore, an embedding dimension of 7 should be enough to obtain a good image of this attractor and a good space portrait of the ictal state. As concluded in Chapter 2, increasing the value of p more than is strictly necessary increases the effect of noise and thus higher values of p are not recommended.

In a new approach [90] it has been shown that the TNM can be applied to the scalp EEGs indirectly. This requires the underlying sources of the brain to be correctly separated from the observed electrode signals without any a priori information about the source signals or the way the signals are combined. The effects of noise and other internal and external artefacts are also highly mitigated by following the same strategy. It is expected to obtain signals similar to the intracranial recordings to which TNM can be applied. To do so, the signal segments are initially separated into their constituent sources using BSS (assuming the sources are independent). Since, for a practical prediction algorithm, the nonlinear dynamics have to be quantified over long-term EEG recordings, after using a block-based BSS algorithm the continuity has to be maintained for the entire recording. This problem turns out not to be easy due to the two inherent ambiguities of BSS: (i) The variances (energies) of the independent components are unknown, and (ii) due to the inherent permutation problem of the BSS algorithms the order of the independent components cannot be determined.

The first ambiguity states that the sources can be estimated up to a scalar factor. Therefore, when moving from one block to another, the amplitude of the sources will be generally different and the signals can be inverted. This ambiguity can be solved as explained below, so its effect can be avoided. The nonlinear dynamics are quantified by the LLE λ_1. Estimation of λ_1 for each block is based on ratios of distances between points within the block. Consequently, as long as λ_1 is estimated for the sources obtained by applying BSS to each block of EEG data individually, there is no need to adjust the energy of the sources.

The second ambiguity, however, severely affects the algorithm. The order in which the estimated sources appear as a result of applying the BSS algorithm changes from block to block. Therefore, we need a procedure to reorder the signals to align the same signal from one block to another and maintain the continuity for the entire recording. The next section explains the approach followed in our algorithm for this purpose.

An overlap window approach is followed to maintain the continuity of the estimated sources, solving both indeterminacies simultaneously. Instead of dividing the EEG recordings into sequential and discontinuous blocks, we employ a sliding window of fixed length, L, with an overlap of $L - N$ samples ($N < L$), and apply the BSS algorithm to the block of data within that window. Therefore, we assume that $\mathbf{x}(n) = [x_1(n), x_2(n), \ldots, x_m(n)]^T$ represents the entire scalp EEG recording, where m is the number of sensors. Two consecutive windows of data are selected as $\mathbf{x}_1(n) = \mathbf{x}(n_0 + n)$ and $\mathbf{x}_2(n) = \mathbf{x}(n_0 + N + n)$ for $t = 1, \ldots, L$, where $n_0 \geq 0$. Therefore,

$$\mathbf{x}_2(n) = \mathbf{x}_1(N + n) \text{ for } n = 1, \ldots, L - N. \tag{15.29}$$

Once the BSS algorithm has been applied to $\mathbf{x}_1(n)$ and $\mathbf{x}_2(n)$, two windows of estimated sources $\hat{\mathbf{s}}_1(n) = [\hat{s}_1(n), \hat{s}_2(n), \ldots, \hat{s}_m(n)]^T$ and $\hat{\mathbf{s}}_2(n) = [\hat{s}'_1(n), \hat{s}'_2(n), \ldots, \hat{s}'_m(n)]^T$ will be obtained, respectively, where m is the number of sources. These two windows overlap within a time interval, but due to the inherent ambiguities of BSS $\hat{\mathbf{s}}_1(n)$ and $\hat{\mathbf{s}}_2(n)$ are not equal in this interval. Instead,

$$\hat{\mathbf{s}}_2(n) = \mathbf{P} \cdot \mathbf{D} \cdot \hat{\mathbf{s}}_1(n + N) \text{ for } n = 1, \ldots, L - N \tag{15.30}$$

where $\mathbf{P}$ is an $n \times n$ permutation matrix and $\mathbf{D} = \text{diag}\{d_1, d_2, \ldots, d_n\}$ is the scaling matrix. Therefore, $\hat{\mathbf{s}}_2(t)$ is just a copy of $\hat{\mathbf{s}}_1(t)$ in the overlap block, with the rows (sources) permuted, and each of them is scaled by a real number d_i that accounts for the scaling ambiguity of BSS. Cross-correlation has been used as a measure of similarity between the rows of $\hat{\mathbf{s}}_1(t)$ and $\hat{\mathbf{s}}_2(t)$ within the overlap region for this purpose. The cross-correlation between two zero mean wide sense stationary random signals $x(t)$ and $y(t)$ is defined as

$$r_{xy}(\tau) = E[x(n)y(n + \tau)] \tag{15.31}$$

where $E[.]$ denotes the expectation operation. This measure gives an idea of the similarity between $x(n)$ and $y(n)$, but its values are not bounded and depend on the amplitude of the signal. Therefore, it is preferable to use a normalization of r_{xy} given by the cross-correlation coefficient defined as:

$$\rho_{xy} = \frac{r_{xy}}{\sigma_x \sigma_y} \tag{15.32}$$

where σ_x and σ_y are the standard deviations of $x(n)$ and $y(n)$, respectively. The cross-correlation coefficient satisfies

$$-1 \leq \rho_{xy} \leq 1 \tag{15.33}$$

Furthermore, if $\rho_{xy} = 1$, then $y = ax$, with $a > 0$, and $x(t)$ and $y(t)$ are perfectly correlated; and if $\rho_{xy} = -1$, then $y = -ax$ and $x(n)$ and $y(n)$ are perfectly anti-correlated. When the two signals have no information in common, $\rho_{xy} = 0$, they are said to be uncorrelated.

Theoretically, the BSS algorithm gives independent sources at the output. Since two independent signals are uncorrelated, if the cross-correlation coefficient ρ is calculated between one row $\hat{s}_i(n)$ of the overlap block of $\hat{S}_1(n)$ and all the rows $\hat{s}'_j(t)$ of the overlap block of $\hat{S}_2(n)$, we should obtain all the values equal to zero except one of them for which $|\rho_{ii}| = 1$. In other words, if we define the matrix $\Gamma = \{\gamma_{ij}\}$ with elements equal to the absolute value of ρ between the ith row of the overlap segment of $\hat{S}_1(n)$ and the jth row of the overlap segment of $\hat{S}_2(n)$, then, $\Gamma = \mathbf{P}^T$ and the permutation problem can be solved.

Once the permutation problem has been solved, each of the signals $\hat{s}_i(n)$ corresponds to only one of the signals $\hat{s}'_j(n)$, but the latter signals are scaled and possibly inverted versions of the former signals, due to the first inherent ambiguity of BSS. The BSS algorithm sets the variances of the output sources to one and, therefore, $\hat{s}_i(n)$ and $\hat{s}'_j(n)$ both have equal variance. Since the signals only share an overlap of $L - N$ samples, the energy of the overlap segment of these signals will generally be different and, therefore, can be used to solve the amplitude ambiguity. In particular

$$\hat{s}_i(n + N) = \text{sign}(\rho_{ij}) \frac{\sigma_i}{\sigma'_j} \hat{s}'_j(n), \text{ for } n = 1, \ldots, L - N \qquad (15.34)$$

where ρ_{ij} is calculated for $\hat{s}_i(n)$ and $\hat{s}'_j(n)$ within the overlap segment, and σ_i and σ'_j are the standard deviations of $\hat{s}_i(n)$ and $\hat{s}'_j(n)$, respectively, within the overlap segment. This should solve the scaling ambiguity of the BSS algorithm.

In practice, the estimated sources are not completely uncorrelated and therefore, $\Gamma \neq \mathbf{P}^T$. However, for each row it is expected to obtain only one of the elements γ_{ij} close to unity corresponding to $j = j_0$, and the rest close to zero. Therefore, the algorithm can still be applied to maintain the continuity of the signals.

After the sources are estimated TNM can be applied to track the dynamics of the estimated source signals.

Using simultaneous scalp and intracranial EEG recordings the performance of the above system has been observed. The intracranial recordings were obtained from multicontact Foramen Ovale (FO) electrodes. Electrode bundles are introduced bilaterally through the FO under fluoroscopic guidance. The deepest electrodes within each bundle lie next to medial temporal structures, whereas the most superficial electrodes lie at or just below the FO [91]. As FO electrodes are introduced via anatomical holes, they provide a unique opportunity to record simultaneously from scalp and medial temporal structures without disrupting the conducting properties of the brain coverings by burr holes and wounds, which can otherwise make simultaneous scalp and intracranial recordings unrepresentative of the habitual EEG [92]. Simultaneously, scalp EEG recordings were obtained from standard silver cup electrodes applied according to the "Maudsley" electrode placement system [93]. The Maudsley system is a modification of the extended 10–20 system. The advantage of the Maudsley system with respect to the standard 10–20 system is that it provides a more extensive coverage of the lower part of the cerebral convexity, increasing the sensitivity of recording from basal sub-temporal structures.

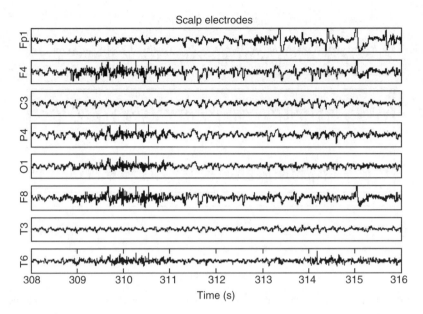

Figure 15.5 Eight seconds of EEG signals from 8 out of 16 scalp electrodes during a seizure (DC removed)

For the scalp EEG the overlap window approach was used to maintain the continuity of the underlying sources and, once the continuity was maintained, the resulting sources were divided into non-overlapping segments of 2048 samples. The largest Lyapunov exponent, λ_1 was estimated for each of these segments. The intracranial signals were also divided into segments of the same size and λ_1 was estimated for each of these segments. In both cases, the parameters used for the estimation of λ_1 were those used by Iasemidis *et al.* for the estimation of STL_{max}, as explained in [81]. After the λ_1s are calculated for different segments, they are included in a time sequence and smoothed by time averaging.

Within the simultaneous intracranial and scalp EEG recording of 5 min and 38 s containing a focal seizure, the seizure is discernible in the intracranial electrodes (Figure 15.5), from around 308 s, and the ictal state lasts throughout the observation. Figure 15.5 shows a segment of the signals recorded by the scalp electrodes during the seizure. The signals are contaminated by noise and artefact signals, such as eye blinking and electrocardiogram (ECG), and the seizure is not clearly discernible. Figure 15.6 shows the signals obtained after applying the BSS-based prediction algorithm to the same segment of scalp EEGs. It is clear that the characteristics of the estimated independent components vary significantly. Some spikes can be seen in IC4. Figure 15.7a and b illustrate the smoothed λ_1 variations for two intracranial electrodes located in the focal area. The smoothed λ_1 is calculated by averaging the current value of λ_1 and the previous two values. These two electrodes show a clear drop in the value of λ_1 at the occurrence of seizure, starting prior to the onset. However, the intracranial EEG was contaminated by a high-frequency activity that causes fluctuations of λ_1 for the entire recording. Figure 15.7c–f illustrate the smoothed λ_1 evolution for four scalp electrodes once the baseline was removed. The value of λ_1 presents large fluctuations that can be due to the presence of noise and artefacts.

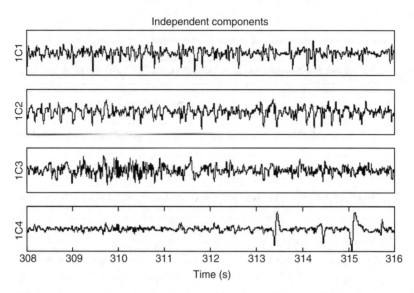

Figure 15.6 The four independent components obtained by applying BSS to the scalp electrode signals shown in Figure 15.5

Although the values seem to be lower as the seizure approaches, there is not a clear trend before seizure in any of the electrodes.

Figure 15.8 shows the results obtained for two of the estimated sources after the application of the proposed BSS algorithm. The algorithm efficiently separates the underlying sources from the eye-blinking artefacts and noise. Both figures show how the BSS algorithm efficiently separates the epileptic components. The value of λ_1 reaches the minimum value more than 1 minprior to seizure, remaining low until the end of the recording. This corresponds with the seizure lasting until the end of the recording.

Figure 15.9–15.11 illustrate the results obtained from a recording of 5 min and 34 s duration. In this particular case the epileptic component was not clearly visible by visual inspection of the intracranial electrode signals. The intracranial electrodes may not have recorded the electrical activity of the epileptic focus because of their location. Figure 15.9a shows a segment of the signals recorded by eight scalp electrodes during the seizure. Although the signals are contaminated by noise and artefacts, the seizure components are discernible in several electrodes. Figure 15.9b illustrates the signals obtained for the same segment of data after the BSS algorithm. In this case the seizure component seems to be separated from noise and artefacts in the third estimated source. Figure 15.10a displays the evolution of the smoothed λ_1 for four different intracranial electrodes. The values fluctuate during the recording but there is a gradual drop in λ_1 starting at the beginning of the recording. A large drop in the value of λ_1 is observed for the four electrodes around 250 s and reaches a minimum value around 275 s. However, the onset of seizure occurs around 225 s and therefore none of the intracranial electrodes is able to predict the seizure. Figure 15.10b shows the variation of smoothed λ_1 for four scalp electrodes. Likewise, for the intracranial electrodes, λ_1 values have large fluctuations but present a gradual drop towards seizure. Similarly, the drop to the lowest value of λ_1 starts after 250 s and, therefore, the signals from these electrodes are not used for seizure prediction.

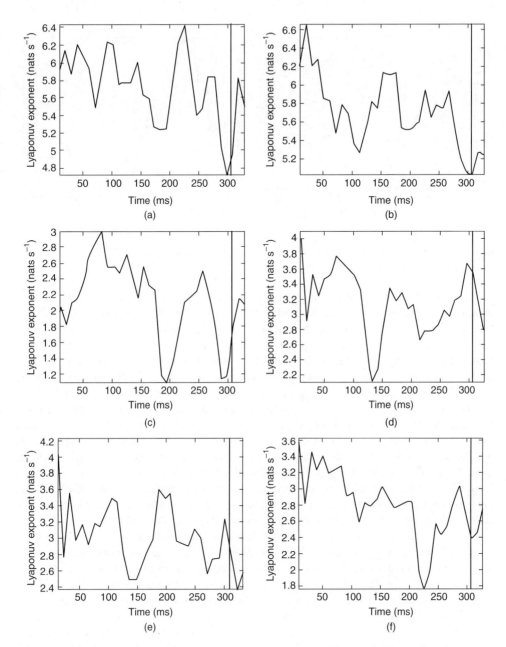

Figure 15.7 The smoothed λ_1 evolution over time for two intracranial electrodes located in the focal area, (a) for the LF4 electrode, (b) for the LF6 electrode. (c)–(f) show the smoothed λ_1 evolutions for four scalp electrodes. The length of the recording is 338 s and the seizure occurs at 306 s (marked by the vertical line)

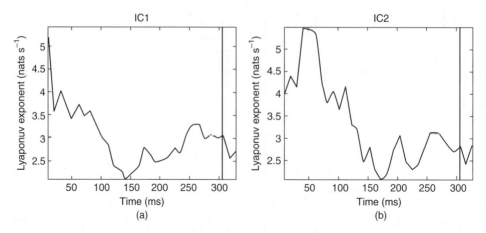

Figure 15.8   Smoothed λ_1 evolution over time for two independent components IC1 and IC2, for which ictal activity is prominent (see Figure 15.5)

Figure 15.11 illustrates the changes in the smoothed λ_1 for the third estimated source obtained after the application of BSS to scalp EEG. λ_1 starts decreasing approximately 2 min before the onset of seizure. The minimum of λ_1 is obtained around the same time as for the intracranial and scalp electrodes. However, a local minimum is clear at the onset of seizure and the values are clearly lower during the seizure than at the beginning of the recording. The BSS algorithm seems to separate the epileptic component in one of the estimated sources allowing the prediction of seizure even when this is not possible from the intracranial electrodes.

Figures 15.12a and b show the results obtained for a third EEG recording lasting 5 min and 37 s. The electrodes recorded a generalized seizure. Figure 15.12a illustrates the results for four intracranial electrodes. The value of λ_1 does not show any clear decrease until 250 s when there is a sudden drop in λ_1 for all the electrodes. The minimum value is obtained several seconds later; however, the onset of seizure was clearly discernible from the intracranial electrodes around 236 s. Therefore, as an important conclusion, the intracranial EEG is not able to predict the onset of seizure in such cases and isonly able to detect the seizure after its onset. There is a clear drop in the value of λ_1 but it does not occur soon enough to predict the seizure.

Figure 15.12b shows the results obtained after the application of BSS. The evolution of λ_1 is similar to the evolution for intracranial recordings. However, the drop in the value of λ_1 for the estimated source seems to start decreasing before it does for the intracranial electrodes. The minimum λ_1 for the estimated source occurs before such a minimum for λ_1 is achieved for the intracranial electrodes. This means that by preprocessing the estimated sources using the BSS-based method, the occurrence time of seizure can be estimated more accurately.

There is no doubt that local epileptic seizures are predictable from the EEGs. Scalp EEG recordings seem to contain enough information about the seizure; however, this information is mixed with the signals from the other sources within the brain, particularly strong cortical potentials. The sources are also affected by noise and artefacts. Incorporating a suitable preprocessing technique, such as the BSS algorithm, separates the seizure signal (long before the seizure) from the rest of the sources, noise, and artefacts within the brain. Therefore, the well known traditional nonlinear methods for evaluation of chaotic behaviour of the EEG signals can be applied. Using BSS, prediction of seizure might be possible from scalp EEG

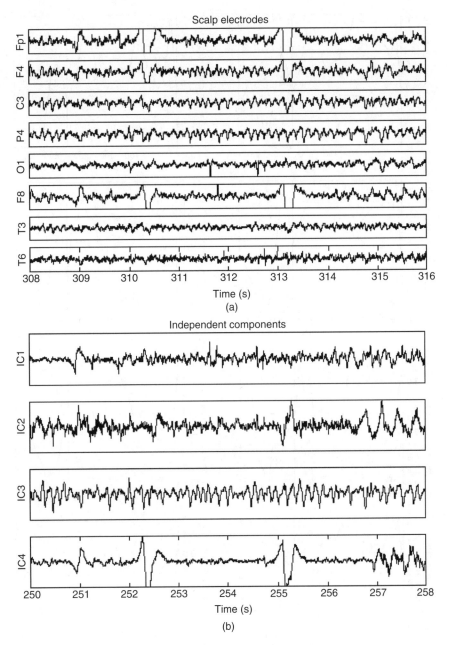

Figure 15.9 (a) A segment of 8 s of EEG signals (with zero mean) for 8 out of 16 scalp electrodes during the seizure, (b) the four independent components obtained by applying BSS to the scalp electrode signals shown in (a). The epileptic activity seems to be discernible in IC3 and its λ_1 evolution is shown in Figure 15.10

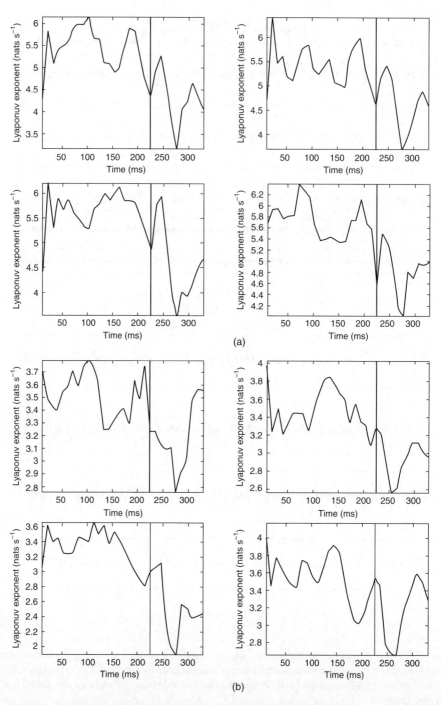

Figure 15.10 (a) Intracranial EEG analysis: 3-point smoothed λ_1 evolution from focal seizure. In this case the electrical activity of the epileptic focus seemed not to be directly recorded by the intracranial electrodes. (b) Scalp EEG analysis: the smoothed λ_1 evolution for four scalp electrodes. The length of the recording is 334 s. The seizure occurs at 225 s (marked by vertical line)

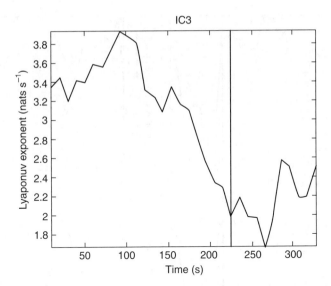

Figure 15.11 Smoothed λ_1 evolution for a focal seizure estimated from the independent component IC3 (i.e. the one where ictal activity is discernible in Figure 15.9(b)). The electrical activity of the epileptic focus was not directly recorded by the intracranial electrodes. The length of the recording is 334 s, the seizure occurs at 225 s (marked by the vertical line)

data, even when this is not possible with intracranial EEG, especially when the epileptic activity is spread over a large portion of the brain in most of the cases. However, the results obtained from three sets with generalized seizure support the idea of unpredictability of this type of seizure, since they are preceded and followed by normal EEG activity.

The results obtained by analysis of the scalp EEGs, although very promising, are subject to several limitations. The principal limitation is due to the length of the recordings. These recordings allow the comparison between scalp and intracranial EEG; however, they were of relatively short duration. Therefore, it is basically assumed that the epileptic component is active during the entire recording. Longer recordings are needed to better examine the value of BSS in the study of seizure predictability. For such recordings it can be hardly assumed that the underlying sources are active during the entire recording and, therefore, the algorithm needs to detect the beginning and end of these activities. Furthermore, the algorithm employed to maintain the continuity fails in some cases where a segment of the scalp EEG is corrupted or the electrical activity is not correctly recorded. The number of corrupted segments increases for longer recordings and, therefore, a new methodology to maintain the continuity of the estimated sources for these particular segments should be combined with the overlap window approach. Another limitation arises from the fixed number of output signals selected for the BSS algorithm.

Generally, a seizure can be predicted if the signals are observed during a previous seizure and by analysing the interictal period. Otherwise, there remains a long way to go before being able to accurately predict the seizure from only EEG. On the other hand, seizure is predictable if the EEG information is combined with other information, such as that gained from a video sequence of the patients, heart rate variability, and respiration.

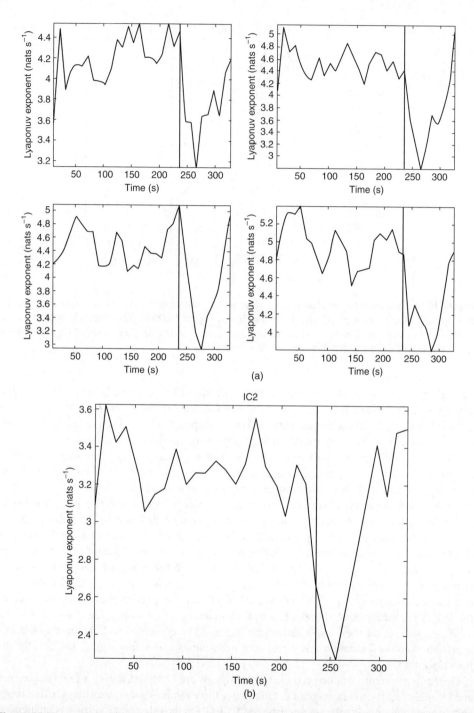

Figure 15.12 (a) Smoothed λ_1 evolution of four intracranial electrodes for a generalized seizure. The length of the recording is 337 s and it records a generalized seizure starting at 236 s (marked by the vertical line). (b) Smoothed λ_1 evolution of IC2 component estimated from the corresponding scalp EEG. The length of the recording is 337 s and it records a generalized seizure at 236 s (marked by the vertical line)

15.6   Fusion of EEG – fMRI Data for Seizure Detection and Prediction

In the above sections it has been verified that some epileptic seizures may be predicted, however, a long-term recording is normally necessary. Processing of the EEGs using the popular techniques such as BSS, however, is carried out on blocks of data. The processed blocks need to be aligned and connected to each other to provide the complete information. The inherent permutation problem of BSS can be solved by incorporating the functional MRI into the prediction system.

Clinical MRI has been of primary importance for visualization/detection of brain tumours, stroke, and multiple sclerosis. In 1990, Ogawa [94] showed that MRI can be sensitized to cerebral oxygenation, using deoxyhemoglobin as an endogenous susceptibility contrast agent. Using gradient-echo imaging, a form of MRI image encoding sensitive to local inhomogeneity of the static magnetic field, he demonstrated (for an animal) that the appearance of the blood vessels of the brain changed with blood oxygenation. In some later papers published by his group they presented the detection of human brain activations using this blood BOLD [95, 96]. Now, it is established that by an increase in neuronal activity, local blood flow increases. The increase in perfusion, in excess of that needed to support the increased oxygen consumption due to neuronal activation, results in a local decrease in the concentration of deoxyhaemoglobin. Since deoxyhaemoglobin is paramagnetic, a reduction in its concentration results in an increase in the homogeneity of the static magnetic field, which yields an increase in the gradient-echo MRI signal. Although the BOLD fMRI does not measure brain activity directly, it relies on neurovascular coupling to encode the information about the brain function into detectible haemodynamic signals. It will clearly be useful to exploit this information in localization of seizure during the ictal period.

As stated in Chapter 14, brain sources can be localized within the brain. The seizure signals, however, are primarily not known either in time or in space. Therefore, if the onset of seizure is detected within one ictal period, the location of the source may be estimated and the source which originates from that location tracked. The FO electrodes were planted within hippocampus. Using fMRI it is easy to detect the BOLD regions corresponding to the seizure sources. Therefore, in separation of the sources using BSS, it can be established that the source signals originating from those regions represent seizure signals. This process can be repeated for all the estimated source segments and, thereby, the permutation problem can be solved and the continuity of the signals maintained.

Unfortunately, in a simultaneous EEG-fMRI recording the EEG signals are highly distorted by the fMRI effects. Figure 15.13 shows a multichannel EEG affected by fMRI. An effective preprocessing technique is then required to remove this artefact. Also, in a long-term recording it is difficult to keep the patient steady under the fMRI system. This also causes severe distortion of the fMRI data and new methods are required for its removal. In Chapter 16 application of simultaneous EEG-fMRI recording and joint analysis to seizure BOLD detection is explained in detail.

15.7   Conclusions

Detection of epileptic seizures using different techniques has been successful, particularly for adult seizures. The false alarm rate for detection of seizure in the newborn is, however, still too high and, therefore, the design of a reliable newborn seizure detection system is still an open

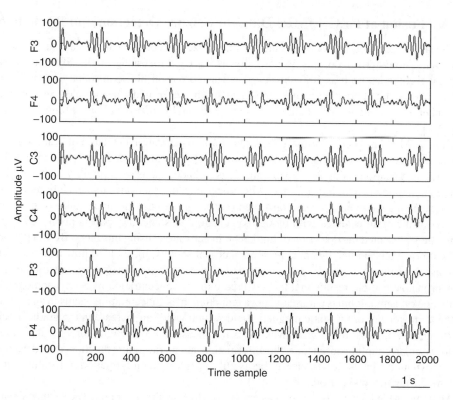

Figure 15.13   A segment of EEG signals affected by the scanner artefact in a simultaneous EEG-fMRI recording

problem. In addition, more has to be done on differentiation between epileptic seizures and seizures caused by other neurological disorders. On the other hand, although predictability of focal seizure from the EEGs (both scalp and intracranial) has been verified, more research is necessary to increase the accuracy of a seizure prediction system and enable a robust prediction of epileptic seizure onset. For the case of neonate seizure, there is much space for research. In addition, seizure source localization using scalp EEG only is difficult to achieve unless an effective brain model is established to map the interictal deep discharges into the scalp electrodes. Currently, subdural electrodes, such as Foreman Ovale electrodes, implanted in the hippocampus are able to locate the seizure sources. Fusion of different neuroimaging modalities may indeed pave the path for more reliable seizure detection and prediction systems.

References

[1] Iasemidis, L.D. (2003) Epileptic seizure prediction and control. *IEEE Trans. Biomed. Eng.*, **50**, 549–558.
[2] Annegers, J.F. (1993) *The Epidemiology of Epilepsy, The Treatment of Epilepsy* (ed. E. Wyllie), Lea and Febiger, Philadelphia, pp. 157–164.
[3] Engel, J. Jr and Pedley, T.A. (1997) *Epilepsy: A Comprehensive Text-Book*, Lippinott-Ravon, Philadelphia.
[4] Lehnertz, K., Mormann, F., Kreuz, T. *et al.* (2003) Seizure prediction by nonlinear EEG analysis. *IEEE Eng. Med. Biol. Mag.*, **22**(1), 57–63.

[5] Conradsen, I., Beniczky, S., Hoppe, K. *et al.* (2012) Automated algorithm for generalized tonic-clonic epileptic seizure onset detection based on sEMG zero-crossing rate. *IEEE Trans. Biomed. Eng.*, **59**(2), 579–585.

[6] Azami, H., Sanei, S. and Mohammadi, K. (2011) A novel signal segmentation method based on standard deviation and variable threshold. *J. Comput. Appl.*, **34**(2), 27–34.

[7] Peitgen, H. (2004) *Chaos and Fractals: New Frontiers of Science*, Springer-Verlag, New York.

[8] Niedermeyer, E. and Da Silva, F.L. (1999) *Electroencephalography, Basic Principles, Clinical Applications, and Retailed Fields*, 4th edn, Lippincott Williams &Wilkins.

[9] Bravais, L.F. (1827) *Researchers sur les Symptoms et le Traitement de l'Épilesie Hémiplégique*, Thése de Paris, Paris, no. 118.

[10] Gavaret, M., Guedj, E., Koessler, L. *et al.* (2010) Reading epilepsy from the dominant temporo-occipital region. *J. Neurol. Neurosurg. Psychiat.*, **81**, 710–715. doi: 10.1136/jnnp.2009.175935

[11] Akay, M. (2001) *Nonlinear Biomedical Signal Processing, Vol. II, Dynamic Analysis and Modelling*, IEEE Press.

[12] Chee, M.W.L., Morris, H.H., Antar, M.A. *et al.* (1993) Presurgical evaluation of temporal lobe epilepsy using interictal temporal spikes and positron emission tomography. *Arch. Neurol.*, **50**, 45–48.

[13] Godoy, J., Luders, H.O., Dinner, D.S. *et al.* (1992) Significance of sharp waves in routine EEGs after epilepsy surgery. *Epilepsia*, **33**, 513–288.

[14] Kanner, A.M., Morris, H.H., Luders, H.O. *et al.* (1993) Usefulness of unilateral interictal sharp waves of temporal lobe origin in prolonged video EEG monitoring studies. *Epilepsia*, **34**, 884–889.

[15] Steinhoff, B.J., So, N.K., Lim, S. and Luders, H.O. (1995) Ictal scalp EEG in temporal lobe epilepsy with unitemporal versus bitemporal interictal epileptiform discharges. *Neurology*, **45**(5), 889–896.

[16] Walter, D.O., Muller, H.F. and Jell, R.M. (1973) Semiautomatic quantification of sharpness of EEG phenomenon. *IEEE Trans. Biomed. Eng.*, BME-**20**, 53–54.

[17] Ktonas, P.Y. and Smith, J.R. (1974) Quantification of abnormal EEG spike characteristics. *Comput. Biol. Med.*, **4**, 157–163.

[18] Gotman, J. and Gloor, P. (1976) Automatic recognition and quantification of interictal epileptic activity in the human scalp EEG. *Electroenceph. Clin. Neurophysiol.*, **41**, 513–529.

[19] Davey, B.L.K., Fright, W.R., Caroll, G.J. and Jones, R.D. (1989) Expert system approach to detection of epileptiform activity in the EEG. *Med. Biol. Eng. Comput.*, **27**, 365–370.

[20] Dingle, A.A., Jones, R.D., Caroll, G.J. and Fright, W.R. (1993) A multistage system to detect epileptiform activity in the EEG. *IEEE Trans. Biomed. Eng.*, **40**, 1260–1268.

[21] Glover, J.R. Jr., Ktonas, P.Y., Raghavan, N. *et al.* (1986) A multichannel signal processor for the detection of epileptogenic sharp transients in the EEG. *IEEE Trans. Biomed. Eng.*, **33**, 1121–1128.

[22] Glover, J.R. Jr., Raghavan, N., Ktonas, P.Y. and Frost, J.D. (1989) Context-based automated detection of epileptogenic sharp transients in the EEG: Elimination of false positives. *IEEE Trans. Biomed. Eng.*, **36**, 519–527.

[23] Gotman, J., Ives, J.R. and Gloor, R. (1979) Automatic recognition of interictal epileptic activity in prolonged EEG recordings. *Electroencephalogr. Clin. Neurophysiol.*, **46**, 510–520.

[24] Gotman, J. and Wang, L.Y. (1991) State-dependent spike detection: concepts and preliminary results. *Electroencephalogr. Clin. Neurophysiol.*, **79**, 11–19.

[25] Gotman, J. and Wang, L.Y. (1992) State-dependent spike detection: validation. *Electroencephalogr. Clin. Neurophysiol.*, **83**, 12–18.

[26] Ozdamar, O., Yaylali, I., Jayakar, P. and Lopez, C.N. (1991) Multilevel neural network system for EEG spike detection, in *Computer-Based Medical Systems, Proceedingsof the 4th Annual IEEE Symposium* (eds I.N. Bankmann and J.E. Tsitlik), IEEE Computer Society, Washington DC.

[27] Webber, W.R.S., Litt, B., Lesser, R.P. *et al.* (1993) Automatic EEG spike detection: what should the computer imitate. *Electrencephalogr. Clin. Neurophysiol.*, **87**, 364–373.

[28] Webber, W.R.S., Litt, B., Wilson, K. and Lesser, R.P. (1994) Practical detection of epileptiform discharges (EDs) in the EEG using an artificial neural network: a comparison of raw and parameterised EEG data. *Electroencephalogr. Clin. Neurophysiol.*, **91**, 194–204.

[29] Wilson, S.B., Harner, R.N., Duffy, B.R. *et al.* (1996) Spike detection I. Correlation and reliability of human experts. *Electroencephalogr. Clin. Neurophysiol.*, **98**, 186–198.

[30] Kurth, C., Gilliam, F. and Steinhoff, B.J. (2000) EEG spike detection with a Kohonen feature map. *Ann. Biomed. Eng.*, **28**, 1362–1369.

[31] Kohonen, T. (1990) The self-organizing map. *Proc. IEEE*, **78**(9), 1464–1480.

[32] Kohonen, T. (1997) *The Self-Organizing Maps*, 2nd edn, Springer, NY.

[33] Nenadic, Z. and Burdick, J.W. (2005) Spike detection using the continuous wavelet transform. *IEEE Trans. Biomed. Eng.*, **52**(1), 74–87.

[34] Boashash, B., Mesbah, M. and Colditz, P. (2003) Time-Frequency Detection of EEG Abnormalities, in *Time–Frequency Signal Analysis and Processing: A Comprehensive Reference*, (ed. B. Boashash), Elsevier Science, Amsterdam, pp. 663–670.

[35] Subasi, A. (2005) Epileptic seizure detection using dynamic wavelet network. *Exp. Syst. Appl.*, **29**, 343–355.

[36] Lutz, A., Lachaux, J-P., Martinerie, J. and Varela, F.J. (2002) Guiding the study of brain dynamics by using first-person data: synchrony patterns correlate with ongoing conscious states during a simple visual task. *Proc. Natl. Acad. Sci.*, **99**(3), 1586–1591.

[37] Osorio, I. (1999) System for the prediction, rapid detection, warning, prevention, or control of changes in activity states in the brain of subject. US patent 5995868.

[38] Osorio, I., Frei, M.G. and Wilkinson, S.B. (1998) Real-time automated detection and quantitative analysis of seizures and short-term prediction of clinical onset. *Epilepsia*, **39**(6), 615–627.

[39] Bhavaraju, N.C., Frei, M.G. and Osorio, I. (2006) Analogue seizure detection and performance evaluation. *IEEE Trans. Biomed Eng.*, **53**(2), 238–245.

[40] Gonzalez-Vellon, B., Sanei, S. and Chambers, J. (1993) Support vector machines for seizure detection. Proceedings of the ISSPIT, Darmstadt, Germany, pp. 126–129.

[41] Raad, A., Antoni, J., Sidahmed, M. (2003) Indicators of cyclostationarity: proposal, statistical evaluation and application to diagnosis. Proceedings of the IEEE ICASSP, vol. VI, pp. 757–760.

[42] Acir, N., Oztura, I., Kuntalp, M. *et al.* (2005) Automatic detection of epileptiform events in EEG by a three-stage procedure based on artificial neural network. *IEEE Trans. Biomed. Eng.*, **52**(1), 30–40.

[43] da Silva, F.H.L., Blanes, W., Kalitzin, S. N. *et al.* (2003) Dynamical diseases of brain systems: different routes to epileptic seizures. *IEEE Trans. Biomed. Eng.*, **59**(5), 540–548.

[44] Adeli, H., Ghosh-Dastidar, S. and Dadmehr, N. (2007) A wavelet-chaos methodology for analysis of EEGs and EEG subbands to detect seizure and epilepsy. *IEEE Trans. Biomed. Eng.*, **54**(2), 205–211.

[45] Cao, L. (1997) Practical method for determining the minimum embedding dimension of a scalar time series. *Physica D*, **110**(1–2), 43–50.

[46] Rana, P., Lipor, J., Lee, H. *et al.* (2012) Seizure detection using the phase-slope index and multichannel ECoG. *IEEE Trans. Biomed. Eng.*, **59**(4), 1125–1134.

[47] Volpe, J.J. (1987) *Neurology of the Newborn*, Saunders, Philadelphia, PA.

[48] Liu, A., Hahn, J.S., Heldt, G.P. and Coen, R.W. (1992) Detection of neonatal seizures through computerized EEG analysis. *Electroencephalogr. Clin. Neurophysiol.*, **82**, 30–37.

[49] Celka, P., Boashash, B. and Colditz, P. (2001) Preprocessing and time-frequency analysis of newborn EEG seizures. *IEEE Eng. Med. Biol. Mag.*, **20**(5), 30–39.

[50] Shoker, L., Sanei, S., Wang, W. and Chambers, J. (2004) Removal of eye blinking artifact from EEG incorporating a new constrained BSS algorithm. *IEEE J. Med. Biol. Eng. Comput.*, **43**, 290–295.

[51] Pfurtscheller, G. and Fischer, G. (1977) A new approach to spike detection using a combination of inverse and matched filter techniques. *Electroenceph. Clin. Neurophysiol.*, **44**, 243–247.

[52] Roessgen, M., Zoubir, A.M. and Boashash, B. (1998) Seizure detection of newborn EEG using a model-based approach. *IEEE Trans. Biomed. Eng.*, **45**(6), 673–685.

[53] Choudhuri, N., Ghosal, S. and Roy, A. (2004) Contiguity of the Whittle measure for a Gaussian time series. *Biometrika*, **91**(1), 211–218.

[54] da Silva, F.H.L., Hoeks, A., Smits, H. and Zetterberg, L.H. (1975) Model of brain rhythmic activity; the alpha rhythm of the thalamus. *Kybernetik*, **15**, 27–37.

[55] O'Toole, J., Mesbah, M. and Boashash, B. (2005) Neonatal EEG seizure detection using a time-frequency matched filter with a reduced template set. Proceedings of the 8th International Symposium on Signal Processing and its Applications, pp. 215–218.

[56] Karayiannis, N.B., Mukherjee, A., Glover, J.R. *et al.* (2006) Detection of pseudosinusoidal epileptic seizure segments in the neonatal EEG by cascading a rule-based algorithm with a neural network. *IEEE Trans. Biomed. Eng.*, **53**(4), 633–641.

[57] Bishop, C.M. (1995) *Neural Networks for Pattern Recognition*, Oxford University Press, New York.

[58] Purushothaman, G. and Karayiannis, N.B. (1997) Quantum neural networks (QNNs): inherently fuzzy feedforward neural networks. *IEEE Trans. Neural. Netw.*, **8**(3), 679–693.

[59] Mitra, J., Glover, J., Ktonas, P. *et al.* (2009) A multistage system for the automated detection of epileptic seizures in neonatal electroencephalography. *J. Clin. Neurophysiol.*, **26**, 1–9.

[60] Celka, P. and Colditz, P. (2002) A computer-aided detection of EEG seizures in infants, a singular-spectrum approach and performance comparison. *IEEE Trans. Biomed. Eng.*, **49**(5), 455–462.

[61] Deburchgraeve, W., Cherian, P., Vos, M. *et al.* (2008) Automated neonatal seizure detection mimicking a human observer reading EEG. *J. Clin. Neurophysiol.*, **119**, 2447–2454.

[62] Temko, A., Lightbody, G., Thomas, E.M. *et al.* (2012) Instantaneous measure of EEG channel importance for improved patient-adaptive neonatal seizure detection. *IEEE Trans. Biomed. Eng.*, **53**(3), 717–727.

[63] Takens, F. (1981) Detecting strange attractors in turbulence, in *Dynamical Systems and Turbulence*, (eds D.A. Rand and L.S. Young) Springer-Verlag, Berlin, pp. 366–381.

[64] Sauer, T., Yurke, J.A. and Casdagli, M. (1991) Embedology. *J. Stat. Phys.*, **65**(3/4), 579–616.

[65] Fraser, A.M. and Swinney, H.L. (1986) Independent coordinates for strange attractors from mutual information. *Phys. Rev. A*, **33**, 1134–1140.

[66] Kennel, M.B., Brown, R. and Abarbanel, H.D.I. (1992) Determining minimum embedding dimension using a geometrical construction. *Phys. Rev. A*, **45**, 3403–3411.

[67] Casdagli, M.C. (1997) Recurrence plots revisited. *Physica D*, **108**(1), 12–44.

[68] Kantz, H. and Schreiber, T. (1995) Dimension estimates and physiological data. *Chaos*, **5**(1), 143–154.

[69] Rosenstein, M.T., Collins, J.J. and Deluca, C.J. (1993) A practical method for calculating largest Lyapunov exponents from small data sets. *Physica D*, **65**, 117–134.

[70] Wolf, A., Swift, J.B., Swinney, H.L. and Vastano, J.A. (1985) Determining Lyapunov exponents from a time series. *Physica D*, **16**, 285–317.

[71] Kawabata, N. (1973) A nonstationary analysis of the electroencephalogram. *IEEE Trans. Biomed. Eng.*, **20**, 444–452.

[72] Iasemidis, L.D., Sackellares, J.C., Zaveri, H.P. and Willians, W.J. (1990) Phase space topography and the Lyapunov exponent of electrocorticograms in partial seizures. *Brain Topogr.*, **2**, 187–201.

[73] Iasemidis, L.D., Shiau, D.-S., Sackellares, J.C. *et al.* (2004) A dynamical resetting of the human brain at epileptic seizures: application of nonlinear dynamics and global optimization techniques. *IEEE Trans. Biomed. Eng.*, **51**(3), 493–506.

[74] Lehnertz, K. and Elger, C.E. (1995) Spatio-temporal dynamics of the primary epileptogenic area in temporal lobe epilepsy characterized by neuronal complexity loss. *Electroencephalogr. Clin. Neurophysiol.*, **95**, 108–117.

[75] Lerner, D.E. (1996) Monitoring changing dynamics with correlation integrals: Case study of an epileptic seizure. *Physica D*, **97**, 563–76.

[76] Osorio, I., Harrison, M.A.F., Lai, Y.C. and Frei, M.G. (2001) Observations on the application of the correlation dimension and correlation integral to the prediction of seizures. *J. Clin. Neurophysiol.*, **18**, 269–274.

[77] Quyen, M.L.V., Martinerie, J., Baulac, M. and Varela, F.J. (1999) Anticipating epileptic seizures in real time by a non-linear analysis of similarity between {EEG} recordings. *NeuroReport*, **10**, 2149–2155.

[78] Moser, H.R., Weber, B., Wieser, H.G. and Meier, P.F. (1999) Electroencephalogram in epilepsy: analysis and seizure prediction within the framework of Lyapunov theory. *Physica D*, **130**, 291–305.

[79] Litt, B., Estellera, R., Echauz, J. *et al.* (2001) Epileptic seizures may begin hours in advance of clinical onset: A report of five patients. *Neuron*, **30**, 51–64.

[80] D'Alessandro, M., Esteller, R., Vachtsevanos, G. *et al.* (2003) Epileptic seizure prediction using hybrid feature selection over multiple intracranial eeg electrode contacts: a report of four patients. *IEEE Trans. Biomed. Eng.*, **50**, 603–615.

[81] Iasemidis, L.D., Principe, J.C. and Sackellares, J.C. (2000) Measurement and quantification of spatio-temporal dynamics of human epileptic seizures, in *Nonlinear Biomedical Signal Processing* (ed. M. Akay), IEEE Press, pp. 296–318.

[82] Sackellares, J.C., Iasemidis, L.D., Shiau, D.S. *et al.* (2000) Epilepsy-when chaos fails, in *Chaos in the Brain?* (eds K. Lehnertz and C.E. Elger), World Scientific, Singapore, pp. 112–133.

[83] Iasemidis, L.D., Shiau, D., Chaovalitwongse, W. *et al.* (2003) Adaptive epileptic seizure prediction system. *IEEE Trans. Biomed. Eng.*, **50**, 616–627.

[84] Hively, L.M., Protopopescu, V.A. and Gailey, P.C. (2000) Timely detection of dynamical change in scalp {EEG} signals. *Chaos*, **10**, 864–875.

[85] Hively, L.M. and Protopopescu, V.A. (2003) Channel-consistent forewarning of epileptic events from scalp EEG. *IEEE Trans. Biomed. Eng.*, **50**, 584–593.

[86] Iasemidis, L., Principe, J., Czaplewski, J. *et al.* (1997) Spatiotemporal transition to epileptic seizures: a nonlinear dynamical analysis of scalp and intracranial EEG recordings, in *Spatiotemporal Models in Biological and Artificial Systems* (eds F. Silva, J. Principe and L. Almeida), IOS Press, Amsterdam, pp. 81–88.

[87] Sackellares, J., Iasemidis, L., Shiau, D. *et al.* (1999) Detection of the preictal transition from scalp EEG recordings. *Epilepsia*, **40**(S7), 176.

[88] Shiau, D., Iasemidis, L., Suharitdamrong, W. *et al.* (2003) Detection of the preictal period by dynamical analysis of scalp EEG. *Epilepsia*, **44**(S9), 233–234.

[89] Takens, F. (1981) Detecting strange attractors in turbulence, in *Lectures Notes in Mathematics, Dynamical Systems and Turbulence, Warwick 1980* (eds D.A. Rand and L.S. Young), Springer-Verlag, Berlin, pp. 366–381.

[90] Corsini, J., Shoker, L., Sanei, S. and Alarcon, G. (2006) Epileptic seizure predictability from scalp EEG incorporating constrained blind source separation. *IEEE Trans. Biomed. Eng.*, **53**(5), 790–799.

[91] Fernandez, J., Alarcon, G., Binnie, C.D. and Polkey, C.E. (1999) Comparison of sphenoidal, foramen ovale and anterior temporal placements for detecting interictal epileptiform discharges in presurgical assessment for temporal lobe epilepsy. *Clin. Neurophysiol.*, **110**, 895–904.

[92] Nayak, D., Valentin, A., Alarcon, G. *et al.* (2004) Characteristics of scalp electrical fields associated with deep medial temporal epileptiform discharges. *Clin. Neurophysiol.*, **115**, 1423–1435.

[93] Margerison, J.H., Binnie, C.D. and McCaul, I.R. (1970) Electroencephalographic signs employed in the location of ruptured intracranial arterial aneurysms. *Electroencephalogr. Clin. Neurophysiol.*, **28**, 296–306.

[94] Menon, R.S. (2002) Postacquisition suppression of large-vessel BOLD signals in high-resolution fMRI. *Magn. Reson. Med.*, **47**(1), 1–9.

[95] Huettel, S.A., Song, A.W. and McCarthy, G. (2004) *Functional Magnetic Resonance Imaging*, Sinauer Assov., Sunderland, MA.

[96] Toga, A.W. and Mazziotta, J.C. (2002) *Brain Mapping: the Methods*, 2nd edn, Academic, San diego, CA.

16

Joint Analysis of EEG and fMRI

A number of neuroimaging tools and systems have been developed over the years to study different aspects of brain function. EEG and functional magnetic resonance imaging (fMRI) have probably been used more than other methods to study the brain function. The advances of acquisition systems in recent years has enabled simultaneous EEG–fMRI recordings of the brain. The information achieved from these two data modalities is important in analysis of brain responses to event-related and movement-related stimulations.

Unfortunately, the EEG recorded in the MRI scanner is highly deteriorated by the magnetic field. Therefore, to ensure that the EEG signal will be useful and informative, a comprehensive preprocessing technique is required.

In this chapter we review the literature and describe the state of the art in new signal processing methods for both fMRI artefact removal from EEG signals and the combination of these two modalities. This is followed by presentation of a number of new approaches for solving these problems.

16.1 Fundamental Concepts

fMRI produces a sequence of low-resolution MRI which is able to detect brain activity by measuring the associated changes in blood flow. Unlike structural MRI, this imaging modality has lower contrast. MRI is produced by a nuclear magnetic resonance (NMR) signal from hydrogen nuclei. An MRI session begins when the subject is placed horizontally inside the bore of a large magnetic field. A typical clinical MRI system requires a field strength of at least 1.5 Tesla. The MRI system requires a radio frequency (RF) coil to generate an oscillating magnetic field [1]. Figure 16.1 represents a block diagram of an MRI scanner. The hydrogen nuclei of the water molecules in the subject's body are the source of the NMR signal. The hydrogen nuclei (protons) have an intrinsic property called nuclear spin. The spin of these particles can be considered as a magnetic moment vector which makes the particle behave like a magnet. Inside the magnetic field, a large fraction of magnetic moments aligns parallel to

Adaptive Processing of Brain Signals, First Edition. Saeid Sanei.
© 2013 John Wiley & Sons, Ltd. Published 2013 by John Wiley & Sons, Ltd.

Magnetic shield

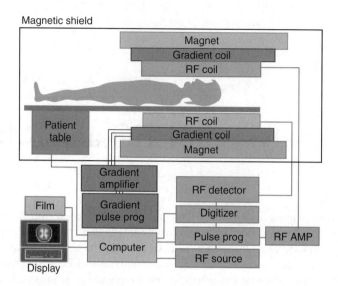

Figure 16.1 Block diagram of an MRI scanner illustrating the components involved in fMRI data recording

the main magnetic field and forms a net magnetization force which oscillates at the Larmor frequency [2]:

$$f = \gamma B \tag{16.1}$$

where B is the strength of the applied magnetic field and γ is a constant related to the type of nuclear particles. Applying an oscillating magnetic field at a frequency equal to the Larmor frequency causes the spins to absorb energy and be excited. The magnetic field is generated by an RF pulse of magnetic energy. Resonance between the magnetic field excitation and equilibrium states induces a current into a coil placed near the subject. The MRI is then performed by a controlled manipulation of the magnetic field [3].

According to (16.1) a spatially varying magnetic field creates different frequencies in different locations. As a result, the detected signal changes based on varying the magnetic field which leads to a signal containing spectral information. An inverse Fourier transform converts the spectrum into a real signal in the spatial domain as fMRI images. The image produced in each scan is called a volume. Each volume is composed of a number of slices through the body, and each slice has a certain thickness and contains a number of 3D unit elements called voxels.

fMRI extends the use of MRI to find information about biological functions as well as anatomical information. Numerous studies during past decades have shown that neural activity causes changes in blood flow and blood oxygenation in the brain. In the early 1990s, Ogawa *et al.* [4] used the MRI technique to measure the haemodynamic response (changes in blood flow) in the brain. When a subject performs a particular task during the MRI scanning, the metabolism of neural cells in a brain region responsible for a task is increased and leads to an increase in the need for oxygen. Existing haemoglobins in the red blood cells deliver oxygen through capillaries to the involved neurons. Oxygenated haemoglobins (oxy-haemoglobin) are diamagnetic. In contrast to oxyhaemoglobins, deoxygenated haemoglobins (deoxyhaemoglobins) are paramagnetic [5]. So, the level of oxygenation affects the NMR signal and, consequently, changes the intensity of the recorded MRI images.

16.1.1 Blood Oxygenation Level Dependent

Blood oxygenation level dependent (BOLD) refers to a phenomenon in fMRI which represents the changes in the NMR signal due to the variation of blood deoxyhaemoglobin concentration. There is a biomedical model for dynamic changes in deoxyhaemoglobin content during brain activation. This model is known as the HRF [6]. HRF presents the temporal properties of brain activation. If the human brain and the fMRI scanner are considered to be linear time invariant systems the variation of the measured intensities of the voxels can be described by a time series $d(n)$ which can be represented as convolution of the stimulus function $h(n)$ and the impulse response $s(n)$. Such a system includes noise $v(n)$ and drift $f(n)$ too:

$$d(n) = h(n) * s(n) + v(n) + f(n) \tag{16.2}$$

where $*$ denotes the convolution operation. The stimulus function $h(n)$ is often called the *time course*. This time course should be known to the model-based system for the construction of BOLD. The HRF function is usually modelled by the difference between two gamma functions [7]:

$$s(n) = \left(\frac{n}{k_1}\right)^{\alpha_1} e^{-(n-k_1)/\beta_1} - c\left(\frac{n}{k_2}\right)^{\alpha_2} e^{-(n-k_2)/\beta_2} \tag{16.3}$$

Where k_j is the time shift of the peak, the constant c determines the contribution of each gamma (often set to $0.30 < c < 0.5$) and α_1, α_2, β_1, and β_2 are the parameters of gamma functions (with typical values of 6, 12, 12, 12 respectively). The haemodynamic response varies for different parts of the brain and different subjects. Figure 16.2 demonstrates the variation of HRF with k and c.

Two common experimental designs are used in fMRI experiments: *block design* and *event-related design* [8]. In block design, the test subjects employ different cognitive processes, alternating between them periodically. This is the most time-efficient approach for comparing brain responses with different tasks. The simplest form of block design experiment includes two states: "rest" and "active". These states are alternated throughout the experiment in order to obtain an optimum experimental design for BOLD detection with sufficient SNR. The block design is an ideal choice for many types of experiments because the shape of the brain response is simple, so it can be very useful in an early exploratory stage. Although in block design measuring the temporal response of the brain is hard to achieve, limiting the flexibility of block design, the robustness of its results still makes it a valuable method.

Unlike in block design, in event-related design discrete stimuli occur randomly during the scanning. In event-related design, the input is modelled by discrete impulses in the stimulus

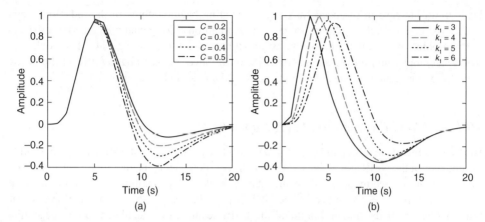

Figure 16.2 Gamma functions with varying parameters; (a) c changes from 0.2 to 0.5 and (b) when the latency (time shift) changes from 3 to 6

time instants. Event-related fMRI is a suitable tool to capture temporal properties of the brain response. Moreover, it provides information on HRF occurrence instants [9].

In event-related fMRI experiments stimuli are separated by inter-stimulus intervals (ISI).

Large ISI causes the HRFs not to overlap due to sequential stimuli and reveals the transient variation in brain response. Several types of noise usually appear in fMRI experiments. First, noise related to data acquisition that can be considered due to object variability. Object variability results from quantum thermodynamics and thermal noise. Thermal noise causes white noise with a constant variance in the image domain [10]. Head movement is another source of noise in fMRI data. This noise is considered as physiological noise. Therefore, fMRI needs to be preprocessed before being further analysed.

16.1.2   *Popular fMRI Data Formats*

The most popular file formats for fMRI are:

> **DICOM** is the standard form of data obtained from the scanner. It is two-dimensional and can be converted to ANALYZE or NIFTI for further analysis. A statistical parametric mapping (SPM) [11] toolbox enables conversion of DICOM to ANALYZE and NIFTI formats.

> **ANALYZE** is another very commonly used software for fMRI analysis using MATLAB. SPM and fMRIB software library (FSL) use ANALYZE 7.5 format. An Analyze 7.5 data format consists of two files, an image file with extension ".img" and a header file with extensions ".hdr". The ".img" file contains the recorded image and the ".hdr" file contains the volume information of the ".img" file, such as voxel size, and the number of pixels in the x, y and z directions (dimensions).

> **NIFTI** (neuroimaging informatics technology initiative) is the most recent standard for fMRI data format. The file extension of NIFTI is ".nii" with arbitrary information. This format can be converted to ANALYZE by the MRIcro toolbox.

16.1.3 Preprocessing of fMRI Data

The first stage in analysing fMRI data is preprocessing with the aim of (i) removing non-task related variability from the data, (ii) increasing SNR and (iii) preparing data for further analysis in order to detect brain active regions.

Motion correction (realignment), coregistration, segmentation, normalization and smoothing are important (spatial) preprocessing approaches. On the other hand, there are temporal preprocessing methods including slice timing correction that is often required.

- *Realignment*: This is to remove the head movement artefact from fMRI data. Any misalignment due to motion can affect brain source localization.
- *Functional–structural coregistration*: structural MRI is a high-resolution image providing static anatomical information. In contrast, functional MRI has low resolution and provides dynamic physiological information [12]. Functional images have little structural information, so it needs a high-resolution anatomical image to determine the activated area. Thus, registration between functional and structural MRI is required to align images produced by these two processes [1].
- *Normalization*: recalling that the data is in the form of two-dimensional images making up a volume, intensity normalization refers to rescaling intensities in all volumes such that at the end they have the same average intensity. Spatial normalization is another important step in preprocessing. This resembles coregistration, however, it corrects dissimilarities in shape between the volumes.
- *Smoothing*: this is to improve SNR in fMRI. In other words, the high-frequency components are removed from fMRI during the smoothing process. Usually a Gaussian kernel with a specified width is used to smooth the fMRI.
- *Slice timing correction*: the differences in image acquisition time between the slices need to be corrected. In fMRI, each slice of a volume is scanned at a time. Since the fMRI slices are acquired separately, the information from different slices comes from varying points in time after the task events and needs to be aligned by performing this step.

16.1.4 Relation between EEG and fMRI

Since the late 1990s, simultaneous EEG–fMRI or more precisely, EEG–correlated fMRI, has emerged as a noninvasive brain imaging technique based on EEG recording [13]. Although EEG provides a direct representation of synaptic activities with high temporal resolution, identification of underlying neuronal sources using EEG is incomplete due to low spatial resolution. In contrast, fMRI has high spatial resolution. The low temporal resolution of fMRI ignores rapid variation of brain responses. Hence, EEG and fMRI complement each other.

Haemodynamic brain response, captured by fMRI, reflects the indirect or secondary effect of neuronal activity. In contrast, EEG is a direct measure of synaptic activity. So, deriving the physiological relation between EEG and fMRI depends on understanding how the changes in neuronal activity captured by EEG affect the fMRI BOLD signal. This relationship, however, is not straightforward.

A basic assumption underlying the interpretation of fMRI maps is that an increase in regional "neuronal activity" (global synaptic product of a local neuronal population in response to inhibitory and excitatory inputs and interneural feed-forward and feedback activity) results in an increase in metabolic demand (of neurons and astrocytes), with increased energy and oxygen consumption. In response to these neurometabolic changes, there is a rise in local brain perfusion that exceeds the metabolic needs. Therefore, although the total oxygen consumption increases, the fraction of oxygen extraction and the percentage of deoxygenated haemoglobin in local venous blood decrease (positive BOLD change as recorded with fMRI). In a neuronal network comprising forward and backward excitatory and inhibitory connections, the BOLD signal changes reflect variation of the network "state" compared to a well and carefully defined baseline. Changes in the local field potential provide the best estimate of the haemodynamic response [13]. The underlying mechanism is depicted in Figure 16.3.

Amongst the differences, there is a time lag difference between synaptic responses and haemodynamic responses [15]. The relation is generally nonlinear too and the SNR in EEG is significantly higher than the SNR in fMRI [16]. Despite this, the relation between the two depends on the type of neuronal activity. Although the relation between the magnitude and spatial scale of the BOLD signal and neuronal physiology is still under debate, some

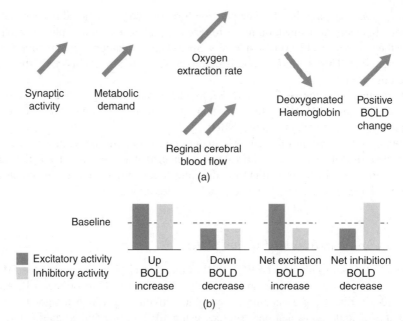

Figure 16.3   The mechanism of the inherent link between EEG and fMRI; (a) an increase in regional "neuronal activity" (global synaptic product of a local neuronal population in response to inhibitory and excitatory inputs and interneural feed-forward and feedback activity) results in an increase in metabolic demand (of neurons and astrocytes), with increased energy and oxygen consumption. In response to these neurometabolic changes, there is a rise in local brain perfusion that exceeds the metabolic needs. Therefore, although the total oxygen consumption increases, both, part of the oxygen extraction and the percentage of deoxygenated haemoglobin in local venous blood decrease (positive BOLD change as recorded with fMRI). (b) In a neuronal network comprising forward and backward excitatory and inhibitory connections, the changes in BOLD signal reflect changes in the network "state" when compared to a carefully defined baseline [13, 14]

researchers have reported a predominantly linear coupling between BOLD and neuronal activity. For example, a linear relationship between somatosensory-evoked potentials and BOLD has been demonstrated in [17]. Rees *et al.* [18] and Heeger *et al.* [19] also declared a linear correlation between the average spike rate in the cortical area of a monkeys's brain and the BOLD signal. This shows EEG and fMRI have an inherent relationship and hence can be analysed jointly. Simultaneous measurement of EEG and fMRI therefore paves the way for providing more information about the exact relation between the two modalities.

16.2 Model-Based Method for BOLD Detection

The general linear model (GLM) is frequently used to analyse fMRI data. The works in [20, 21] are good examples of BOLD detection from fMRI using GLM. Based on this model the stimulus time course is used to construct a HRF to be convolved with the fMRI sequence to highlight the BOLD. The GLM is the most common approach in model-based fMRI analyses. The GLM approach is used to find a good solution for a variety of research questions within an infinite number of different experimental designs. This model can be presented as:

$$X = YB + E \tag{16.4}$$

where X is a matrix containing a series of multivariate measurements, Y is called a design matrix, B is a matrix containing unknown parameters that are usually to be estimated and E represents errors or noise, assumed to have normal distribution.

In GLM, each voxel of the collected fMRI image volume is considered as a linear combination of the haemodynamic responses of stimuli and its corresponding weighted parameters, that is, each column of matrix $X \in R^{M \times P}$ corresponds to the fMRI time series for each voxel. Hence, GLM interprets the time course variation for each voxel. Figure 16.4 shows the time series for one sample voxel in an fMRI experiment. M refers to the number of scans in the experiment and P is the number of all the voxels in one scanned volume. In fMRI analysis using GLM, the design matrix is constructed using a predicted brain response to a given task. Each column of $Y \in R^{M \times N}$ shows one of the responses obtained by convolving the stimulus function and HRF. N refers to the number of available predictors obtained based on some prior knowledge. The stimuli function is determined based on the type of experimental design during the fMRI recording session which can be block design or event-related design. The HRF can be chosen amongst known functions such as the gamma function.

Figure 16.5 represents the schematics of the constructed model for a sample fMRI time series.

In this model there are M parameters $\beta_1, \beta_2, \ldots, \beta_M$ to be estimated. These parameters are considered as the coefficients of predefined regressors obtained based on some prior knowledge. After the model is identified the weight parameters, B, should be estimated. This can be done by using techniques such as maximum likelihood estimation (MLE) or Bayesian estimation. The parameter values that fit the data and minimize the squared error can be estimated by:

$$B = (Y^T Y)^{-1} Y^T X \tag{16.5}$$

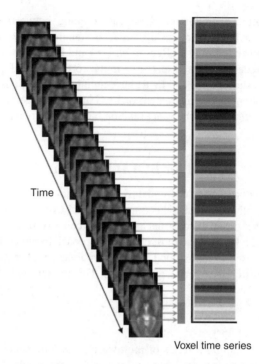

Voxel time series

Figure 16.4   fMRI time series for a voxel sample

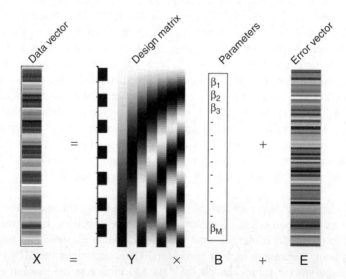

Figure 16.5   A schematic presentation of a GLM model for a sample fMRI time series

After parameter estimation, the activated regions are detected by evaluating the statistical significance of the whole brain voxels [22]. Common techniques for analysing.

fMRI data evaluate the statistical characteristics in each voxel of interest by t-statistic [22]. Then, the activated areas are detected by selecting the voxels with higher statistical significance than a certain threshold level [23]. The SPM [11] and FSL [24] toolboxes are commonly used for analysing fMRI data modelled using GLM. The main drawback of GLM is that the exact time course has to be known.

16.3 Simultaneous EEG-fMRI Recording: Artefact Removal from EEG

In simultaneous EEG and fMRI data recordings, the EEG signals are severely affected by magnetic field effects inducing strong artefacts. Without removing the fMRI-induced artefacts the underlying information within the EEG can hardly be realised. EEG signals recorded in the magnetic field suffer from two major artefacts namely gradient and ballistocardiogram (BCG) artefacts. Prior to any joint processing of these data, the artefacts should be effectively removed. In the following sections the main characteristics of these artefacts are summarised and the common and recent approaches for removing them from the EEG signals are provided.

16.3.1   Gradient Artefact Removal

Gradient artefact or imaging artefact results from the changes in magnetic field of an MRI scanner during image acquisition. This artefact is characterised by its high frequency and amplitude up to 100 times larger than the EEG average amplitude and, therefore, makes any visual data inspection almost impossible [25]. Figure 16.6 presents a segment of the EEG corrupted by gradient artefact. Eleven fMRI slices are recorded during this segment. The enlarged frame on the top right-hand side of the figure presents the high-frequency contents of the gradient artefact.

The easiest way to avoid gradient artefact is to use interleaved EEG–fMRI protocols such as periodic interleaved scanning [26, 27] or EEG-triggered fMRI [28]. These techniques are not flexible enough for general use and are often less efficient than continuous recording.

The most common technique for gradient removal is average artefact subtraction (AAS). Since this artefact does not show significant variability over time it can be subtracted from the EEG signal using an average template approach. Following this method, first, an artefact template is created by averaging successive artefact cycles in each channel. Then the created template is subtracted from the data.

In an effective work by Allen *et al.* [29] AAS and an adaptive noise cancelation technique have been combined. In order to generate the artefact template, the EEG signal should be segmented based on certain timing cues or triggers. These triggers are either volume timing triggers or slice timing triggers. These triggers are recorded by the scanner and denote time points when the MRI scanner starts to scan each volume/slice. It is obvious that any misalignment between the actual data and the artefact template removes some information within the EEG signals. In order to avoid this problem, slice timing triggers are used. Since these triggers are not recorded by the system due to shortage of memory a correlation-based algorithm is

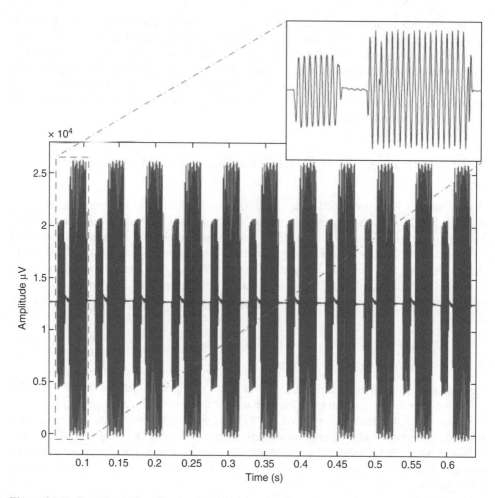

Figure 16.6   Gradient artefact for a sample segment of EEG signal recorded simultaneously with fMRI

developed to detect the triggers. Figure 16.7 presents a 5 s segment of data from 13 channels obtained after applying Allen's method for removing the gradient artefact. BCG is evident in these signals.

16.3.2   Ballistocardiogram Artefact Removal

Unlike gradient artefact, BCG has a more complex morphology and characterization. This artefact can be seen after the EEG is restored from the gradient artefact. BCG has severe destructive influence on the EEG signals. BCG is caused by movements of EEG electrodes in the magnetic field during the fMRI scans. There is a small movement in each electrode during the cardiac pulsation and, as a result, a variable signal is induced into each electrode. The BCG artefact obscures EEG at alpha frequencies (8–13 Hz) and below, with amplitudes around

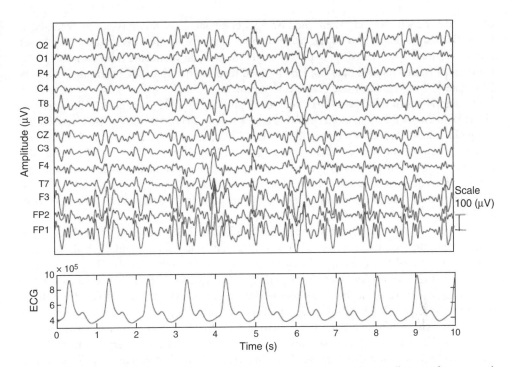

Figure 16.7 A set of 13-channel EEG data covering the entire head, after gradient artefact removal, contaminated with BCG artefact. Electrocardiogram (ECG) trend has also been illustrated at the bottom

150 μV inside a magnetic field with strength of 1.5 T [30]. BCG is quasi-periodic with varying morphology. This is because of irregularities in the heart beat, body movement, and the fact that during cardiac pulsation the electrodes are not affected by the magnetic field uniformly.

There have been intensive researches for the removal of this artefact. For example, Allen *et al.* [25] suggested an efficient technique to reduce BCG by firmly bandaging the electrodes and wires to the subject.

The AAS, more often used for gradient artefact removal, was one of the first algorithms used for cancelation of the BCG artefact [25]. Using this method, BCG is subsequently reduced by template subtraction from each trial. One of the major drawbacks of this widely used method is incompatibility with the artefact as it changes over time. Therefore, the assumption of having a similar artefact in all trials is not always valid. Moreover, in all the methods based on averaging, the reference ECG channel is needed.

In order to modify the method to deal with heart beat variation, a dynamic template with adaptive amplitude estimated by sliding and averaging has been proposed [27]. In another research direction, Bonmassar *et al.* [30] proposed an algorithm based on adaptive filtering to remove both ECG and motion artefact. Although this method is popular, it requires acquiring another reference signal using a motion sensor attached to the subject. Niazy *et al.* [31] developed an algorithm using principal component analysis (PCA) known as an optimal basis set (OBS). In their method, first a mixture of signals obtained from trials affected by artefact is formed for each channel. Then, the principal components of this mixture are calculated. In the third step, only a few of the first principal components are selected as the basis set. Finally, a template is created using the selected basis set and subtracted from each BCG trial. Unfortunately, this method does not lead to good results when there are other artefacts in the EEG data which are mainly due to subject movement.

Discrete Hermite transform (DHT) has been proposed for BCG removal [32]. The main objective in this method is to model the BCG artefact using DHT. The shape of BCG is modelled using Gaussian functions which are initial Hermite functions. These Gaussian functions are eigenvectors of a centred or shifted Fourier matrix. In [32], the DHT of the EEG signal is obtained by computing the inner product between the signal and the Gaussian functions. This provides a set of transformed values corresponding to a particular shape within the EEG signal. Then, the artefact template is built using some of the transformed values and subtracted from the EEG signal.

Another class of artefact removal methods is based on BSS mainly based on ICA. Blind approaches often are more useful when no reference signal, such as ECG in this case, is available for template matching. Moreover, they do not consider that BCG is predictable. Methods using ICA assume that the recorded EEG signals can be represented as a linear mixture of independent neural activities inside the brain, and artefacts caused by muscles and noise. On the other hand, ICA decomposes the EEG signals into a set of independent components (ICs). Removing the ICs containing the BCG artefact and back-projecting the remaining ICs to the electrode space results in clean EEG signals. In several studies ICA has been used to remove the BCG artefact [33]. In a research the performances of different ICA algorithms for removing BCG from EEG data have been evaluated [34]. They also used two different post-processing methods to improve the results of ICA. In a recent method, Ghaderi *et al.* [35] proposed a blind source extraction (BSE) technique with cyclostationarity constraint to extract the BCG sources. In another attempt, Leclercq *et al.* [36] proposed a constrained ICA (cICA) to remove the BCG artefact. In the method proposed in [36], first, a template for BCG artefact is estimated for each channel. Then, the artefact-related components based on these constraints are extracted using the cICA algorithm. In the next step, the estimated artefact components are clustered and averaged over each cluster to have better estimation of the components related to BCG artefact. Then, the Gram–Schmidt algorithm is used for orthogonalisation and computing the separation matrix. In the last step, clean EEG signals are recovered by deflating (recursively subtracting) the sources selected as artefact and back-projecting the remaining sources to the electrode space.

An important issue in BCG removal using ICA, is selection of the correct number of extracted components that should be deflated. Different numbers of ICs, for example, 3 [34], 3–6 [33], 1 [37], considered as the BCG artefacts, have been reported by different researchers. Selecting and removing only a small number of sources as BCG may leave out some artefacts in the signals whereas selecting and removing a large number of sources may eliminate useful information from the EEG signals.

In a new approach by Ferdowsi *et al.* [38] ICA and DHT have been combined. This can be achieved by applying DHT to those ICA sources labelled as BCG artefacts. After detecting the peaks in BCG by applying a simple peak detection method, each BCG source is divided into time segments centred at the detected peaks. Hence, the segments have one peak in each segment. Then, an adaptive DHT is applied to each segment in order to model the artefact. Finally, the obtained model is subtracted from each segment which gives a clean component with no artefact. All the obtained components based on this procedure will be projected back to the electrode space giving a clean EEG signal.

The main advantage of this approach is its robustness against changes in the shape of the artefact over time. The proposed method alleviates the uncertainty in choosing the right number of sources to be deflated in ICA-based methods. Moreover, an adaptive parameter selection strategy is proposed to decrease the sensitivity of DHT-based methods to variations of the model parameters.

Methods based solely on DHT have some drawbacks. For example, a weakness of DHT in BCG removal is its disability to model the whole artefact with a minimum number of basis functions. Based on such weaknesses, large numbers of coefficients are needed to reconstruct the BCG template having the same amplitude and duration. Using a large number of basis functions also leads to eliminating some details in the EEG data. The hybrid ICA-DHT algorithm may be summarized in the following steps:

1. Apply ICA to the EEG data contaminated by BCG artefact.
2. Select six ICA components which are more correlated with the ECG channel.
3. Apply DHT on each selected BCG source to find a template for the artefact.
4. Subtract the template from each BCG source and back-project the residuals together with the remaining sources.

In an experiment natural EEG data were recorded from five healthy men. All the subjects were right-handed and their ages ranged from 18 to 50 years. The EEG was acquired using the Neuroscan Maglink RT system (with impedances kept within 10–20 kΩ), providing a 64-channel comprising 62 scalp electrodes, one ECG electrode and one EOG electrode.

The sampling rate of raw EEG data was set at 10 kHz. During fMRI acquisition 300 volumes including 38 slices ($3.2969 \times 3.2969 \times 3.3$ mm resolution, repetition time, TR = 2000 ms, and echo time, TE = 25 ms) were acquired.

After preprocessing, the Infomax ICA is applied to 10 s of EEG data. The number of sources is chosen to be equal to the number of sensors. After applying ICA, the extracted sources are clustered based on their correlations with the ECG channel. DHT is then used for modelling and removing these spikes from the selected sources. For this purpose, these sources are segmented such that the spikes fall in the centre of each segment. In order to have one artefact in each segment, with no overlap, the length of each segment is selected to be 256. This is because the period of BCG artefact is approximately 1 s, and the sampling rate of data is 250 Hz [38].

The residual contains brain rhythms retrieved by the proposed method. In this work, 15 DHT coefficients (5% of the total coefficients) are selected to model the artefact. The reason for selecting such a small fraction is to avoid losing useful EEG information while removing BCG. Figure 16.8 a shows the results of artefact removal from one "EEG channel" using ICA and ICA-DHT. The highlighted (circled by dotted lines) areas show some peaks of BCG artefact which have not been removed by ICA. The number of deflated sources in ICA is five, while six sources have been selected to deflate in ICA-DHT method. Figure 16.8 b shows the BCG source which is not deflated in ICA.

Figure 16.9 shows the results of applying different artefact removal methods for a segment of EEG signal labelled as the C_Z channel. The threshold and dilation parameter in DHT need to be set appropriately and this is often done using an adaptive approach. In this

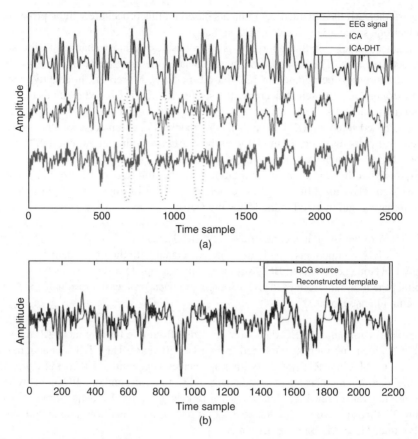

Figure 16.8 (a) Comparison between ICA and ICA-DHT for BCG removal from EEG and (b) The BCG source, not deflated in ICA, together with corresponding model obtained using ICA-DHT (see Plate 19 for the coloured version)

experiment the averaged dilation parameter and the threshold have been set to 4.25 and 0.001, respectively [38].

In another attempt at BCG removal the structure of BCG has been exploited to develop a source extraction method based on both short and long term prediction [39]. The predictability is therefore used as a criterion for source separation. Looking at Figure 16.10 a sample at time t may be predicted using its previous K_1 samples and K_2 samples of the previous cycle within one cycle interval denoted as τ.

The prediction error resulting from both short and long term prediction can be expressed as:

$$e_s(t) = y(t) - \sum_{p=1}^{K_1} b_p y(t - p)$$

$$e_l(t) = y(t) - \sum_{q=1}^{K_2} d_q y(t - \tau - q)$$

(16.6)

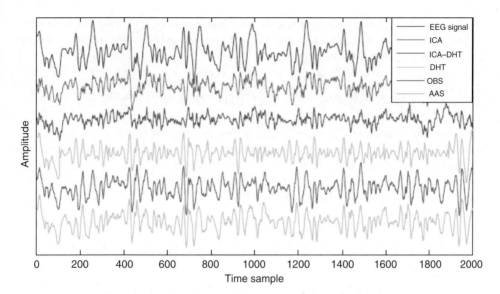

Figure 16.9 Results of artefact removal from C_Z channel using ICA, combined ICA-DHT, DHT, OBS, and AAS (see key on figure) (see Plate 20 for the coloured version)

Where b_p and d_q are, respectively, the short and long term prediction coefficients, K_1 and K_2 are, respectively, the short and long term prediction orders and $\mathbf{y}(t) = \mathbf{W}\mathbf{x}(t)$ are the separated sources at time instant t. Here, $\mathbf{x}(t) = \mathbf{A}\mathbf{s}(t) + \mathbf{v}(t)$ are the mixtures of EEG signals and artefacts measured at the electrodes and $\mathbf{v}(t)$ is additive noise. Equation (16.6) can be written in matrix/vector form as:

$$e_s(t) = \mathbf{w}^T\mathbf{x}(t) - \mathbf{b}\bar{\mathbf{y}}(t)$$
$$e_l(t) = \mathbf{w}^T\mathbf{x}(t) - \mathbf{d}\bar{\mathbf{y}}_l(t) \tag{16.7}$$

where $\mathbf{y}(t) = [y_1(t), \ldots, y_n(t)]^T$ are the $n \times 1$ estimated source signals at time t, $\mathbf{W}$ is the separating matrix, $\mathbf{w} = [w_1, w_2, \ldots, w_m]^T$, $\mathbf{b} = [b_1, b_2, \ldots, b_{K1}]^T$, $\bar{\mathbf{y}}(t) = [y(t-1),$

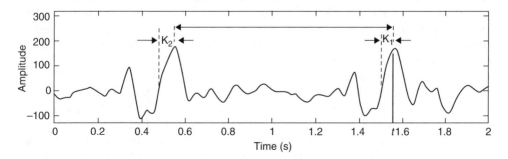

Figure 16.10 A cycle of BCG artefact in a sample segment of EEG signal

$y(t-2), \ldots, y(t-K_1)]^{\mathrm{T}}$, K_1 is the short-term prediction order, $\bar{y}_l(t) = [y(t-\tau-1), y(t-\tau-2), \ldots, y(t-\tau-K_1)]^{\mathrm{T}}$ and K_2 is the long-term prediction order, (for notational simplicity we omit index j from $y_j(t)$ and $\mathbf{w}_j$).

$$J(\mathbf{w}, \mathbf{b}, \mathbf{d}) = \frac{1}{2}E\left\{e_s^2\right\} + \frac{1}{2}E\left\{e_l^2\right\} = J_s(\mathbf{w}, \mathbf{b}) + J_l(\mathbf{w}, \mathbf{d}) \qquad (16.8)$$

The optimum values for the source extractor filter $\mathbf{w}$ (a column vector of matrix $\mathbf{W}$ related to the source of interest) and short and long term coefficient vectors $\mathbf{b}$ and $\mathbf{d}$ can be estimated by minimising the cost function $J(\mathbf{w},\mathbf{b},\mathbf{d})$ in (16.8) using alternating least squares (ALS) [40]. The update equations for estimating these parameters can then be derived through some exhaustive mathematical manipulation and concluded as [39]:

$$\mathbf{w}^{(\kappa+1)} = \mathbf{w}^{(\kappa)} - \eta_{\mathbf{w}}\left(\mathbf{w}^{(\kappa)}\right)^{\mathrm{T}}\left\{2\mathbf{R}_{\tilde{\mathbf{x}}}(0) - 2\sum_{p=1}^{K_1}b_p^{(\kappa)}\mathbf{R}_{\tilde{\mathbf{x}}}(p) - 2\sum_{p=1}^{K_1}d_p^{(\kappa)}\mathbf{R}_{\tilde{\mathbf{x}}}(q+\tau)\right.$$
$$\left. + \sum_{p=1}^{K_1}\sum_{q=1}^{K_1}b_p^{(\kappa)}d_q^{(\kappa)}\mathbf{R}_{\tilde{\mathbf{x}}}(p-q) + \sum_{p=1}^{K_2}\sum_{q=1}^{K_2}b_p^{(\kappa)}d_q^{(\kappa)}\mathbf{R}_{\tilde{\mathbf{x}}}(p-q)\right\} \qquad (16.9)$$

$\forall k = 1, \cdots K_1$

$$b_k^{(\kappa+1)} = b_k^{(\kappa)} - \eta_b\left\{-\left(\mathbf{w}^{(\kappa)}\right)^{\mathrm{T}}\mathbf{R}_{\tilde{\mathbf{x}}}(k)\mathbf{w}^{(\kappa)} + \sum_{p=1}^{K_1}b_p^{(\kappa)}\left(\mathbf{w}^{(\kappa)}\right)^{\mathrm{T}}\mathbf{R}_{\tilde{\mathbf{x}}}(p-k)\mathbf{w}^{(\kappa)}\right\} \qquad (16.10)$$

$\forall k = 1, \cdots K_1$

$$d_k^{(\kappa+1)} = d_k^{(\kappa)} - \eta_b\left\{-\left(\mathbf{w}^{(\kappa)}\right)^{\mathrm{T}}\mathbf{R}_{\tilde{\mathbf{x}}}(k)\mathbf{w}^{(\kappa)} + \sum_{p=1}^{K_2}d_p^{(\kappa)}\left(\mathbf{w}^{(\kappa)}\right)^{\mathrm{T}}\mathbf{R}_{\tilde{\mathbf{x}}}(p-k)\mathbf{w}^{(\kappa)}\right\} \qquad (16.11)$$

where $\eta_{\mathbf{w}}$, η_b and $\eta_{\mathbf{d}}$ are step sizes and $\mathbf{w}^{(u+1)}$, $b_k^{(u+1)}$ and $d_k^{(u+1)}$ are the updated parameter values and $\mathbf{R}_x$ is the autocorrelation function of $\tilde{\mathbf{x}}$ (whitened $\mathbf{x}$). The step sizes are kept fixed here. The updates continue for all the sources from 1 to n and the values of $\mathbf{w}$, $\mathbf{b}$ and $\mathbf{d}$ are updated to minimize (16.8). $\mathbf{w}$ is normalized after each iteration to preserve a unity norm for each column of the separation matrix. In this approach K_1 and K_2 have been selected carefully for optimum solutions [39]. The update process ends when the changes in the parameters remain below a predefined threshold.

To investigate the quality of results one way is to determine how the detected BCG is correlated with the corresponding ECG. Higher correlation values indicate a better separation. Another method is to find out how much of the artefact is still remaining in the original mixtures. This can be achieved by measuring the correlation between the detected artefacts and the mixtures. Better results correspond to smaller correlation values. Figure 16.11 represent the topographic maps illustrating the μ (mu) rhythm after the artefact is removed by employing

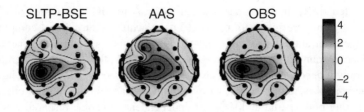

Figure 16.11   Topographic maps illustrating mu rhythm after the artefact is removed by employing AAS, OBS, and the prediction based BSE (SLTP-BSE) approach

AAS, OBS, and the prediction-based BSE (SLTP-BSE) approaches [39]. The measure of effectiveness of a method is how well it can localise the mu rhythm.

16.4   BOLD Detection in fMRI

Detection of BOLD is the prime objective in the analysis of fMRI data. Non-negative matrix factorization (NMF), as a BSS approach, is able to detect the BOLD signal without relying on any prior knowledge about the instants of stimulus onset. This is in contrast to GLM which is the most widely used technique for fMRI analysis. NMF decomposes a mixture of fMRI data into a set of time courses and their corresponding spatial sources. Extracted time courses represent the brain temporal response to stimuli or artefacts. Each time course is related to a source known as an active map and represents the active area in the brain. Ulfarsson *et al.* [41] state that a suitable fMRI analysis technique should provide sparse sources. That means having a small number of active (non-zero) voxels in each source. Brain networks of interest, such as the motor or visual cortex, typically have sparse spatial structure [41]. NMF is able to produce sparse results as a consequence of allowing only non-negative decomposition. These advantages of NMF make it suitable for analysis of fMRI which is inherently positive. To improve the results and ensure uniqueness of the solution, some spatial or temporal constraints are defined and incorporated into the factorisation formulation.

The most common model-free methods applied to fMRI data are ICA and NMF. These methods detect BOLD by decomposing a mixture containing measured fMRI data into a set of time courses and their corresponding spatial maps [10, 42]. Generally, separated sources and time courses are divided into two groups: signal of interest and signal not of interest. Signals of interest include task-related, function-related and transiently task-related ones [10]. BOLD is a task-related signal and is the main outcome of fMRI. If the brain response to a given task dies off before turning off the stimulation or change due to repeated stimuli, it will lead to a transiently task-related signal [10]. Function-related signals reveal the similarities between voxels inside a particular functional domain [43].

Signals not of interest include physiology-related, motion-related and scanner-related signals [10]. Breathing and heart beat are considered to be those producing the physiology-related signals. Brain ventricles and areas which contain large blood vessels are the origin of physiology-related signals.

The proposed method by McKeown *et al.* [44] was the first application of the spatial ICA for fMRI analysis. In spatial ICA (SICA), each row of the mixture matrix refers to the vectorized form of collected fMRI image in one scan. In contrast to SICA, in temporal ICA (TICA) each column of a mixture refers to an fMRI image acquired during one complete scan. In SICA, the algorithm attempts to find a set of spatially independent components and their associated (unconstrained) time courses. However, in TICA, the algorithm attempts to find a set of temporally independent time courses and their associated spatial maps. The decision to choose either spatial or temporal ICA or even both depends on the desired criteria [45]. In spite of exploiting both spatial and temporal independence in some research works [45, 46], most approaches still rely on the assumption of spatial independence because of a lack of good understanding of the unknown brain activities. Moreover, spatial ICA has lower computational complexity than TICA. This favours SICA for decomposing fMRI data. ICA employs different algorithms to separate independent sources and their associated time courses. Performance evaluation of these algorithms is an important issue which has been studied in some researches [47–49]. Source distribution imposes a suitable criterion to find a suitable ICA algorithm. The Infomax BSS algorithm often leads to reliable separation results for fMRI.

Although ICA has been found very useful and effective in fMRI analysis, it is not able to extract the required sparse sources effectively; mainly due to the number of zero-valued components which make higher order averages hard to handle. In [42] constrained NMF has been proposed for detection of the active area in the brain. They use sparsity and uncorrelatedness as constraints on decomposed factors and in another work they developed their method using the K-mean clustering algorithm to improve the initialization of NMF [50]. In these works, spatial NMF is used to decompose a set of fMRI images into a set of non-negative sources and a set of non-negative time-courses. We also employ spatial NMF throughout this chapter. Although constraints such as sparsity and uncorrelatedness make NMF more reliable, they do not incorporate any information about sources of interest.

16.4.1   Implementation of Different NMF Algorithms for BOLD Detection

The squared Euclidean distance and generalized KL divergence are the best known and the most frequently used cost functions for NMF. Csiszar's divergence [51], Bregman divergence [52], generalized divergence measure [53] and α or β divergences [54] are other alternatives. A suitable cost function can be determined based on the assumption about noise distribution. When the noise is normally distributed, the squared Euclidean distance is an optimal choice. In some applications, such as pattern recognition, image processing and statistical learning, the noise is not necessarily Gaussian and the cost functions based on information divergence are often used.

In the following discussion the performances of different NMF algorithms for BOLD detection are compared. These includes the α-divergence [55] algorithm. Given a non-negative matrix $\mathbf{X} \in \mathbb{R}^{M \times N}$, containing the input data, two non-negative matrices $\mathbf{A} \in \mathbb{R}^{M \times J}$ and $\mathbf{S} \in \mathbb{R}^{J \times N}$ are estimated such that $\mathbf{X} = \mathbf{AS} + \mathbf{V}$, where $\mathbf{V} \in \mathbb{R}^{M \times N}$ is the factorization error and J denotes rank of factorization and is assumed to be known or estimated by an information-theoretic criterion.

The Euclidean norm is the most common NMF cost function:

$$J_{\mathrm{F}} = \|\mathbf{X} - \mathbf{AS}\|_F^2 = \sum_i \sum_t (x_{it} - \{\mathbf{AS}\}_{it})^2 \qquad (16.12)$$

Multiplicative update rules for the above optimization problem have been derived by Lee and Seung [56]. They proved that the referred cost functions would converge to a local minimum under these update rules. As another algorithm, the basic α-divergence between $\mathbf{X}$ and $\mathbf{AS}$ is defined as:

$$J_\alpha = \frac{1}{\alpha(\alpha - 1)} \sum_i \sum_t \left(x_{it}^\alpha \{\mathbf{AS}\}_{it}^{1-\alpha} - \alpha x_{it} + (\alpha - 1)\{\mathbf{AS}\}_{it} \right) \tag{16.13}$$

where $0 \le \alpha \le 2$. Special cases for the α-divergence algorithm are also defined as follows [55]:
KL I-divergence ($\alpha \to 1$):

$$J_{KL} = \sum_i \sum_t \left(x_{it} \ln \frac{x_{it}}{\{\mathbf{AS}\}_{it}} - x_{it} + \{\mathbf{AS}\}_{it} \right) \tag{16.14}$$

Dual KL I-divergence ($\alpha \to 0$):

$$J_{dKL} = \sum_i \sum_t \left(\{\mathbf{AS}\}_{it} \ln \frac{\{\mathbf{AS}\}_{it}}{x_{it}} + x_{it} - \{\mathbf{AS}\}_{it} \right) \tag{16.15}$$

Squared Hellinger divergence ($\alpha = 0.5$):

$$J_{SH} = \sum_i \sum_t \left(\{\mathbf{AS}\}_{it}^{1/2} - x_{it}^{1/2} \right)^2 \tag{16.16}$$

Pearson divergence ($\alpha = 2$):

$$J_p = \sum_i \sum_t \frac{(x_{it} - \{\mathbf{AS}\}_{it})^2}{\{\mathbf{AS}\}_{it}} \tag{16.17}$$

The choice of optimal α depends on the application and the data being analysed. The following equations present the main learning rules for α-divergence algorithms. These update rules are suitable for large scale NMF [54] and can be expressed as:

$$a_{ij} \leftarrow \left(a_{ij} \left(\sum s_{jt} \left(\frac{x_{it}}{\{AS\}_{it}} \right)^\alpha \right)^{\omega/\alpha} \right)^{1-\alpha_{sa}} \tag{16.18}$$

$$s_{ij} \leftarrow \left(s_{ij} \left(\sum a_{jt} \left(\frac{x_{it}}{\{AS\}_{it}} \right)^\alpha \right)^{\omega/\alpha} \right)^{1-\alpha_{ss}} \tag{16.19}$$

where ω is the over relaxation parameter and is typically selected to be between 0.5 and 2. The over-relaxation parameter accelerates the convergence and stabilizes the algorithm. α_{sa}

and α_{ss} are small positive parameters which are used to enforce the sparsity constraint on the algorithm.

The α-divergence-based NMF algorithm has been used to detect brain activation in a set of synthetic and real fMRI data. In order to find the optimal value of α, the source separation procedure was repeated with different α values. Moreover, the performances of groups of algorithms were compared with those of more common NMF algorithms such as Euclidean distance-based methods.

16.4.2   BOLD Detection Experiments

The NMF algorithms have been examined by employing a set of test data and using the software in the machine learning for signal processing (MLSP) laboratory [57] and also a set of EEG–fMRI recorded data. The simulated fMRI data included eight sources and their corresponding time courses, as depicted in Figure 16.12. $\mathbf{s}_1$ shows the simulated task-related source (BOLD), $\mathbf{s}_2$ and $\mathbf{s}_6$ are transient task-related, and the rest represent the artefact-related sources. Using the ICA and NMF algorithms the task-related source or BOLD is separated from the fMRI sequence.

In order to generate the mixture, a matrix of time courses is multiplied by the matrix of sources. SIR, defined as $\text{SIR}_i = \|\hat{\mathbf{s}}_i\|_2 / \|\hat{\mathbf{s}}_i - \mathbf{s}_i\|_2$, for each source i, has been used for evaluation of the algorithm performance. $\| \cdot \|_2$ indicates the Euclidean norm. In this equation $\hat{\mathbf{s}}_i$ and $\mathbf{s}_i$ are respectively the extracted and actual task-related (BOLD) sources.

The performances of Euclidean distance, KL I-divergence, dual KL I-divergence, square Hellinger divergence and Pearson divergence have been evaluated for this data set a number of times and averaged. Our experiments show that the best convergence requires the over-relaxation parameter to be $\omega = 1.9$ and sparsity regularization parameters $\alpha_{sa} = \alpha_{ss} = 0.001$.

Figure 16.13 shows the computed average SIR for the results of Euclidean distance and different α-divergence-based NMF algorithms. It is seen from the figure that the best SIR is related to a Euclidean distance of 29.33 dB. This value is much higher than the results of α-divergence-based methods. Amongst α-divergence-based NMF algorithms, KL I-divergence and square Hellinger divergence show higher SIR than dual KL I-divergence and Pearson divergence.

It should be noted that the sparsity constraint is used in the Euclidean distance-based algorithm to obtain more accurate results. The sparsity regularization parameter has been set to 0.1 for this experiment.

Real data sets taken from the SPM website [11] are used next. The first data set is auditory fMRI data from a single subject experiment. This dataset has been recorded using a 2T Siemens MAGNETOM Vision scanner with the scan to scan repeat time (TR) of 7 s. The auditory stimuli are bi-syllabic words presented binaurally at a rate of 60 per min. The data set contains 96 scans and each scan consists of 64 contiguous slices ($3 \times 3 \times 3$mm voxel size). The 96 scans include 8 blocks of size 12, each of which contains 6 scans under rest and 6 scans under auditory stimulations.

The second data set is a visual fMRI data and was collected using a 2T Siemens MAGNE-TOM vision system. There are 360 scans during four runs, each consisting of four conditions which are *fixation, attention, no attention* and *stationary*. In the *attention* condition the subject should detect the changes in scene and during the *no attention* condition the subject was

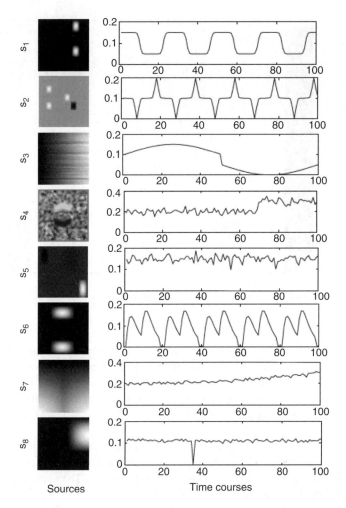

Figure 16.12 Simulated fMRI including the sources and corresponding time courses (using MLSP Laboratory tools at http://mlsp.umbc.edu/resources.html)

instructed to just keep their eyes open. During *attention* and *no attention* the subjects fixated centrally while white dots emerged from the fixation point to the edge of the screen.

The given data sets need preprocessing for subsequent processing for data separation. All the preprocessing steps including realignment, slice timing, coregistration, normalization and smoothing have been performed using SPM software [11].

In the experiments on real fMRI data, we applied KL I-Divergence and Square Hellinger to both data sets. Visual comparison of the results for different methods is difficult. Figure 16.14 shows the extracted BOLD and its corresponding time course of the first data set. As is seen, the activated region is correctly detected for different brain slices.

Figure 16.15 presents the results of applying KL I-divergence to the visual fMRI data set.

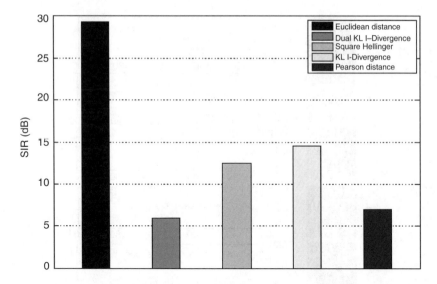

Figure 16.13   Computed SIR of source of interest for different methods

BOLD has been detected in the occipital lobe which is responsible for visual processing tasks. The extracted time course also verifies the temporal behaviour of activation.

In order to compare the results, the normalized correlation between the extracted time-course and the predicted temporal response of the brain has been calculated. The brain temporal response to a specific task can be modelled by convolving the task-waveform and the HRF [7].

The results show that the normalized correlation between the extracted time-course and the predicted temporal response of the brain for the results of Euclidean distance has a higher value than those for the two other α-divergence based methods. The numerical results of this comparison for both data sets are given in Table 16.1.

Any temporal or spatial pre-knowledge about the BOLD can enhance the effectiveness of its detection. As an example, auditory BOLD is expected to be seen in lateral lobes, visual BOLD in fronto-temporal lobes, and focal seizure BOLD in fronto-parietal lobes. Such information can be incorporated into the separation algorithm to enhance its accuracy.

Table 16.1   Normalized correlation between the extracted BOLD time course and predicted brain temporal response

Dataset	Auditory	Visual
Algorithm		
KL I – Divergence	0.6950	0.8736
Square Hellinger	0.6371	0.8152
Euclidean Distance	0.8689	0.9102

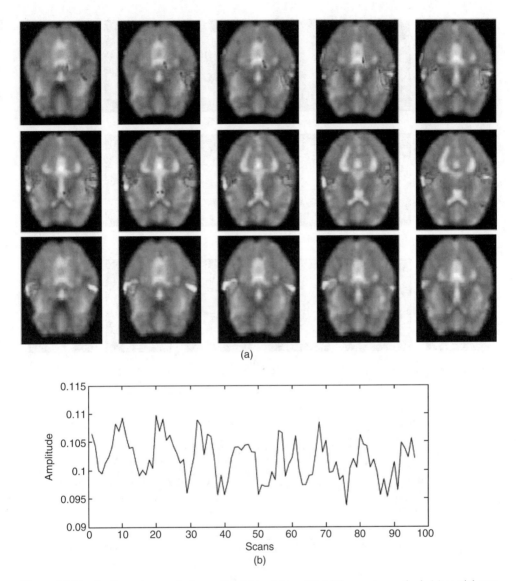

Figure 16.14 Auditory data analysis results obtained from KL I-Divergence method; (a) spatial map (see Plate 21 for the coloured version) and (b) corresponding time-course

16.5 Fusion of EEG and fMRI

16.5.1 *Extraction of fMRI Time-Course from EEG*

To construct, enhance, or detect BOLD, GLM is the most robust technique if the time course is available. In cases where the recordings are not time-locked, there is no cue, there is voluntary movement, or the BOLD is due to a neuro-physiological disorder, such as seizure, the time

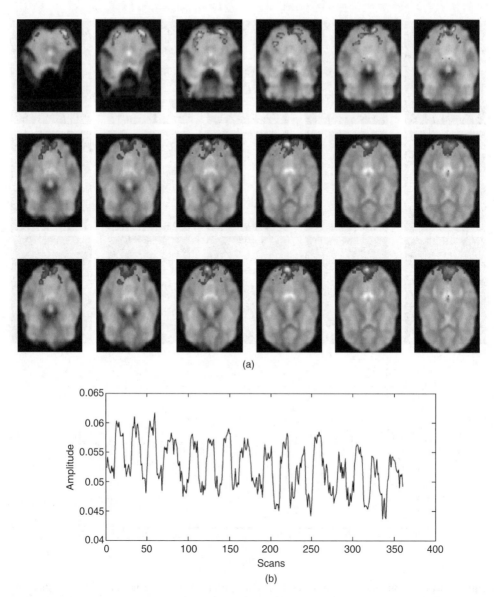

Figure 16.15 Visual data analysis results obtained from KL I-Divergence method; (a) spatial map (see Plate 22 for the coloured version) and (b) corresponding time-course

course is not known. Although blind techniques, such as ICA or NMF, are able to estimate the spatial and temporal information, they often introduce errors leading to misinterpretation of the results and, therefore, estimation of an effective time-course allows application of model-based approaches, such as GLM, by accurate construction of HRF.

In the cases of movement-related brain activity, the changes in synchronous activity of the neuronal population cause event-related variations in the amplitude of EEG oscillations

[58]. In earlier chapters it was mentioned that there are two strong event-related changes in brain oscillations: event-related desynchronization (ERD) in alpha rhythm and event-related synchronization (ERS) in beta rhythm. The act of preparation for a movement in a subject suppresses EEG oscillations in both alpha and beta rhythms over the sensory motor area [59–61]. This is the ERD signal and starts approximately 1 s before the movement. Following movement termination, while the alpha rhythm returns to the baseline, beta power also returns to the baseline and exceeds the pre-movement level [62, 63]. This sharp increase in the signal power of the beta band is the ERS signal or post-movement beta rebound (PMBR). Using the information obtained from EEG rhythms as a regressor for fMRI analysis enables localization of BOLD which is correlated with a specific neuronal rhythm. Concurrent recording of EEG and fMRI is useful for this purpose.

16.5.2   Fusion of EEG and fMRI, Blind Approach

EEG–fMRI integration reveals more complete information about brain functionality which cannot be observed using EEG and fMRI separately. Generally, despite the information provided at the beginning of this chapter the relationship between synchronous neuronal activity and BOLD is not totally clear. Simultaneous recording of EEG and fMRI provides the opportunity to identify different areas of the brain involved during EEG events. Using the extracted time course referred to in Section 16.5.1, the correlation between post-movement beta rebound and BOLD can be investigated. The information extracted from EEG analysis is used as a constraint to incorporate information derived from the EEG signal into the fMRI analysis procedure.

Tensor factorization, explained in previous chapters of this book, may be used for this purpose. In a novel approach PARAFAC2 as a tensor factorization algorithm is selected for EEG–fMRI fusion. The main advantage of using PARAFAC2 is the capability to analyze data in multiple modes rather than just two. This feature leads to obtaining valuable information about BOLD and its time course. Moreover, the multiway data analysis enables processing of the data in a systematic way. This approach can effectively detect the area in the brain which is responsible for PMBR. This can be empirically proved when comparing the results with those obtained from GLM.

Current approaches for EEG–fMRI integration are divided into two main groups: model driven and data driven. The computational biophysical model forms the basis of model-driven approaches [64]. In these methods, the assumptions about the neural activities are used to model the relation between EEG and fMRI. Despite the fact that model-driven approaches provide deeper understanding about the neuronal mechanisms, they require an explicit description of mutual neuronal substrates to annotate the measured EEG and fMRI data. However, the demand for the application of model-driven methods has decreased due to lack of knowledge about the neuronal substrates [64].

Data-driven approaches are based on common interactions between EEG and fMRI. These approaches are classified into two main categories: EEG–fMRI fusion based on constraints and EEG–fMRI fusion based on prediction. In methods based on constraint, the fMRI active map obtained by fMRI analysis is used as a priori information for electromagnetic source localization. Since the number of EEG sensors is generally smaller than the number of sources within the brain, the EEG inverse problem is often underdetermined and ill-posed. Therefore, additional constraints or a priori information are needed in order to obtain a unique and stable solution for the location of sources in theEEG.

There are two approaches for EEG–fMRI integration based on constraint, namely fMRI-constraint dipole fitting and distributed source modelling. Fujimaki *et al.* [65] proposed a method based on dipole fitting. In their method the neural activity at each fMRI hotspot is modelled as an equivalent regional current dipole. Then, the dipole locations are fixed to fMRI hotspots or the fMRI hotspots are used as seed points for dipole fitting while a maximum distance constraint is applied. After the dipole locations are estimated, the dipole moments can be determined by fitting the electric current dipole (ECD) model to the EEG data. The temporal dynamics of the regional neural activity is determined after estimating the dipole time courses. Although the methods give unique solutions, they are only able to detect the active dipoles at fMRI activation areas. Some techniques have also been developed based on distributed source modelling [66, 67]. In these works, the geometrical information of BOLD obtained from the fMRI active map is used to derive the covariance prior for source reconstruction. Weighted minimum norm frameworks and Wiener filter are two techniques used for this purpose in [66, 67], respectively. In contrast to methods based on dipole fitting, the fMRI constrained distributed source imaging methods are capable of obtaining dipoles, not only at fMRI activation areas but also at areas where fMRI fails to show any activation. However, the performance of these methods is affected by some drawbacks, such as identification of fMRI weighting factors and problems which result from the temporal resolution difference between the EEG and fMRI. These problems are "fMRI extra sources", "fMRI invisible sources" and "fMRI discrepancy sources" [68]. In [69–71], the authors have suggested some empirical values for weighting factors to overcome the problem. In another research, Phillips *et al.* [72] used expectation maximization (EM) as a data-driven technique to select the fMRI weighting factors.

Integration through prediction refers to incorporation of EEG features as additional regressors for fMRI analysis. The main objective of these techniques is to explore the correlation between the fMRI time-series and event-related potentials or EEG rhythm oscillations. The main superiority of these algorithms is the ability to use neural responses (measured by EEG) directly instead of using regressors relying only on the timings of the stimuli or tasks. One of the best examples is in the investigation of the interictal and ictal epileptic activity to localize epileptic foci and characterize the relationship between epileptic activity and the haemo-dynamic response [73, 74]. In another research, Horovitz *et al.* [75] investigated the neural activations underlying the brain rhythm modulations in the rest or pathological brain. Formaggio *et al.* [76] studied the correlation between the changes of mu rhythm and BOLD signal peak. In both researches, the regressors are derived from the power of a specific frequency band.

Recently, BSS and its variants have been applied to combine EEG and fMRI. This class of methods works based on measuring the mutual dependence between the two modalities. In these approaches, first the original EEG and fMRI data are decomposed into several components. Then, they are cross-matched. Calhoun *et al.* [77] proposed a multivariate technique using ICA to analyze the features extracted from EEG and fMRI. In another research [78] an algorithm using parallel ICA has been developed for EEG–fMRI fusion. In that method the modalities have been integrated using a pair-wise matching across trial modulation. Joint ICA is another BSS variant proposed by Moosmann *et al.* [79] to link the components from the two modalities.

Tensor factorization as another variant of BSS may be used for the same purpose. PARAFAC2 has been employed to analyze EEG–fMRI data obtained from a simultaneous recording. Beckmann *et al.* [80] have used this technique to perform multisubject and multi-session fMRI analysis. The main advantage of multiway as compared to two-way data analysis techniques in fMRI applications is to extract meaningful features in more than two modes. In this application PARAFAC2 is first used to analyze fMRI with the aim of extracting the signal of interest in the temporal, spatial and slice modes. The main reason for using PARAFAC2 instead of PARAFAC is that we are dealing with a non-trilinear data mixture due to changing the size of the brain area scanned at each slice. Then, a semi-blind method based on PARAFAC2 is proposed to integrate EEG and fMRI. In the proposed method a constrained technique is developed to incorporate the time course obtained from the Rolandic beta rhythm into the separation procedure. Imposing such a constraint leads to being able to separate the active area inside the brain which is highly correlated with the time course derived from the EEG signals. The proposed technique can be categorized as a data-driven technique in the family of EEG–fMRI fusion techniques based on prediction.

As part of our work in this area a partially constrained multiway BSS has been developed as follows. Recalling the PARAFAC2 model, the cost function

$$J(\mathbf{F}_q, \mathbf{A}, \mathbf{D}_1, \dots, \mathbf{D}_q) = \sum_{q=1}^{Q} \left\| \mathbf{X}_q - \mathbf{F}_q \mathbf{D}_q \mathbf{A}^{\mathrm{T}} \right\|_{\mathrm{F}}^2 \qquad (16.20)$$

is subject to the constraint $\mathbf{F}_q^{\mathrm{T}} \mathbf{F}_q = \mathbf{F}_p^{\mathrm{T}} \mathbf{F}_p$ for all pairs $p, q = 1, \dots, Q$. Here, $\mathbf{X}_q = \mathbf{X}(:, :, q)^{\mathrm{T}}$ is the transposed qth frontal slice of the tensor $\underline{\mathbf{X}}$ for $q = 1, \dots, Q$. $\mathbf{A}$ is the component matrix in the first mode which is fixed for all slabs, $\mathbf{F}_q$ is the component matrix in the second mode corresponding to the qth frontal slice of $\mathbf{X}$ and $\mathbf{D}_q$ is a diagonal matrix holding the qth row of the component matrix $\mathbf{C}$. The above cost function is reformulated as follows to allow us to impose the constraint:

$$J(\mathbf{P}_1, \dots, \mathbf{P}_q, \mathbf{F}, \mathbf{A}, \mathbf{D}_1, \dots, \mathbf{D}_q) = \sum_{q=1}^{Q} \left\| \mathbf{X}_q - \mathbf{P}_q \mathbf{F} \mathbf{D}_q \mathbf{A}^{\mathrm{T}} \right\|_{\mathrm{F}}^2 \qquad (16.21)$$

subject to $\mathbf{P}_q^{\mathrm{T}} \mathbf{P}_q = \mathbf{I}_R$ and $\mathbf{D}q$ is diagonal for all q. Using the method proposed by Kiers *et al.* [81], the PARAFAC2 problem is changed to a PARAFAC problem when $\mathbf{X}q$ is replaced by $\mathbf{P}_q^{\mathrm{T}} \mathbf{X}_q$.

Assuming the data tensor $\mathbf{X} \in \mathfrak{R}^{I \times J \times Q}$ contains the recorded fMRI images for one subject. Each fMRI volume recorded in one scan is composed of a number of slices. In order to arrange the fMRI data in a multiway tensor, first the slices are converted to a vector. Then, they are inserted as rows of the tensor. Hence, $\mathbf{X}(i, :, :)$ holds the recorded volume in the ith scan, $\mathbf{X}(:, :, q)$ holds the qth slice of all recorded volumes during all scans and $\mathbf{X}(:, j, :)$ holds the recorded voxel in the jth spatial location. Therefore, matrices $\mathbf{A}$, $\mathbf{F}$ and $\mathbf{C}$ denote the loading factors in the temporal, spatial and slice domains.

The available information about the loading factor in the temporal domain, extracted from the EEG signals, can then be used as a temporal constraint. In order to incorporate the prior information obtained by analysis of the EEG signals, the following constrained optimization problem is proposed:

$$J_{\text{new}}(\mathbf{A}, \mathbf{F}, \mathbf{C}, \mathbf{M}_{uk}, \mathbf{R}) = \|\mathbf{Y} - [\mathbf{A}, \mathbf{F}, \mathbf{C}]\|_F^2 + \lambda \|\mathbf{M} - \mathbf{A}\mathbf{R}^T\|_F^2 \text{ subject to } \mathbf{R}^T\mathbf{R} = \mathbf{I} \quad (16.22)$$

where $\mathbf{Y} = \mathbf{P}_q^T \mathbf{X}_q$, $\mathbf{M} \in \Re^{l \times L}$ is a matrix containing the prior information about the temporal signature of $\mathbf{Y}$ including known, $\mathbf{M}_k$, and unknown, $\mathbf{M}_{uk}$, parts, and $\mathbf{R} \in \Re^{L \times L}$ is the permutation matrix. $\| . \|_F$ denotes the Frobenius norm and λ is the regularization parameter which stabilizes the trade-off between the main part of the cost function and the constraint.

The constraint matrix $\mathbf{M}$ is designed such that its columns hold the regressors derived as the result of simultaneously recorded EEG analysis. Consider that K out of L columns of $\mathbf{M}$ are known. These columns indicate the available regressors to be used in the fMRI analysis. So, the constraint matrix is defined as $\mathbf{M} = [\mathbf{M}_k \vdots \mathbf{M}_{uk}]$ such that $\mathbf{M}_k \in \Re^{l \times K}$ and $\mathbf{M}_{uk} \in \Re^{l \times (L-K)}$ are the known and unknown sub-matrices respectively. Hence, $\mathbf{M}$ may be written in the following form:

$$\mathbf{M} = [\mathbf{M}_k \vdots \mathbf{M}_{uk}] = \begin{bmatrix} m_{11}^k \cdots m_{1k}^k & m_{1K+1}^{uk} \cdots m_{1L}^{uk} \\ \vdots & \vdots \\ m_{I1}^k \cdots m_{Ik}^k & m_{IK+1}^{uk} \cdots m_{IL}^{uk} \end{bmatrix} \quad (16.23)$$

This means each column of $\mathbf{M}$ refers to the time course of a stimulus. This allows detection of BOLDs for a number of stimuli at the same time. In this particular experiment there is only one stimulus and, therefore, the first column of $\mathbf{M}$ is known (extracted from the EEG). This regressor is built up by convolving the extracted time-course, denoting the onset of the Rolandic beta rhythm, and HRF. Matrix $\mathbf{R}$ matches the constraint matrix $\mathbf{M}$ with the estimated factor for temporal mode $\mathbf{A}$. ALS is used to estimate the factors. Following ALS, the gradients of the cost function with respect to all the factors are calculated. For this purpose, the unfolded version of data array is used. The factors are derived as follows.

$$\mathbf{A} \leftarrow \left(\mathbf{Y}_{(1)}(\mathbf{C} \odot \mathbf{F}) + \lambda \mathbf{M}\mathbf{R}\right)((\mathbf{C}^T\mathbf{C}) \otimes (\mathbf{F}^T\mathbf{F}) + \lambda \mathbf{R}^T\mathbf{R})^\dagger \quad (16.24)$$

$$\mathbf{F} \leftarrow \mathbf{Y}_{(2)}(\mathbf{C} \odot \mathbf{A})((\mathbf{C}^T\mathbf{C}) \otimes (\mathbf{A}^T\mathbf{A}))^\dagger \quad (16.25)$$

$$\mathbf{C} \leftarrow \mathbf{Y}_{(3)}(\mathbf{F} \odot \mathbf{A})((\mathbf{F}^T\mathbf{F}) \otimes (\mathbf{A}^T\mathbf{A}))^\dagger \quad (16.26)$$

$$\mathbf{M}_{uk} \leftarrow ([\mathbf{A}\mathbf{R}^T] :, \ K+1 : L) \quad (16.27)$$

where $\odot$ and $\otimes$ are respectively Khatri–Rao and Hadamard products. In order to calculate the permutation matrix, $\mathbf{R}$, the same procedure as was to compute $\mathbf{P}_k$ is performed. Since $\mathbf{R}$ is orthonormal, minimizing (16.22) over $\mathbf{R}$ is reduced to

$$\mathbf{R} \leftarrow \arg \max_{\mathbf{R}} \{\text{tr}(\mathbf{R}\mathbf{A}^T\mathbf{M})\} \quad (16.28)$$

Let $\mathbf{A}^T\mathbf{M} = \mathbf{U}\mathbf{\Sigma}\mathbf{V}^T$, so that the unique minimum of (16.28) is obtained as $\mathbf{R} = \mathbf{V}\mathbf{U}^T$. The resulting matrix $\mathbf{R}$ is an orthonormal matrix. The performance improves if λ decreases with the number of iterations.

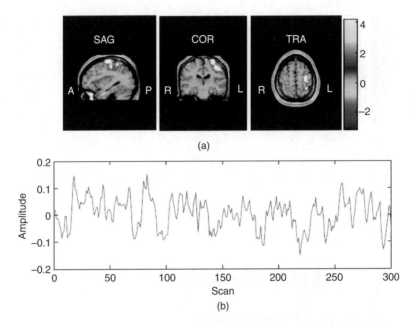

(a)

(b)

Figure 16.16   Detected BOLD (a) with its corresponding time-course (see Plate 23 for the coloured version) (b) using PARAFAC2. SAG, COR, and TRA refer, respectively, to sagittal, coronal, and transverse planes

   This method has been applied to the same real data in the previous section and the results shown in Figure 16.16 present the extracted spatial pattern of BOLD with its corresponding columns in matrices **A** and **C**. The results show activity in the left area of the primary motor cortex. The weights obtained in matrix **C** imply that the BOLD has maximum contribution in slice no. 62 for the subject. Using GLM on similar data shows comparable results.

16.5.3   Fusion of EEG and fMRI, Model-Based Approach

The correlation between the post-movement beta rebound and fMRI has been exploited in establishing a data-driven-based EEG–fMRI fusion which uses the extracted information from the EEG signal as a predictor for the BOLD signal. For this purpose, the power time-course of the Rolandic beta rhythm obtained by the proposed method in the previous section has been utilized. The calculated power time-course represents the instantaneous interaction between the EEG activity in the beta-band and the motor task. The power time-course is then convolved with HRF to make the regressor for fMRI analysis. This regressor is used to predict the BOLD response in fMRI data which are collected simultaneously with the EEG. A schematic illustration of data analysis is given in Figure 16.17.

16.6   Application to Seizure Detection

A growing number of engineering and clinical studies have transformed EEG–fMRI into a powerful tool to investigate not only the haemodynamic changes associated with spontaneous

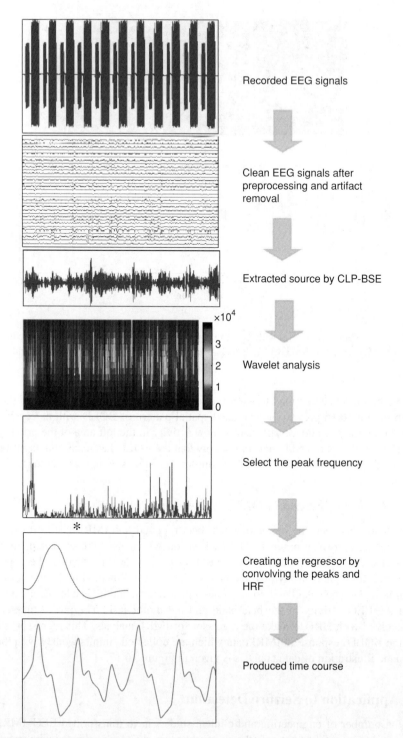

Figure 16.17 Schematic of different steps of model-based EEG–fMRI analysis (see Plate 24 for the coloured version)

brain activity in epileptic networks, including subcortical changes, but also to study endogenous brain rhythms of wakefulness and sleep, as well as evoked brain activity [13].

The interactions between neuronal populations and brain regions (macroscale, extrinsic connectivity) form the foundation for dynamic causal modelling (DCM). The EEG signal comes from the instantaneous electrical potentials generated by the pyramidal cells, spread through the head volume. At the cellular level, the fMRI signal is the end result of a metabolic and haemodynamic cascade occurring over a time scale of the order of seconds. Symmetric fusion models rely on finding inverse solutions of these models to estimate the neuronal activity and connectivity of the underlying neuronal population [13].

As a direct imaging tool of neuroelectric activity, the presurgical localization of epilepsy has been validated by several intracranial EEG studies and surgical series. Recent studies also suggest a good concordance between EEG–fMRI findings and similar gold standard localization tools. Studies comparing or combining localization and EEG–fMRI reveal that the combination of these techniques can provide new localizing and temporal information, with potential clinical relevance. The ongoing biophysical modelling and experimental developments attempt to solve the combined "electrovascular inverse problem" and to uncover the precise nature of microscopic and macroscopic neurovascular coupling. The future translation from group studies of cognitive evoked-related potentials to individual patients with spontaneously occurring epileptic discharges is far from simple but will likely help us to better understand the underlying mechanisms of the generation, propagation, and termination of epileptic discharges. These techniques may also have wider application in the investigation of the relationship between epileptic activity and other brain networks involved in the resting state, cognitive processes, and sleep [13].

A number of research works, including [82, 83], attempted to develop new signal processing tools and algorithms to combine EEG and fMRI in epilepsy studies. However, developing a robust technique in the detection of the intracranial spikes will be a big step for enhancing the robustness of these algorithms.

16.7 Conclusions

During the past two decades a tremendous amount of research has been carried out to combine and utilize the information achieved by simultaneous EEG and fMRI recordings. After the EEGs are restored from various artefacts, three major lines of research have been developed within biomedical signal processing and machine learning communities. These methods may be referred to as EEG-driven fMRI, fMRI-driven EEG, and joint EEG–fMRI signal processing. Although there are a number of distinguished findings in these areas, unfortunately, none of these techniques has been widely used for clinical diagnosis and patient treatment. Evoked potentials and their influence on both EEG and fMRI modalities have been the main applications of the algorithms. On the other hand, more precise restoration of EEG from BCG artefact needs more research.

References

[1] Jezzard, P., Matthews, P.M. and Smith, S.M. (1997) Functional MRI: Methods and applications. PhD Thesis.
[2] Wright, G.A. (1997) Magnetic resonance imaging. *Signal Process. Mag., IEEE*, **14**(1), 56–66.

[3] Lauterbur, P.C. (1973) Image formation by induced local interactions: Examples employing nuclear magnetic resonance. *Nature*, **242**(5394), 190–191.

[4] Ogawa, S., Lee, T.M., Key, A.R. and Tank, D.W. (1990) Brain magnetic resonance imaging with contrast dependent on blood oxygenation. *Proc. Natl. Acad. Sci. U. S. A.*, **87**(24), 9868–9872.

[5] Pauling, L. and Coryell, C.D. (1936) The magnetic properties and structure of hemoglobin, oxyhemoglobin and carbonmonoxyhemoglobin. *Proc. Natl. Acad. Sci. U. S. A.*, **22**(4), 210–216.

[6] Anders, M.D. and Randy, L.B. (1997) Selective averaging of rapidly presented individual trials using fMRI. *Hum. Brain Mapp.*, **5**(5), 329–340.

[7] Friston, K.J., Fletcher, P., Josephs, O. *et al.* (1998) Event-related fMRI: Characterizing differential responses. *NeuroImage*, **7**(1), 30–40.

[8] Jezzard, P., Matthews, P.M. and Smith, S.M. (2003) *Functional MRI: An Introduction to Methods*, Oxford University Press, USA.

[9] Huettel, S.A., Song, A.W. and McCarthy, G. (2004) *Functional Magnetic Resonance Imaging*, Sinauer Associates Inc., Sunderland, MA.

[10] Calhoun, V.D., Adali, T., Hansen, L.K. *et al.* (2003) ICA of functional MRI data: An overview. Proceedings of the International Workshop on Independent Component Analysis and Blind Signal Separation, pp. 281–288.

[11] Institute of Neuroimaging, University College London, Statistical parameter mapping (spm). http://www.fil.ion.ucl.ac.uk/spm/ accessed 12th January 2013.

[12] Symms, M., Jager, H., Schmierer, K. and Yousry, T. (2004) A review of structural magnetic resonance neuroimaging. *J. Neurol. Neurosurg. Psychiat.*, **75**(9), 1235–1244.

[13] Vulliemoz, S., Lemieux, L., Daunizeau, J. *et al.* (2009) The combination of EEG source imaging and EEG-correlated functional MRI to map epileptic networks. *Epilepsia*, 1–15. doi: 10.1111/j.1528-1167.2009.02342.x

[14] Logothetis, N.K. (2008) What we can do and what we cannot do with fMRI. *Nature*, **453**, 869–878.

[15] Shmuel, A. (2010) *Locally Measured Neuronal Correlates of Functional MRI Signals*, Springer Berlin Heidelberg.

[16] Logothetis, N.K., Pauls, J., Augath, M. *et al.* (2001) Neurophysiological investigation of the basis of the fMRI signal. *Nature*, **412**(6843), 150–157.

[17] Ogawa, S., Lee, T.M., Stepnoski, R. *et al.* (2000) An approach to probe some neural systems interaction by functional MRI at neural time scale down to milliseconds. *Proc. Natl. Acad. Sci. U. S. A.*, 11026–11031.

[18] Rees, G., Friston, K. and Koch, Ch. (2000) A direct quantitative relationship between the functional properties of human and macaque V5. *Nat. Neurosci.*, **3**(7), 716–723.

[19] Heeger, D.J., Hukl, A.C., Geisler, W.S. and Albrech, D.G. (2000) Spikes versus BOLD: what does neuroimaging tell us about neuronal activity? *Nat. Neurosci.*, **3**(7), 631–633.

[20] Price, C.J. and Friston, K.J. (1997) Cognitive conjunction: A new approach to brain activation experiments. *NeuroImage*, **5**(4), 261–270.

[21] Calhoun, V.D., Stevens, M.C., Pearlson, G.D. and Kiehl, K.A. (2004) fMRI analysis with the general linear model: removal of latency-induced amplitude bias by incorporation of hemodynamic derivative terms. *NeuroImage*, **22**(1), 252–257.

[22] Friston, K.J., Holmes, A.P., Worsley, K.J. *et al.* (1994) Statistical parametric maps in functional imaging: A general linear approach. *Hum. Brain Mapp.*, **2**(4), 189–210.

[23] Jing, M. (2008) Predictability of epileptic seizures by fusion of scalp EEG and fMRI. PhD thesis, Cardiff University, UK.

[24] FMRIB Software Library v5.0 September 2012, created by the Analysis Group, FMRIB, Oxford, UK. http://www.fmrib.ox.ac.uk/fsl/ accessed 12th January 2013

[25] Allen, P.J., Polizzi, G., Krakow, K. *et al.* (1998) Identification of EEG events in the MR scanner: The problem of pulse artefact and a method for its subtraction. *NeuroImage*, **8**(3), 229–239.

[26] Goldman, R.I., Stern, J.M., Engel, J. and Cohen, M.S. (2000) Acquiring simultaneous EEG and functional MRI. *Clin. Neurophysiol.*, **111**, 1974–1980.

[27] Kruggel, F., Wiggins, C.J., Herrmann, C.S. and von Cramon, D.Y. (2000) Recording of the event-related potentials during functional MRI at 3.0 Tesla field strength. *Magn. Resonan. Med.*, **44**(2), 277–282.

[28] Seeck, M., Lazeyras, F., Michel, C.M. *et al.* (1998) Non-invasive epileptic focus localization using EEG-triggered functional MRI and electromagnetic tomography. *Electroencephalogr. Clin. Neurophysiol.*, **106**(6), 508–512.

[29] Allen, P.J., Josephs, O. and Turner, R. (2000) A Method for removing imaging artefact from continuous EEG recorded during functional MRI. *NeuroImage*, **12**(2), 230–239.

[30] Bonmassar, G., Purdon, P.L., Jskelinen, I.P. *et al.* (2002) Belliveau. Motion and ballistocardiogram artefact removal for interleaved recording of EEG and EPs during MRI. *NeuroImage*, **16**(4), 1127–1141.

[31] Niazy, R.K., Beckmann, C.F., Iannetti, G.D. *et al.* (2005) Removal of fMRI environment artefacts from EEG data using optimal basis sets. *NeuroImage*, **28**(3), 720–737.

[32] Mahadevan, A., Acharya, A., Sheffer, S. and Mugler, D.H. (2008) Ballistocardiogram artefact removal in EEG-fMRI signals using discrete Hermite transforms. *Select. Top. Signal Process.*, **2**(6), 839–853.

[33] Mantini, D., Perrucci, M.G., Cugini, S. *et al.* (2007) Complete artefact removal for EEG recorded during continuous fMRI using independent component analysis. *NeuroImage*, **34**(2), 698–607.

[34] Nakamura, W., Anami, K., Mori, T. *et al.* (2006) Removal of ballistocardiogram artefacts from simultaneously recorded EEG and fMRI data using independent component analysis. *IEEE Trans. Biomed. Eng.*, **53**(7), 1294–1308.

[35] Ghaderi, F., Nazarpour, K., McWhirter, J.G. and Sanei, S. (2010) Removal of ballistocardiogram artefacts using the cyclostationary cource extraction method. *IEEE Trans. Biomed. Eng.*, **57**(11), 2667–2676.

[36] Leclercq, Y., Balteau, E., Dang-Vu, T. *et al.* (2009) Rejection of pulse related artefact (PRA) from continuous electroencephalographic (EEG) time series recorded during functional magnetic resonance imaging (fMRI) using constraint independent component analysis (cICA). *NeuroImage*, **44**(3), 679–691.

[37] Dyrholm, M., Goldman, R., Sajda, P. and Brown, T.R. (2009) Removal of BCG artefacts using a non-kirchhoffian overcomplete representation. *IEEE Trans. Biomed. Eng.*, **56**(2), 200–204.

[38] Ferdowsi, S., Sanei, S., Nottage, J. *et al.* (2012) A hybrid ICA-Hermite transform for removal of ballistocardiogram from EEG. Proceedings of the European Signal Processing Conference, EUSIPCO, Romania, 2012.

[39] Ferdowsi, S., Abolghasemi, V. and Sanei, S. (2012) Blind separation of balistocardiogram from EEG via short- and long-term linear prediction filtering. Proceedings of the Machine Learning and Signal Processing, MLSP 2012, Spain.

[40] Haykin, S., (2001) *Adaptive Filter Theory*, 4th edn, Prentice Hall.

[41] Ulfarsson, M.O. and Solo, V. (2007) Sparse variable principal component analysis with application to fMRI. 4th IEEE International Symposium on Biomedical Imaging: From Nano to Macro, pp. 460–463.

[42] Wang, X., Tian, J., Li, X. *et al.* (2004) Detecting brain activations by constrained non-negative matrix factorization from task-related BOLD fMRI. Proceedings of the SPIE 5369, Medical Imaging 2004: Physiology, Function, and Structure from Medical Images, 675, April, 2004.

[43] Biswal, B., Yetkin, F.Z., Haughton, V.M. and Hyde, J.S. (1995) Functional connectivity in the motor cortex of resting human brain using echo-planar MRI. *Magn. Reson. Med.*, **34**(4), 537–541.

[44] Mckeown, M.J., Makeig, S., Brown, G.G. *et al.* (1998) Analysis of fMRI data by blind separation into independent spatial components. *Hum. Brain Mapp.*, **6**, 160–188.

[45] Calhoun, V.D., Adali, T., Pearlson, G.D. and Pekar, J.J. (2001) Spatial and temporal independent component analysis of functional MRI data containing a pair of task-related waveforms. *Hum. Brain Mapp.*, **13**, 43–53.

[46] Stone, J.V., Porrill, J., Porter, N.R. and Wilkinson, I.D. (2002) Spatiotemporal independent component analysis of event-related fMRI data using skewed probability density functions. *NeuroImage*, **15**(2), 407–421.

[47] Cichocki, A. (2002) ICALAB, http://www.bsp.brain.riken.jp/icalab, released 2002, accessed 12th January 2013.

[48] Correa, N., Adali, T., Li, Y. and Calhoun, V.D. (2005) Comparison of blind source separation algorithms for fMRI using a new Matlab toolbox: GIFT. Proceedings of the IEEE International Conference on Acoustics, Speech, and Signal Processing, ICASSP, vol. 5, pp. v/401–v/404.

[49] Calhoun, V.D. (2004) Group ICA of fMRI toolbox (GIFT).

[50] Wang, X., Tian, J., Yang, L. and Hu, J. (2005) Clustered cNMF for fMRI data analysis. *Med. Images*, **5746**, 631–638.

[51] Cichocki, A., Zdunek, R. and Amari, S.I. (2006) *Csiszrs Divergences for Non-Negative Matrix Factorization: Family of New Algorithms*, LNCS, Springer, pp. 32–39.

[52] Dhillon, I.S. and Sra, S. (2005) Generalized nonnegative matrix approximations with Bregman divergencesin Neural Information Processing Systems Conference (NIPS) vol.17, pp. 283–290.

[53] Kompass, R. (2007) A generalized divergence measure for nonnegative matrix factorization. *Neural Computing*, **19**, 780–791.

[54] Cichocki, A., Amari, S.I., Zdunek, R. *et al.* (2006) Extended smart algorithms for non-negative matrix factorization. Artificial Intelligence and Soft Computing ICAISC 2006, vol. 4029 of Lecture Notes in Computer Science, pp. 548–562.

[55] Cichocki, A., Lee, H., Kim, Y.D. and Choi, S. (2008) Non-negative matrix factorization with α-divergence. *Pattern Recog. Lett.*, **29**(9), 1433–1440.

[56] Lee, D.D. and Seung, H.S. (2001) Algorithms for non-negative matrix factorization in Neural Information Processing Systems Conference (NIPS)vol. 13, pp. 556–562.

[57] MLSP-Lab, provided by the University of Maryland, Baltimore County, http://mlsp.umbc.edu, accessed 12th January 2013.

[58] Pfurtscheller, G. and Lopes da Silva, F.H. (1999) Event-related EEG/MEG synchronization and desynchronization: basic principles. *Clin. Neurophysiol.*, **110**(11), 1842–1857.

[59] Jasper, H. and Penfield, W. (1949) Electrocorticograms in man: Effect of voluntary movement upon the electrical activity of the precentral gyrus. *Eur. Arch. Psych. Clin. Neuro.*, **183**, 163–174.

[60] Jong, R., Gladwin, T.E. and Hart, B.M. (2006) Movement-related EEG indices of preparation in task switching and motor control. *Brain Res.*, **1105**(1), 73–82.

[61] Krusienski, D.J., Schalk, G., McFarland, D.J. and Wolpaw, J.R. (2007) A μ-rhythm matched filter for continuous control of a brain-computer interface. *IEEE Trans. Biomed. Eng.*, **54**(2), 273–280.

[62] Pfurtscheller, G., Stanck, A. Jr. and Edlinger, G. (1997) On the existence of different types of central beta rhythms below 30 Hz. *Electroencephalogr. Clin. Neurophysiol.*, **102**(4), 316–325.

[63] Stevenson, C.M., Brookes, M.J. and Morris, P.G. (2011) β-band correlates of the fMRI bold response. *Hum. Brain Mapp.*, **32**(2), 182–197.

[64] Valdes-Sosa, P.A., Sanchez-Bornot, J.M., Sotero, R.C. *et al.* (2009) Model driven EEG/fMRI fusion of brain oscillations. *Hum. Brain Mapp.*, **30**(9), 2701–2721.

[65] Fujimaki, N., Hayakawa, T., Nielsen, M. *et al.* (2002) An fMRIConstrained MEG Source Analysis with Procedures for Dividing and Grouping Activation. *NeuroImage*, **17**(1), 324–343.

[66] Ahlfors, S.P. and Simpson, G.V. (2004) Geometrical interpretation of fMRI-guided MEG/EEG inverse estimates. *NeuroImage*, **22**(1), 323–332.

[67] Liu, Z. and He, B. (2008) fMRI-EEG integrated cortical source imaging by use of time variant spatial constraints. *NeuroImage*, **39**(3), 1198–1214.

[68] Liu, Z., Kecman, F. and He, B. (2006) Effects of fMRIEEG mismatches in cortical current density estimation integrating fMRI and EEG: A simulation study. *Clin. Neurophysiol.*, **117**(7), 1610–1622.

[69] Babiloni, F., Babiloni, C., Carducci, F. *et al.* (2003) Multimodal integration of high-resolution EEG and functional magnetic resonance imaging data: a simulation study. *NeuroImage*, **19**(1), 1–15.

[70] Wagner, M., Fuchs, M. and Kastner, J. (2000) fMRI-constrained dipole fits and current density reconstructions. Proceedings of 12th International Conference on Biomagnetism, pp. 785–788.

[71] Liu, A.K., Belliveau, J.W. and Dale, A.M. (1998) Spatiotemporal imaging of human brain activity using functional MRI constrained magnetoencephalography data: Monte Carlo simulations. *Proc. Natl. Acad. Sci. U. S. A.*, **95**, 8945–8950.

[72] Phillips, C., Mattout, J., Rugg, M.D. *et al.* (2005) An empirical Bayesian solution to the source reconstruction problem in EEG. *Neuroimage*, **24**(4), 997–1011.

[73] Al-Asmi, A., Benar, C.G., Gross, D.W. *et al.* (2003) fMRI activation in continuous and spike-triggered EEG-fMRI studies of epileptic spikes. *Epilepsia*, **44**(10), 1328–1339.

[74] Gotman, J., Grova, C., Bagshaw, A. *et al.* (2005) Generalized epileptic discharges show thalamocortical activation and suspension of the default state of the brain. *Proc. Nat. Acad. Sci. U. S. A.*, **102**(42), 15236–15240.

[75] Horovitz, S.G., Fukunaga, M., de Zwart, J.A. *et al.* (2008) Low frequency BOLD fluctuations during resting wakefulness and light sleep: A simultaneous EEG-fMRI study. *Hum. Brain Mapp.*, **29**(6), 671–682.

[76] Formaggio, E., Storti, S., Avesani, M. *et al.* (2008) EEG and fMRI coregistration to investigate the cortical oscillatory activities during finger movement. *Brain Topogr.*, **21**(2), 100–111.

[77] Calhoun, V.D. and Adali, T. (2009) Feature-based fusion of medical imaging data. *IEEE Trans. Inf. Technol. Biomed.*, **13**(5), 711–720.

[78] Eichele, T., Calhoun, V.D., Moosmann, M. *et al.* (2008) Unmixing concurrent EEG-fMRI with parallel independent component analysis. *Int. J. Psychophysiol.*, **67**(3), 222–234.

[79] Moosmann, M., Eichele, T., Nordby, H. *et al.* (2008) Joint independent component analysis for simultaneous EEG-fMRI: Principle and simulation. *Int. J. Psychophysiol.*, **67**(3), 212–221.

[80] Beckmann, C.F. and Smith, S.M. (2005) Tensorial extensions of independent component analysis for multisubject fMRI analysis. *NeuroImage*, **25**(1), 294–311.

[81] Kiers, H.A.L., ten Berge, J.M.F. and Bro, R. (1999) PARAFAC2-Part I. A direct fitting algorithm for the PARAFAC2 model. *J. Chemomet.*, **13**(3–4), 275–294.

[82] Jing, M. and Sanei, S. (2009) Simultaneous EEG-fMRI analysis with application to detection and localizaion of seizure signal sources, in *Recent Advances in Signal Processing* (ed. A.A. Zaher), IN-TECH Pub., 978-953-307-002-5.

[83] Jing, M., Sanei, S. and Hamandi, K. (2008) A novel ICA approach for separation of seizure BOLD from fMRI, constrained by the simultaneously recorded EEG signals. Proceedings of the European Signal Processing Conference, EUSIPCO 2008, Switzerland.

Index